Operator Theory: Advances and Applications
Vol. 162

Operator Theory in Krein Spaces and Nonlinear Eigenvalue Problems

Karl-Heinz Förster
Peter Jonas
Heinz Langer
Editors

Birkhäuser Verlag
Basel · Boston · Berlin

Editors:

Karl-Heinz Förster
Institut für Mathematik
Technische Universität Berlin
D-10623 Berlin
Germany
e-mail: foerster@math.tu-berlin.de

Heinz Langer
Institut für Analysis und Scientific Computing
Technische Universität Wien
A-1040 Wien
Austria
e-mail: hlanger@mail.zserv.tuwien.ac.at

Peter Jonas
Institut für Mathematik
Technische Universität Berlin
D-10623 Berlin
Germany
e-mail: jonas@math.tu-berlin.de

2000 Mathematics Subject Classification: Primary: 47xx; Secondary: 46xx, 15xx, 30xx, 49R50

A CIP catalogue record for this book is available from the
Library of Congress, Washington D.C., USA

Bibliographic information published by Die Deutsche Bibliothek
Die Deutsche Bibliothek lists this publication in the Deutsche Nationalbibliografie; detailed bibliographic data is available in the Internet at <http://dnb.ddb.de>.

ISBN 3-7643-7452-7 Birkhäuser Verlag, Basel – Boston – Berlin

Member of the BertelsmannSpringer Publishing Group
Printed on acid-free paper produced from chlorine-free pulp. TCF∞
Cover design: Heinz Hiltbrunner, Basel
Printed in Germany
ISBN-10: 3-7643-7452-7
ISBN-13: 978-3-7643-7452-5
e-ISBN: 3-7643-7453-5

9 8 7 6 5 4 3 2 1

www.birkhauser.ch

Contents

Preface

This volume contains papers written by the participants of the 3rd Workshop on Operator Theory in Krein spaces and Nonlinear Eigenvalue Problems, held at the Technische Universität Berlin, Germany, December 12 to 14, 2003.

The workshop covered topics from spectral, perturbation and extension theory of linear operators in Krein spaces. They included generalized Nevanlinna functions and related classes of functions, boundary value problems for differential operators, spectral problems for matrix polynomials, and perturbation problems for second order evolution equations. All these problems are reflected in the present volume. The workshop was attended by 46 participants from 12 countries.

It is a pleasure to acknowledge the substantial financial support received from the

- Research Training Network HPRN-CT-2000-00116 "Analysis and Operators" by the European Community,
- DFG-Forschungszentrum MATHEON "Mathematik für Schlüsseltechnologien",
- Institute of Mathematics of the Technische Universität Berlin.

We would also like to thank Petra Grimberger for her great help. Last but not least, special thanks are due to Jussi Behrndt, Christian Mehl and Carsten Trunk for their excellent work in the organisation of the workshop and the preparation of this volume. Without their assistance the workshop might not have taken place.

The Editors

Operator Theory:
Advances and Applications, Vol. 162, 1–17

Partial Non-stationary Perturbation Determinants for a Class of J-symmetric Operators

Vadim Adamyan, Peter Jonas and Heinz Langer

Abstract. We consider the partial non-stationary perturbation determinant

$$\Delta^{(1)}_{H/A}(t) := \det\left(e^{itA}P_1e^{-itH}\Big|_{\mathcal{H}_1}\right), \quad t \in \mathbb{R}.$$

Here H is a self-adjoint operator in some Krein space $\mathcal{K}$ and A is a self-adjoint operator in the Hilbert space $\mathcal{H}_1$, which is the positive component of a fundamental decomposition of $\mathcal{K}$ with corresponding orthogonal projection P_1. The asymptotic behavior of $\Delta^{(1)}_{H/A}(t)$ for $t \to \infty$ and the spectral shift function for H and its diagonal part are studied. Analogous results for the case if the underlying space is a Hilbert space were obtained in [1].

Mathematics Subject Classification (2000). 47A55, 47B50; 47A15, 47B10.

Keywords. Perturbation determinant, trace class perturbation, positive operator in Krein space, skew symmetric operator, block matrix operator, spectral shift function.

1. Introduction

Let H_0 and H be self-adjoint operators in some Hilbert or Krein space $\mathcal{H}$, which generate groups $\left(e^{-itH_0}\right)_{t\in\mathbb{R}}$ and $\left(e^{-itH}\right)_{t\in\mathbb{R}}$ of unitary operators and are such that the closure N of the difference $H - H_0$ belongs to the set $\mathcal{S}_1$ of trace class operators. We call the function

$$\Delta_{H/H_0}(t) := \det\left(e^{itH_0}e^{-itH}\right), \qquad t \in \mathbb{R},$$

V. Adamyan gratefully acknowledges support by the USA Civil Research and Development Foundation, the Ministry of Education and Science of Ukraine (grant UM1-2567-OD-03), and from the Deutsche Akademische Austauschdienst (DAAD) (during his visit to TU-Berlin in December 2003). P. Jonas was supported by the Hochschul- und Wissenschaftsprogramm des Bundes und der Länder of Germany.

the *non-stationary perturbation determinant* (of the pair H_0, H). It is known (see, e.g., [1]) that it satisfies the relation

$$\Delta_{H/H_0}(t) = e^{-it\,\mathrm{tr}\,N}, \quad t \in \mathbb{R}.$$

A *partial non-stationary perturbation determinant* is defined by the relation

$$\Delta^{(1)}_{H/A}(t) := \det\left(e^{itA}P_1 e^{-itH}|_{\mathcal{H}_1}\right), \quad t \in \mathbb{R},$$

where A is a self-adjoint operator in some Hilbert space $\mathcal{H}_1$, H is a self-adjoint operator in a larger Hilbert or Krein space $\mathcal{K} \supset \mathcal{H}_1$ with $\mathrm{dom}\,A \subset \mathrm{dom}\,H$, P_1 is the orthogonal projection in $\mathcal{K}$ onto $\mathcal{H}_1$ and the closure of the operator $P_1 H|_{\mathrm{dom}\,A} - A$ is a trace class operator in $\mathcal{H}_1$.

If all the underlying spaces are Hilbert spaces, these notions were introduced in [1], motivated by applications from physics. Some results of [1] can be summarized as follows.

Theorem 1.1. *Let $\mathcal{H} = \mathcal{H}_1 \oplus \mathcal{H}_2$ with orthogonal projections P_1, P_2 onto $\mathcal{H}_1$, $\mathcal{H}_2$, respectively. Consider in $\mathcal{H}$ the self-adjoint operator H given by the matrix representation*

$$H = \begin{pmatrix} A+V & B \\ B^* & D \end{pmatrix}, \tag{1.1}$$

where A and D are self-adjoint operators in $\mathcal{H}_1$ and $\mathcal{H}_2$, respectively, such that the spectra of A and D are weakly separated by a real point α, that is,

$$\sigma(A) \le \alpha \le \sigma(D) \quad \text{and } \alpha \text{ is not an eigenvalue of } A \text{ and of } D,$$

V is a symmetric trace class operator in $\mathcal{H}_1$, and B is a trace class operator from $\mathcal{H}_2$ into $\mathcal{H}_1$. Assume further that the spectrum of H is absolutely continuous in at least one of the intervals $(-\infty, \alpha]$, $[\alpha, +\infty)$. Then

$$\Delta^{(1)}_{H/A}(t) = e^{-b-iat}\left(1 + o(1)\right), \quad t \to \infty,$$

with $b \ge 0$; if $V \le 0$ then $a \le 0$ and

$$a = 0 \iff V = B = 0.$$

The aim of this note is to generalize this result to the situation where $\mathcal{H}$ is a Krein space $\mathcal{K}$ and $\mathcal{H}_1 \subset \mathcal{K}$ is a maximal uniformly positive subspace of $\mathcal{K}$, that is, it is a component in a fundamental decomposition of the Krein space $\mathcal{K}$, and the operator H is self-adjoint in this Krein space. In other words, instead of the operator H of the form (1.1) we consider an operator

$$H = \begin{pmatrix} A+V & B \\ -B^* & D \end{pmatrix}, \tag{1.2}$$

where again A and D are self-adjoint operators in the Hilbert spaces $\mathcal{H}_1$ and $\mathcal{H}_2$, respectively.

In the following section we consider a pair of strongly continuous groups of unitary operators in Krein spaces, the generators of which differ by a trace class operator, and study the corresponding non-stationary perturbation determinant. The maximal semidefinite invariant subspaces of a special class of self-adjoint operators in Krein spaces and corresponding Riccati equations are considered in Section 3. These self-adjoint operators are very special in the sense that in their matrix representation (1.2) with $V = 0$ the diagonal entries A and D have strongly separated spectrum and the bounded 'perturbation' B is so small that for some $\alpha \in \mathbb{R}$ the operator $\alpha I - H$ is invertible and nonnegative in the Krein space. This implies that H is even similar to a self-adjoint operator in a Hilbert space, see Lemma 3.2 below. In Section 4 we consider the partial non-stationary perturbation determinants for this class of operators in Krein spaces and study their asymptotic properties. Finally, in Section 5 we prove some definiteness properties of the spectral shift function for these operators in Krein spaces. These properties explain, e.g., the fact (see also [3]), that if a diagonal self-adjoint operator $H_0 = \begin{pmatrix} A & 0 \\ 0 & D \end{pmatrix}$, for which the spectra of the diagonal entries are strongly separated by a real number α, is perturbed to an operator

$$H = \begin{pmatrix} A & B \\ -B^* & D \end{pmatrix}, \tag{1.3}$$

then, as long as B is sufficiently small, the spectrum of H_0 to the right of α moves to the left whereas the spectrum to the left of α moves to the right. Recall that in the Hilbert space case, that is if in (1.3) the operator $-B^*$ is replaced by B^*, then these parts of the spectrum move in the opposite directions.

By $\mathcal{B}_{\mathcal{H}}$ we denote the algebra of all bounded operators in the Hilbert space $\mathcal{H}$, $\mathcal{S}_1$ is the class of trace class operators in a Hilbert space or between two Hilbert spaces (the spaces should be always clear from the context), and $\|\cdot\|_1$ is the trace class norm.

2. Non-stationary perturbation determinants in Krein spaces

Let $\mathcal{K}$ be a Krein space, and $(U_0(t))_{t\in\mathbb{R}}$ be a strongly continuous group of unitary operators in $\mathcal{K}$. Recall that the operators $U_0(t)$, $t \in \mathbb{R}$, are not necessarily uniformly bounded, but if they have this property then, according to a theorem of Sz.-Nagy (comp. [5, Theorem II.5.18]), the group $(U_0(t))_{t\in\mathbb{R}}$ is similar to a group of unitary operators in a Hilbert space. Clearly, the *generator* H_0 of a strongly continuous group $(U_0(t))_{t\in\mathbb{R}}$ of unitary operators in $\mathcal{K}$, which is defined by the relation

$$H_0 x = -\lim_{t\to 0} \frac{1}{\mathrm{i}t} (U_0(t) - I)\, x, \quad x \in \operatorname{dom} H_0,$$

is a self-adjoint operator in $\mathcal{K}$.

Theorem 2.1. *Let H_0 be the generator of the strongly continuous group $(U_0(t))_{t\in\mathbb{R}}$ of unitary operators in the Krein space $\mathcal{K}$, and let V be a trace class operator in $\mathcal{K}$: $V \in \mathcal{S}_1$. Then the operator $H := H_0 + V$ is the generator of a strongly continuous group $(U(t))_{t\in\mathbb{R}}$ in $\mathcal{K}$ such that $U_0(-t)U(t) - I \in \mathcal{S}_1$ and*

$$\Delta_{H/H_0}(t) \equiv \det\,(U_0(-t)U(t)) = e^{-\mathrm{i}t\,\mathrm{tr}V}, \quad t \in \mathbb{R}. \tag{2.1}$$

If the operator V is self-adjoint in $\mathcal{K}$, then the operators $U(t)$, $t \in \mathbb{R}$, are unitary in $\mathcal{K}$.

Proof. The fact that $H_0 + V$ is the generator of a strongly continuous group $(U(t))_{t\in\mathbb{R}}$ in $\mathcal{K}$ follows from [10, Theorem IX.2.1], and according to [10, (IX.2.3)] the operators $U(t)$ satisfy the integral equation

$$U(t) = U_0(t) - \mathrm{i}\int_0^t U_0(t-s)VU(s)\,\mathrm{d}s, \quad t \in \mathbb{R}. \tag{2.2}$$

Since also $(U(t)^+)_{t\in\mathbb{R}}$ is a strongly continuous group in $\mathcal{K}$ and $V^+ \in \mathcal{S}_1$ it follows that $VU(s) = (U(s)^+V^+)^+$ depends continuously on s with respect to the norm of $\mathcal{S}_1$. Then the same is true for $s \longmapsto U_0(t-s)VU(s)$ with fixed $t \in \mathbb{R}$. Therefore, the integral in (2.2) and $U_0(t)^{-1}U(t)$ belong to $\mathcal{S}_1$.

The formula (2.1) can be proved as the corresponding formula in [1, Theorem 2.1]. It follows also more directly from the general formula

$$\frac{d}{dt}\mathrm{tr}\,\{\ln(I - A(t))\} = -\mathrm{tr}\left\{(I - A(t))^{-1}\frac{dA(t)}{dt}\right\}, \tag{2.3}$$

which holds for a real differentiable $\mathcal{S}_1$-valued operator function $A(\cdot)$, see [8, footnote on p. 164]. Indeed, if in this formula we set

$$A(t) := I - U_0(-t)U(t),$$

then from (2.2) we find

$$A(t) = \mathrm{i}\int_0^t U_0(-s)VU(s)\,ds,$$

hence

$$\frac{d}{dt}A(t) = \mathrm{i}\,U_0(-t)VU(t),$$

and the relation (2.3) yields

$$\frac{d}{dt}\ln\left(\det U_0(-t)U(t)\right) = -\mathrm{i}\,\mathrm{tr}\,\{U(-t)U_0(t)U_0(-t)VU(t)\} = -\mathrm{i}\,\mathrm{tr}V.$$

The last claim of the theorem is a consequence of the fact that $H_0 + V$ is self-adjoint. □

Corollary 2.2. *If in Theorem* 2.1 *the operator V is self-adjoint in the Krein space $\mathcal{K}$, then*

$$\left|\Delta_{H/H_0}(t)\right| = 1. \tag{2.4}$$

Indeed, if V is self-adjoint in $\mathcal{K}$, then $\mathrm{tr}\,V$ is real since with any fundamental symmetry J of $\mathcal{K}$ we have

$$\mathrm{tr}\,V = \mathrm{tr}\,JV^*J = \mathrm{tr}\,V^* = \overline{\mathrm{tr}\,V}.$$

Now (2.4) follows from (2.1).

3. Skew symmetric perturbations of a self-adjoint block diagonal operator matrix

Let $(\mathcal{H}, (\cdot,\cdot))$ be a Hilbert space, let $\mathcal{H}_1$ be a nontrivial subspace of $\mathcal{H}$ with orthogonal complement $\mathcal{H}_2$:

$$\mathcal{H} = \mathcal{H}_1 \oplus \mathcal{H}_2. \tag{3.1}$$

In this section we consider an operator H in $\mathcal{H}$ which, with respect to the decomposition (3.1), is given by a block operator matrix of the form

$$H = \begin{pmatrix} A & B \\ -B^* & D \end{pmatrix}, \tag{3.2}$$

where A is a self-adjoint operator in $\mathcal{H}_1$, D is a self-adjoint operator in $\mathcal{H}_2$ (both operators can be unbounded), and B is a bounded operator from $\mathcal{H}_2$ into $\mathcal{H}_1$. If the space $\mathcal{H}$ is equipped with the indefinite inner product

$$[x,y] := (x_1,y_1) - (x_2,y_2), \quad x = x_1 + x_2,\ y = y_1 + y_2,\ x_1,y_1 \in \mathcal{H}_1,\ x_2,y_2 \in \mathcal{H}_2,$$

then $\mathcal{K} := (\mathcal{H}, [\cdot,\cdot])$ becomes a Krein space. Evidently, with respect to the decomposition (3.1) of $\mathcal{K}$ the corresponding fundamental symmetry J has the representation

$$J = \begin{pmatrix} I & 0 \\ 0 & -I \end{pmatrix},$$

that is $[x,y] = (Jx,y)$, $x,y \in \mathcal{K}$. The assumptions about the operators on the right-hand side of (3.2) mean that H is a self-adjoint operator in this Krein space $\mathcal{K}$ with domain $\operatorname{dom} H = \operatorname{dom} A \oplus \operatorname{dom} D$.

We assume additionally, that the spectra of A and of D are separated, that is that A is bounded from above, D is bounded from below, and

$$\max \sigma(A) < \min \sigma(D).$$

Without loss of generality we can assume that for some $\delta > 0$

$$\max \sigma(A) \le -\delta < 0 < \delta \le \min \sigma(D). \tag{3.3}$$

Let $(-A)^{1/2}, D^{1/2}$ denote the positive square roots of the positive operators $-A$ and D, respectively, and introduce also the operator

$$T := (-A)^{-1/2} B D^{-1/2}.$$

Lemma 3.1. *Under the assumption* (3.3), *if* $q := \|T\| < 1$ *then the operator* $-H$ *is positive in the Krein space* $\mathcal{K}$ *and*

$$\sigma(H) \subset \big(-\infty, -(1-q)\delta\big] \cup \big[(1-q)\delta, \infty\big).$$

Proof. Since $\|T\| < 1$ the operator

$$W := \begin{pmatrix} I & -T \\ -T^* & I \end{pmatrix}$$

is strictly positive in $(\mathcal{H}, (\cdot,\cdot))$, in fact $W \geq (1-q)I$. On the other hand,

$$-JH = \begin{pmatrix} -A & -B \\ -B^* & D \end{pmatrix} = \begin{pmatrix} (-A)^{1/2} & 0 \\ 0 & D^{1/2} \end{pmatrix} W \begin{pmatrix} (-A)^{1/2} & 0 \\ 0 & D^{1/2} \end{pmatrix},$$

hence $-JH \geq (1-q)\delta I$, and $-H$ is a positive operator in the Krein space $\mathcal{K}$. Therefore $\sigma(H)$ is real, see [13]. Since from $0 \leq (-JH)^{-1} \leq \delta^{-1}(1-q)^{-1}I$ we obtain that

$$\|H^{-1}\| \leq \frac{1}{\delta(1-q)},$$

the interval $(-(1-q)\delta, (1-q)\delta)$ belongs to $\rho(H)$. □

Lemma 3.2. *If the assumption* (3.3) *holds and* $q = \|(-A)^{-\frac{1}{2}}BD^{-\frac{1}{2}}\| < 1$, *then in the Krein space* $\mathcal{K}$ *there exist a uniformly positive subspace* $\mathcal{L}_+$ *and a uniformly negative subspace* $\mathcal{L}_-$ *such that the following holds.*

(i) $\mathcal{K} = \mathcal{L}_+[\dot{+}]\mathcal{L}_-$, $\operatorname{dom} H = (\operatorname{dom} H \cap \mathcal{L}_+)[\dot{+}](\operatorname{dom} H \cap \mathcal{L}_-)$, *and* $\mathcal{L}_+$ *and* $\mathcal{L}_-$ *are invariant under* H, *that is,* $H(\operatorname{dom} H \cap \mathcal{L}_\pm) \subset \mathcal{L}_\pm$.

(ii) *The operators* $H|_{\mathcal{L}_+}$ *and* $H|_{\mathcal{L}_-}$ *are self-adjoint in the Hilbert spaces* $(\mathcal{L}_+, [\cdot,\cdot])$ *and* $(\mathcal{L}_-, -[\cdot,\cdot])$, *respectively, and we have*

$$\sigma(H|_{\mathcal{L}_-}) = \sigma(H) \cap [(1-q)\delta, \infty), \quad \sigma(H|_{\mathcal{L}_+}) = \sigma(H) \cap (-\infty, -(1-q)\delta].$$

The subspaces $\mathcal{L}_+$ *and* $\mathcal{L}_-$ *are uniquely determined by the properties in* (i).

Proof. If $B = 0$, then the operator in (3.2) becomes

$$\widehat{H} = \begin{pmatrix} A & 0 \\ 0 & D \end{pmatrix},$$

and, if the corresponding subspaces from Lemma 3.2 are denoted by $\widehat{\mathcal{L}}_\pm$, we have, evidently,

$$\widehat{\mathcal{L}}_+ = \mathcal{H}_1, \quad \widehat{\mathcal{L}}_- = \mathcal{H}_2.$$

This means that if ∞ is a critical point of $\widehat{H}$ then it is a regular critical point (for the definition see [13] or [7]). Since the domains of $\widehat{H}$ and H coincide, according to a theorem of B. Ćurgus [7], ∞ is also a regular critical point of H. If $\mathcal{L}_+$ denotes the spectral subspace of H corresponding to the interval $(-\infty, -(1-q)\delta]$ and $\mathcal{L}_-$ denotes the spectral subspace of H corresponding to the interval $[(1-q)\delta, \infty)$, then $\mathcal{L}_+$ and $\mathcal{L}_-$ satisfy the conditions (i) and (ii).

If $\mathcal{L}'_+$ is a uniformly positive subspace of $\mathcal{K}$ and $\mathcal{L}'_-$ is a uniformly negative subspace of $\mathcal{K}$ such that (i) holds with $\mathcal{L}_\pm$ replaced by $\mathcal{L}'_\pm$, then also (ii) holds with $\mathcal{L}_\pm$ replaced by $\mathcal{L}'_\pm$. This implies $\mathcal{L}'_\pm = \mathcal{L}_\pm$. □

We mention that the existence of the subspaces $\mathcal{L}_\pm$ can also be shown in a more direct way by proving that the integral

$$\frac{1}{\pi \mathrm{i}} \int_{\mathrm{i}\mathbb{R}} (H-z)^{-1} \,\mathrm{d}z \tag{3.4}$$

(along the imaginary axis and with Cauchy principal value at $\pm i\infty$), exists in the strong operator topology, and that it equals the difference of the orthogonal projections onto the subspaces $\mathcal{L}_-$ and $\mathcal{L}_+$, see [16], and also [2], [14].

As uniformly definite and mutually orthogonal subspaces, $\mathcal{L}_\pm$ admit graph representations by means of an angular operator K which is a strict contraction from $\mathcal{H}_1$ into $\mathcal{H}_2$:

$$\mathcal{L}_+ = \left\{ \begin{pmatrix} x_1 \\ Kx_1 \end{pmatrix} : x_1 \in \mathcal{H}_1 \right\}, \quad \mathcal{L}_- = \left\{ \begin{pmatrix} K^* x_2 \\ x_2 \end{pmatrix} : x_2 \in \mathcal{H}_2 \right\}. \tag{3.5}$$

If $z \neq \bar{z}$ we have

$$\begin{aligned} &P_1(H-z)^{-1}\mathcal{L}_+ \\ &= (A-z)^{-1}(I + B(D-z)^{-1}B^*(A-z)^{-1})^{-1}(I - B(D-z)^{-1}K)\,\mathcal{H}_1. \end{aligned} \tag{3.6}$$

For sufficiently large $|\operatorname{Im} z|$ the set on the right-hand side of (3.6) coincides with dom A (cf. [15, proof of Theorem 4.1]). This implies the relation

$$\operatorname{dom} H \cap \mathcal{L}_+ = \left\{ \begin{pmatrix} x_1 \\ Kx_1 \end{pmatrix} : x_1 \in \operatorname{dom} A \right\};$$

the analogous relation

$$\operatorname{dom} H \cap \mathcal{L}_- = \left\{ \begin{pmatrix} K^* x_2 \\ x_2 \end{pmatrix} : x_2 \in \operatorname{dom} D \right\}$$

follows in the same way. In particular,

$$K \operatorname{dom} A \subset \operatorname{dom} D, \qquad K^* \operatorname{dom} D \subset \operatorname{dom} A. \tag{3.7}$$

Theorem 3.3. *Let H be the self-adjoint operator* (3.2) *in the Krein space $\mathcal{K}$ such that A and D are self-adjoint and satisfy* (3.3)*, that B is a bounded operator from $\mathcal{H}_2$ into $\mathcal{H}_1$, and that $\|(-A)^{-1/2}BD^{-1/2}\| < 1$. Then there exists a unique strict contraction K from $\mathcal{H}_1$ into $\mathcal{H}_2$ such that the spectral subspaces $\mathcal{L}_+$ and $\mathcal{L}_-$ admit the representations* (3.5) *and:*

(i) $K \operatorname{dom} A \subset \operatorname{dom} D$ *and K satisfies the Riccati equation*

$$KBKx + B^*x - DKx + KAx = 0, \quad x \in \operatorname{dom} A. \tag{3.8}$$

(ii) *The restriction of H to $(\mathcal{L}_+, [\cdot,\cdot])$ is unitarily equivalent to $A + BK$, which is a self-adjoint and negative operator in the Hilbert space $\widehat{\mathcal{H}}_1 := (\mathcal{H}_1, [\cdot,\cdot]_1)$ with*

$$[x_1, x_1]_1 := ((I - K^*K)\,x_1, x_1)\,, \quad x_1 \in \mathcal{H}_1.$$

(iii) *The restriction of H to $(\mathcal{L}_-, -[\cdot,\cdot])$ is unitarily equivalent to the operator $D - B^*K^*$, which is self-adjoint and positive in the Hilbert space $\widehat{\mathcal{H}}_2 := (\mathcal{H}_2, [\cdot,\cdot]_2)$, with*

$$[x_2, x_2]_2 := ((I - KK^*)\,x_2, x_2)\,, \quad x_2 \in \mathcal{H}_2.$$

(iv) *If B is compact then K is compact, and $B \in \mathcal{S}_1$ implies $K \in \mathcal{S}_1$.*

Proof. By Lemma 3.2 and the remarks before Theorem 3.3, the proof of (i–iii) is straightforward (cf. [2]) and is therefore left to the reader.

To verify (iv) assume that $B \in \mathcal{S}_1$. If B is only compact, a similar reasoning applies. Let E_+ and E_- be the orthogonal projections in $\mathcal{K}$ onto $\mathcal{L}_+$ and $\mathcal{L}_-$, respectively. Then (cf. (3.4))

$$\begin{aligned} E_- - E_+ - (P_2 - P_1) &= (\pi \mathrm{i})^{-1} \int_{\mathrm{i}\mathbb{R}} \left((H-z)^{-1} - (\hat{H}-z)^{-1}\right) \mathrm{d}z \\ &= -(\pi \mathrm{i})^{-1} \int_{\mathrm{i}\mathbb{R}} (H-z)^{-1}(H-\hat{H})(\hat{H}-z)^{-1}\, \mathrm{d}z =: T. \end{aligned} \tag{3.9}$$

Since the closure of $H - \hat{H}$ belongs to $\mathcal{S}_1$, the integral on the right-hand side of (3.9) converges in $\mathcal{S}_1$: $T \in \mathcal{S}_1$. In view of $E_+ + E_- = P_1 + P_2 = I$, the relation (3.9) implies $P_1 - E_+ = \frac{1}{2} T \in \mathcal{S}_1$. Hence

$$P_1 - E_+ = \begin{pmatrix} I & 0 \\ 0 & 0 \end{pmatrix} - \begin{pmatrix} (I-K^*K)^{-1} & -(I-K^*K)^{-1}K^* \\ K(I-K^*K)^{-1} & -K(I-K^*K)^{-1}K^* \end{pmatrix} \in \mathcal{S}_1$$

and $K \in \mathcal{S}_1$ follows. □

Remark 3.4. The same argument as in [4] shows that the operator K in Theorem 3.3 is the unique contractive solution of the Riccati equation (3.8).

Below we use the identities:

$$(I - F^*F)^{1/2}F^* = F^*(I-FF^*)^{1/2}, \; (I-FG)^{-1}F = F(I-GF)^{-1} \tag{3.10}$$

which hold for arbitrary bounded operators F and G between appropriate Hilbert spaces if for the second relation only the inverses in it exist.

Corollary 3.5. *Under the assumptions of Theorem 3.3 the operator H admits the following block diagonalization:*

$$H \equiv \begin{pmatrix} A & B \\ -B^* & D \end{pmatrix} = S^{-1} \begin{pmatrix} \widetilde{A} & 0 \\ 0 & \widetilde{D} \end{pmatrix} S = S^{-1}\widetilde{H}S, \tag{3.11}$$

where

$$\begin{aligned} \widetilde{A} &:= (I-K^*K)^{1/2}(A+BK)(I-K^*K)^{-1/2}, \\ \widetilde{D} &:= (I-KK^*)^{1/2}(D-B^*K^*)(I-KK^*)^{-1/2}, \end{aligned}$$

$$\widetilde{H} := \begin{pmatrix} \widetilde{A} & 0 \\ 0 & \widetilde{D} \end{pmatrix},$$

and

$$S = \begin{pmatrix} (I-K^*K)^{-1/2} & -(I-K^*K)^{-1/2}K^* \\ -K(I-K^*K)^{-1/2} & (I-KK^*)^{-1/2} \end{pmatrix}. \tag{3.12}$$

The operators $\widetilde{A}$, $\widetilde{D}$ *and* $\widetilde{H}$ *are self-adjoint in the Hilbert spaces* $\mathcal{H}_1$, $\mathcal{H}_2$ *and* $\mathcal{H}$, *respectively. If, in addition,* B *is compact then*

$$\operatorname{dom} \widetilde{A} = \operatorname{dom} A, \quad \operatorname{dom} \widetilde{D} = \operatorname{dom} D, \quad \operatorname{dom} \widetilde{H} = \operatorname{dom} H. \tag{3.13}$$

Proof. The first two assertions are consequences of Theorem 3.3. It remains to verify the relations (3.13). The inclusions (3.7) imply

$$(I - K^*K) \operatorname{dom} A \subset \operatorname{dom} A, \quad (I - K^*K) \operatorname{dom} D \subset \operatorname{dom} D. \tag{3.14}$$

Moreover, as the projection E_+ on $\mathcal{L}_+$ in $\mathcal{H}$ corresponding to the decomposition $\mathcal{H} = \mathcal{L}_+[\dot{+}]\mathcal{L}_-$,

$$E_+ = \begin{pmatrix} (I - K^*K)^{-1} & -(I - K^*K)^{-1}K^* \\ (I - KK^*)^{-1}K & I - (I - KK^*)^{-1} \end{pmatrix},$$

maps $\operatorname{dom} H$ into itself, we obtain

$$(I - K^*K)^{-1} \operatorname{dom} A \subset \operatorname{dom} A, \quad (I - KK^*)^{-1} \operatorname{dom} D \subset \operatorname{dom} D. \tag{3.15}$$

In view of $HE_+ = E_+H$ and $H(I - E_+) = (I - E_+)H$ we find

$$\begin{aligned} (A + BK)(I - K^*K)^{-1} &= (I - K^*K)^{-1}(A + K^*B^*), \\ (D - B^*K^*)(I - KK^*)^{-1} &= (I - KK^*)^{-1}(D - KB). \end{aligned} \tag{3.16}$$

These relations together with (3.14) and (3.15) imply that $(I - K^*K)^{-1}|_{\operatorname{dom} A}$ $\big((I - KK^*)^{-1}|_{\operatorname{dom} D}\big)$ is an automorphism of $\operatorname{dom} A$ (resp. $\operatorname{dom} D$) with respect to the graph norm $\|\cdot\|_A$ (resp. $\|\cdot\|_D$).

Suppose now that B is compact. Then by Theorem 3.3, (iv), K is compact. Therefore, $I - (I - K^*K)^{-1}$ and $I - (I - KK^*)^{-1}$ are compact in $\mathcal{H}_1$ and $\mathcal{H}_2$, respectively. Then the relations (3.16) imply that $I - (I - K^*K)^{-1}|_{\operatorname{dom} A}$ and $I - (I - KK^*)^{-1}|_{\operatorname{dom} D}$ are compact in $\operatorname{dom} A$ and $\operatorname{dom} D$, respectively, with respect to the corresponding graph norms. It follows that

$$\begin{aligned} \sigma\big((I - K^*K)^{-1}|_{\operatorname{dom} A}\big) \setminus \{1\} &\subset \sigma\big((I - K^*K)^{-1}\big), \\ \sigma\big((I - KK^*)^{-1}|_{\operatorname{dom} D}\big) \setminus \{1\} &\subset \sigma\big((I - KK^*)^{-1}\big). \end{aligned}$$

Therefore, the square roots

$$\big((1 - K^*K)^{-1}|_{\operatorname{dom} A}\big)^{\frac{1}{2}} \quad \text{and} \quad \big((1 - KK^*)^{-1}|_{\operatorname{dom} D}\big)^{\frac{1}{2}}$$

defined with the help of the Riesz-Dunford functional calculus are bijections of $\operatorname{dom} A$ and $\operatorname{dom} D$, respectively, which implies (3.13). □

Proposition 3.6. *Under the assumptions of Theorem* 3.3 *the operator* $P_1 e^{-itH}\big|_{\mathcal{H}_1}$ *is invertible for all* $t \in \mathbb{R}$.

Proof. The inverse of the operator S in (3.12) has the matrix representation

$$S^{-1} = \begin{pmatrix} (I - K^*K)^{-1/2} & (I - K^*K)^{-1/2}K^* \\ K(I - K^*K)^{-1/2} & (I - KK^*)^{-1/2} \end{pmatrix}.$$

Then, according to (3.11) we have

$$\begin{aligned} P_1 e^{-\mathrm{i}tH}\big|_{\mathcal{H}_1} &= P_1 S^{-1} e^{-\mathrm{i}t\widetilde{H}} S\big|_{\mathcal{H}_1} \\ &= \left((I-K^*K)^{-1/2} \quad K^*(I-KK^*)^{-1/2}\right)\begin{pmatrix} e^{-\mathrm{i}t\widetilde{A}} & 0 \\ 0 & e^{-\mathrm{i}t\widetilde{D}} \end{pmatrix}\begin{pmatrix} (I-K^*K)^{-1/2} \\ -K(I-K^*K)^{-1/2} \end{pmatrix} \\ &= e^{-\mathrm{i}t(A+BK)}\left(I-\widetilde{Q}(t)\right)(I-K^*K)^{-1}, \end{aligned}$$

where

$$\widetilde{Q}(t) := e^{\mathrm{i}t(A+BK)} K^* e^{-\mathrm{i}t(D-B^*K^*)} K. \tag{3.17}$$

The operator $A+BK$ is similar to a self-adjoint operator, therefore $e^{-\mathrm{i}t(A+BK)}$ is similar to a unitary operator and hence invertible. Now the invertibility of $P_1 e^{-\mathrm{i}tH}\big|_{\mathcal{H}_1}$ follows if the invertibility of $I-\widetilde{Q}(t)$ is shown. The operators

$$\begin{aligned} e^{-\mathrm{i}\widetilde{A}t} &= (I-K^*K)^{1/2} e^{-\mathrm{i}t(A+BK)} (I-K^*K)^{-1/2}, \\ e^{-\mathrm{i}\widetilde{D}t} &= (I-KK^*)^{1/2} e^{-\mathrm{i}t(D-B^*K^*)} (I-KK^*)^{-1/2} \end{aligned}$$

are unitary in the Hilbert spaces $\mathcal{H}_1$ and $\mathcal{H}_2$, respectively, and therefore the relation

$$I-\widetilde{Q}(t) = (I-K^*K)^{-1/2}\left(I-e^{\mathrm{i}\widetilde{A}t}K^*e^{-\mathrm{i}\widetilde{D}t}K\right)(I-K^*K)^{1/2}.$$

shows the invertibility of $I-\widetilde{Q}(t)$ since K is a strict contraction. □

4. Partial non-stationary perturbation determinants for skew symmetric perturbations of self-adjoint operators

Let $\mathcal{H}$ be a Hilbert space as in (3.1), let A, V, and D be self-adjoint operators in $\mathcal{H}_1$ and $\mathcal{H}_2$, respectively, let B be a an operator from $\mathcal{H}_2$ into $\mathcal{H}_1$ and suppose that $B,\, V \in \mathcal{S}_1$. We consider the following operator in $\mathcal{H}$:

$$H = \begin{pmatrix} A+V & B \\ -B^* & D \end{pmatrix}. \tag{4.1}$$

Moreover, it is assumed that the spectra of the diagonal components of H are separated such that there exists a $\delta > 0$ with $A+V \le -\delta I$ and $D \ge \delta I$, and that $\left\|(-A-V)^{-1/2} B D^{-1/2}\right\| < 1$. These assumptions imply that H is strictly negative in the Krein space $\mathcal{K}$, i.e., $-JH \gg 0$ in $\mathcal{H}$, see Lemma 3.1.

For the operator H in (4.1) all conditions of Theorem 3.3 are satisfied with A replaced by $A+V$. Therefore for H in (4.1) the assertion of Corollary 3.5 holds with a strict contraction $K \in \mathcal{S}_1$. It follows that

$$e^{-\mathrm{i}tH} = \begin{pmatrix} I & K^* \\ K & I \end{pmatrix}\begin{pmatrix} e^{-\mathrm{i}t\widehat{A}} & 0 \\ 0 & e^{-\mathrm{i}t\widehat{D}} \end{pmatrix}\begin{pmatrix} (I-K^*K)^{-1} & -(I-K^*K)^{-1}K^* \\ -(I-KK^*)^{-1}K & (I-KK^*)^{-1} \end{pmatrix},$$

where

$$\widehat{A} := A+V+BK, \qquad \widehat{D} := D-B^*K^*.$$

If P_1 denotes again the orthogonal projection onto $\mathcal{H}_1$, then

$$P_1 e^{-\mathrm{i}tH}\big|_{\mathcal{H}_1} = e^{-\mathrm{i}t\widehat{A}}(I - K^*K)^{-1} - K^* e^{-\mathrm{i}t\widehat{D}}(I - KK^*)^{-1}K. \tag{4.2}$$

From (4.2), (3.10) and Theorem 2.1 it follows as in [1] that

$$\Delta^{(1)}_{H/A}(t) \equiv \det\left(e^{\mathrm{i}tA} P_1 e^{-\mathrm{i}tH}\big|_{\mathcal{H}_1}\right) = e^{-\mathrm{i}t\,a} \frac{\widetilde{\Delta}(t)}{\det(I - K^*K)}, \tag{4.3}$$

where

$$a = \operatorname{tr}(V + BK) \tag{4.4}$$

and

$$\widetilde{\Delta}(t) := \det\left(I - e^{\mathrm{i}t\widetilde{A}_1} K^* e^{-\mathrm{i}t\widetilde{D}} K\right).$$

with the self-adjoint operators

$$\begin{aligned} \widetilde{A}_1 &:= (I - K^*K)^{1/2}\widehat{A}\,(I - K^*K)^{-1/2}, \\ \widetilde{D} &:= (I - KK^*)^{1/2}\widehat{D}\,(I - KK^*)^{-1/2}. \end{aligned} \tag{4.5}$$

Theorem 4.1. *Suppose that the operator H in $\mathcal{H}$, given by (4.1), satisfies all the assumptions from the first paragraph of this section, and suppose that at least one of the self-adjoint operators $\widetilde{A}_1$ and $\widetilde{D}$ from (4.5) has absolutely continuous spectrum. Then in the relation (4.3) we have $\lim_{t\to\infty}\widetilde{\Delta}(t) = 1$, that is*

$$\Delta^{(1)}_{H/A}(t) = e^{-b-\mathrm{i}\,at}(1 + \mathrm{o}(1)),$$

with

$$a = \operatorname{tr}(V + BK), \qquad b = \ln\det(I - K^*K).$$

The proof of this theorem follows immediately from (4.3) and [1, Lemma 3.2]; the latter we formulate here for the convenience of the reader.

Lemma 4.2. *Let $(U(t))_{t\in\mathbb{R}}$ be a strongly continuous group of unitary operators in the Hilbert space $\mathcal{H}_1$, let $Y \in \mathcal{S}_1$ be an operator from $\mathcal{H}_1$ into the Hilbert space $\mathcal{H}_2$, and let $W(t)$, $t > 0$, be a bounded function with values in $\mathcal{B}_{\mathcal{H}_2}$: $\|W(t)\| \le c$, $t > 0$. If the infinitesimal generator of the group $(U(t))_{t\in\mathbb{R}}$ has absolutely continuous spectrum, then*

$$\lim_{t\to\infty} \det\left(I + U(t)Y^*W(t)Y\right) = 1.$$

5. The spectral shift function for skew symmetric perturbations

If H_0, H is a pair of self-adjoint operators in a Hilbert space $\mathcal{H}$ such that $H = H_0 + N$ for some $N \in \mathcal{S}_1$, we denote by $D_{H/H_0}(z)$, $\operatorname{Im} z \ne 0$, the (*stationary*) *perturbation determinant* of H and H_0:

$$D_{H/H_0}(z) := \det\left((H - z)(H_0 - z)^{-1}\right) = \det\left(I + (H - H_0)(H_0 - z)^{-1}\right).$$

Recall that according to a result of M.G. Krein (see [12],[6]) the perturbation determinant $D_{H/H_0}(z)$ admits the representation

$$\log D_{H/H_0}(z) = \int_{-\infty}^{\infty} \frac{\xi\,(\lambda; H, H_0)}{\lambda - z}\,\mathrm{d}\lambda, \quad \operatorname{Im} z \neq 0, \tag{5.1}$$

where $\xi(\lambda; H, H_0)$ is the *spectral shift function* of the pair H_0, H, which is a real summable function on $\mathbb{R}$. The logarithm is determined by the property

$$\lim_{\eta\uparrow\infty} \log D_{H/H_0}(\pm \mathrm{i}\eta) = 0,$$

hence

$$\xi(\lambda; H, H_0) = \frac{1}{\pi} \lim_{\eta\downarrow 0} \arg D_{H/H_0}(\lambda + \mathrm{i}\eta), \quad \text{a.e. } \lambda \in \mathbb{R}. \tag{5.2}$$

The spectral shift function has the property

$$\int_{-\infty}^{+\infty} \xi(\lambda; H, H_0)\,\mathrm{d}\lambda = \operatorname{tr} N \tag{5.3}$$

and for all C^∞-functions f with compact support the so-called trace formula

$$\operatorname{tr}\left(f(H) - f(H_0)\right) = \int_{-\infty}^{+\infty} f'(\lambda)\,\xi(\lambda; H, H_0)\,\mathrm{d}\lambda \tag{5.4}$$

holds.

If the condition $H = H_0 + N$, $N \in \mathcal{S}_1$, is replaced by the more general assumption that for some nonreal z the difference of the resolvents $(H - z)^{-1} - (H_0 - z)^{-1}$ is a trace class operator, then there exists still a real locally summable function $\widetilde{\xi}$ on $\mathbb{R}$ with

$$\int_{-\infty}^{+\infty} |\widetilde{\xi}(\lambda)|(1 + \lambda^2)^{-1}\,\mathrm{d}\lambda < \infty \tag{5.5}$$

such that (5.4) holds for the same functions f as above. $\widetilde{\xi}$ is uniquely determined by (5.4) only up to a constant function.

It was shown in [9, Satz 4.2.1 and Satz 4.2.5] that these statements hold true if the Hilbert space $\mathcal{H}$ is replaced by a Krein space and H_0, H are nonnegative (or nonpositive) self-adjoint operators in this Krein space with the following properties: (i) $0 \in \rho(H_0) \cap \rho(H)$, (ii) H_0 and H are similar to self-adjoint operators in some Hilbert space, (iii) $H = H_0 + N$ for some $N \in \mathcal{S}_1$. In fact, also for such a pair H_0, H there exists a spectral shift function $\xi(\cdot; H, H_0) \in L^1(\mathbb{R})$ such that the relations (5.1)–(5.4) hold. If instead of (iii) only the difference of the resolvents of H and H_0 is of trace class, there exists a spectral shift function $\widetilde{\xi}$ satisfying (5.5) such that (5.4) holds. These results are also related to [11, Theorem 2] and the remarks at the end of this paper.

Now, as in the preceding sections, we consider the Krein space $\mathcal{K} = \mathcal{H}_1[\dot{+}]\mathcal{H}_2$, where $(\mathcal{H}_1, [\cdot,\cdot])$ and $(\mathcal{H}_2, -[\cdot,\cdot])$ are Hilbert spaces. Suppose that A and V are self-adjoint operators in $(\mathcal{H}_1, [\cdot,\cdot])$, $V \in \mathcal{S}_1$, D is a self-adjoint operator in $(\mathcal{H}_2, -[\cdot,\cdot])$, and B is a trace class operator from $(\mathcal{H}_2, -[\cdot,\cdot])$ into $(\mathcal{H}_1, [\cdot,\cdot])$ such that for some $\delta > 0$

$$A + V \leq -\delta I \qquad D \geq \delta I \tag{5.6}$$

and

$$\|B\| < \delta \tag{5.7}$$

Observe that the assumption (5.7) on B is slightly stronger than that of Section 4. Let

$$H_0 := \begin{pmatrix} A & 0 \\ 0 & D \end{pmatrix}, \quad H_1 := \begin{pmatrix} A+V & 0 \\ 0 & D \end{pmatrix}, \quad H := \begin{pmatrix} A+V & B \\ -B^* & D \end{pmatrix}. \tag{5.8}$$

Then $\xi(\cdot; H_1, H_0) = \xi(\cdot; A+V, A) \in L^1(\mathbb{R})$ and $\xi(\cdot; H_1, H_0)$ satisfies the relations (5.1)–(5.4) with H replaced by H_1, in particular,

$$\int_{-\infty}^{\infty} \xi(\lambda; H_1, H_0)\, \mathrm{d}\lambda = \operatorname{tr} V. \tag{5.9}$$

Since H arises from H_1 by a trace class perturbation, H_1 and H are negative in $\mathcal{K}$ and H_1 and H are similar to self-adjoint operators in Hilbert spaces, the spectral shift function $\xi(\cdot; H, H_1)$ exists, is summable on $\mathbb{R}$ and fulfills (5.1)–(5.4) with H_0 replaced by H_1 ([9, Satz 4.2.1 and Satz 4.2.5]).

In [3] the spectral shift function for the pair H_1, H under the above assumptions was defined by the relation (5.2), and it was shown that it has the properties

$$\begin{aligned} \xi(\lambda; H, H_1) &\leq 0 \quad \text{for a.e. } \lambda > -\delta + \|B\|, \\ \xi(\lambda; H, H_1) &\geq 0 \quad \text{for a.e. } \lambda < \delta - \|B\|. \end{aligned} \tag{5.10}$$

Theorem 5.1. *Let H, H_0, H_1 be given by (5.8), where A, V, B, D satisfy the assumptions formulated before (5.8), and let K be the strict contraction from Theorem 3.3 with A replaced by $A+V$. Then the following statements hold:*

(i) $\xi(\lambda; H, H_1) = 0$ *for a. e.* $\lambda \in (-\delta + \|B\|, \delta - \|B\|)$.

(ii) $\int_{-\infty}^{0} \xi(\lambda; H, H_1)\, \mathrm{d}\lambda = \operatorname{tr}(BK)$.

(iii) $\int_{0}^{\infty} \xi(\lambda; H, H_1)\, \mathrm{d}\lambda = -\operatorname{tr}(BK)$.

(iv) $\xi(\lambda; H, H_1) \leq 0$ *for a. e.* $\lambda \geq 0$, *and* $\xi(\lambda; H, H_1) \geq 0$ *for a. e.* $\lambda \leq 0$.

(v) *If* $V \geq 0$ *then* $\xi(\lambda; H, H_0) \leq 0$ *for a. e.* $\lambda \geq 0$, *and* $\xi(\lambda; H, H_0) \geq 0$ *for a. e.* $\lambda \leq 0$.

First we prove a lemma which will be needed in the proof of Theorem 5.1.

Lemma 5.2. *If, in some Hilbert space, C is a bounded self-adjoint and T is a trace class operator such that $I+T$ is boundedly invertible, then*

$$\operatorname{tr}\left((I+T)C(I+T)^{-1} - C\right) = 0.$$

Proof. The claim follows from the relations

$$(I+T)\,C\,(I+T)^{-1} \;-\; C \;=\; (TC-CT)\,(I+T)^{-1} \;\in\; \mathcal{S}_1$$

and

$$\operatorname{tr} TC(I+T)^{-1} = \operatorname{tr} C(I+T)^{-1}T = \operatorname{tr} CT(I+T)^{-1},$$

where for the first equality we have used [8, Theorem III.8.2]. □

Proof of Theorem 5.1. 1. First we verify the assertion (i). If

$$H_s := H_1 + (s-1)\begin{pmatrix} 0 & B \\ -B^* & 0 \end{pmatrix}, \qquad s\in[1,2],$$

then by (5.2) $\xi(\cdot\,; H_s, H_1)$ restricted to $(-\delta+\|B\|, \delta-\|B\|)$ is an integer constant n_s. Since n_s depends continuously on $s\in[1,2]$ and $n_1=0$, we have $n_s=0$ for all $s\in[1,2]$, hence $\xi(\cdot\,;H_2,H_1)=\xi(\cdot\,;H,H_1)$ is zero on $(-\delta+\|B\|,\delta-\|B\|)$.

2. As in Theorem 4.1 we consider the operators

$$\begin{aligned}\widetilde{A}_1 &:= (I-K^*K)^{\frac12}(A+V+BK)(I-K^*K)^{-\frac12},\\ \widetilde{D} &:= (I-KK^*)^{\frac12}(D-B^*K^*)(I-KK^*)^{-\frac12}\end{aligned}$$

in $\mathcal{H}_1$ and $\mathcal{H}_2$, respectively. Since the difference of the resolvents of $A+V$ and $\widetilde{A}_1$ (of D and $\widetilde{D}$, respectively,) is of trace class, there exists a spectral shift function $\widetilde{\xi}_1 \in L^1(\mathbb{R},(1+t^2)^{-1})$ of the pair $A+V$, $\widetilde{A}_1$ ($\widetilde{\xi}_2\in L^1(\mathbb{R},(1+t^2)^{-1})$ of the pair D, $\widetilde{D}$, respectively) which is zero on $(0,\infty)$ (resp. on $(-\infty,0)$) and on some neighborhood of 0, and these functions $\widetilde{\xi}_1$ and $\widetilde{\xi}_2$ are uniquely determined. Then $\widetilde{\xi} := \widetilde{\xi}_1+\widetilde{\xi}_2 \in L^1(\mathbb{R},(1+t^2)^{-1})$ is a spectral shift function for the pair H_1, $\widetilde{H}_1$ where

$$\widetilde{H}_1 := \begin{pmatrix} \widetilde{A}_1 & 0 \\ 0 & \widetilde{D}\end{pmatrix}.$$

We have

$$\widetilde{H}_1 = SHS^{-1}, \qquad I-S\in\mathcal{S}_1, \tag{5.11}$$

where

$$S = \begin{pmatrix} (I-K^*K)^{-\frac12} & -(I-K^*K)^{-\frac12}K^* \\ -K(I-K^*K)^{-\frac12} & (I-KK^*)^{-\frac12}\end{pmatrix}.$$

In view of (5.11) and Lemma 5.2, for every $f\in C_0^\infty(\mathbb{R})$, we find

$$\begin{aligned}\operatorname{tr}\{f(H)-f(H_1)\} &= \operatorname{tr}\{f(S^{-1}\widetilde{H}_1S)-f(H_1)\}\\ &= \operatorname{tr}\{S^{-1}f(\widetilde{H}_1)S - f(\widetilde{H}_1)+f(\widetilde{H}_1)-f(H_1)\}\\ &= \operatorname{tr}\{f(\widetilde{H}_1)-f(H_1)\} = \int_{-\infty}^{\infty}\widetilde{\xi}(\lambda)f'(\lambda)\,d\lambda.\end{aligned}$$

Since $\widetilde{\xi}$ and $\xi(\cdot; H, H_1)$ are zero in some neighborhood of 0, these functions coincide, and it follows that $\widetilde{\xi}$, $\widetilde{\xi}_1$, $\widetilde{\xi}_2 \in L^1(\mathbb{R})$. Moreover, using again Lemma 5.2, we find

$$\begin{aligned}\int_{-\infty}^{0} \xi(\lambda; H, H_1)\,\mathrm{d}\lambda &= \int_{-\infty}^{\infty} \widetilde{\xi}_1(\lambda)\,\mathrm{d}\lambda = \lim_{\eta\uparrow\infty}\left\{-\eta^2\int_{-\infty}^{\infty}\widetilde{\xi}_1(\lambda)(\lambda-\mathrm{i}\eta)^{-2}\,\mathrm{d}\lambda\right\}\\ &= \lim_{\eta\uparrow\infty}\left\{\eta^2\mathrm{tr}\left((\widetilde{A}_1-\mathrm{i}\eta)^{-1}-(A+V-\mathrm{i}\eta)^{-1}\right)\right\}\\ &= \lim_{\eta\uparrow\infty}\left\{\eta^2\mathrm{tr}\left((I-K^*K)^{-\frac{1}{2}}(\widetilde{A}_1-\mathrm{i}\eta)^{-1}(I-K^*K)^{\frac{1}{2}}-(A+V-\mathrm{i}\eta)^{-1}\right)\right\}\\ &= \lim_{\eta\uparrow\infty}\left\{\eta^2\mathrm{tr}\left((A+V+BK-\mathrm{i}\eta)^{-1}-(A+V-\mathrm{i}\eta)^{-1}\right)\right\}\\ &= \lim_{\eta\uparrow\infty}\left\{-\eta^2\mathrm{tr}\left((A+V-\mathrm{i}\eta)^{-1}(A+V+BK-\mathrm{i}\eta)^{-1}BK\right)\right\}\\ &= \mathrm{tr}\,(BK).\end{aligned}$$

This relation together with

$$\int_{-\infty}^{\infty} \xi(\lambda; H, H_1)\,\mathrm{d}\lambda = \mathrm{tr}\,(H-H_1) = 0$$

yields (iii).

3. Assertion (iv) is a consequence of (i) and (5.10). If $V \geq 0$ then in addition to (5.6) we have $A \leq -\delta I$. Then $\xi(\cdot; H_1, H_0) = \xi(\cdot; A+V, A)$ is zero on $(-\delta, \infty)$ and nonnegative on $(-\infty, -\delta)$. In this case H_0 and H are nonpositive operators in $\mathcal{K}$ which satisfy the assumptions of [9, Satz 4.2.5] mentioned above, and we have

$$\xi(\cdot; H, H_0) = \xi(\cdot; H_1, H_0) + \xi(\cdot; H, H_1)$$

Therefore, (iv) implies (v). □

We need one more lemma.

Lemma 5.3. *Let A, B and D be as in* (5.6) *and* (5.7). *Then for the pair H_1, H as in* (5.8) *we have*

$$\xi(\lambda; H, H_1) = 0 \text{ for a.e. } \lambda \in \mathbb{R} \iff B = 0.$$

Proof. As in the proof of the corresponding result [1, Theorem 3.3] for a symmetric trace class perturbation of H_1 it follows that $\mathrm{tr}(H-H_1)^2 = 0$, and the relation

$$(H-H_1)^2 = \begin{pmatrix} -BB^* & 0 \\ 0 & -B^*B \end{pmatrix}$$

yields $B = 0$. □

Theorem 5.4. *Suppose that the operator H in $\mathcal{H}$, given by* (4.1), *satisfies all the assumptions from the first paragraph of Section* 4 *and assume additionally that $\|B\| < \delta$ holds. If $V \geq 0$, then the coefficient a in* (4.3) *is nonnegative, and we have*

$$a = 0 \iff V = B = 0.$$

Proof. By (4.4), Theorem 5.1, and (5.9) the assumption $V \geq 0$ implies that $a = \operatorname{tr} V + \operatorname{tr} BK \geq 0$. If $a = 0$, then $\operatorname{tr} V = 0$ and $\operatorname{tr} BK = 0$. It follows that $V = 0$ and, again by Theorem 5.1, $\xi(\lambda; H, H_1) = 0$ for a. e. $\lambda \in \mathbb{R}$. It remains to apply Lemma 5.3. □

Note. We use this opportunity to point out that in [1, Theorem 3.4] the first relation after (3.19) should read $a = \operatorname{tr}(V+BX)$ (instead of $a = \operatorname{tr}(V+BX) \leq 0$).

References

[1] V. Adamyan and H. Langer: Partial non-stationary perturbation determinants. *Operator Theory: Advances and Applications* **154** (2004), 1–18.

[2] V. Adamyan and H. Langer: Spectral properties of a class of rational operator-valued functions *J. Operator Theory* **33** (1995), 259–277.

[3] V. Adamjan and H. Langer: The spectral shift function for certain block operator matrices. *Math. Nachr.* **211** (2000), 5–24.

[4] V. Adamyan, H. Langer and C. Tretter: Existence and uniqueness of contractive solutions of some Riccati equations. *J. Funct. Anal.* **179** (2001), 448–473

[5] T.Ya. Azizov, I.S. Iokhvidov: Linear Operators in Spaces with an Indefinite Metric. Wiley, Chichester, 1989.

[6] M.S. Birman, D.R. Yafaev: The spectral shift function. The papers of M.G. Krein and their further development. *Algebra i Analiz* **4,5** (1992), 1–44 (Russian).

[7] B. Ćurgus: On the regularity of the critical point infinity of definitizable operators. *Integral Equations Operator Theory* **8** (1985), 462–488.

[8] I.C. Gohberg, M.G. Krein: Introduction to the Theory of Linear Nonselfadjoint Operators. *Translations of Mathematical Monographs* **18**, AMS: Providence, R.I., 1969.

[9] P. Jonas: Die Spurformel der Störungstheorie für einige Klassen unitärer und selbstadjungierter Operatoren im Kreinraum. *Report* R-MATH-**06/86**, Karl-Weierstrass-Institut für Mathematik, Berlin 1986.

[10] T. Kato: Perturbation Theory for Linear Operators, Springer-Verlag, New York 1966.

[11] L.S. Koplienko: The trace formula in abstract scattering theory for a pair of spaces. (Russian) *Scattering theory. Theory of oscillations. Probl. Mat. Fiz.* **9**, pp. 41–53, Leningrad. Univ., Leningrad, 1979.

[12] M.G. Krein: On the trace formula in perturbation theory. *Mat. Sb.* **33**(**75**) (1953), 597–626. (Russian)

[13] H. Langer: Spectral functions of definitizable operators in Krein spaces. *Lecture Notes in Mathematics* **948**, pp. 1–46, Springer-Verlag, Berlin 1982.

[14] H. Langer and C. Tretter: Spectral decomposition of some non-selfadjoint operator matrices. *J. Operator Theory* **39** (1998), 1–20.

[15] H. Langer and C. Tretter: Diagonalization of certain block operator matrices and application to Dirac operators. *Operator Theory: Advances and Applications* **122**, (2001), 331–358.

[16] K. Veselić: On spectral properties of a class of J-selfadjoint operators. I, II. *Glasnik Mat.* Ser. III **7(27)** (1972), 229–248; ibid. **7(27)** (1972), 249–254.

Vadim Adamyan
Department of Theoretical Physics
Odessa National University,
65026 Odessa
Ukraine
e-mail: vadamyan@paco.net

Peter Jonas
Institut für Mathematik
Technische Universität Berlin
Str. d. 17. Juni 136
D-10623 Berlin
Germany
e-mail: jonas@math.berlin.de

Heinz Langer
Institut für Analysis und Scientific Computing
Technische Universität Wien,
Wiedner Hauptstr. 8–10
A-1040 Wien
Austria
e-mail: hlanger@mail.zserv.tuwien.ac.at

Operator Theory:
Advances and Applications, Vol. 162, 19–45

Reproducing Kernel Spaces of Series of Fueter Polynomials

Daniel Alpay, Michael Shapiro and Dan Volok

Abstract. We study reproducing kernel spaces of power series of Fueter polynomials and their multipliers. In particular we prove a counterpart of Beurling–Lax theorem in the quaternionic Arveson space and we define and characterize counterparts of the Schur–Agler classes. We also address the notion of rationality in the hyperholomorphic setting.

Introduction

Reproducing kernel Hilbert spaces of analytic functions in one complex variable play an important role in operator theory, in particular in operator models and in interpolation theory, to name two instances. An important case is that of reproducing kernels of the form $c(z\overline{w})$ where c is a function analytic in a neighborhood of the origin with power series expansion $c(t) = \sum_{n=0}^{\infty} c_n t^n$ such that $c_n \geq 0$ for all $n \in \mathbb{N}$. The function $K(z, w) = c(z\overline{w})$ is positive and the associated reproducing kernel Hilbert space is the set of functions

$$f(z) = \sum_{\substack{n=0 \\ c_n \neq 0}}^{\infty} f_n z^n,$$

with norm

$$\|f\|^2 = \sum_{\substack{n=0 \\ c_n \neq 0}}^{\infty} \frac{|f_n|^2}{c_n} < \infty.$$

The research of Michael Shapiro was partially supported by CONACYT projects as well as by Instituto Politécnico Nacional in the framework of COFAA and CGPI programs. The stay of Michael Shapiro at Ben–Gurion University and the postdoctoral fellowship of Dan Volok were supported by the Center for Advanced Studies in Mathematics of Ben Gurion University of the Negev.

Typical examples include the Hardy space and the Bergman space, corresponding respectively to the functions

$$c(t) = \frac{1}{1-t} \quad \text{and} \quad c(t) = \frac{1}{(1-t)^2}.$$

These spaces of power series also have counterparts in the setting of several complex variables: $K(z,w) = c(\langle z,w\rangle)$ with $z,w \in \mathbb{C}^N$ and

$$t = \langle z,w\rangle = \sum_{i=1}^{N} z_i\overline{w_i}.$$

Examples include the Hardy space of the unit ball $\mathbb{B}_N$, the Bergman space and the Arveson space and the corresponding functions c are respectively

$$c(t) = \frac{1}{(1-t)^N}, \qquad c(t) = \frac{1}{(1-t)^{N+1}}, \qquad \text{and} \quad c(t) = \frac{1}{(1-t)}.$$

Other spaces of interest correspond to kernels of the form

$$K(z,w) = \sum_{\substack{n\in\mathbb{N}^N \\ c_n\neq 0}} c_n z^n \overline{w}^n,$$

where we have used the multi-index notation and where the $c_n \geq 0$. The corresponding reproducing kernel Hilbert spaces are sometimes called *weighted power series spaces*. The spaces under consideration will now include the Hardy space of the polydisk $\mathbb{D}^N$, corresponding to $c_n \equiv 1$.

We study in the present paper the counterparts of these weighted power series spaces in the quaternionic setting; power series are now replaced by series of Fueter polynomials. See the discussion at the beginning of Section 3.

This paper is written with two audiences in mind and is at the intersection of two different fields; on the one hand, people familiar with the theory (or one should say, theories) of reproducing kernel Hilbert spaces of power series in one and several complex variables and on the other hand, people familiar with hypercomplex analysis.

The paper intends to be of a review nature and also to contain new results: among the new results presented we mention:

1. Another approach to quaternionic rational functions; see Theorem 2.8.
2. The fact that the quaternionic Cauchy kernel is rational. See Corollary 2.10.
3. A characterisation of the Leibenson shift operators. See Theorem 3.5.
4. A Beurling type theorem in the quaternionic Arveson space.
5. A definition and study of Schur–Agler type classes in the quaternionic setting. See Section 4.2.

The setting which we present contains both a non-commutative and an analytic aspects. It is different from the non-commutative theory (but some formulas are quite similar; see, e.g., formula (3.12) for the realization of a Schur multiplier) and it is also quite different from the analytic setting. The Fueter polynomials (see

Definition 2.1) play now the role of the usual monomials $z_1^{n_1} \cdots z_N^{n_N}$ and, although similar in notation, have quite different properties.

Before considering the hyperholomorphic case we discuss briefly in the next section the case of several complex variables.

1. The case of several complex variables

1.1. Rational functions

Quaternionic hyperholomorphic rational functions play an important role in this paper and we begin by reviewing some facts for the corresponding objects in the setting of one and several complex variables. A *rational function* of one complex variable is just a quotient of polynomials with complex coefficients. A matrix-valued function is rational if its entries are rational. Equivalently, it is rational if it is the quotient of a matrix polynomial (a matrix function with polynomial entries) with a scalar polynomial. Originating with the theory of linear systems, another representation of rational function proved to be very useful:

Proposition 1.1. *A matrix-valued function $r(z)$ analytic in a neighborhood of the origin is rational if and only if it can be written as*

$$r(z) = D + C(I_n - zA)^{-1}zB \tag{1.1}$$

where I_n denotes the identity matrix of order n and where A, B, C and D denote matrices of appropriate sizes.

An expression of the form (1.1) is called a *realization*. See, e.g., [9], where functions analytic at infinity rather than at the origin are considered.

In several complex variables, rational functions are also defined as quotient of polynomials, and the realization result extend: a matrix-valued function of several complex variables analytic in a neighborhood of the origin is rational if and only if it can be written as

$$r(z) = D + C(I_n - \sum_{i=1}^{N} z_i A_i)^{-1} \left(\sum_{i=1}^{N} z_i B_i \right) \tag{1.2}$$

where $A_1, \ldots, A_N, B_1, \ldots, B_N, C$ and D are matrices of appropriate sizes. We can rewrite (1.2) as (1.1) by setting

$$z = \begin{pmatrix} z_1 I_n & \cdots, z_N I_n \end{pmatrix}, \quad A = \begin{pmatrix} A_1 \\ \vdots \\ A_N \end{pmatrix} \quad B = \begin{pmatrix} B_1 \\ \vdots \\ B_N \end{pmatrix}.$$

For a recent proof of this well-known realization result, see [3]. We refer to the papers [18] and [29] for connections with the theory of linear systems (note that the paper [29] considers a different kind of realization).

1.2. Some Hilbert spaces of power series

In this section we review some results from the theory of reproducing kernel Hilbert spaces of power series in several complex variables.

Definition 1.2. *Let* $\mathbf{c} = \{c_\nu\}$ *be a sequence of positive numbers indexed by* $\mathbb{N}^N$ *and let its support* supp $(\mathbf{c})$ *be defined by*

$$\text{supp }(\mathbf{c}) = \left\{\nu \in \mathbb{N}^N \mid c_\nu \neq 0\right\}. \tag{1.3}$$

We denote by $\mathcal{H}(\mathbf{c})$ *the space of power series of the form*

$$f(x) = \sum_{\nu \in \text{supp }(\mathbf{c})} z^\nu f_\nu$$

where the $f_\nu \in \mathbb{C}$ *are such that*

$$\|f\|_{\mathbf{c}}^2 := \sum_{\nu \in \text{supp }(\mathbf{c})} \frac{|f_\nu|^2}{c_\nu} < \infty. \tag{1.4}$$

In the sequel we use the notion of *lower inclusive sets*. These sets were introduced in the work [8] of Ball, Li, Timotin and Trent (the term itself was coined in Woerderman's paper [37]) and used in [37] to solve the Carathéodory–Féjer interpolation problem in the polydisk.

Define a partial order $\leq_p$ on $\mathbb{N}^N$ as follows: For $k = (k_1, k_2, \ldots, k_N) \in \mathbb{N}^N$ and $\ell = (\ell_1, \ell_2, \ldots, \ell_N) \in \mathbb{N}^N$, we say that $k \leq_p \ell$ if and only if $k_i \leq \ell_i \quad i = 1, 2, \ldots, N$.

Definition 1.3. *A set* $\mathcal{K} \subseteq \mathbb{N}^N$ *is said to be lower inclusive if the following condition holds:*
if $k \in \mathcal{K}$ *and* $\ell \leq_p k$, *then* $\ell \in \mathcal{K}$.

1.3. Gleason's problem and the Leibenson's shift operators

What is now called Gleason's problem was considered by Hefer; the paper [25] was published in 1950. In a footnote, Behnke and Stein state that the author died in 1941 and that the paper is part of his 1940 Munster dissertation. For a related result, see also [11].

Problem 1.4. *Let* $\mathcal{M}$ *be a set of functions analytic in a subset* $\Omega \subset \mathbb{C}^N$ *and let* $a \in \Omega$. *Given* $f \in \mathcal{M}$; *to find functions* $g_1(z, a), \ldots, g_N(z, a) \in \mathcal{M}$ *such that*

$$f(z) - f(a) = \sum_{j=1}^N (z_j - a_j) g_j(z, a).$$

A more restrictive requirement is to ask that there are bounded operators $T_{j,a}$ such that $g_j(z, a) = (T_{j,a} f)(z)$. We then say that the $T_{j,a}$ solve Gleason's problem. When $N = 1$ and $a = 0$, we get back to the well-known notion of backward-shift invariance: is $\mathcal{M}$ invariant under the backward-shift operator $\frac{f(z) - f(0)}{z}$?

Let us take $a = 0$ and $N \geq 1$. Differentiating the function $t \mapsto f(tz)$ (with $t \in [0,1]$) and integrating back one obtains

$$f(z) - f(0) = \sum_{\ell=1}^{N} z_j(\mathcal{R}_j f)(z)$$

where

$$\mathcal{R}_j f(z) = \int_0^1 \frac{\partial f}{\partial z_j}(tz)dt.$$

One has

$$\mathcal{R}_j z^\alpha = \begin{cases} \frac{\alpha_j}{|\alpha|} z^{\alpha - e_j} & \text{if } \alpha_j > 0, \\ 0 & \text{if } \alpha_j = 0, \end{cases}$$

where e_j denotes the row vector with all components equal to 0, besides the jth one equal to 1, and so:

Lemma 1.5. *A necessary condition for a space $\mathcal{H}(\mathbf{c})$ to be $\mathcal{R}_j$-invariant is that the set* supp $(\mathbf{c})$ *is lower inclusive.*

The $\mathcal{R}_j$ are the generalized backward-shift operators introduced by Leibenson and one version of Gleason's problem is to ask whether the space $\mathcal{M}$ is invariant under the $\mathcal{R}_j$. Of course a negative answer does not mean that Gleason's problem is not solvable in $\mathcal{M}$.

The functions $g_j(z,a)$ are not uniquely defined in general; on the other hand, when $a = 0$, the choice $g_j(z,0) = R_j f(z)$ is unique under appropriate hypothesis. A first set of such hypothesis was given in [15], where E. Doubtsov proved that the Leibenson solution is a minimal solution (in an appropriate sense). In [2] it is shown that if the space is $\mathcal{R}_j$-invariant the $\mathcal{R}_j$ are the only commutative solution to Gleason's problem.

Theorem 1.6. *Let $\mathcal{P}$ be a space of $\mathbb{C}^p$-valued functions analytic on a domain $\Omega \subset \mathbb{C}^N$ containing the origin, and which is invariant under the multiplication operators M_{z_j} for $j = 1, \ldots, N$. The set of commuting, bounded operators solving Gleason's problem in $\mathcal{P}$, if it exists, is unique, and is given by*

$$T_j := T_{j,0} \colon f(z) \mapsto g_j(z)$$

where $g_j(z)$ is the uniquely determined element of $\mathcal{P}$ having Taylor expansions with center point at the origin given by

$$g_j(z) = \sum_{\alpha \in \mathbb{N}^N \colon \alpha_j \geq 1} z^{\alpha - e_j} \frac{\alpha_j}{|\alpha|} f_\alpha$$

if $f(z)$ has Taylor expansion at the origin given by

$$f(z) = \sum_{\alpha \in \mathbb{N}^N} z^\alpha f_\alpha.$$

In Section 3.2 we prove a similar result in the setting of hyperholomorphic functions.

1.4. Schur multipliers

Definition 1.7. *Let $K(z,w)$ be a function positive on the set Ω. The function $s : \Omega \longrightarrow \mathbb{C}$ is called a Schur multiplier if the operator M_s of multiplication by s is a contraction from $\mathcal{H}(K)$ (the reproducing kernel Hilbert space with reproducing kernel K) into itself.*

It is well known (but, as we will see, the hyperholomophic counterparts of these formulas are more involved; see, e.g., formula (3.2)) that

$$M_s^* K(\cdot,w) = K(\cdot,w)s(w)^* \tag{1.5}$$

and that s is a Schur multiplier if and only if the function

$$(1 - s(z)s(w)^*)K(z,w)$$

is positive in Ω.

1.5. The Hardy space of the ball

The Hardy space of the ball $\mathbf{H}_2(\mathbb{B}_N)$ is the reproducing kernel Hilbert space with reproducing kernel $\frac{1}{(1-\langle z,w\rangle)^N}$. Since

$$\frac{1}{(1-\langle z,w\rangle)^N} = \sum_{\alpha\in\mathbb{N}^N} \frac{(N+|\alpha|-1)!}{\alpha!(N-1)!} z^\alpha \overline{w^\alpha}$$

the space $\mathbf{H}_2(\mathbb{B}_N)$ is a weighted power series space and its elements can be characterized via (1.4). A function s analytic in the ball is a contractive multiplier for the Hardy space if and only if the kernel

$$\frac{1 - s(z)s(w)^*}{(1-\langle z,w\rangle)^N} \tag{1.6}$$

is positive in $\mathbb{B}_N$.

The norm (1.4) has also a geometric interpretation as

$$\|f\|^2 = \sup_{0<r<1} \int_{|z|=1} |f(rz)|^2 d\lambda(z).$$

Thanks to this interpretation, Schur multipliers of the Hardy space are readily seen (as in the case $N=1$) to be exactly the set of functions analytic and contractive in $\mathbb{B}_N$. For $N=1$ interpolation theory and realization theory of these functions is a very well developed topic, known as Schur analysis; see, e.g., [19], [21] for some references. For $N>1$ these same questions (interpolation theory and realization theory) seem beyond the scope of current methods of several complex variables and operator theory.

1.6. The Arveson space and de Branges–Rovnyak spaces

The kernel $\frac{1}{1-\langle z,w\rangle}$ is positive in the open unit ball. When $N > 1$ the associated reproducing kernel Hilbert space is strictly and contractively included in the Hardy space of the ball. This space was introduced by Drury in [16] and studied further by Arveson [7]. Following other authors we will call it the Arveson space.

A function s analytic in the ball is a contractive multiplier for the Arveson space if and only if the kernel

$$\frac{1 - s(z)s(w)^*}{1 - \langle z, w\rangle}$$

is positive in $\mathbb{B}_N$. Note the difference with (1.6). We note that there are functions analytic and contractive in the ball and which are not Schur multipliers of the Arveson space. In the statement (and in the sequel of the paper), a co-isometric operator is an operator whose adjoint is isometric.

Theorem 1.8. *A function s analytic in the ball is a Schur multiplier of the Arveson space if and only if there exists a Hilbert space $\mathcal{H}$ and a co-isometric operator*

$$\begin{pmatrix} A & B \\ C & D \end{pmatrix} = \begin{pmatrix} A_1 & B_1 \\ \vdots & \vdots \\ A_N & B_N \\ C & D \end{pmatrix} : \mathcal{H}^N \oplus \mathbb{C} \Longrightarrow \mathcal{H} \oplus \mathbb{C}$$

such that

$$s(z) = D + C(I_{\mathcal{H}} - zA)^{-1}zB \tag{1.7}$$

where

$$zA = z_1A_1 + \cdots + z_NA_N, \quad zB = z_1B_1 + \cdots + z_NB_N.$$

Remark 1.9. *It follows from formula* (1.7) *that we have the power series expansion*

$$s(z) = D + \sum_{k=1}^{N} \sum_{\nu\in\mathbb{N}^N} \frac{|\nu|!}{\nu!} z_k z^\nu C A^\nu B_k,$$

where

$$A^\nu = A_1^{\times\nu_1} \times \cdots \times T_N^{\times\nu_N}$$

and

$$A_1 \times A_2 \times \cdots \times A_n = \frac{1}{n!} \sum_{\sigma\in S_n} A_{\sigma(1)}A_{\sigma(2)} \cdots A_{\sigma(n)}.$$

Remark 1.10. *The knowledgeable reader will have noticed that in the above statements it is not necessary to assume that s is* analytic *in the ball. It is enough to assume that s is defined on a uniqueness set in the ball. This is one instance of a general principle where positivity forces analyticity.*

We presented the definition and characterization of Schur multipliers in the scalar case, but these also make sense in the case of operator-valued functions.

1.7. The polydisk and the Schur–Agler classes

The Hardy space of the polydisk is the reproducing kernel Hilbert space with reproducing kernel

$$k(z,w)=\frac{1}{\prod_1^N(1-z_j\overline{w_j})}.$$

It would seem natural to define in the setting of the polydisk the class of Schur multipliers, that is functions analytic in the polydisk and such that the kernel

$$\frac{1-s(z)\overline{s(w)}}{\prod_1^N(1-z_j\overline{w_j})} \tag{1.8}$$

is positive in $\mathbb{D}^N$. Unfortunately, as soon as $N>2$, these are not classes for which there is a nice characterization in terms of realization. J. Agler introduced (see [1]) the class of functions s such that

$$1-s(z)\overline{s(w)}=\sum_{j=1}^N(1-z_j\overline{w_j})K_j(z,w)$$

for some (in general not uniquely defined) functions $K_1,\ldots,K_N$ positive in $\mathbb{D}^N$.

Dividing both sides of the above equality by $\prod_1^N(1-z_j\overline{w_j})$ we see that the kernel (1.8) is positive. In particular the function s is a contractive multiplier of the Hardy space of the polydisk and is thus automatically analytic there. For $N>2$ the Schur–Agler class is strictly smaller than the class of contractive multipliers of the Hardy space of the polydisk. As in Section 1.6 we focus on the scalar case.

Theorem 1.11. *A function s analytic in the polydisk is in the Schur–Agler class if and only if it can be written as*

$$s(z)=D+C(I-d(z)A)^{-1}d(z)B,$$

where in the expression $d(z)=\operatorname{diag}\,(z_jI_{\mathcal{H}_j})$ for some Hilbert spaces $\mathcal{H}_j$ and

$$\begin{pmatrix}A & B\\ C & D\end{pmatrix}:\begin{pmatrix}\mathcal{H}_1\\ \vdots\\ \mathcal{H}_N\\ \mathbb{C}\end{pmatrix}\longrightarrow\begin{pmatrix}\mathcal{H}_1\\ \vdots\\ \mathcal{H}_N\\ \mathbb{C}\end{pmatrix}$$

is a co-isometric operator.

One of the main results of this paper is the definition and characterization of the Schur–Agler classes in the quaternionic setting; see Section 4.2.

2. Hyperholomorphic functions

2.1. Quaternions and quaternionic hyperholomorphic functions

The building of the skew-field of quaternions has a fascinating history; see for instance [14]. For our present purposes it is enough to define directly the quaternions as

$$\mathbb{H} = \left\{ q = \begin{pmatrix} z & w \\ -\overline{w} & \overline{z} \end{pmatrix}, \quad z, w \in \mathbb{C} \right\}.$$

This is readily seen to be a skew-field. Writing $z = x_0 + ix_1$ and $w = x_2 + ix_3$ we have that

$$x = x_0\mathbf{e}_0 + x_1\mathbf{e}_1 + x_2\mathbf{e}_2 + x_3\mathbf{e}_3 \tag{2.1}$$

where

$$\mathbf{e}_0 = \begin{pmatrix} 1 & 0 \\ 0 & 1 \end{pmatrix}, \quad \mathbf{e}_1 = \begin{pmatrix} i & 0 \\ 0 & -i \end{pmatrix}, \quad \mathbf{e}_2 = \begin{pmatrix} 0 & 1 \\ -1 & 0 \end{pmatrix}, \quad \text{and} \quad \mathbf{e}_3 = \begin{pmatrix} 0 & i \\ i & 0 \end{pmatrix}.$$

We will denote $\mathbf{e}_0 = 1$ and note that the $\mathbf{e}_j$ satisfy the Cayley multiplication table

	$\mathbf{e_0}$	$\mathbf{e_1}$	$\mathbf{e_2}$	$\mathbf{e_3}$
$\mathbf{e_0}$	$\mathbf{e_0}$	$\mathbf{e_1}$	$\mathbf{e_2}$	$\mathbf{e_3}$
$\mathbf{e_1}$	$\mathbf{e_1}$	$-\mathbf{e_0}$	$\mathbf{e_3}$	$-\mathbf{e_2}$
$\mathbf{e_2}$	$\mathbf{e_2}$	$-\mathbf{e_3}$	$-\mathbf{e_0}$	$\mathbf{e_1}$
$\mathbf{e_3}$	$\mathbf{e_3}$	$\mathbf{e_2}$	$-\mathbf{e_1}$	$-\mathbf{e_0}$

. (2.2)

In the sequel we identify the space $\mathbb{R}^3$ with the set of purely vectorial quaternions, that is quaternions x such that $x_0 = 0$.

The function $f : \Omega \subset \mathbb{R}^4 \to \mathbb{H}$ is called *left-hyperholomorphic* if

$$D\ f := \frac{\partial}{\partial x_0} f + \mathbf{e_1}\frac{\partial}{\partial x_1} f + \mathbf{e_2}\frac{\partial}{\partial x_2} f + \mathbf{e_3}\frac{\partial}{\partial x_3} f = 0. \tag{2.3}$$

Write $f = f_0 + \mathbf{e}_1 f_1 \mathbf{e}_2 + f_2 + \mathbf{e}_3 f_3$. The components f_j of f satisfy the system

$$\begin{aligned}
&\frac{\partial f_0}{\partial x_0} - \frac{\partial f_1}{\partial x_1} - \frac{\partial f_2}{\partial x_2} - \frac{\partial f_3}{\partial x_3} = 0,\\
&\frac{\partial f_0}{\partial x_1} + \frac{\partial f_1}{\partial x_0} - \frac{\partial f_2}{\partial x_3} + \frac{\partial f_3}{\partial x_2} = 0,\\
&\frac{\partial f_0}{\partial x_2} + \frac{\partial f_1}{\partial x_3} + \frac{\partial f_2}{\partial x_0} - \frac{\partial f_3}{\partial x_1} = 0,\\
&\frac{\partial f_0}{\partial x_3} - \frac{\partial f_1}{\partial x_2} + \frac{\partial f_2}{\partial x_1} + \frac{\partial f_3}{\partial x_0} = 0.
\end{aligned} \tag{2.4}$$

See, e.g., [20, equations (2a) p. 76]. The system of equations (2.4) when there is no dependence on x_0 appears in [27, (5) p. 985]; nowadays it bears the name of the Moisil–Theodoresco system and there is a long list of works about its properties. A curious reader can find it useful to look into the books [17] and [23] as well as into the papers [22], [31], [33] and [36].

The case where $f_0 \equiv 0$ and where f_1, f_2 and f_3 do not depend on x_0 is of special interest; see [20, p. 78]. The system (2.4) can be re-written now as

$$\begin{aligned}
&\operatorname{div} \vec{f} = 0,\\
&\operatorname{rot} \vec{f} = 0,
\end{aligned} \tag{2.5}$$

where $\vec{f} = (f_1, f_2, f_3)$; hence (2.5) being a particular case of (2.4) has both purely mathematical and physical developments. Again, a long list of references could be composed from which we indicate a few instances: [10, pp. 81–96], [38], [24], [30].

A solution of (2.5), and more generally of its generalization to any dimension, is called a system of conjugate harmonic functions; see [28, p. 18]. The paper [32] can be useful for a first acquaintance and to understand the main ideas.

We now introduce a family of hyperholomorphic polynomials, and describe the counterpart of the Taylor series at the origin. Let f be left-hyperholomorphic. The chain rule gives

$$\frac{\mathrm{d}}{\mathrm{d}\,t} f(tx) = \sum_{\ell=0}^{3} x_\ell \frac{\partial f}{\partial x_\ell}(tx). \tag{2.6}$$

Since the function is left-hyperholomorphic we have

$$\frac{\partial f}{\partial x_0} = -\mathbf{e_1}\frac{\partial}{\partial x_1} f - \mathbf{e_2}\frac{\partial}{\partial x_2} f - \mathbf{e_3}\frac{\partial}{\partial x_3} f.$$

Replacing $\frac{\partial f}{\partial x_0}$ by this expression in (2.6) we obtain

$$\begin{aligned}\frac{\mathrm{d}}{\mathrm{d}\,t} f(tx) &= x_0\left(-\mathbf{e_1}\frac{\partial}{\partial x_1} f(tx) - \mathbf{e_2}\frac{\partial}{\partial x_2} f(tx) - \mathbf{e_3}\frac{\partial}{\partial x_3} f(tx)\right) + \\ &\quad + \sum_{\ell=1}^{3} x_\ell \frac{\partial f}{\partial x_\ell}(tx) \\ &= \sum_{\ell=1}^{3} (x_\ell - x_0\mathbf{e}_\ell)\frac{\partial f}{\partial x_\ell}(tx).\end{aligned}$$

Integrating with respect to t we obtain

$$f(x) - f(0) = \sum_{\ell=1}^{3} (x_\ell - x_0\mathbf{e}_\ell)\int_0^1 \frac{\partial f}{\partial x_\ell}(tx)dt. \tag{2.7}$$

It remains to show that the functions $g_\ell(x) = \int_0^1 \frac{\partial f}{\partial x_\ell}(tx)dt$ are left-hyperholomorphic. This follows from the fact that, for a given t, Df evaluated at the point tx is equal to 0 and that we can interchange integration and derivation when computing Dg_ℓ.

We note that the functions $\zeta_\ell(x) = x_\ell - x_0\mathbf{e}_\ell$ are hyperholomorphic. Iterating formula (2.7) we get

$$f(x) - f(0) = \sum_{\nu\in\mathbb{N}^3} \zeta^\nu f_\nu$$

where $f_\nu \in \mathbb{H}$ and ζ^ν are non-commutative homogeneous hyperholomorphic polynomials in the ζ_j given by the formula

$$\zeta^\nu(x) = \zeta_1(x)^{\times\nu_1} \times \zeta_2(x)^{\times\nu_2} \times \zeta_3(x)^{\times\nu_3}, \tag{2.8}$$

where $\nu = (\nu_1, \nu_2, \nu_3)$ and where the symmetrized product of $a_1, \ldots, a_n \in \mathbb{H}$ is defined by

$$a_1 \times a_2 \times \cdots \times a_n = \frac{1}{n!} \sum_{\sigma \in S_n} a_{\sigma(1)} a_{\sigma(2)} \cdots a_{\sigma(n)}, \tag{2.9}$$

where S_n is the set of all permutations of the set $\{1, \ldots, n\}$.

Definition 2.1. *The polynomials $\zeta^\nu(x)$ defined by (2.8) are called the Fueter polynomials.*

2.2. The Cauchy–Kovalevskaya extension and product

The pointwise product of two hyperholomorphic functions is not in general hyperholomorphic. The Cauchy–Kovalevskaya product allows to remedy this situation. Let $\varphi(x_1, x_2, x_3)$ be a real analytic function from some open domain of $\mathbb{R}^3$ into $\mathbb{H}$, that is φ is given by four coordinate real analytic real-valued functions

$$\varphi(x_1, x_2, x_3) = \varphi_0(x_1, x_2, x_3) + \sum_1^3 \mathbf{e}_i \varphi_i(x_1, x_2, x_3).$$

The Cauchy–Kovalevskaya theorem (in fact, in its simplest form; see [26, Section 1.7]; see also [26, Section 1.10] and [13, Section I.7] for the general version) implies that the system of equations (2.4) with initial conditions

$$f_i(0, x_1, x_2, x_3) = \varphi_i(x_1, x_2, x_3)$$

admits a unique real analytic solution in a neighborhood of the origin in $\mathbb{R}^4$. This solution

$$f(x_0, x_1, x_2, x_3) = f_0(x_0, x_1, x_2, x_3) + \sum_1^3 \mathbf{e}_i f_i(x_0, x_1, x_2, x_3)$$

is hyperholomorphic by definition and is called the Cauchy–Kovalevskaya extension of the function φ.

Example 2.2. *The Cauchy–Kovalevskaya extension of the polynomial x^α is the Fueter polynomial ζ^α.*

More generally, the Cauchy–Kovalevskaya extension of the $\mathbb{H}$-valued real-analytic function $\sum_{k=0}^\infty \sum_{|\nu|=k} x^\alpha f_\nu$ (where $f_\nu \in \mathbb{H}$) is the function

$$f(x) = \sum_{k=0}^\infty \sum_{|\nu|=k} \zeta^\nu f_\nu.$$

Consider now two hyperholomorphic functions f and g and let φ and ψ be their restrictions to $\mathbb{R}^3$ (that is, when setting $x_0 = 0$). The functions φ and ψ are real analytic and so is their (pointwise) product. The Cauchy–Kovalevskaya extension of $\varphi\psi$ is called the *Cauchy–Kovalevskaya product* of f and g. It was first introduced by F. Sommen in [35].

Consider now a function f hyperholomorphic in a neighborhood of the origin. The function $f(0,x_1,x_2,x_3)$ is real analytic in a neighborhood of the origin of $\mathbb{R}^3$ and thus we can write

$$f(0,x_1,x_2,x_3)-f(0,0,0,0)=x_1h_1(x_1,x_2,x_3)+x_2h_2(x_1,x_2,x_3)+x_3h_3(x_1,x_2,x_3),$$

where the h_j are $\mathbb{H}$-valued and real analytic in a neighborhood of the origin of $\mathbb{R}^3$. Taking the Cauchy–Kovaleskaya extension of this expression we get to

$$f(x)-f(0)=\zeta_1(x)\odot g_1(x)+\zeta_2(x)\odot g_2(x)+\zeta_3(x)\odot g_3(x), \tag{2.10}$$

where the g_j are hyperholomorphic in a neighborhood of the origin.

The Cauchy–Kovalevskaya product can be defined also directly in terms of the power series expansions at the origin of the two functions. More precisely we have:

Theorem 2.3. *Let f and g be two functions hyperholomorphic in a neighborhood of the origin, with power series expansions*

$$f(x)=\sum_{k=0}^{\infty}\sum_{|\nu|=k}\zeta^{\nu}f_{\nu}\quad \textit{and}\quad f(x)=\sum_{k=0}^{\infty}\sum_{|\nu|=k}\zeta^{\nu}g_{\nu}.$$

Then,

$$(f\odot g)(x)=\sum_{n=0}^{\infty}\sum_{|\eta|=n}\zeta^{\eta}\sum_{0\leq\nu\leq\eta}f_{\nu}g_{\eta-\nu}.$$

The proof of Theorem 2.3 can be found in [12]. In view of Example 2.2, it follows from the right $\mathbb{H}$-linearity of the equation (2.3).

2.3. The quaternionic Cauchy kernel

Neither the quaternionic variable (2.1) nor its powers x^n (with $n=\pm1,\pm2,\dots$) are hyperholomorphic. As noted by Fueter [20, p. 77] the functions $\Delta_{\mathbb{R}^4}x^n$ are both left- and right-hyperholomorphic.

The quaternionic Cauchy kernel is defined by the formula: for $x\neq 0$,

$$K(x):=-\frac{1}{2\,\mathrm{vol}\,\mathbb{S}^3}\overline{D}\frac{1}{|x|^2}=\frac{1}{\mathrm{vol}\,\mathbb{S}^3}\frac{\overline{x}}{|x|^4}=-\frac{1}{4\,\mathrm{vol}\,\mathbb{S}^3}\Delta_{\mathbb{R}^4}x^{-1}.$$

For a function $f(x)$ left hyperholomorphic in a neighborhood of the ball $B(0,R)$ the following Cauchy formula holds:

$$f(x)=\int_{|y|=R}K(y-x)d\sigma f(y).$$

Finally we note the expansion

$$\Delta\,(y-x)^{-1}=y^{-1}\sum_{n=0}^{\infty}\Delta\,(xy^{-1})^{n+2}=\sum_{n\in\mathbb{N}^3}\alpha_n(y)\beta_n(x)$$

valid for $|x|<|y|$. See [20, p. 81]. The β_n are hyperholomorphic polynomials, and are in fact the Fueter polynomials defined above.

2.4. Rational functions

In [4] we defined matrix-valued rational hyperholomorphic functions as functions obtained from Fueter polynomials after a finite number of operations of the following type: addition, Cauchy–Kovalevskaya multiplication, and if the the function is invertible at the origin, say $R(x) = R(0)(I_p - T(x))$, with $T(0) = 0$ and $R(0)$ invertible, then the Cauchy–Kovalevskaya inversion is defined by

$$R(x)^{-1} = \{I_p + T(x) + T(x) \odot T(x) + T(x) \odot T(x) \odot T(x) + \cdots\} R(0)^{-1}.$$

We proved that:

Proposition 2.4. *A matrix-valued function hyperholomorphic in a neighborhood of the origin is rational if and only if it can be written as*

$$\begin{aligned} R(x) = D + C \odot (I - (\zeta_1(x)A_1 + \zeta_2(x)A_2 + \zeta_3(x)A_3))^{-\odot} \odot \\ \odot (\zeta_1(x)B_1 + \zeta_2(x)B_2 + \zeta_3(x)B_3) \end{aligned} \tag{2.11}$$

where $A_1, A_2, A_3, B_1, B_2, B_3, C$ and D are matrices of appropriate sizes.

Compare (2.11) with (1.2).

We now give another characterization of hyperholomorphic rational functions. As a corollary we will obtain that the quaternionic Cauchy kernel is rational see also [5].

We first define rational functions of three real variables and whose values are matrices with quaternionic entries. Since the variables and the coefficients commute the notion of polynomials makes no difficulty. We call matrix-polynomial any finite sum

$$p(x) = \sum x^{\alpha_1} x^{\alpha_2} x^{\alpha_3} p_{(\alpha_1,\alpha_2,\alpha_3)} \tag{2.12}$$

where the $p_\alpha \in \mathbb{H}^{p\times q}$.

Definition 2.5. *A rational function of three real variables and with quaternionic coefficients is any function obtained from polynomials of the form* (2.12) *after a finite number of the following operations: addition, pointwise multiplication and inversion.*

In the sequel we focus on the case of functions which are real analytic in a neighborhood of the origin.

Proposition 2.6. *A function of three real variables, real analytic in a neighborhood of the origin and with quaternionic coefficients is rational if and only if it can be represented as*

$$r(x_1, x_2, x_3) = D + C(I - (x_1A_1 + x_2A_2 + x_3A_3))^{-1}(x_1B_1 + x_2B_2 + x_3B_3) \tag{2.13}$$

where $A_1, A_2, A_3, B_1, B_2, B_3, C$ and D are matrices of appropriate sizes.

Proof. The proof follows a classical argument and proceeds in a number of steps (we omit the proofs):

Step 1. Constant matrices and monomials of the form x_iM have realizations of the form (2.13).

Step 2. If r and s admit realizations of the form (2.13) and if the product rs (resp. the sum $r+s$) makes sense, then the product (resp. the sum) admits also a realization of the form (2.13).

We note that the result on the sum follows from the result on the product since (with identities of appropriate sizes)

$$r+s = \begin{pmatrix} r & I \end{pmatrix} \begin{pmatrix} I \\ s \end{pmatrix}.$$

Step 3. If r admits a realization and $r(0)$ is invertible, then r^{-1} also admits a realization.

Proposition 2.7. *A function of three real variables, real analytic in a neighborhood of the origin and with quaternionic coefficients is rational if and only if it can be represented as*

$$r(x_1, x_2, x_3) = \frac{q(x_1, x_2, x_3)}{p(x_1, x_2, x_3)},$$

where q is a polynomial with quaternionic coefficients and p is a polynomial with real coefficients, such that $p(0) \neq 0$.

Proof. This is an immediate consequence of the inversion formula

$$q^{-1} = \frac{\overline{q}}{|q|^2} \quad \forall q \in \mathbb{H} \setminus \{0\}.$$

□

We now turn to the main result of this section:

Theorem 2.8. *A function defined in an open set Ω of $\mathbb{R}^4$ containing the origin is hyperholomorphic rational if and only if its restriction to $\Omega \cap \mathbb{R}^3$ is a rational $\mathbb{H}$-valued function of the three real variables x_1, x_2, x_3.*

Proof. Assume that R is hyperholomorphic and rational, that is, admits a realization of the form (2.11). Then, setting $x_0 = 0$ in (2.11), the Cauchy–Kovalevskaya products become usual products and we obtain

$$R(0, x_1, x_2, x_3) = D + C(I - (x_1A_1 + x_2A_2 + x_3A_3))^{-1}(x_1B_1 + x_2B_2 + x_3B_3),$$

and so the restriction $R(0, x_1, x_2, x_3)$ of R to $\Omega \cap \mathbb{R}^3$ is rational.

Conversely, (2.13) defines a function which is real analytic in a neighborhood of the origin. Taking the Cauchy–Kovalevskaya extension of both sides of (2.13) we obtain by definition on the left a hyperholomorphic function. On the right, by definition of the Cauchy–Kovalevskya product we obtain an expression of the form (2.11). □

Remark 2.9. *There is a fundamental difference between the expressions* (2.11) *and* (2.13). *The former is local (that is valid only in a neighborhood of the origin), while the latter is global. It makes sense for every* $(x_1, x_2, x_3) \in \mathbb{R}^3$ *where the matrix* $I - (x_1A_1 + x_2A_2 + x_3A_3)$ *is invertible.*

Corollary 2.10. *Let* $y \neq 0 \in \mathbb{H}$. *The quaternionic Cauchy kernel* $x \mapsto \frac{1}{2\pi^2}\frac{\overline{x}-\overline{y}}{|x-y|^4}$ *is rational.*

Proof. It suffices to note that the restriction of the function $x \mapsto \frac{\overline{x}-\overline{y}}{|x-y|^4}$ to $\mathbb{R}^3$ is rational in x_1, x_2, x_3 and has no singularities in a neighborhood of 0 when $y \neq 0$. □

In a similar way the quaternionic Bergman kernel for the unit ball is rational; indeed, this kernel is shown in [34, p. 10] to be equal to

$$K(x,y) = \frac{2}{\pi^2}\frac{(1-2\langle y,x\rangle + |y|^2|x|^2)(1-2\overline{x}y) + (\overline{y} - \overline{x}|y|^2)(x - y|x|^2)}{(1-2\langle y,x\rangle + |y|^2|x|^2)^3}.$$

3. Reproducing kernel spaces of power series of Fueter polynomials

We now define the counterparts of the spaces $\mathcal{H}(\mathbf{c})$ in the setting of hyperholomorphic functions. Our motivation for studying such spaces came from the quaternionic Arveson space, which we defined and studied in [4], [6].

3.1. Generalities

Theorem 3.1. *Let* $\mathbf{c} = \{c_\nu\}$ *be a sequence of positive numbers indexed by* $\mathbb{N}^3$ *with support* supp $(\mathbf{c})$ *(defined by* (1.3)*). Let*

$$k_{\mathbf{c}}(x,y) = \sum_{k=0}^{\infty} \sum_{\substack{\nu \in \text{supp } (\mathbf{c}) \\ |\nu| = k}} c_\nu \zeta^\nu(x)\overline{\zeta^\nu(y)}.$$

and

$$\Omega_{\mathbf{c}} = \left\{ x \in \mathbb{R}^4 \mid \sum_{k=0}^{\infty} \sum_{\substack{\nu \in \text{supp } (\mathbf{c}) \\ |\nu| = k}} c_\nu |\zeta^\nu(x)|^2 < \infty \right\}.$$

Then $k_{\mathbf{c}}$ *is positive for* $x, y \in \Omega_{\mathbf{c}}$ *and the associated reproducing kernel Hilbert space of left hyperholomorphic functions is the set of functions*

$$f(x) = \sum_{\nu \in \text{supp } (\mathbf{c})} \zeta^\nu(x) f_\nu$$

where the $f_\nu \in \mathbb{H}$ *are such that*

$$\|f\|_{\mathbf{c}}^2 := \sum_{\nu \in \text{supp } (\mathbf{c})} \frac{|f_\nu|^2}{c_\nu} < \infty. \tag{3.1}$$

We will denote by $\mathcal{H}(\mathbf{c})$ the reproducing kernel Hilbert space of left hyperholomorphic functions with reproducing kernel $k_{\mathbf{c}}$. Its norm is given by (3.1). We note that $\mathcal{H}(\mathbf{c})$ contains the span of the ζ^ν where $\nu \in \text{supp}\,(\mathbf{c})$.

We will say that the function s hyperholomorphic in $\Omega_{\mathbf{c}}$ is a multiplier (resp. a Schur multiplier) if the operator of Cauchy–Kovalevskaya multiplication by s on the left is bounded (resp. is a contraction) from $\mathcal{H}(\mathbf{c})$ into itself. We now present the counterpart of formula (1.5).

Proposition 3.2. *Let s be a multiplier of $\mathcal{H}(\mathbf{c})$. Then it holds that:*

$$\left(M_s^*\left(k_y a\right)\right)(x) = \sum_{k=0}^{\infty}\sum_{|\nu|=k} c_\nu \zeta^\nu(x)\overline{(s\odot\zeta^\nu(y))}a. \tag{3.2}$$

Proof. Let $a, b \in \mathbb{H}$ and $x, y \in \Omega_{\mathbf{c}}$. We have:

$$\begin{aligned}\langle M_S^* k_y a, k_x b\rangle_{\mathcal{H}(\mathbf{c})} &= \langle k_y a, s\odot(k_x b)\rangle_{\mathcal{H}(\mathbf{c})}\\ &= \langle k_y a, \sum_{k=0}^{\infty}\sum_{|\nu|=k} c_\nu (s\odot\zeta^\nu)\overline{\zeta^\nu(x)}b\rangle_{\mathcal{H}(\mathbf{c})}\\ &= \sum_{k=0}^{\infty}\sum_{|\nu|=k} c_\nu \overline{\left(\overline{a}\,(s\odot\zeta^\nu)\,(y)\overline{\zeta^\nu(x)}b\right)}\\ &= \sum_{k=0}^{\infty}\sum_{|\nu|=k} c_\nu \overline{b}\zeta^\nu(x)\overline{(s\odot\zeta^\nu(y))}a\end{aligned}$$

and hence we obtain the formula (3.2). □

As a corollary we obtain that the function s is a Schur multiplier if and only if the kernel

$$\sum_{k=0}^{\infty}\sum_{|\nu|=k} c_\nu\left(\zeta^\nu(x)\overline{\zeta^\nu(y)} - (s\odot\zeta^\nu)(x)\overline{(s\odot\zeta^\nu)(y)}\right) \tag{3.3}$$

is positive in $\Omega_{\mathbf{c}}$.

3.2. Gleason's problem and Leibenson's shift operators

Proposition 3.3. *The operator*

$$\mathcal{R}_j f(z) = \sum_{k=0}^{\infty}\sum_{|\nu|=k} \zeta^{\nu-e_j}\frac{\nu_j}{|\nu|}f_\nu \tag{3.4}$$

is bounded in $\mathcal{H}(\mathbf{c})$ if and only if the following two conditions hold: the set supp $(\mathbf{c})$ *is lower inclusive and it holds that*

$$\sup_j \left(\frac{\nu_j}{|\nu|}\right)^2\frac{c_\nu}{c_{\nu_j}} < \infty. \tag{3.5}$$

When it is bounded it is equal to the Leibenson backward-shift operator

$$\mathcal{R}_j f(x) = \int_0^1 \frac{\partial f}{\partial x_j}(tx)dt. \tag{3.6}$$

Proof. The operator $\mathcal{R}_j$ is bounded if and only if there exists a constant $K > 0$ such that

$$\sum_{k=0}^{\infty} \sum_{|\nu|=k} \left(\frac{\nu_i}{|\nu|}\right)^2 \frac{|f_\nu|^2}{c_{\nu-e_i}} \leq K \left(\sum_{k=0}^{\infty} \sum_{|\nu|=k} \frac{|f_\nu|^2}{c_\nu}\right).$$

The result follows easily. □

The Leibenson backward-shift operator operators (3.6) have previously appeared in (2.7) which can be viewed as Gleason problem (see Problem 1.4) with respect to hyperholomorphic variables. However, we are mainly interested in the following version of this problem, formulated in terms of the Cauchy–Kovalevskaya product:

Problem 3.4. *Let $\mathcal{M}$ be a set of functions left hyperholomorphic in a neighborhood Ω of the origin. Given $f \in \mathcal{M}$; to find functions $p_1, \ldots, p_3 \in \mathcal{M}$ such that*

$$f(x) - f(0) = \sum_{j=1}^{3} (\zeta_j \odot p_j)(x).$$

Theorem 3.5. *Under hypothesis* (3.5) *Problem* 3.4 *is solvable in the spaces $\mathcal{H}(\mathbf{c})$ and the Leibenson type operators* (3.6) *are the only commutative solution of the problem.*

Proof. The proof parallels the proof of the similar fact in $\mathbb{C}^N$ presented in [2]. Here we consider the special case of power series expansions at the origin.

First of all, we note that the operators $\mathcal{R}_j$ given by (3.4) commute:

$$\mathcal{R}_j \mathcal{R}_\ell f = \sum_{k=0}^{\infty} \sum_{|\nu|=k} \zeta^\nu \frac{(\nu_j+1)(\nu_\ell+1)}{(|\nu|+1)(|\nu|+2)} f_{\nu+e_j+e_\ell} = \mathcal{R}_\ell \mathcal{R}_j f,$$

and solve the Gleason problem:

$$\sum_{j=1}^{3} \zeta_j \odot (\mathcal{R}_j f) = \sum_{j=1}^{3} \zeta_j \odot \sum_{k=0}^{\infty} \sum_{|\nu|=k} \zeta^\nu \frac{\nu_j+1}{|\nu|+1} f_{\nu+e_j} = \sum_{k=1}^{\infty} \sum_{|\nu|=k} \sum_{j=1}^{3} \frac{\nu_j}{|\nu|} \zeta^\nu f_\nu = f - f(0).$$

Furthermore, let us assume that T_1, T_2, T_3 are some commuting bounded operators on $\mathcal{H}(\mathbf{c})$ which solve the Gleason problem, as well. Then we have for $f \in \mathcal{H}(\mathbf{c})$ and x in a neighborhood of the origin

$$\begin{aligned} f(x) &= f(0) + \sum_{j=1}^{3} (\zeta_j \odot (T_j f))(x) \\ &= f(0) + \sum_{j=1}^{3} \zeta_j(x) T_j f(0) + \sum_{j,\ell=1}^{3} (\zeta_j \odot \zeta_\ell \odot (T_\ell T_j f))(x). \end{aligned}$$

Continuing to iterate this formula and taking into account that T_j commute, we obtain the Taylor series for $f(x)$ in the form

$$f(x)=\sum_{k=0}^{\infty}\sum_{|\nu|=k}\frac{|\nu|!}{\nu!}\zeta^{\nu}(x)(T^{\nu}f)(0),$$

where

$$T^{\nu}=T_1^{\nu_1}T_2^{\nu_2}T_3^{\nu_3}.$$

In particular,

$$(T^{\nu}f)(0)=\frac{|\nu|!}{\nu!}f_{\nu}.$$

Now we write the Taylor expansion for T_jf:

$$(T_jf)(x)=\sum_{k=0}^{\infty}\sum_{|\nu|=k}\frac{|\nu|!}{\nu!}\zeta^{\nu}(x)(T^{\nu+e_j}f)(0)=\sum_{k=0}^{\infty}\sum_{|\nu|=k}\frac{|\nu|!}{\nu!}\zeta^{\nu}(x)f_{nu+e_j}=(\mathcal{R}_jf)(x).$$

□

3.3. The quaternionic Arveson space

The quaternionic Arveson space $\mathcal{A}$ corresponds to the choice $c_\nu=\frac{|\nu|!}{\nu!}$.

Proposition 3.6. *Assume that* $c_\nu=\frac{|\nu|!}{\nu!}$. *Then* $\Omega_{\mathbf{c}}$ *is the ellipsoid*

$$\Omega_{\mathbf{c}}=\left\{x\in\mathbb{H}\mid 3x_0^2+x_1^2+x_2^2+x_3^2<1\right\}$$

and

$$K_{\mathcal{A}}(x,y):=K_{\mathbf{c}}(x,y)=(1-\zeta_1(x)\overline{\zeta_1(y)}-\zeta_2(x)\overline{\zeta_2(y)}-\zeta_3(x)\overline{\zeta_3(y)})^{-\odot}.$$

A proof can be found in [6].

The choices $s(x)=\zeta_j(x)$ for $j=1,2,3$ in (3.3) leads to:

Theorem 3.7. *Let C be the operator of evaluation at the origin. It holds that*

$$I-\sum_{j=1}^{3}M_{\zeta_j}M_{\zeta_j}^{*}=C^{*}C \tag{3.7}$$

if and only if $c_\nu=\frac{|\nu|!}{\nu!}$, *that is, if and only if we are in the setting of the quaternionic Arveson space* $\mathcal{A}$.

Proof. Applying on both sides of the operator identity (3.7) to the kernel $k_{\mathbf{c}}$ we obtain

$$c_{\nu-e_1}+c_{\nu-e_2}+c_{\nu-e_3}=c_\nu.$$

The only solution of this equation with $c_0=1$ is $c_\nu=\frac{|\nu|!}{\nu!}$. □

Theorem 3.8. *The operators M_j defined by $f\mapsto f\odot\zeta_j$ are continuous in the quaternionic Arveson space $\mathcal{A}$ and their adjoints are given by $M_j^*=\mathcal{R}_j$. The Arveson space is the only space of hyperholomorhpic functions with these two properties.*

Proof. The result follows from (3.7) and from Theorem 3.5. □

3.4. $\mathcal{H}^*$-valued hyperholomorphic functions

Let $\mathcal{H}$ be a right linear quaternionic Hilbert space and let $\mathcal{H}^*$ denote the (left) dual space of bounded $\mathbb{H}$-linear functionals on $\mathcal{H}$. Let Ω be a domain in $\mathbb{R}^4$ containing the origin and let $f : \Omega \mapsto \mathcal{H}^*$ be a mapping such that $\forall h \in \mathcal{H}$, $f(\cdot)h$ is a left-hyperholomorphic function in Ω. Such a mapping f is said to be an $\mathcal{H}^*$-valued left-hyperholomorphic function in Ω.

Theorem 3.9. *Let $f(x)$ be an $\mathcal{H}^*$-valued left-hyperholomorphic function in a ball $B(0,R)$. Then f can be represented as the series*

$$f(x) = \sum_{k=0}^{\infty} \sum_{|\nu|=k} \zeta^{\nu}(x) f_{\nu}, \quad f_{\nu} \in \mathcal{H}^*,$$

which converges normally in $B(0,R)$ with respect to the operator norm.

Proof. First we note that for every $R' \in (0,R)$ the family of functionals $\{f(x) : |x| \leq R'\}$ is uniformly bounded: $\sup_{|x|\leq R'} \|f(x)\| < \infty$. Let $h \in \mathcal{H}$ and let

$$f(x)h = \sum_{k=0}^{\infty} P_k(x,h)$$

be the expansion of $f(x)h$ into the series of homogeneous polynomials of x. Then it follows from the Cauchy formula for the hyperholomorphic functions that for $x < R'$

$$|P_k(x,h)| \leq C_1(k+2)(k^2+1)\left(\frac{|x|}{R'}\right)^k \|h\| \sup_{|x|\leq R'} \|f(x)\|,$$

which can be proved like in [12, p.82], and where C_1 is a positive real constant independent of k. By uniqueness of the Taylor expansion, $P_k(x,h)$ is linear with respect to h, hence we can write $P_k(x,h) = P_k(x)h$ where $P_k(x) \in \mathcal{H}^*$ and the series $\sum_{k=0}^{\infty} P_k(x)$ converges to $f(x)$ normally in $B(0,R)$ with respect to the operator norm. Furthermore, since the polynomial $P_k(x)h$ is hyperholomorphic, we have

$$P_k(x)h = \sum_{|\nu|=k} \zeta^{\nu}(x) f_{\nu}(h),$$

where $f_{\nu}(h)$ is linear with respect to h and satisfies for $x < R'$ the estimate

$$|f_{\nu}(h)| \leq C_2 \frac{1}{(R')^k} \frac{(k+2)!}{\nu!} \|h\| \sup_{|x|\leq R'} \|P_k(x)\| \leq C(\nu, R')\|h\| \sup_{|x|\leq R'} \|f(x)\|.$$

Hence $f_{\nu} \in \mathcal{H}^*$. □

Corollary 3.10. *A positive kernel $k(x,y)$ can be represented as*

$$k_{\mathbf{c}}(x,y) = g(x)g(y)^*,$$

where $g(x)$ is an $\mathcal{H}(k)^*$-valued hyperholomorphic function.

3.5. de Branges–Rovnyak spaces

We shall say that an $\mathcal{H}^*$-valued hyperholomorphic function $s(x)$ is a Schur multiplier if the kernel

$$K_s(x,y) = \sum_{k=0}^{\infty} \sum_{|\nu|=k} \frac{|\nu|!}{\nu!} \left(\zeta^\nu(x)\overline{\zeta^\nu(y)} - (\zeta^\nu \odot s)(x)(\zeta^\nu \odot s)(y)^* \right)$$

is positive (compare with (3.3)). Note that, in view of Corollary 3.10, this condition is in force if and only if there exist a quaternionic Hilbert space $\mathcal{G}$ and a $\mathcal{G}^*$-valued hyperholomorphic function $g(x)$ such that

$$1 - s(x)s(y)^* = g(x)g(y)^* - \sum_{\ell=1}^{3} (\zeta_\ell \odot g)(x)(\zeta_\ell \odot g)(y)^*.$$

Our terminology can be explained as follows. Let us denote by $\ell^2(\mathcal{H})$ the quaternionic Hilbert space of sequences $(f_\nu : \nu \in \mathbb{N}^3, f_\nu \in \mathcal{H})$ such that $\sum \frac{\nu!}{|\nu|!}|f_\nu|^2 < \infty$. Then $s = \sum \zeta^\nu s_\nu$ is a Schur multiplier if and only if the operator M_s defined by

$$M_s(f_\nu) = \sum_\nu \zeta^\nu \left(\sum_{\mu \le \nu} s_\mu f_{\nu-\mu} \right)$$

is a contraction from $\ell^2(\mathcal{H})$ into the quaternionic Arveson space $\mathcal{A}$. In this case $K_s(\cdot, y) = (I - M_s M_s^*)K_{\mathcal{A}}$ and we denote by

$$\mathcal{H}(s) := (I - M_s M_s^*)^{\frac{1}{2}} \mathcal{A}$$

the quaternionic Hilbert space with the reproducing kernel K_s. This is the de Branges–Rovnyak space in the present setting.

Theorem 3.11. *Let s be an $\mathcal{H}^*$-valued hyperholomorphic Schur multiplier. Then there exists a co-isometry*

$$V = \begin{pmatrix} T_1 & F_1 \\ T_2 & F_2 \\ T_3 & F_3 \\ G & H \end{pmatrix} : \begin{pmatrix} \mathcal{H}(s) \\ \mathcal{H} \end{pmatrix} \mapsto \begin{pmatrix} \mathcal{H}(s)^3 \\ \mathbb{H} \end{pmatrix}$$

such that

$$\left(\sum_{k=1}^{3} \zeta_k \odot (T_k f) \right)(x) = f(x) - f(0); \tag{3.8}$$

$$\left(\sum_{k=1}^{3} \zeta_k \odot (F_k h) \right)(x) = (s(x) - s(0))h; \tag{3.9}$$

$$Gf = f(0); \tag{3.10}$$

$$Hh = s(0)h. \tag{3.11}$$

Furthermore, $s(x)$ admits the representation

$$s(x)h = Hh + \sum_{k=1}^{3}\sum_{\nu\in\mathbb{N}^3} \frac{|\nu|!}{\nu!}(\zeta_k \odot \zeta^\nu)(x)GT^\nu F_k h, \quad x \in \Omega,\ h \in \mathcal{H}, \tag{3.12}$$

where we use the notation

$$T^\nu = T_1^{\times\nu_1} \times T_2^{\times\nu_2} \times T_3^{\times\nu_3}.$$

Proof. Let us denote by $\mathcal{H}(s)_3$ the closure in $\mathcal{H}(s)^3$ of the linear span of the elements of the form

$$w_y = \begin{pmatrix} (I - M_s M_s^*)\mathcal{R}_1 K_{\mathcal{A}}(\cdot, y) \\ (I - M_s M_s^*)\mathcal{R}_2 K_{\mathcal{A}}(\cdot, y) \\ (I - M_s M_s^*)\mathcal{R}_3 K_{\mathcal{A}}(\cdot, y) \end{pmatrix}, \quad y \in \Omega.$$

Define

$$\left(\hat{T} w_y q\right)(x) = \left(K_s(x,y) - K_s(x,0)\right) q, \quad \hat{F} w_y q = \left(s(y)^* - s(0)^*\right) q, \tag{3.13}$$

$$\left(\hat{G} q\right)(x) = K_s(x,0) q, \quad \hat{H} q = s(0)^* q, \tag{3.14}$$

then it follows from Theorems 3.7, 3.8 that

$$\left\langle \begin{pmatrix} \hat{T} w_{y_1} q_1 + \hat{G} p_1 \\ \hat{F} w_{y_1} q_1 + \hat{H} p_1 \end{pmatrix}, \begin{pmatrix} \hat{T} w_{y_2} q_2 + \hat{G} p_2 \\ \hat{F} w_{y_2} q_2 + \hat{H} p_2 \end{pmatrix} \right\rangle = \left\langle \begin{pmatrix} w_{y_1} q_1 \\ p_1 \end{pmatrix}, \begin{pmatrix} w_{y_2} q_2 \\ p_2 \end{pmatrix} \right\rangle$$

for any $y_1, y_2 \in \Omega$ and $p_1, p_2, q_1, q_2 \in \mathbb{H}$. Hence the operator matrix $\hat{V} = \left(\begin{smallmatrix} \hat{T} & \hat{G} \\ \hat{F} & \hat{H} \end{smallmatrix}\right)$ can be extended as an isometry from $\left(\begin{smallmatrix} \mathcal{H}(s)_3 \\ \mathbb{H} \end{smallmatrix}\right)$ into $\left(\begin{smallmatrix} \mathcal{H}(s) \\ \mathcal{H} \end{smallmatrix}\right)$. Let us set $V = \left(\begin{smallmatrix} T & F \\ G & H \end{smallmatrix}\right) = \hat{V}^*$. Then the relations (3.13), (3.14) imply (3.8)–(3.11). Now, iterating (3.8) as in the proof of Theorem 3.5 we obtain (3.12). $\square$

Theorem 3.12. *Let $\mathcal{G}, \mathcal{H}$ be right quaternionic Hilbert spaces and let*

$$V = \begin{pmatrix} T_1 & F_1 \\ T_2 & F_2 \\ T_3 & F_3 \\ G & H \end{pmatrix} : \begin{pmatrix} \mathcal{G} \\ \mathcal{H} \end{pmatrix} \mapsto \begin{pmatrix} \mathcal{G}^3 \\ \mathbb{H} \end{pmatrix}$$

be a co-isometry. Then

$$s_V(x) = H + \sum_{k=1}^{3}\sum_{\nu\in\mathbb{N}^3} \frac{|\nu|!}{\nu!}(\zeta_k \odot \zeta^\nu)(x)GT^\nu F_k$$

is an $\mathcal{H}^*$-valued Schur multiplier.

Proof. Define

$$A_\mu(x) = \sum_{\nu\in\mathbb{N}^3} \frac{|\nu|!}{\nu!}\times$$
$$\times\left((\zeta^{\mu+\nu}\odot\zeta_1)(x)GT^\nu \quad (\zeta^{\mu+\nu}\odot\zeta_2)(x)GT^\nu \quad (\zeta^{\mu+\nu}\odot\zeta_3)(x)GT^\nu\right),$$

$$B_\mu(x) = \sum_{\nu\in\mathbb{N}^3} \frac{|\nu|!}{\nu!}\zeta^{\mu+\nu}(x)GT^\nu, \quad C(x) = \sum_{\nu\in\mathbb{N}^3} \frac{|\nu|!}{\nu!}\zeta^\nu(x)GT^\nu.$$

Then

$$\begin{aligned} A_\mu(x)A_\mu(y)^* + \zeta^\mu(x)\overline{\zeta^\mu(y)} &= \left(A_\mu(x) \quad \zeta^\mu(x)\right)VV^*\left(A_\mu(y) \quad \zeta^\mu(y)\right)^* \\ &= \left(B_\mu(x) \quad (\zeta^\mu\odot s_V)(x)\right)\left(B_\mu(y) \quad (\zeta^\mu\odot s_V)(y)\right)^* \\ &= B_\mu(x)B_\mu(y)^* + (\zeta^\mu\odot s_V)(x)(\zeta^\mu\odot s_V)(y)^*. \end{aligned}$$

Hence

$$\begin{aligned} K_{s_V}(x,y) &= \sum_{\mu\in\mathbb{N}^3}\frac{|\mu|!}{\mu!}\left(\zeta^\mu(x)\overline{\zeta^\mu(y)} - (\zeta^\mu\odot s_V)(x)(\zeta^\mu\odot s_V)(y)^*\right) \\ &= \sum_{\mu\in\mathbb{N}^3}\frac{|\mu|!}{\mu!}\left(B_\mu(x)B_\mu(y)^* - A_\mu(x)A_\mu(y)^*\right). \end{aligned}$$

Furthermore,

$$\begin{aligned} &\sum_{\mu\in\mathbb{N}^3}\frac{|\mu|!}{\mu!}\sum_{n=1}^3(\zeta^{\mu+\nu}\odot\zeta_n)(x)GT^\nu(T^\eta)^*G^*\overline{(\zeta^{\mu+\eta}\odot\zeta_n)(y)} \\ &\quad = \sum_{n=1}^3\sum_{\mu:\mu_n>0}\frac{\mu_n}{|\mu|}\frac{|\mu|!}{\mu!}\zeta^{\mu+\nu}(x)GT^\nu(T^\eta)^*G^*\overline{\zeta^{\mu+\eta}(y)} \\ &\quad = \sum_{|\mu|>0}\zeta^{\mu+\nu}(x)GT^\nu(T^\eta)^*G^*\overline{\zeta^{\mu+\eta}(y)}, \end{aligned}$$

and thus $K_{s_V}(x,y) = C(x)C(y)^*$ is positive. □

Theorem 3.13. *Let $\mathcal{H}$ be a right linear quaternionic reproducing kernel Hilbert space of functions, left-hyperholomorphic in a neighborhood of the origin. Assume that there exist bounded operators T_1, T_2 T_3 from $\mathcal{H}$ into itself such that*

$$\left(\sum_{k=1}^3\zeta_k\odot(T_kf)\right)(x) = f(x) - f(0)$$

and

$$\sum_{k=1}^3\|T_kf\|^2 \le \|f\|^2 - |f(0)|^2.$$

Then there exist a quaternionic Hilbert space $\mathcal{G}$ and a $\mathcal{G}^*$-valued Schur multiplier s such that $\mathcal{H} = \mathcal{H}(s)$.

Proof. Since $T^*T + G^*G \leq I$, there exist F, H such that $V = \begin{pmatrix} T & F \\ G & H \end{pmatrix}$ is a co-isometry. Hence s_V is a Schur multiplier, and in particular

$$K_{s_V}(x,y) = C(x)C(y)^*$$

where

$$C(x) = \sum_{\nu \in \mathbb{N}^3} \frac{|\nu|!}{\nu!} \zeta^\nu(x) G T^\nu.$$

But then for $f \in \mathcal{H}$ we have $C(x)f = f(x)$, hence $\mathcal{H} = \mathcal{H}(s_V)$. □

4. The analogue of the Hardy space of the polydisk

4.1. The Hardy type space

The counterpart of Hardy space here corresponds to the case $c_\nu \equiv 1$. Now

$$\Omega_{\mathbf{c}} = \left\{ x \in \mathbb{H} \mid \sup_{i=1,2,3} (|x_0|^2 + |x_i|^2) < 1 \right\}.$$

4.2. Quaternionic Schur–Agler spaces and realization theory

Definition 4.1. *Let s be hyperholomorphic in a neighborhood of the origin. Then s is said to belong to the Schur–Agler class, if there exist hyperholomorphic operator-valued $g_1(x), g_2(x), g_3(x)$ such that*

$$1 - s(x)\overline{s(y)} = \sum_{\ell=1}^{3} \left(g_\ell(x) g_\ell(y)^* - \zeta_\ell \odot g_\ell(x) (\zeta_\ell \odot g_\ell(y))^* \right).$$

Theorem 4.2. *Let s be in the Schur–Agler class and for $\ell = 1,2,3$ let $\mathcal{H}_\ell$ be the Hilbert space with the reproducing kernel $g_\ell(x)g_\ell(y)^*$. Then there exists a co-isometry*

$$\begin{pmatrix} A & B \\ C & D \end{pmatrix} : \begin{pmatrix} \bigoplus_{\ell=1}^3 \mathcal{H}_\ell \\ \mathbb{H} \end{pmatrix} \mapsto \begin{pmatrix} \bigoplus_{\ell=1}^3 \mathcal{H}_\ell \\ \mathbb{H} \end{pmatrix},$$

such that for arbitrary $q \in \mathbb{H}$ and $h_\ell \in \mathcal{H}_\ell$, $\ell = 1,2,3$

$$\zeta \odot Ah = \sum_{\ell=1}^{3} h_\ell - h_\ell(0), \quad \zeta \odot Bq = (s - s(0))q,$$

$$Ch = \sum_{\ell=1}^{3} h_\ell(0), \quad Dq = s(0)q.$$

In terms of the operators A, B, C, D the function s admits the realization

$$s = D + \sum_{\nu \in \mathbb{N}^3} \frac{|\nu|!}{\nu!} \zeta^\nu \odot \sum_{\ell=1}^{3} \zeta_\ell C A^{[\nu]} \pi_\ell B, \tag{4.1}$$

where π_ℓ is the orthogonal projection onto $\mathcal{H}_\ell$ in $\bigoplus_{\ell=1}^3 \mathcal{H}_\ell$ and

$$A^{[\nu]} = (\pi_1 A)^{\times \nu_1} \times (\pi_2 A)^{\times \nu_2} \times (\pi_3 A)^{\times \nu_3}.$$

Proof. The proof is analogous to that of Theorem 3.11. We consider

$$\mathcal{H} = \overline{\text{span}} \left\{ \begin{pmatrix} (\zeta_1 \odot g_1)(y)^* a \\ (\zeta_2 \odot g_2)(y)^* a \\ (\zeta_3 \odot g_3)(y)^* a \\ b \end{pmatrix} \right\}.$$

and define

$$\begin{aligned} \hat{A}\left(\zeta_\ell \odot g_\ell(y)^* a\right) &= \left(g_\ell(y)^* a - g_\ell(0)^* a\right), \quad \hat{C} b = g_\ell(0)^* b, \\ \hat{B}\left(\zeta_\ell \odot g_\ell(y)^* a\right) &= \overline{s(y)} a - \overline{s(0)} a, \quad \hat{D} b = \overline{s(0)} b. \end{aligned}$$

Then $\begin{pmatrix} \hat{A} & \hat{C} \\ \hat{B} & \hat{D} \end{pmatrix}$ can be extended as an isometry from $\begin{pmatrix} \mathcal{H} \\ \mathbb{H} \end{pmatrix}$ into $\begin{pmatrix} \oplus_{\ell=1}^3 \mathcal{H}_\ell \\ \mathbb{H} \end{pmatrix}$ and the adjoint operator matrix $\begin{pmatrix} A & B \\ C & D \end{pmatrix} = \begin{pmatrix} \hat{A} & \hat{C} \\ \hat{B} & \hat{D} \end{pmatrix}^*$ possesses the desire properties. □

Theorem 4.3. *Let s be of the form (4.1), where $\begin{pmatrix} A & B \\ C & D \end{pmatrix}$ is a co-isometry. Then s belongs to the Schur–Agler class.*

Proof. Note that

$$\left(\sum_{\nu\in\mathbb{N}^3} \tfrac{|\nu|!}{\nu!} \zeta^\nu \odot \sum_{\ell=1}^3 \zeta_\ell C A^{[\nu]} \pi_\ell \quad 1\right) \begin{pmatrix} A & B \\ C & D \end{pmatrix} = \left(\sum_{\nu\in\mathbb{N}^3} \tfrac{|\nu|!}{\nu!} \zeta^\nu C A^{[\nu]} \quad s\right).$$

Hence

$$\begin{aligned} \sum_{\ell=1}^3 \left(\zeta_\ell \odot \sum_{\nu\in\mathbb{N}^3} \frac{|\nu|!}{\nu!} \zeta^\nu C A^{[\nu]} \pi_\ell\right)(x) \left(\zeta_\ell \odot \sum_{\nu\in\mathbb{N}^3} \frac{|\nu|!}{\nu!} \zeta^\nu C A^{[\nu]} \pi_\ell\right)(y)^* + 1 = \\ + \left(\sum_{\nu\in\mathbb{N}^3} \frac{|\nu|!}{\nu!} \zeta^\nu C A^{[\nu]} \pi_\ell\right)(x) \left(\sum_{\nu\in\mathbb{N}^3} \frac{|\nu|!}{\nu!} \zeta^\nu C A^{[\nu]} \pi_\ell\right)(y)^* + s(x)\overline{s(y)}. \end{aligned}$$

□

References

[1] J. Agler. *On the representation of certain holomorphic functions defined on a polydisk*, volume 48 of *Operator Theory: Advances and Applications*, pages 47–66. Birkhäuser Verlag, Basel, 1990.

[2] D. Alpay and C. Dubi. On commuting operators solving Gleason's problem. *PAMS*, 133:3285–3293, 2005.

[3] D. Alpay and C. Dubi. A realization theorem for rational functions of several complex variables. *System & Control Letters*, 49:225–229, 2003.

[4] D. Alpay, B. Schneider, M. Shapiro, and D. Volok. Fonctions rationnelles et théorie de la réalisation: le cas hyper-analytique. *Comptes Rendus Mathématiques*, 336:975–980, 2003.

[5] D. Alpay, M. Shapiro, and D Volok. Rational hyperholomorphic functions in $\mathbb{R}^4$. *Journal of Functional Analysis*, 221:122–149, 2005.
Available at http://arxiv.org/pdf/math.FA/0312504.

[6] D. Alpay, M. Shapiro, and D. Volok. Espaces de de Branges Rovnyak: le cas hyperanalytique. *Comptes Rendus Mathématiques*, 338:437–442, 2004.

[7] W. Arveson. The curvature of a Hilbert module over $\mathbf{c}[z_1, \ldots, z_d]$. *Proc. Natl. Acad. Sci. USA*, 96(20):11096–11099 (electronic), 1999.

[8] J.A. Ball, W.S. Li, D. Timotin, and T.T. Trent. A commutant lifting theorem on the polydisc. *Indiana Univ. Math. J.*, 48(2):653–675, 1999.

[9] H. Bart, I. Gohberg, and M. Kaashoek. *Minimal factorization of matrix and operator functions*, volume 1 of *Operator Theory: Advances and Applications*. Birkhäuser Verlag, Basel, 1979.

[10] S. Bergman. *Integral operators in the theory of linear partial differential equations.* Second revised printing. Ergebnisse der Mathematik und ihrer Grenzgebiete, Band 23. Springer-Verlag New York Inc., New York, 1969.

[11] M.F. Bessmertnyĭ. On realizations of rational matrix functions of several complex variables. In *Interpolation theory, systems theory and related topics (Tel Aviv/Rehovot, 1999)*, volume 134 of *Oper. Theory Adv. Appl.*, pages 157–185. Birkhäuser, Basel, 2002. Translated from the Russian by D. Alpay and V. Katsnelson.

[12] F. Brackx, R. Delanghe, and F. Sommen. *Clifford analysis*, volume 76. Pitman research notes, 1982.

[13] R. Courant and D. Hilbert. *Methods of mathematical physics. Vol. II: Partial differential equations.* (Vol. II by R. Courant.). Interscience Publishers (a division of John Wiley & Sons), New York-London, 1962.

[14] M.J. Crowe. *A history of vector analysis.* Dover Books on Advanced Mathematics. Dover Publications Inc., New York, 1985. The evolution of the idea of a vectorial system, Reprint of the 1967 edition.

[15] E. Doubtsov. Minimal solutions of the Gleason problem. *Complex Variables Theory Appl.*, 36(1):27–35, 1998.

[16] S.W. Drury. A generalization of von Neumann's inequality to the complex ball. *Proc. Amer. Math. Soc.*, 68(3):300–304, 1978.

[17] A. Dzhuraev. *Methods of singular integral equations*, volume 60 of *Pitman Monographs and Surveys in Pure and Applied Mathematics.* Longman Scientific & Technical, Harlow, 1992. Translated from the Russian; revised by the author.

[18] E. Fornasini and G. Marchesini. State-space realization theory of two-dimensional filters. *IEEE Trans. Automatic Control*, AC–21(4):484–492, 1976.

[19] B. Fritzsche and B. Kirstein, editors. *Ausgewählte Arbeiten zu den Ursprüngen der Schur-Analysis*, volume 16 of *Teubner-Archiv zur Mathematik.* B.G. Teubner Verlagsgesellschaft, Stuttgart–Leipzig, 1991.

[20] R. Fueter. Die Theorie der regulären Funktionen einer quaternionen Variablen. In *Comptes rendus du congrès international des mathématiciens, Oslo 1936, Tome I*, pages 75–91, 1937.

[21] I. Gohberg, editor. *I. Schur methods in operator theory and signal processing*, volume 18 of *Operator theory: Advances and Applications*. Birkhäuser Verlag, Basel, 1986.

[22] Yu.M. Grigor′ev and V.V. Naumov. Approximation theorems for a Moisil-Theodorescu system. *Sibirsk. Mat. Zh.*, 25(5):9–19, 1984.

[23] K. Gürlebeck and W. Sprössig. *Quaternionic analysis and elliptic boundary value problems*, volume 56 of *Mathematical Research*. Akademie-Verlag, Berlin, 1989.

[24] K. Gürlebeck and W. Sprössig. *Quaternionic and Clifford calculus for physicists and engineers*, volume 1 of *Mathematical methods in practice*. John Wiley and Sons, 1997.

[25] H. Hefer. Zur Funktionentheorie mehrerer Veränderlichen. Über eine Zerlegung analytischer Funktionen und die Weilsche Integraldarstellung. *Math. Ann.*, 122:276–278, 1950.

[26] S.G. Krantz and H.P. Parks. *A primer of real analytic functions*, volume 4 of *Basler Lehrbücher [Basel Textbooks]*. Birkhäuser Verlag, Basel, 1992.

[27] G. Moisil. Sur l'équation $\Delta u = 0$. *C. R. Acad. Sci. Paris Sér. I Math.*, 191:984–986, 1930.

[28] B. Muckenhoupt and E.M. Stein. Classical expansions and their relation to conjugate harmonic functions. *Trans. Amer. Math. Soc.*, 118:17–92, 1965.

[29] R. Roesser. A discrete state-space model for linear image processing. *IEEE Trans. Automatic Control*, AC–20:1–10, 1975.

[30] M.V. Shapiro. Structure of the quaternionic modules and some properties of the involutive operators. *J. Natur. Geom.*, 1(1):9–37, 1992.

[31] M.V. Shapiro. Some remarks on generalizations of the one-dimensional complex analysis: hypercomplex approach. In *Functional analytic methods in complex analysis and applications to partial differential equations (Trieste, 1993)*, pages 379–401. World Sci. Publishing, River Edge, NJ, 1995.

[32] M.V. Shapiro. On the conjugate harmonic functions of M. Riesz–E. Stein–G. Weiss. In S. Dimiev and K. Sevigawa, editor, *Topics in complex analysis, differential geometry and mathematical physics (St. Konstantin, 1996)*, pages 8–32. World Sci. Publishing, River Edge, NJ, 1997.

[33] M.V. Shapiro and N.L. Vasilevski. Quaternionic ψ-hyperholomorphic functions, singular integral operators and boundary value problems. I. ψ-hyperholomorphic function theory. *Complex Variables Theory Appl.*, 27(1):17–46, 1995.

[34] M.V. Shapiro and N.L. Vasilevski. On the Bergman kernel function in hyperholomorphic analysis. *Acta Appl. Math.*, 46(1):1–27, 1997.

[35] F. Sommen. A product and an exponential function in hypercomplex function theory. *Applicable Anal.*, 12(1):13–26, 1981.

[36] N.N. Tarkhanov. A remark on the Moisil-Teodorescu system. *Sibirsk. Mat. Zh.*, 28(3):208–213, 225, 1987.

[37] H. Woerdeman. Positive Carathéodory interpolation on the polydisc. *Integral Equations Operator Theory*, 42(2):229–242, 2002.

[38] M. Zhdanov. *Integral transforms in geophysics*. Springer-Verlag, Berlin, 1988. Translated from the Russian by Tamara M. Pyankova.

Daniel Alpay
Department of Mathematics
Ben–Gurion University of the Negev
Beer-Sheva 84105
Israel
e-mail: dany@math.bgu.ac.il

Michael Shapiro
Departamento de Matemáticas
E.S.F.M del I.P.N.
07300 México, D.F.
México
e-mail: shapiro@esfm.ipn.mx

Dan Volok
Department of Mathematics
Ben–Gurion University of the Negev
Beer-Sheva 84105
Israel
e-mail: volok@math.bgu.ac.il

Operator Theory:
Advances and Applications, Vol. 162, 47–69

Extremal Extensions of a $C(\alpha)$-suboperator and Their Representations

Yury Arlinskiĭ

Abstract. The operator and the block-operator matrix forms for extremal rigid and soft extensions of a $C(\alpha)$-suboperator in a Hilbert space are given. Representations of the Friedrichs and Kreĭn–von Neumann maximal sectorial extensions of a sectorial linear relation space as strong resolvent limits of a family maximal accretive extensions are obtained.

Mathematics Subject Classification (2000). Primary 47A06, 47A20, 47A63, 47B34; Secondary 47B65.

Keywords. Contraction, contractive extension, $C(\alpha)$-suboperator, sectorial linear relation, rigid and soft extensions, Friedrichs extension, Kreĭn-von Neumann extension.

1. Introduction

Let H be a complex Hilbert space with the inner product $(\cdot,\cdot)$ and the norm $||\cdot||$, let I be the identity operator in H, and let $\alpha \in (0,\pi/2)$.

Definition 1.1. [4], [7]. Let a linear operator A in H be defined on the subspace H_0 and let A satisfy the condition

$$||A\sin\alpha \pm i\cos\alpha I|| \le 1, \tag{1.1}$$

Then in the case $H_0 = H$, we say that A belongs to the class $C(\alpha)$ and in the case $H_0 \neq H$ the operator A is called a $C(\alpha)$-suboperator.

Clearly, $A \in C(\alpha) \iff A^* \in C(\alpha)$. It is easy to see that the condition (1.1) is equivalent to

$$\tan\alpha\left(||f||^2 - ||Af||^2\right) \ge 2|\mathrm{Im}\,(Af,f)| \quad \text{for all} \quad f \in H_0. \tag{1.2}$$

It follows that operators from the class $C(\alpha)$ and $C(\alpha)$-suboperators are contractions and it is reasonable to define the class $C(0)$ as the set of all selfadjoint contractions in H and a nondensely defined Hermitian contraction we will call a $C(0)$-suboperator.

$C(\alpha)$-suboperators (operators of the class $C(\alpha)$) naturally arise as the fractional-linear transformations of the form

$$K(\mathbf{S}) = \{\langle u + u', u - u' \rangle, \langle u, u' \rangle \in \mathbf{S}\} = (I - \mathbf{S})(I + \mathbf{S})^{-1} \tag{1.3}$$

of sectorial (maximal sectorial) linear relations (l.r.) $\mathbf{S}$ with the vertex at the origin and the semiangle α [18], [23], [9], [22]. As it was proved in [18], [23], every α-sectorial l.r. $\mathbf{S}$ has $m-\alpha$-sectorial Friedrichs extension $\mathbf{S}_F$ which is associated with the closure $\mathbf{S}_F[u, v]$ of the sesquilinear form $(\mathbf{S}\cdot, \cdot)$. In the case of densely defined nonnegative symmetric operator S ($\alpha = 0$) M.G. Kreĭn [19] established that the set of all its nonnegative and selfadjoint extensions has minimal and maximal elements in the sense of quadratic forms. Fractional-linear transformations of the form (1.3) of these extremal nonnegative selfadjoint extensions are contractive selfadjoint extensions A_M and A_μ of the nondensely defined Hermitian contraction $A = (I - S)(I + S)^{-1}$. Operators A_M and A_μ possess properties:

$$\begin{aligned} &\inf \{((I - A_M)(h - \varphi), h - \varphi),\ \varphi \in \mathrm{dom}\,(A)\} = 0, \\ &\inf \{((I + A_\mu)(h - \varphi), h - \varphi),\ \varphi \in \mathrm{dom}\,(A)\} = 0 \quad \text{for all} \quad h \in H. \end{aligned}$$

The operator $S_\mu = (I - A_\mu)(I + A_\mu)^{-1}$ is exactly the Friedrichs extension S_F of S and the operator $S_N = (I - A_M)(I + A_M)^{-1}$ is the minimal among all nonnegative selfadjoint extensions of S. In addition, the operator S_N coincides with the extension constructed by J. von Neumann when S is a positive definite symmetric operator. Definitions of the minimal nonnegative selfadjoint extension for a nondensely defined nonnegative Hermitian operator and a nonnegative l.r. were given in [2] and [14]. Representations of the nonnegative extensions S_F and S_N as strong resolvents limits of certain family of selfadjoint extensions were established in [2] and [25]. The expressions of endpoints A_μ and A_M were given by T. Ando [1] in the block-matrix form and in [11] in the operator form.

In the general case of a non Hermitian but sectorial l.r. $\mathbf{S}$ the definition of the analog of $\mathbf{S}_N$ was given in [6], [8] similarly to the nonnegative case [14]: $\mathbf{S}_N = \left(\left(\mathbf{S}^{-1}\right)_F\right)^{-1}$. In the sequel $\mathbf{S}_N$ is called the Kreĭn–von Neumann extension. Properties of $\mathbf{S}_N$ and $\mathbf{S}_F$ and their strong resolvent limit representations are established in [8]. When $\alpha \neq 0$ we will denote by the same symbols A_μ and A_M the fractional-linear transformations $(I - \mathbf{S}_F)(I + \mathbf{S}_F)^{-1}$ and $(I - \mathbf{S}_N)(I + \mathbf{S}_N)^{-1}$, correspondingly. These operators are extensions of the $C(\alpha)$-suboperator $A = (I - \mathbf{S})(I + \mathbf{S})^{-1}$, belong to the class $C(\alpha)$, and possess properties for all $h \in H$ [6], [7]:

$$\begin{aligned} &\inf \{\mathrm{Re}\,((I + A_\mu)(h - \varphi), h - \varphi), \varphi \in \mathrm{dom}\,(A)\} = 0, \\ &\inf \{\mathrm{Re}\,((I - A_M)(h - \varphi), h - \varphi), \varphi \in \mathrm{dom}\,(A)\} = 0 \quad \text{for all} \quad h \in H. \end{aligned}$$

Operators A_μ and A_M are called the *rigid* and *soft* extensions, respectively.

Let A be a nondensely defined contraction in the Hilbert space H with the domain $\mathrm{dom}\,(A) = H_0$. M.G. Crandall [15] gave a parametrization of all contractive extensions on H of the operator A in the following operator form:

$$\widetilde{A}_K = AP_{H_0} + (I - AA^*)^{1/2} K P_{\mathfrak{N}}, \tag{1.4}$$

where $A^* : H \to H_0$ is the adjoint to $A : H_0 \to H$, $\mathfrak{N} := H \ominus H_0$, P_{H_0} and $P_{\mathfrak{N}}$ are orthogonal projections in H onto H_0 and $\mathfrak{N}$, respectively, and $K : \mathfrak{N} \to \overline{\mathrm{ran}}(I - AA^*)^{1/2}$ is a contractive parameter. Later in [24], [13], [16] the description of all contractive extensions of a nondensely defined contraction A was given in the form of block-operator matrices.

In [6], [7], [10] the expressions for the contractive parameters K_μ and K_M in (1.4) corresponding to the rigid and soft extensions of a $C(\alpha)$-suboperator A were obtained.

In this paper we consider the holomorphic family on the unit disk of contractive extensions of A of the form

$$\widetilde{A}(z) = AP_{H_0} + (I - AA^*)^{1/2} \widetilde{\mathfrak{X}}_0(z) P_{\mathfrak{N}},$$

where $\widetilde{\mathfrak{X}}_0(z)$ is B. Sz.-Nagy and C. Foias [26] characteristic function of the contraction $\widetilde{A}_0 := AP_{H_0}$. We establish properties of this family and prove that

$$A_\mu = \text{s-}\lim_{z \to -1} \widetilde{A}(z), \ A_M = \text{s-}\lim_{z \to 1} \widetilde{A}(z),$$

and, moreover,

$$\begin{aligned} A_\mu &= AP_{H_0} + (I - AA^*)^{1/2} \widetilde{\mathfrak{X}}_0(-1) P_{\mathfrak{N}}, \\ A_M &= AP_{H_0} + (I - AA^*)^{1/2} \widetilde{\mathfrak{X}}_0(1) P_{\mathfrak{N}}, \end{aligned}$$

where $\widetilde{\mathfrak{X}}_0(\pm 1)$ are strong nontangential limit values of $\widetilde{\mathfrak{X}}_0(z)$. Thus, new representations of A_μ an A_M in Crandall's form (1.4) are obtained. Another goal of the present paper is to give the block-matrix form of the rigid and soft extensions of a $C(\alpha)$-suboperator A similar to [24], [13], [16], i.e., to obtain the explicit expressions for the corresponding parameters. For this purpose we give in Section 2 one more (relatively short) proof of the main result in [24], [13], [16]. Our approach enables us to reduce the Crandall's form of contractive extensions to a block-matrix form. Note that the block-matrix representations of A_μ and A_M close to the case of a Hermitian contraction [1] were given in [22].

If $\mathbf{S}$ is an α-sectorial l.r. and $A = (I - \mathbf{S})(I + \mathbf{S})^{-1}$ then the fractional-linear transformations $\widetilde{\mathbf{S}}_\lambda = (I - \widetilde{A}(z))(I + \widetilde{A}(z))^{-1}$, $\lambda = (z-1)(1+z)^{-1}$, are m-accretive extensions of $\mathbf{S}$ and take the form

$$\widetilde{\mathbf{S}}_\lambda = \mathbf{S} \dot{+} \left\{ \langle \varphi_\lambda, -\lambda \varphi_\lambda \rangle, \ \varphi_\lambda \in \mathfrak{N}_\lambda(S) \right\},$$

where $\mathfrak{N}_\lambda(S) := H \ominus \mathrm{ran}\,(\mathbf{S} - \bar{\lambda} I)$ are the defect subspaces of $\mathbf{S}$. We prove that $\mathbf{S}_F$ and $\mathbf{S}_N$ are strong resolvent limits of the family $\widetilde{\mathbf{S}}_\lambda$:

$$\mathbf{S}_N = \text{s-R-}\lim_{\lambda \to 0} \widetilde{\mathbf{S}}_\lambda, \ \mathbf{S}_F = \text{s-R-}\lim_{\lambda \to \infty} \widetilde{\mathbf{S}}_\lambda.$$

We will keep the following notations. The class of all continuous linear operators defined on a complex Hilbert space $\mathfrak{H}_1$ and taking values in a complex Hilbert space

$\mathfrak{H}_2$ is denoted by $\mathcal{L}(\mathfrak{H}_1, \mathfrak{H}_2)$ and $\mathcal{L}(\mathfrak{H}) := \mathcal{L}(\mathfrak{H}, \mathfrak{H})$. By $P_{\mathfrak{K}}$ we always denote the orthogonal projection in a Hilbert space $\mathfrak{H}$ onto its subspace $\mathfrak{K}$. For a contraction $T \in \mathcal{L}(\mathfrak{H}_1, \mathfrak{H}_2)$ the nonnegative square root $D_T = (I - T^*T)^{1/2}$ is called the defect operator of T and $\mathfrak{D}_T$ stands for the closure of the range $\operatorname{ran}(D_T)$. It is well known that the defect operators satisfy the following commutation relation: $TD_T = D_{T^*}T$, cf. [26]. By $\operatorname{dom}(T)$, $\operatorname{ran}(T)$, and $\ker(T)$ we denote the domain, the range and the null-space of a linear operator T, respectively. We will often use the following well-known theorem of R.G. Douglas [17].

Theorem 1.2. [17]. *For every $A, B \in L(\mathfrak{H})$ the following statements are equivalent:*

(i) $\operatorname{ran} A \subset \operatorname{ran} B$;
(ii) $A = BC$ *for some* $C \in L(\mathfrak{H})$;
(iii) $AA^* \le \lambda BB^*$ *for some* $\lambda \ge 0$.

In this case there is a unique C satisfying $\|C\|^2 = \inf\{\lambda : AA^* \le \lambda BB^*\}$ *and* $\operatorname{ran} C \subset \overline{\operatorname{ran}}\, B^*$, *in which case* $\ker C = \ker A$.

Finally we note that two different descriptions in the operator form of all extensions of the class $C(\beta)$, $\beta \in [\alpha, \pi/2)$ of a given $C(\alpha)$-suboperator were obtained in [10], and in [21], [22] (see Remark 3.11).

2. Nondensely defined contractions and their contractive extensions

2.1. M.G. Crandall's theorem

Let H and H' be two Hilbert spaces. Suppose that H_0 is a subspace of H and $A : H_0 \to H'$ is a contraction. The operator $\widetilde{A}$ defined on H is called a contractive extension of A if $\widetilde{A} \supset A$ and $\|\widetilde{A}\| \le 1$.

Consider A as an operator from $\mathcal{L}(H_0, H')$. Then A has the adjoint $A^* \in \mathcal{L}(H', H_0)$. Let $\mathfrak{N} = H \ominus H_0$. The following result belongs to M.G. Crandall [15].

Theorem 2.1. [15]. *The formula*

$$\widetilde{A} = AP_{H_0} + D_{A^*}KP_{\mathfrak{N}} \tag{2.1}$$

establishes a one-to-one correspondence between all contractive extensions of A and all contractions $K \in \mathcal{L}(\mathfrak{N}, \mathfrak{D}_{A^*})$.

Proof. Let the operator $\widetilde{A}$ be given by (2.1), where $K \in \mathcal{L}(\mathfrak{N}, \mathfrak{D}_{A^*})$ is a contraction. Then

$$\widetilde{A}^* = A^* + K^*D_{A^*}.$$

It follows that for all $f \in H'$

$$\|\widetilde{A}^*f\|^2 = \|A^*f\|^2 + \|K^*D_{A^*}f\|^2 \le \|A^*f\|^2 + \|D_{A^*}f\|^2 = \|f\|^2.$$

Thus $\widetilde{A}^*$ is a contraction. Hence the operator $\widetilde{A}$ is contraction. Moreover $\widetilde{A}\restriction H_0 = A$.

Conversely, if $\widetilde{A}$ is a contractive extension of A, then its adjoint $\widetilde{A}^* : H \to H$ is a contraction. Because $A \subset \widetilde{A}$, we get $P_{H_0}\widetilde{A}^* = A^*$. Therefore the operator $\widetilde{A}^*$ takes the form

$$\widetilde{A}^* = A^* + L,$$

where the range of the operator L is contained in $\mathfrak{N}$. It follows that $||\widetilde{A}^* f||^2 = ||A^* f||^2 + ||Lf||^2$ for all $f \in H'$. Since $\widetilde{A}^*$ is a contraction, we obtain

$$||Lf||^2 \leq ||f||^2 - ||A^* f||^2, \ f \in H'.$$

By R.G. Douglas's theorem 1.2 we get

$$L^* = D_{A^*}K,$$

where $K : \mathfrak{N} \to \mathfrak{D}_{A^*}$ is a contraction. □

As a consequence for $\widetilde{A} = AP_{H_0} + D_{A^*}KP_{\mathfrak{N}}$ with a contraction $K \in \mathcal{L}(\mathfrak{N}, \mathfrak{D}_{A^*})$ one has the following relations

$$\begin{aligned} ||D_{\widetilde{A}} f||^2 &= ||(D_A P_{H_0} - A^* K P_{\mathfrak{N}}) f||^2 + ||D_K P_{\mathfrak{N}} f||^2, \ f \in H, \\ ||D_{\widetilde{A}^*} g||^2 &= ||D_{K^*} D_{A^*} g||^2, \ g \in H'. \end{aligned} \tag{2.2}$$

Because $A^* \mathfrak{D}_{A^*} \subset \mathfrak{D}_A$, the first relation in (2.2) gives

$$\inf\left\{||D_{\widetilde{A}} f - D_{\widetilde{A}}\varphi||^2, \ \varphi \in H_0\right\} = ||D_K P_{\mathfrak{N}} f||^2 \quad \text{for all} \quad f \in H.$$

This means that (see [19])

$$\operatorname{ran}(D_{\widetilde{A}}) \cap \mathfrak{N} = \operatorname{ran}(D_K). \tag{2.3}$$

The second equality in (2.2) yields

$$\operatorname{ran}(D_{\widetilde{A}^*}) = D_{A^*} \operatorname{ran}(D_{K^*}). \tag{2.4}$$

2.2. Block-matrix form of contractive extensions

Suppose now that the Hilbert space H' is decomposed as $H' = H_0' \oplus \mathfrak{M}$. Then $A = A_0 + C$, where $A_0 := P_{H_0'} A \in \mathcal{L}(H_0, H_0')$ and $C = P_{\mathfrak{M}} A \in \mathcal{L}(H_0, \mathfrak{M})$. We can rewrite A in the block-matrix form

$$A = \begin{pmatrix} A_0 \\ C \end{pmatrix}.$$

Since A is a contraction, we have $||A_0 g||^2 + ||Cg||^2 \leq ||g||^2$ for all $g \in H_0$. It follows that

$$C = K_0 D_{A_0}, \tag{2.5}$$

where $K_0 \in \mathcal{L}(\mathfrak{D}_{A_0}, \mathfrak{M})$ is a contraction. A bounded extension $\widetilde{A}$ of A also has the block-matrix form

$$\widetilde{A} = \begin{pmatrix} A_0 & B \\ C & D \end{pmatrix} : \begin{pmatrix} H_0 \\ \mathfrak{N} \end{pmatrix} \to \begin{pmatrix} H_0' \\ \mathfrak{M} \end{pmatrix}.$$

The description of blocks B and D of all contractive extensions $\widetilde{A}$ was obtained in [13], [16], and [24]. Here we propose another approach based on the Crandall's form (2.1).

Theorem 2.2. [13], [16], [24]. *The formula*

$$\widetilde{A} = \begin{pmatrix} A_0 & D_{A_0^*}N \\ K_0 D_{A_0} & -K_0 A_0^* N + D_{K_0^*} X D_N \end{pmatrix} : \begin{pmatrix} H_0 \\ \mathfrak{N} \end{pmatrix} \to \begin{pmatrix} H_0' \\ \mathfrak{M} \end{pmatrix} \tag{2.6}$$

establishes a bijective correspondence between all contractive extensions $\widetilde{A}$ *of the contraction* $A = A_0 + K_0 D_{A_0}$ *and all pairs*

$$\left\langle N \in \mathcal{L}(\mathfrak{N}, \mathfrak{D}_{A_0^*}),\ X \in \mathcal{L}(\mathfrak{D}_N, \mathfrak{D}_{K_0^*})\right\rangle$$

of contractive operators.

Proof. From (2.5) it follows that

$$A^* = A_0^* P_{H_0'} + D_{A_0} K_0^* P_{\mathfrak{M}}.$$

Therefore for all $f \in H'$:

$$\begin{aligned} ||f||^2 - ||A^* f||^2 &= ||P_{H_0'} f||^2 + ||P_{\mathfrak{M}} f||^2 - ||(A_0^* P_{H_0'} + D_{A_0} K_0^* P_{\mathfrak{M}}) f||^2 \\ &= ||P_{\mathfrak{M}} f||^2 + ||P_{H_0'} f||^2 - ||A_0^* P_{H_0'} f||^2 - ||D_{A_0} K_0^* P_{\mathfrak{M}} f||^2 \\ &\quad - 2\mathrm{Re}\,(A_0^* P_{H_0'} f, D_{A_0} K_0^* P_{\mathfrak{M}} f) = ||D_{A_0^*} P_{H_0'} f||^2 - ||K_0^* P_{\mathfrak{M}} f||^2 \\ &\quad + ||A_0 K_0^* P_{\mathfrak{M}} f||^2 - 2\mathrm{Re}\,(D_{A_0^*} P_{H_0'} f, A_0 K_0^* P_{\mathfrak{M}} f) + ||P_{\mathfrak{M}} f||^2 \\ &= ||D_{A_0^*} P_{H_0'} f - A_0 K_0^* P_{\mathfrak{M}} f||^2 + ||D_{K_0^*} P_{\mathfrak{M}} f||^2. \end{aligned}$$

Thus,

$$||D_{A^*} f||^2 = ||D_{A_0^*} P_{H_0'} f - A_0 K_0^* P_{\mathfrak{M}} f||^2 + ||D_{K_0^*} P_{\mathfrak{M}} f||^2,\ f \in H'. \tag{2.7}$$

In view of the equality $A_0 D_{A_0} = D_{A_0^*} A_0$ we get that $A_0 \mathfrak{D}_{A_0} \subset \mathfrak{D}_{A_0^*}$ and since $\mathrm{ran}\,(K_0^*) \subset \mathfrak{D}_{A_0}$, from (2.7) it follows that

$$\inf\left\{||D_{A^*}(f - \varphi)||^2,\ \varphi \in H_0\right\} = ||D_{K_0^*} P_{\mathfrak{M}} f||^2,\ f \in H'. \tag{2.8}$$

Let $\mathfrak{H}_0' := \overline{D_{A^*} H_0'}$ and let $\mathfrak{M}_0' = \mathfrak{D}_{A^*} \ominus \mathfrak{H}_0'$. Observe that (see [19])

$$\mathfrak{M}_0' = \left\{f \in \mathfrak{D}_{A^*} : D_{A^*} f \in \mathfrak{M}\right\}.$$

From (2.7) and (2.8) we get the equalities

$$\begin{aligned} ||P_{\mathfrak{H}_0'} D_{A^*} f||^2 &= ||D_{A_0^*} P_{H_0'} f - A_0 K_0^* P_{\mathfrak{M}} f||^2, \\ ||P_{\mathfrak{M}_0'} D_{A^*} f||^2 &= ||D_{K_0^*} P_{\mathfrak{M}} f||^2,\ f \in H'. \end{aligned} \tag{2.9}$$

In particular,

$$||P_{\mathfrak{H}_0'} D_{A^*} \varphi||^2 = ||D_{A_0^*} \varphi||^2,\ \varphi \in H_0'.$$

From (2.9) it follows that there are a unitary operator U_0 from $\mathfrak{H}_0'$ onto $\mathfrak{D}_{A_0^*}$ and a unitary operator Z_0 from $\mathfrak{M}_0'$ onto $\mathfrak{D}_{K_0^*}$ such that

$$\begin{aligned} U_0 P_{\mathfrak{H}_0'} D_{A^*} f &= D_{A_0^*} P_{H_0'} f - A_0 K_0^* P_{\mathfrak{M}} f, \\ Z_0 P_{\mathfrak{M}_0'} D_{A^*} f &= D_{K_0^*} P_{\mathfrak{M}} f,\ f \in H'. \end{aligned} \tag{2.10}$$

Let $U_0^* = U_0^{-1} \in \mathcal{L}(\mathfrak{D}_{A_0^*}, \mathfrak{H}_0')$ and $Z_0^* = Z_0^{-1} \in \mathcal{L}(\mathfrak{M}_0', \mathfrak{D}_{K_0^*})$ be the adjoint operators. Then from (2.10) we have

$$D_{A^*} = U_0^* \left(D_{A_0^*} P_{H_0'} - A_0 K_0^* P_{\mathfrak{M}}\right) + Z_0^* D_{K_0^*} P_{\mathfrak{M}}$$

and

$$D_{A^*} = \left(D_{A_0^*} - K_0 A_0^*\right) U_0 P_{\mathfrak{H}_0'} + D_{K_0^*} Z_0 P_{\mathfrak{M}_0'}. \tag{2.11}$$

Let $K \in \mathcal{L}(\mathfrak{N}, \mathfrak{D}_{A^*})$. Then $K = P_{\mathfrak{H}_0'} K + P_{\mathfrak{M}_0'} K$. Put

$$N = U_0 P_{\mathfrak{H}_0'} K,\ Y = Z_0 P_{\mathfrak{M}_0'} K,\ \widetilde{K} = N + Y.$$

It follows that

$$K = U_0^* N + Z_0^* Y$$

and $||Kh||^2 = ||\widetilde{K}h||^2 = ||Nh||^2 + ||Yh||^2$ for all $h \in \mathfrak{N}$. Clearly, $K \in \mathcal{L}(\mathfrak{N}, \mathfrak{D}_{A^*})$ is a contraction $\iff$ the operator $\widetilde{K} \in \mathcal{L}(\mathfrak{M}, \mathfrak{D}_{A_0^*} \oplus \mathfrak{D}_{K_0^*})$ is a contraction and $\widetilde{K}$ is a contraction $\iff$ $Y = XD_N$, where $X \in \mathcal{L}(\mathfrak{D}_N, \mathfrak{D}_{K_0^*})$ is a contraction. Further for a contraction $K \in \mathcal{L}(\mathfrak{N}, \mathfrak{D}_{A^*})$ from (2.11) and for all $h \in \mathfrak{N}$ we get

$$D_{A^*} Kh = \left(D_{A_0^*} - K_0 A_0^*\right) Nh + D_{K_0^*} X D_N h. \tag{2.12}$$

Let $\widetilde{A} = AP_{H_0} + D_{A^*} K P_{\mathfrak{N}}$. Then (2.1) and (2.12) yield (2.6). If the operator $\widetilde{A}$ is given by (2.2) with contractions $N \in \mathcal{L}(\mathfrak{N}, \mathfrak{D}_{A_0^*})$ and $X \in \mathcal{L}(\mathfrak{D}_N, \mathfrak{D}_{K_0^*})$ then the operators $\widetilde{K} = N + XD_N$ and $K = U_0^* N + Z_0^* X D_N$ are contractions. The operator K belongs to $\mathcal{L}(\mathfrak{N}, \mathfrak{D}_{A^*})$. Hence from (2.2) and (2.11) we obtain

$$\begin{aligned}\widetilde{A} &= (A_0 + K_0 D_{A_0}) P_{H_0} + \left((D_{A_0^*} - K_0 A_0^*) N + D_{K_0^*} X D_N\right) P_{\mathfrak{N}} \\ &= AP_{H_0} + \left((D_{A_0^*} - K_0 A_0^*) U_0 P_{\mathfrak{H}_0'} + D_{K_0^*} Z_0 P_{\mathfrak{M}_0'}\right) K P_{\mathfrak{N}} = AP_{H_0} + D_{A^*} K P_{\mathfrak{N}}.\end{aligned}$$

Thus, $\widetilde{A}$ is a contractive extension of A. □

Remark 2.3. Other proofs of Theorem 2.2 can also be found in [20] and [22].

Suppose that $H_0 \subset H$, $H_0' \subset H'$, $A : H_0 \to H'$, $B : H_0' \to H$, and the operators A and B forms a dual pair, i.e.,

$$(Af, h)_{H'} = (f, Bh)_H \quad \text{for all} \quad f \in H_0,\ g \in H_0'.$$

The operator $\widetilde{A} \in \mathcal{L}(H, H')$ is called an extension of the dual pair $\langle A, B \rangle$ if $\widetilde{A} \supset A$ and $\widetilde{A}^* \supset B$. Theorem 2.2 enables us to give the block-matrix form of all contractive extensions of the dual pair of contractions $\langle A, B \rangle$. The Crandall's operator form (2.1) of contractive extensions of a dual pair of contractions is given in [3].

3. $C(\alpha)$-suboperators and its extremal extensions

3.1. Operators of the class $\widetilde{C}$

From (1.2) it follows that $T \in C(\alpha)$ if and only if the operator $(I - T^*)(I + T)$ is a sectorial operator with the vertex at the origin and the semiangle α.

Let

$$\widetilde{C} = \cup\{C(\alpha),\ \alpha \in [0, \pi/2)\}.$$

Properties of operators of the class $\widetilde{C}$ were studied in [4], [5], [20]. In [4] it was proved that if $T \in \widetilde{C}$ then

(i) $\operatorname{ran}(D_{T^n}) = \operatorname{ran}(D_{T^{*n}}) = D_{T_R}$ for all natural numbers n,

where $T_R = (T + T^*)/2$ is the real part of T,

(ii) the subspace $\mathfrak{D}_T$ reduces the operator T, and, moreover, the operator $T{\restriction}\ker(D_T)$ is a selfadjoint and unitary, and $T{\restriction}\mathfrak{D}_T$ is a completely nonunitary contraction of the class C_{00} [26], i.e.,

$$\lim_{n\to\infty} T^n f = \lim_{n\to\infty} T^{*n} f = 0 \quad \text{for all} \quad f \in \mathfrak{D}_T.$$

Let

$$\Theta_T(z) = \left[-T + zD_{T^*}(I - zT^*)^{-1}D_T\right]{\restriction}\mathfrak{D}_T,\ |z| < 1$$

be the characteristic function [26] of the operator T from the class $\widetilde{C}$. Then $\Theta(z)$ is bi-inner [26] and there exist unitary strong nontangential limits [5]

$$\Theta_T(\pm 1) = \text{s-}\lim_{z\to\pm 1} \Theta_T(z).$$

Observe that if T is a selfadjoint contraction (i.e., belongs to the class $C(0)$) then $\Theta_T(\pm 1) = \pm I{\restriction}\mathfrak{D}_T$.

Let T belong to the class $C(\alpha)$ in the Hilbert space H. Then $T^* \in C(\alpha)$. By (1.1) we obtain

$$2|\operatorname{Im}(Tf, f)| \le \tan\alpha ||D_T f||^2,\ 2|\operatorname{Im}(Tf, f)| \le \tan\alpha ||D_{T^*} f||^2,\ f \in H.$$

It follows that

$$T - T^* = 2iD_T \Phi D_T,\ T - T^* = 2iD_{T^*}\Phi_* D_{T^*},$$

where Φ and Φ_* are selfadjoint operators in the subspace $\mathfrak{D}_T = \mathfrak{D}_{T^*}$ and

$$||\Phi|| \le \frac{\tan\alpha}{2},\ ||\Phi_*|| \le \frac{\tan\alpha}{2}.$$

Since $\operatorname{ran}(D_T) = \operatorname{ran}(D_{T^*})$, by R.G. Douglas Theorem 1.2 there exists a bounded and boundedly invertible operator $L_T \in \mathcal{L}(\mathfrak{D}_T)$ such that

$$D_T = D_{T^*} L_T$$

The connections between operators Φ, Φ_*, L_T, and limit values $\Theta(\pm 1)$ of the characteristic function $\Theta(z)$ of T are given by the following relations ([4], [5], [10]):

$$\begin{aligned}
&\Phi_* = L_T \Phi L_T^*,\\
&\Theta_T(\pm 1)\Phi = \Phi_* \Theta_T(\pm 1),\\
&\Theta_T(\pm 1) = \pm (L_T^{*-1} - 2iT\Phi)(I \pm 2i\Phi)^{-1},\\
&(I - 2i\Phi)(I + 2i\Phi)^{-1} = -\Theta_T^{-1}(-1)\Theta_T(1),\\
&2i\Phi = \left(I - \Theta_T^{-1}(-1)\Theta_T(1)\right)^{-1}\left(I + \Theta_T^{-1}(-1)\Theta_T(1)\right).
\end{aligned}$$

These connections yield, for instance, that

$$\begin{aligned}
&I - \Theta_T^{-1}(-1)\Theta_T(1) \quad \text{is a sectorial operator with the semiangle } \alpha,\\
&\ker\left(\Theta_T(-1) + \Theta_T(1)\right) = \ker \Phi,\\
&\left\| \frac{\Theta_T(-1) + \Theta_T(1)}{2} \right\| \le \sin\alpha.
\end{aligned} \tag{3.1}$$

3.2. $C(\alpha)$-suboperators

Let A be a nondensely defined contraction in a Hilbert space H with $\operatorname{dom}(A) = H_0$, let $\mathfrak{N}$ be the orthogonal complement to H_0 in H, and let $A^* \in \mathcal{L}(H, H_0)$ be the adjoint operator to A. The subspace $\mathfrak{N}_z(A) = H \ominus \operatorname{ran}(A - \bar{z}I)$ is called the defect subspace of A. It is evident that

$$\varphi_z \in \mathfrak{N}_z(A) \iff A^*\varphi_z = zP_{H_0}\varphi_z. \tag{3.2}$$

Because A is a contraction, (3.2) yields the direct sum decomposition

$$H = H_0 \dot{+} \mathfrak{N}_{\frac{1}{z}}, \ |z| < 1. \tag{3.3}$$

Let $A_0 := P_{H_0}A$. Then A_0 is a contraction in the subspace H_0. It follows from (2.5) that

$$P_{\mathfrak{N}}A = K_0 D_{A_0},$$

where $K_0 \in \mathcal{L}(\mathfrak{D}_{A_0}, \mathfrak{N})$ is a contraction. Thus,

$$A - A_0 + K_0 D_{A_0}. \tag{3.4}$$

Hence $A^* = A_0^* + D_{A_0}K_0^*P_{\mathfrak{N}}$. Moreover,

$$\left\|D_A\varphi\right\|^2 = \left\|D_{K_0}D_{A_0}\varphi\right\|^2, \ \varphi \in H_0 \tag{3.5}$$

and therefore, by R.G. Douglas's Theorem 1.2 we obtain

$$D_A = D_{A_0}D_{K_0}V_0 \tag{3.6}$$

where V_0 is an isometry from $\mathfrak{D}_A$ onto $\mathfrak{D}_{K_0}$. The first equality in (2.9) gives in this case

$$||D_{A^*}f||^2 = ||(D_{A_0^*}P_{H_0} - A_0K_0^*P_{\mathfrak{N}})f||^2 + ||D_{K_0^*}P_{\mathfrak{N}}f||^2, \ f \in H.$$

It follows once again (see (2.8)) that for all $f \in H$

$$\inf\left\{||D_{A^*}f - D_{A^*}\varphi||^2, \ \varphi \in H_0\right\} = ||D_{K_0^*}P_{\mathfrak{N}}f||^2.$$

Hence [19]

$$\operatorname{ran}(D_{A^*})\cap\mathfrak{N}=\operatorname{ran}(D_{K_0^*}). \tag{3.7}$$

From the equality $\|D_{A^*}\varphi\|^2=\|D_{A_0^*}\varphi\|^2$, $\varphi\in H_0$, it follows that

$$U_0D_{A^*}=D_{A_0^*}, \tag{3.8}$$

where U_0 is an isometry from the subspace $\mathfrak{H}_0':=\overline{D_{A^*}H_0}$ onto the subspace $\mathfrak{D}_{A_0^*}$ (see (2.10)).

Let A be a $C(\alpha)$-suboperator. Because

$$\operatorname{Im}(A\varphi,\varphi)=\left(\frac{A_0-A_0^*}{2i}\varphi,\varphi\right),\ \varphi\in H_0,$$

the relation (1.2) yields that there exists a selfadjoint contraction F in the subspace $\mathfrak{D}_A$ such that

$$A_0-A_0^*=i\tan\alpha D_AFD_A. \tag{3.9}$$

From (3.6) we get

$$A_0-A_0^*=2iD_{A_0}G_0D_{A_0},$$

where

$$G_0:=\frac{\tan\alpha}{2}D_{K_0}V_0FV_0^*D_{K_0}.$$

Therefore, the operator A_0 belongs to the class $C(\alpha)$ in the subspace H_0. It implies the equality $\operatorname{ran}(D_{A_0})=\operatorname{ran}(D_{A_0^*})$.

Let

$$\mathfrak{X}_0(z)=\left[-A_0+zD_{A_0^*}(I-zA_0^*)^{-1}D_{A_0}\right]\restriction\mathfrak{D}_{A_0} \tag{3.10}$$

be the characteristic function of the contraction A_0. Because $A_0\in C(\alpha)$, there exist strong nontangential unitary limit values $\mathfrak{X}_0(-1)$ and $\mathfrak{X}_0(1)$ and $\ker(\mathfrak{X}_0(-1)+\mathfrak{X}_0(1))=\ker G_0$. Therefore, from (3.1) we get

$$\ker(\mathfrak{X}_0(-1)+\mathfrak{X}_0(1))\subset\ker D_{K_0}. \tag{3.11}$$

Theorem 3.1. *Let A be a $C(\alpha)$-suboperator in H with* $\operatorname{dom}(A)=H_0$. *Define the contractive extension $\widetilde{A}_0:=AP_{H_0}$ of A and let*

$$\widetilde{\mathfrak{X}}_0(z)=\left[-\widetilde{A}_0+zD_{\widetilde{A}_0^*}(I-z\widetilde{A}_0^*)^{-1}D_{\widetilde{A}_0}\right]\restriction\mathfrak{D}_{\widetilde{A}_0}. \tag{3.12}$$

be the characteristic function [26] of $\widetilde{A}_0$. Then there exist nontangential strong limits

$$\widetilde{\mathfrak{X}}_0(\pm1)=\text{s-}\lim_{z\to\pm1}\widetilde{\mathfrak{X}}_0(z),\ \widetilde{\mathfrak{X}}_0^*(\pm1)=\text{s-}\lim_{z\to\pm1}\widetilde{\mathfrak{X}}_0^*(z).$$

Moreover, the operators $\widetilde{\mathfrak{X}}_0(\pm1)\restriction\mathfrak{N}$ are isometries.

Proof. Because $\widetilde{A}_0^*=A^*$, we have $D_{A^*}=D_{\widetilde{A}_0^*}$, $D_{\widetilde{A}_0}=D_AP_{H_0}+P_{\mathfrak{N}}$, and hence

$$\begin{aligned}\mathfrak{D}_{\widetilde{A}_0^*}&=\mathfrak{D}_{A^*},\\ \mathfrak{D}_{\widetilde{A}_0}&=\mathfrak{D}_A\oplus\mathfrak{N}.\end{aligned} \tag{3.13}$$

From (3.6) and (3.8) we get

$$D_{A^*}(I - zA_0^*)^{-1}D_A\varphi = U_0^* D_{A_0^*}(I - zA_0^*)^{-1}D_{A_0}D_{K_0}V_0\varphi,\ \varphi \in \mathfrak{D}_A.$$

Consequently,

$$\widetilde{\mathfrak{X}}_0(z)\varphi = -A\varphi + U_0^*\big(\mathfrak{X}_0(z) + A_0\big)D_{K_0}V_0\varphi,\ \varphi \in \mathfrak{D}_A. \tag{3.14}$$

Let $h \in \mathfrak{N}$. Then

$$\begin{aligned} &D_{A^*}(I - zA^*)^{-1}h = D_{A^*}(I - zA^*)^{-1}(h - zA^*h + zA^*h)\\ &= D_{A^*}h + zD_{A^*}(I - zA^*)^{-1}A^*h = D_{A^*}h + zD_{A^*}(I - zA^*)^{-1}D_{A_0}K_0^*h\\ &= D_{A^*}h + U_0^*(\mathfrak{X}_0(z) + A_0)K_0^*h. \end{aligned}$$

Therefore,

$$\widetilde{\mathfrak{X}}_0(z)h = zD_{A^*}h + zU_0^*(\mathfrak{X}_0(z) + A_0)K_0^*h,\ h \in \mathfrak{N}. \tag{3.15}$$

Since A_0 and A_0^* belong to the class $C(\alpha)$ in the subspace H_0, there exist strong unitary nontangential limit values $\mathfrak{X}_0(\pm1)$ and $\mathfrak{X}_0^*(\pm1)$ of $\mathfrak{X}_0(z)$ and $\mathfrak{X}_0^*(z)$ [5]. This statement and (3.14), (3.15) imply that there exist strong nontangential limit values $\widetilde{\mathfrak{X}}_0(\pm1)$, $\widetilde{\mathfrak{X}}_0^*(\pm1)$ and

$$\widetilde{\mathfrak{X}}_0(\pm1) = \big[-A + U_0^*(\mathfrak{X}_0(\pm1) + A_0)D_{K_0}V_0\big]P_{H_0} \pm \big[D_{A^*} + U_0^*(\mathfrak{X}_0(\pm1) + A_0)K_0^*\big]P_{\mathfrak{N}}.$$

Let us show that $\widetilde{\mathfrak{X}}_0(\pm1)\restriction\mathfrak{N}$ are isometries. One can easily check that

$$||h||^2 - ||\widetilde{\mathfrak{X}}_0(z)h||^2 = (1 - |z|^2)||(I - zA^*)^{-1}h||^2,\ h \in \mathfrak{N}.$$

For $h \in \mathfrak{N}$ from the equality $A^*P_{\mathfrak{N}} = D_{A_0}K_0^*P_{\mathfrak{N}}$ we have

$$\begin{aligned} &\sqrt{1 - |z|^2}\,(I - zA^*)^{-1}h = \sqrt{1 - |z|^2}\,h + z\sqrt{1 - |z|^2}\,(I - zA^*)^{-1}A^*h\\ &= \sqrt{1 - |z|^2}\,h + z\sqrt{1 - |z|^2}\,(I - zA^*)^{-1}D_{A_0}K_0^*h. \end{aligned}$$

Since

$$||f||^2 - ||\mathfrak{X}_0(z)f||^2 = (1 - |z|^2)||(I - zA^*)^{-1}D_{A_0}f||^2,\ f \in \mathfrak{D}_{A_0}$$

and operators $\mathfrak{X}_0(\pm1)$ are unitary in $\mathfrak{D}_{A_0}$, we have

$$\text{s-}\lim_{z\to\pm1}\sqrt{1 - |z|^2}\,(I - zA^*)^{-1}D_{A_0} = 0.$$

Therefore, for all $h \in \mathfrak{N}$ we get

$$\begin{aligned} &\lim_{z\to\pm1}\Big(||h||^2 - ||\widetilde{\mathfrak{X}}_0(z)h||^2\Big)\\ &= \lim_{z\to\pm1}\Big\|\sqrt{1 - |z|^2}\,h + z\sqrt{1 - |z|^2}\,(I - zA^*)^{-1}D_{A_0}K_0^*h\Big\|^2 = 0. \end{aligned}$$

Thus, operators $\widetilde{\mathfrak{X}}_0(\pm1)\restriction\mathfrak{N}$ are isometries. □

3.3. Holomorphic family of contractive extensions of a $C(\alpha)$-suboperator

Let again A be a $C(\alpha)$-suboperator in H with $\operatorname{dom}(A) = H_0$. Recall that the direct sum decomposition (3.3) holds. Consider the family of operators defined as follows

$$\widetilde{A}(z)f = \begin{cases} Af, \; f \in H_0 \\ zf, \; f \in \mathfrak{N}_{1/z}(A) \end{cases}, \quad |z| < 1 \tag{3.16}$$

and $\widetilde{A}(0) = AP_{H_0}$.

Proposition 3.2. *Operators $\widetilde{A}(z)$ are contractions for all $z \in \mathbb{D}$ and take the form*

$$\widetilde{A}(z) = AP_{H_0} + D_{A^*}\widetilde{\mathfrak{X}}_0(z)P_{\mathfrak{N}}. \tag{3.17}$$

Moreover, the function $\widetilde{A}(z)$ is the transfer function of the linear passive discrete-time system with the state space H, input space H, and output space H [12] given by the block-matrix

$$T = \begin{pmatrix} AP_{H_0} & D^2_{A^*} \\ P_{\mathfrak{N}} & A^* \end{pmatrix} : \begin{pmatrix} H \\ H \end{pmatrix} \to \begin{pmatrix} H \\ H \end{pmatrix}. \tag{3.18}$$

Proof. Let us show that $\widetilde{A}(z)$ is a contraction. Let $f = \varphi + \varphi_z$, where $\varphi \in H_0$, $\varphi_z \in \mathfrak{N}_{1/z}$. Then (3.16) yields

$$\begin{aligned} ||f||^2 - ||\widetilde{A}(z)f||^2 &= ||\varphi + \varphi_z||^2 - ||A\varphi + z\varphi_z||^2 \\ &= ||\varphi||^2 - ||A\varphi||^2 + (1 - |z|^2)||\varphi_z||^2 + 2\mathrm{Re}\left((I - \bar{z}A)\varphi, \varphi_z\right) \\ &= ||\varphi||^2 - ||A\varphi||^2 + (1 - |z|^2)||\varphi_z||^2 \geq 0. \end{aligned}$$

Using (3.2) from (3.16) we obtain

$$\widetilde{A}(z) = A(I - zA^*)^{-1}P_{H_0}(I - zA^*) + z(I - zA^*)^{-1}P_{\mathfrak{N}}, \tag{3.19}$$

and hence

$$\widetilde{A}(z) = \widetilde{A}_0 + zD^2_{A^*}(I - zA^*)^{-1}P_{\mathfrak{N}}.$$

Using (3.12) we get (3.17). Observe that $||\widetilde{\mathfrak{X}}_0(z)|| \leq 1$ for $|z| < 1$ [26]. Making use of Theorem 2.1 we get once again that $\widetilde{A}(z)$ is a contractive extension of the operator A for every $|z| < 1$. Moreover, $\widetilde{\mathfrak{X}}_0(0)\upharpoonright\mathfrak{N} = 0$, therefore from Schwartz's lemma we obtain $\left\|\widetilde{\mathfrak{X}}_0(z)\upharpoonright\mathfrak{N}\right\| \leq |z|$, $|z| < 1$.

Let us show that the operator T in $H \oplus H$ defined by (3.18) is a contraction. Let $\mathbf{f} = \begin{pmatrix} f_1 \\ f_2 \end{pmatrix}$, where $f_1, f_2 \in H$. Then

$$\begin{aligned} ||\mathbf{f}||^2 - ||T\mathbf{f}||^2 &= ||f_1||^2 + ||f_2||^2 - ||AP_{H_0}f_1 + D^2_{A^*}f_2||^2 - ||P_{\mathfrak{N}}f_1 + A^*f_2||^2 \\ &= ||D_AP_{H_0}f_1||^2 + ||D_{A^*}f_2||^2 - ||D^2_{A^*}f_2||^2 - 2\mathrm{Re}\,(D_AP_{H_0}f_1, D_AA^*f_2) \\ &= ||D_AP_{H_0}f_1||^2 + ||A^*D_{A*}f_2||^2 - 2\mathrm{Re}\,(D_AP_{H_0}f_1, A^*D_{A^*}f_2) \\ &= ||D_A(P_{H_0}f_1 - A^*f_2)||^2 \geq 0. \end{aligned}$$

Thus, the operator T can be considered as the linear passive discrete-time system with the state space H, input space H, and output space H [12]. The transfer function of this system takes the form

$$U_T(z) = AP_{H_0} + zD^2_{A^*}(I - zA^*)^{-1}P_{\mathfrak{N}},\ z \in \mathbb{D}$$

and belongs to the Schur class $\mathbf{S}(H, H)$ of holomorphic on the unit disk $\mathbb{D}$ and contractive operator valued functions. Taking into account relations (3.12) and (3.17) we get the equality $U_T(z) = \widetilde{A}(z)$ for all $|z| < 1$. Hence once again $||\widetilde{A}(z)|| \leq 1$, $|z| < 1$. □

Proposition 3.3. *With respect to the decomposition $H = H_0 \oplus \mathfrak{N}$ the operator $\widetilde{A}(z)$ takes the following block-matrix form:*

$$\widetilde{A}(z) = \begin{pmatrix} A_0 & zD_{A_0^*}\mathfrak{X}_0(z)K_0^* \\ K_0D_{A_0} & -zK_0A_0^*\mathfrak{X}_0(z)K_0^* + zD^2_{K_0^*} \end{pmatrix}. \tag{3.20}$$

Proof. Using (3.4), (3.17), (3.15), and (2.11) we get

$$\begin{aligned}\widetilde{A}(z) &= AP_{H_0} + D_{A^*}\widetilde{\mathfrak{X}}(z)P_{\mathfrak{N}}\\ &= AP_{H_0} + z\left(D^2_{A^*} + (D_{A_0^*} - K_0A_0^*)(\mathfrak{X}_0(z) + A_0)K_0^*\right)P_{\mathfrak{N}}\\ &= AP_{H_0} + z\Big(I - (A_0 + K_0D_{A_0})D_{A_0}K_0^* + D_{A_0^*}\mathfrak{X}_0(z)K_0^* + D_{A_0^*}A_0K_0^*\\ &\quad - K_0A_0^*\mathfrak{X}_0(z)K_0^* - K_0A_0^*A_0K_0^*\Big)P_{\mathfrak{N}}\\ &= AP_{H_0} + z\left(D_{A_0^*}\mathfrak{X}(z)K_0^* - K_0A_0^*\mathfrak{X}_0(z)K_0^* + I - K_0K_0^*\right)P_{\mathfrak{N}}\end{aligned}$$

This yields the block-matrix representation (3.20). □

Theorem 3.4. *Let $A = A_0 + K_0D_{A_0}$ be a $C(\alpha)$-suboperator. Then the following conditions are equivalent:*

(i) $||K_0|| < 1$;
(ii) $\operatorname{ran}(D_{A^*}) \supset \mathfrak{N}$;
(iii) $\operatorname{ran}(D_A) = \operatorname{ran}(D_{A_0})$;
(iv) *for all $h \in \mathfrak{N}$ holds*

$$\sup_{\varphi \in H_0} \frac{|(A\varphi, h)|^2}{||D_A\varphi||^2} < \infty; \tag{3.21}$$

(v) *for at least one $z \in \mathbb{D}$ the operator $\widetilde{A}(z)$ belongs to the class $\widetilde{C}$;*
(vi) *for all $z \in \mathbb{D}$ the operator $\widetilde{A}(z)$ belongs to the class $\widetilde{C}$.*

Proof. By (3.7) and (3.6) conditions (i), (ii), and (iii) are equivalent for every nondensely defined contraction A.

As it is well known, condition (3.21) is equivalent to

$$A^*\mathfrak{N} \subset \operatorname{ran}(D_A).$$

The last condition by R.G. Douglas's theorem 1.2 is equivalent to

$$||P_{\mathfrak{N}}A\varphi|| \le \gamma||D_A\varphi|| \quad \text{for all} \quad \varphi \in H_0,$$

where $\gamma > 0$. Since $P_{\mathfrak{N}}A = K_0 D_{A_0}$ and (3.5) holds, the last inequality is equivalent to

$$||K_0\varphi||^2 \le \frac{\gamma^2}{1+\gamma^2}||\varphi||^2,\ \varphi \in \mathfrak{D}_{A_0} \iff ||K_0|| < 1.$$

Suppose that for some $z \in \mathbb{D}$ the operator $\widetilde{A}(z)$ belongs to the class $\widetilde{C}$. Then holds the equality $\operatorname{ran}(D_{\widetilde{A}(z)}) = \operatorname{ran}(D_{\widetilde{A}^*(z)})$. In particular,

$$\operatorname{ran}(D_{\widetilde{A}(z)}) \cap \mathfrak{N} = \operatorname{ran}(D_{\widetilde{A}^*(z)}) \cap \mathfrak{N}.$$

Taking into account (2.3) with $K = \widetilde{\mathfrak{X}}_0(z)\restriction\mathfrak{N}$ and because $||\widetilde{\mathfrak{X}}_0(z)\restriction\mathfrak{N}|| \le |z| < 1$ we get

$$\operatorname{ran}(D_{\widetilde{A}(z)}) \cap \mathfrak{N} = \mathfrak{N}.$$

Now (2.4) gives $\operatorname{ran}(D_{A^*}) \supset \mathfrak{N}$.

Let $\operatorname{ran}(D_{A^*}) \supset \mathfrak{N}$. Let us show that the operator $\widetilde{A}_0 = \widetilde{A}(0) = AP_{H_0}$ belongs to the class $\widetilde{C}$ in H. Since $||K_0|| < 1$, it follows from (3.6) that $\operatorname{ran}(D_A) = \operatorname{ran}(D_{A_0})$, $\operatorname{ran}(D_{\widetilde{A}_0}) = \operatorname{ran}(D_{A_0}) \oplus \mathfrak{N}$, $D_{\widetilde{A}_0} = D_A P_{H_0} + P_{\mathfrak{N}}$. Further from (3.6) and (3.9):

$$\begin{aligned}
\widetilde{A}_0 - \widetilde{A}_0^* &= AP_{H_0} - A^* = (A_0 - A_0^*)P_{H_0} + K_0 D_{A_0} P_{H_0} - D_{A_0} K_0^* P_{\mathfrak{N}} \\
&= i\tan\alpha D_A F D_A P_{H_0} + P_{\mathfrak{N}} K_0 D_{A_0} P_{H_0} - D_{A_0} K_0^* P_{\mathfrak{N}} \\
&= i\tan\alpha D_{\widetilde{A}_0} F P_{H_0} D_{\widetilde{A}_0} + D_{\widetilde{A}_0} K_0 D_{K_0}^{-1} V_0 P_{H_0} D_{\widetilde{A}_0} - D_{\widetilde{A}_0} V_0^* D_{K_0}^{-1} K_0^* P_{\mathfrak{N}} D_{\widetilde{A}_0} \\
&= D_{\widetilde{A}_0} \widetilde{G}_0 D_{\widetilde{A}_0},
\end{aligned}$$

where

$$\widetilde{G}_0 = i\tan\alpha F P_{H_0} + K_0 D_{K_0}^{-1} V_0 P_{H_0} - V_0^* D_{K_0}^{-1} K_0^* P_{\mathfrak{N}}$$

is a skew selfadjoint bounded operator in $\mathfrak{D}_{\widetilde{A}_0}$. Thus $\widetilde{A}_0$ belongs to the class $\widetilde{C}$. It follows that

$$\operatorname{ran}(D_{A^*}) = \operatorname{ran}(D_{\widetilde{A}_0}) = \operatorname{ran}(D_{A_0}) \oplus \mathfrak{N}. \tag{3.22}$$

Let us show that under the condition (ii) the operator $\widetilde{A}(z)$ belongs to the class $\widetilde{C}$ for any $z \in \mathbb{D}$. Actually, from the second relation in (2.2) with the operator $K = \widetilde{\mathfrak{X}}_0(z)\restriction\mathfrak{N}$ and taking into account that $||\widetilde{\mathfrak{X}}_0(z)\restriction\mathfrak{N}|| < 1$ we get that $\operatorname{ran}(D_{\widetilde{A}^*(z)}) = \operatorname{ran}(D_{A^*})$. This equality and (3.22) imply that there are two bounded and boundedly invertible operators L_1 and L_2 in $\mathfrak{D}_{A^*}$ such that

$$D_{A^*} = L_1 D_{\widetilde{A}^*(z)} \quad \text{and} \quad D_{\widetilde{A}_0} = L_2 D_{\widetilde{A}^*(z)}.$$

Hence,

$$\begin{aligned}\operatorname{Im}(\widetilde{A}(z)f,f) &= \operatorname{Im}(AP_{H_0}f,f) + \operatorname{Im}(D_{A^*}\widetilde{\mathfrak{X}}_0(z)P_{\mathfrak{N}}f,f)\\ &= \operatorname{Im}(\widetilde{G}_0 D_{\widetilde{A}_0}f, D_{\widetilde{A}_0}f) + \operatorname{Im}(\widetilde{\mathfrak{X}}_0(z)P_{\mathfrak{N}}f, D_{A^*}f)\\ &= \operatorname{Im}(\widetilde{G}_0 L_2 D_{\widetilde{A}^*(z)}f, L_2 D_{\widetilde{A}^*(z)}f) + \operatorname{Im}(\widetilde{\mathfrak{X}}_0(z)P_{\mathfrak{N}}D_{\widetilde{A}_0}f, D_{A^*}f)\\ &= \operatorname{Im}(\widetilde{G}_0 L_2 D_{\widetilde{A}^*(z)}f, L_2 D_{\widetilde{A}^*(z)}f) + \operatorname{Im}(\widetilde{\mathfrak{X}}_0(z)P_{\mathfrak{N}}L_2 D_{\widetilde{A}^*(z)}f, L_1 D_{\widetilde{A}^*(z)}f)\\ &= (\widetilde{G}(z)D_{\widetilde{A}^*(z)}f, D_{\widetilde{A}^*(z)}f),\ f\in H,\end{aligned}$$

where $\widetilde{G}(z)$ is some selfadjoint operator in $\mathfrak{D}_{\widetilde{A}^*(z)}$. This means that $\widetilde{A}(z)\in\widetilde{C}$ for all $z\in\mathbb{D}$. □

3.4. Rigid and soft extensions of $C(\alpha)$-suboperator and their representations in operator and block-matrix forms

Definition 3.5. [4], [7]. Let A be a $C(\alpha)$-suboperator defined on the subspace H_0 of the Hilbert space H. The extension $T\in\widetilde{C}$ of A is called the rigid if for all $h\in H$

$$\inf\{\operatorname{Re}((I+T)(h-\varphi),h-\varphi),\ \varphi\in H_0\}=0,$$

and the soft if for all $h\in H$

$$\inf\{\operatorname{Re}((I-T)(h-\varphi),h-\varphi),\ \varphi\in H_0\}=0.$$

For a Hermitian contraction these notions were introduced by M.G. Kreĭn in [19].

Denote by A_μ and A_M the rigid and soft extension, respectively. Clearly, $A_M=-(-A)_\mu$. The equalities for an arbitrary contraction A in H

$$2\operatorname{Re}\big((I\pm A)\phi,\phi\big)=||(I\pm A)\phi||^2+||\phi||-||A\phi||^2\qquad\forall\phi\in H_0\tag{3.23}$$

imply that if A is $C(\alpha)$-suboperator then operators $I\pm A$ are α-sectorial. As was shown in [8] that the following relations hold

$$A_\mu=(I+A)_N-I,\quad A_M=I-(I-A)_N,$$

where $(I\pm A)_N$ are Kreĭn–von Neumann extensions of operators $I\pm A$.

Theorem 3.6. [8]. *Let A be a $C(\alpha)$-suboperator defined on the subspace H_0 of the Hilbert space H and let*

$$\begin{aligned}&A_M(z)f=\begin{cases}\overline{z}Af,\ f\in H_0\\ f,\ f\in\mathfrak{N}_{1/z}(A)\end{cases},\\ &A_\mu(z)=-A_M(z),\ \ |z|<1.\end{aligned}\tag{3.24}$$

Then the operators $A_\mu(z)$ and $A_M(z)$ are the rigid and the soft extensions of the $C(\alpha)$-suboperators $-\overline{z}A$ and $\overline{z}A$, respectively and the following equalities hold:

$$\begin{aligned}A_M&=\operatorname{s-lim}_{z\to1}A_M(z),\\ A_\mu&=\operatorname{s-lim}_{z\to-1}A_\mu(z),\end{aligned}\tag{3.25}$$

where limits are nontangential to the unit circle. Operators A_M and A_μ belong to the class $C(\alpha)$.

Using (3.2) (3.25) one can derive the relation

$$A_M(z) = \bar{z}A(I - zA^*)^{-1}P_{H_0}(I - zA^*) + (I - zA^*)^{-1}P_{\mathfrak{N}}. \tag{3.26}$$

$$A_M(z) - \bar{z}\widetilde{A}(z) = -\left(A_\mu(z) + \bar{z}\widetilde{A}(z)\right) = (1 - |z|^2)(I - zA^*)^{-1}P_{\mathfrak{N}}. \tag{3.27}$$

Theorem 3.7. *Let A be a $C(\alpha)$-suboperator and let $\widetilde{A}(z)$ be the family of operators on the unit disk defined by (3.16). Then*

$$\begin{aligned} \text{s-}\lim_{z\to 1} \widetilde{A}(z) &= A_M, \\ \text{s-}\lim_{z\to -1} \widetilde{A}(z) &= A_\mu, \end{aligned} \tag{3.28}$$

where limits are nontangential to the unit circle.

Operators A_M and A_μ take the operator form

$$\begin{aligned} A_M &= AP_{H_0} + D_{A^*}\widetilde{\mathfrak{X}}_0(1)P_{\mathfrak{N}}, \\ A_\mu &= AP_{H_0} + D_{A^*}\widetilde{\mathfrak{X}}_0(-1)P_{\mathfrak{N}}, \end{aligned} \tag{3.29}$$

and the block-matrix form

$$\begin{aligned} A_M &= \begin{pmatrix} A_0 & D_{A_0^*}\mathfrak{X}_0(1)K_0^* \\ K_0D_{A_0} & -K_0A_0^*\mathfrak{X}_0(1)K_0^* + D^2_{K_0^*} \end{pmatrix}, \\ A_\mu &= \begin{pmatrix} A_0 & -D_{A_0^*}\mathfrak{X}_0(-1)K_0^* \\ K_0D_{A_0} & K_0A_0^*\mathfrak{X}_0(-1)K_0^* - D^2_{K_0^*} \end{pmatrix} \end{aligned} \tag{3.30}$$

with respect to the decomposition $H = H_0 \oplus \mathfrak{N}$.

Proof. Because A_0 belongs to the class $C(\alpha)$ in H_0, holds the equality $\mathfrak{D}_{A_0} = \mathfrak{D}_{A_0^*}$ and linear manifolds $\operatorname{ran}(I \pm A_0^*)$ are dense in $\mathfrak{D}_{A_0}$. Since $A^*\upharpoonright H_0 = A_0^*$, one has

$$(I - zA^*)^{-1}P_{\mathfrak{N}}h = P_{\mathfrak{N}}h + z(I - zA_0^*)^{-1}A^*P_{\mathfrak{N}}. \tag{3.31}$$

Hence

$$(1 - |z|^2)(I - zA^*)^{-1}P_{\mathfrak{N}}h = (1 - |z|^2)P_{\mathfrak{N}}h + z(1 - |z|^2)(I - zA_0^*)^{-1}A^*P_{\mathfrak{N}}h.$$

Since $(1 - |z|^2)||(I - zA_0^*)^{-1}|| \le 1 + |z|$ for $|z| < 1$ and

$$(1 - |z|^2)(I - zA_0^*)^{-1}(I \pm A_0^*)\varphi = (1 - |z|^2)\varphi \mp (1 - |z|^2)(1 \pm z)(I - zA_0^*)^{-1}A_0^*\varphi$$

for every $\varphi \in H_0$, we have for $x \in \mathfrak{D}_{A_0}$:

$$\lim_{z\to\pm 1}(1 - |z|^2)(I - zA_0^*)^{-1}x = 0.$$

This relation and (3.25), (3.27), (3.31) yield (3.28). The relations (3.29), (3.30) follow from (3.17), (3.20) and Theorem 3.1. □

Observe that from Theorems 3.7 and 3.1 it follows that hold the relations

$$A_\mu^* = \text{s-}\lim_{z\to -1} \widetilde{A}^*(z), \quad A_M^* = \text{s-}\lim_{z\to 1} \widetilde{A}^*(z). \tag{3.32}$$

Corollary 3.8. *The following conditions are equivalent:*

(i) $\operatorname{ran}(D_{A^*})\cap\mathfrak{N}=\{0\}$;

(ii) *for all* $h\in\mathfrak{N}\setminus\{0\}$ *the relation*

$$\sup_{\varphi\in H_0}\frac{|(A\varphi,h)|^2}{||D_A\varphi||^2}=\infty \tag{3.33}$$

holds;

(iii) *the operator* K_0^* *is isometric;*

(iv) *the* $C(\alpha)$*-suboperator* A *has a unique extension of the class* $\widetilde{C}$.

Proof. The equivalence of conditions (i) and (iii) follows from (3.7). As it is well known, condition (3.33) is equivalent to

$$A^*\mathfrak{N}\cap\operatorname{ran}(D_A)=\{0\}.$$

By $A^*h=D_{A_0}K_0^*h$, $h\in\mathfrak{N}$ and (3.6), we get

$$A^*\mathfrak{N}\cap\operatorname{ran}(D_A)=\{0\}\iff\operatorname{ran}(K_0^*)\cap\operatorname{ran}(D_{K_0})=\{0\}.$$

Since $K_0^*D_{K_0^*}=D_{K_0}K_0^*$, we have the equivalence

$$\operatorname{ran}(K_0^*)\cap\operatorname{ran}(D_{K_0})=\{0\}\iff\quad\text{the operator}\quad K_0^*\quad\text{is isometric.}$$

Let us prove that (iii) $\iff$ (iv). From (3.30) it follows that

$$A_M-A_\mu=\begin{pmatrix}0 & D_{A_0^*}(\mathfrak{X}_0(1)+\mathfrak{X}_0(-1))K_0^*\\ 0 & -K_0A_0^*(\mathfrak{X}_0(1)+\mathfrak{X}_0(-1))K_0^*+2D_{K_0^*}^2\end{pmatrix}.$$

Therefore,

$$A_M=A_\mu\iff\begin{cases}(\mathfrak{X}_0(1)+\mathfrak{X}_0(-1))K_0^*=0,\\ D_{K_0^*}=0.\end{cases}$$

The equality $D_{K_0^*}=0$ is equivalent to K_0^* being isometric. If K_0^* is an isometry then $\ker(D_{K_0})=\operatorname{ran}(K_0^*)$ and (3.11) yields the equality $(\mathfrak{X}_0(1)+\mathfrak{X}_0(-1))K_0^*=0$, i.e., $A_M=A_\mu$. Finally, the condition $A_M=A_\mu$ is equivalent to that the operator A has a unique extension of the class $\widetilde{C}$ [6], [7], [10]. □

Let us make a few remarks.

Remark 3.9. Let us define the following subspaces in $\mathfrak{D}_{A^*}$:

$$\begin{aligned}\mathfrak{K}_0&:=\overline{(D_{A^*}-i\tan\alpha AFD_A)H_0},\\ \mathfrak{H}_0&:=\overline{D_{A^*}H_0},\\ \mathfrak{L}_0'&:=\mathfrak{D}_{A^*}\cap\ker\left(P_{H_0}(D_{A^*}+i\tan\alpha D_AFA^*)\right)=\mathfrak{D}_{A^*}\ominus\mathfrak{K}_0.\end{aligned}$$

Then holds the direct sum decomposition [6], [7], [10]:

$$\mathfrak{D}_{A^*}=\mathfrak{H}_0\dot{+}\mathfrak{L}_0'.$$

Note that subspaces $\mathfrak{H}_0$ and $\mathfrak{L}'_0$ are orthogonal iff $\alpha = 0$. Let $\mathcal{P}_{\mathfrak{H}_0}$ and $\mathcal{P}_{\mathfrak{L}'_0}$ be the skew projections in $\mathfrak{D}_{A^*}$ onto $\mathfrak{H}_0$ and $\mathfrak{L}'_0$ corresponding this decomposition. The operator $N_0 \in \mathcal{L}(\mathfrak{K}_0, \mathfrak{N})$ given by the equality

$$N_0(D_{A^*} - i\tan\alpha AFD_A)\varphi = P_{\mathfrak{N}}A\varphi,\ \varphi \in H_0$$

is well defined and is a contraction. Let $N_0^* \in \mathcal{L}(\mathfrak{N}, \mathfrak{K}_0)$ be the adjoint of N_0 and let

$$M_0 = \mathcal{P}_{\mathfrak{H}_0}N_0^* \in \mathcal{L}(\mathfrak{N}, \mathfrak{H}_0).$$

As was proved in [6], [10] and earlier in [11] for the case $\alpha = 0$, the operators

$$K_M = M_0 + \mathcal{P}_{\mathfrak{L}'_0}D_{A^*},\ K_\mu = M_0 - \mathcal{P}_{\mathfrak{L}'_0}D_{A^*}$$

are isometries in $\mathcal{L}(\mathfrak{N}, \mathfrak{D}_{A^*})$ and the soft and rigid extensions of A take the form

$$A_M = AP_{H_0} + D_{A^*}K_M P_{\mathfrak{N}},\ A_\mu = AP_{H_0} + D_{A^*}K_\mu P_{\mathfrak{N}}.$$

Comparing now with (3.29) we get the equalities

$$\widetilde{\mathfrak{X}}_0(1)\!\upharpoonright\!\mathfrak{N} = \left(M_0 + \mathcal{P}_{\mathfrak{L}'_0}D_{A^*}\right)\upharpoonright\mathfrak{N},\ \widetilde{\mathfrak{X}}_0(-1)\!\upharpoonright\!\mathfrak{N} = \left(M_0 - \mathcal{P}_{\mathfrak{L}'_0}D_{A^*}\right)\upharpoonright\mathfrak{N}.$$

When A is a Hermitian contraction one has $\mathfrak{X}_0(\pm 1) = \pm I\!\upharpoonright\!\mathfrak{D}_{A_0}$ and therefore the soft and rigid extensions takes the form

$$\begin{aligned} A_M &= \begin{pmatrix} A_0 & D_{A_0}K_0^* \\ K_0 D_{A_0} & -K_0A_0K_0^* + D^2_{K_0^*}\end{pmatrix}, \\ A_\mu &= \begin{pmatrix} A_0 & D_{A_0}K_0^* \\ K_0 D_{A_0} & -K_0A_0K_0^* - D^2_{K_0^*}\end{pmatrix}. \end{aligned} \tag{3.34}$$

Remark 3.10. In [22] the following family of noncontractive extensions of A was considered

$$\widehat{A}(\lambda)(\varphi + \varphi_\lambda) = A\varphi + \lambda\varphi_\lambda,\ \varphi \in H_0, \varphi_\lambda \in \mathfrak{N}_\lambda,\ |\lambda| > 1.$$

It was proved in [22] that

$$A_M = \operatorname{s-lim}_{\lambda\downarrow 1}\widehat{A}(\lambda),\ A_\mu = \operatorname{s-lim}_{\lambda\downarrow 1}\widehat{A}(-\lambda)$$

and with respect to the orthogonal decomposition $H = H_0 \oplus \mathfrak{N}$ the operators A_μ and A_M have the following block-matrix representations

$$\begin{aligned} A_\mu &= \begin{pmatrix} A_0 & D_{A_0}K_0^* + (A_0 - A_0^*)(A_0^* + I)^{-1}D_{A_0}K_0^* \\ K_0D_{A_0} & -I + K_0D_{A_0}(A_0^* + I)^{-1}D_{A_0}K_0^*\end{pmatrix}, \\ A_M &= \begin{pmatrix} A_0 & D_{A_0}K_0^* + (A_0 - A_0^*)(A_0^* - I)^{-1}D_{A_0}K_0^* \\ K_0D_{A_0} & I + K_0D_{A_0}(A_0^* - I)^{-1}D_{A_0}K_0^*\end{pmatrix}, \end{aligned}$$

where by the definition

$$\begin{aligned} (A_0 - A_0^*)(A_0^* \pm I)^{-1}D_{A_0} &= \lim_{r\downarrow 1}(A_0 - A_0^*)(A_0^* \pm rI)^{-1}D_{A_0}, \\ D_{A_0}(A_0^* \pm I)^{-1}D_{A_0} &= \lim_{r\downarrow 1}D_{A_0}(A_0^* \pm rI)^{-1}D_{A_0}. \end{aligned}$$

Using the relations in Subsection 3.1 it can be proved that

$$D_{A_0}K_0^* + (A_0 - A_0^*)(A_0^* \pm I)^{-1}D_{A_0}K_0^* = \pm D_{A_0}\mathfrak{X}_0(\pm 1)K_0^*,$$
$$K_0D_{A_0}(A_0^* \pm I)^{-1}D_{A_0}K_0^* \mp I = \mp D^2_{K_0^*} \pm K_0A_0^*\mathfrak{X}_0(\mp 1)K_0^*.$$

Thus, these blocks coincide with corresponding blocks in (3.30).

Remark 3.11. 1) Let A be a Hermitian contraction. In [11] were parametrized all contractive extensions T possessing the property $T^* \supset A$ (quasi-selfadjoint contractive extensions). In particular, the following result has been established.

Theorem. [11]. *Let A be a Hermitian contraction and let $\beta \in [0, \pi/2)$. Then the formula*

$$T = \frac{A_M + A_\mu}{2} + \frac{1}{2}(A_M - A_\mu)^{1/2}Z(A_M - A_\mu)^{1/2} \tag{3.35}$$

gives a bijective correspondence between all quasi-selfadjoint extensions T of A from the class $C(\beta)$ and all $Z \in C(\beta)$ in the subspace $\overline{\operatorname{ran}}\,(A_M - A_\mu)$.

From (3.34) it follows for a Hermitian A contraction the equalities

$$\frac{A_M + A_\mu}{2} = \begin{pmatrix} A_0 & D_{A_0}K_0^* \\ K_0D_{A_0} & -K_0A_0K_0^* \end{pmatrix}, \quad \frac{A_M - A_\mu}{2} = \begin{pmatrix} 0 & 0 \\ 0 & D_{K_0^*} \end{pmatrix}.$$

Hence from (3.35) we obtain the block-matrix representation of all quasi-selfadjoint extensions T of the class $C(\beta)$ for a Hermitian contraction A:

$$T = \begin{pmatrix} A_0 & D_{A_0}K_0^* \\ K_0D_{A_0} & -K_0A_0K_0^* + D_{K_0^*}XD_{K_0^*} \end{pmatrix},$$

where $X \in C(\beta)$ in the subspace $\mathfrak{D}_{K_0^*}$.

2) In [10] a description of all extensions $T \in C(\beta)$ of a $C(\alpha)$ suboperator A was given in the following form.

Theorem. [10]. *Let $T_0 = (A_\mu + A_M)/2$ and let $\Theta_{T_0}(z)$ be the characteristic function of the contraction T_0. Then the formula*

$$T = T_0 + D_{T_0^*}(I + \Theta_{T_0}(1)YT_0^*)^{-1}\Theta_{T_0}(1)YD_{T_0} \tag{3.36}$$

gives a bijective correspondence between all extensions T of the class $C(\beta)$ of $C(\alpha)$-suboperator A and all contractions $Y \in \mathcal{L}(\mathfrak{D}_{T_0})$ that possess the properties:

$$\ker Y \supset \overline{D_{T_0}H_0}, \ (I + \Theta_{T_0}(1)YT_0^*)^{-1} \in \mathcal{L}(\mathfrak{D}_{T_0}),$$
$$|((I - Y^*)\left(\Theta_{T_0}(-1) - \Theta_{T_0}(1)\right)^{-1}\left(\Theta_{T_0}(-1) + \Theta_{T_0}(+1)\right)(I - Y) + Y - Y^*)h, h)|$$
$$\le \tan\beta \left\|D_Y h\right\|^2, \ h \in \mathfrak{D}_{T_0}.$$

The operator $Y = P_{\mathfrak{H}_0}$ corresponds to A_M and $Y = \Theta_{T_0}^{-1}(1)\Theta_{T_0}(-1)P_{\mathfrak{H}_0}$ corresponds to A_μ, where $P_{\mathfrak{H}_0}$ is the orthogonal projection in $\mathfrak{D}_{T_0}$ onto the subspace $\mathfrak{H}_0 = \{f \in \mathfrak{D}_{T_0} : D_{T_0}f \in \mathfrak{N}\}$. When A is a Hermitian contraction ($\alpha = 0$) the formula (3.36) *takes the form*

$$T = T_0 + D_{T_0}(I + YT_0)^{-1}YD_{T_0} \tag{3.37}$$

where $Y \in C(\beta)$ *in the subspace* $\mathfrak{D}_{T_0}$, $(I + YT_0)^{-1} \in \mathcal{L}(\mathfrak{D}_{T_0})$, *and* $\ker Y \supset \overline{D_{T_0}H_0}$. *If in additional* $\ker Y^* \supset \overline{D_{T_0}H_0}$ *then* (3.37) *can be transformed into* (3.35).

3) The next result has been established in [22].

Theorem. [22]. *The following equivalence holds for an extension* T *of the* $C(\alpha)$-*suboperator* A:

$$T \in C(\beta) \iff T = A_\mu + KP_{\mathfrak{N}} \quad \text{and} \quad R_\pm^{-1}(KP_{\mathfrak{N}} + Q_\pm) \quad \text{is a contraction,}$$

where

$$R_\pm = D_{A^*} \pm i \cot\beta(AP_{H_0} - A^*) + \cot^2\beta P_{\mathfrak{N}},$$

and

$$Q_\pm = (A_\mu \pm i\cot\beta I)P_{\mathfrak{N}}.$$

Finally we note that the problem of a block-matrix representation of all extensions of the class $C(\beta)$, $\beta \in [\alpha, \pi/2)$ is still open.

4. Limit representations of the Friedrichs and Kreĭn–von Neumann extensions of a sectorial linear relation

Let $\mathbf{S}$ be a sectorial l.r. with the vertex at the origin and let $\mathfrak{N}_\lambda(\mathbf{S}) = H \ominus \operatorname{ran}(\mathbf{S} - \bar\lambda I)$ be the defect subspaces of $\mathbf{S}$. Let us define the following family of extensions of $\mathbf{S}$:

$$\widetilde{\mathbf{S}}_\lambda = \mathbf{S} \dot{+} \left\{\langle\varphi_\lambda, -z\varphi_\lambda\rangle,\ \varphi_\lambda \in \mathfrak{N}_\lambda\right\},\ \operatorname{Re}\lambda < 0. \tag{4.1}$$

L.r. $\widetilde{\mathbf{S}}_\lambda$ are m-accretive. In fact, for $x \in \operatorname{dom}(\mathbf{S}), \varphi_\lambda \in \mathfrak{N}_\lambda$ we have

$$\left(\widetilde{\mathbf{S}}_\lambda(x + \varphi_\lambda), x + \varphi_\lambda\right) = \left(\mathbf{S}(x), x\right) - \lambda(\varphi_\lambda, \varphi_\lambda) - 2i\operatorname{Im}\left[\lambda(\varphi_\lambda, x)\right]$$

and

$$\operatorname{Re}\left(\widetilde{\mathbf{S}}_\lambda(x + \varphi_\lambda), x + \varphi_\lambda\right) = \operatorname{Re}\left(\mathbf{S}(x), x\right) - \operatorname{Re}\lambda\left\|\varphi_\lambda\right\|^2 \geq 0.$$

Besides,

$$\widetilde{\mathbf{S}}_\lambda(x + \varphi_\lambda) - \bar\lambda(x + \varphi_\lambda) = \mathbf{S}(\varphi) - \bar\lambda x - 2\operatorname{Re}\lambda\,\varphi_\lambda.$$

The definition of the defect subspace implies $\operatorname{ran}(\widetilde{\mathbf{S}}_\lambda - \bar\lambda I) = H$. Thus, $\widetilde{\mathbf{S}}_\lambda$ is m-accretive. Note that in the case of a closed sectorial operator S the extensions $\widetilde{S}_\lambda$ take the form

$$\begin{aligned}
&\operatorname{dom}(\widetilde{S}_\lambda) = \operatorname{dom}(S) \dot{+} \mathfrak{N}_\lambda(S),\\
&\widetilde{S}_\lambda(x + \varphi_\lambda) = Sx - \lambda\varphi_\lambda,\ x \in \operatorname{dom}(S),\ \varphi_\lambda \in \mathfrak{N}_\lambda(S).
\end{aligned}$$

Recall (see Introduction) that the Friedrichs and Kreĭn–von Neumann extensions of $\mathbf{S}$ are given by the the fractional-linear transformations of A_μ and A_M:

$$\mathbf{S}_F = (I - A_\mu)(I + A_\mu)^{-1},\ \mathbf{S}_N = (I - A_M)(I + A_M)^{-1}.$$

Theorem 4.1. *Let* $\mathbf{S}$ *be a sectorial l.r. with the vertex at the origin. Then for its Friedrichs and Kreĭn–von Neumann extensions the following strong resolvent limits representations*

$$\mathbf{S}_N = \text{s-R-}\lim_{\lambda\to 0} \widetilde{\mathbf{S}}_\lambda,\ \mathbf{S}_F = \text{s-R-}\lim_{\lambda\to\infty} \widetilde{\mathbf{S}}_\lambda,$$
$$\mathbf{S}_N^* = \text{s-R-}\lim_{\lambda\to 0} \widetilde{\mathbf{S}}_\lambda^*,\ \mathbf{S}_F^* = \text{s-R-}\lim_{\lambda\to\infty} \widetilde{\mathbf{S}}_\lambda^*$$

hold. The limits are nontangential to the imaginary axis.

Proof. Consider the linear fractional transformation $A = (I - \mathbf{S})(I + \mathbf{S})^{-1}$. The operator A is a $C(\alpha)$-suboperator for some α. One can readily check the equalities

$$\mathfrak{N}_\lambda(\mathbf{S}) = \mathfrak{N}_{1/z}(A),\ \widetilde{\mathbf{S}}_\lambda = \big(I - \widetilde{A}(z)\big)\big(I + \widetilde{A}(z)\big)^{-1},\ \lambda = (z-1)(1+z)^{-1},$$

where $\widetilde{A}(z)$ is defined by (7). This implies the relation

$$(\widetilde{\mathbf{S}}_\lambda + I)^{-1} = \frac{1}{2}(I + \widetilde{A}(z)).$$

Taking into account that

$$(\mathbf{S}_F + I)^{-1} = \frac{1}{2}(I + A_\mu),\ (\mathbf{S}_N + I)^{-1} = \frac{1}{2}(I + A_M),$$
$$(\mathbf{S}_F^* + I)^{-1} = \frac{1}{2}(I + A_\mu^*),\ (\mathbf{S}_N^* + I)^{-1} = \frac{1}{2}(I + A_M^*)$$

and applying (3.28) and (3.32) we get the assertion of Theorem 4.1. □

Remark 4.2. In [9] it is proved that the following conditions for $\widetilde{\mathbf{S}}_\lambda$ defined by (4.1) are equivalent:

(i) the linear relation $\widetilde{\mathbf{S}}_\lambda$ is m-sectorial for some λ, $\operatorname{Re}\lambda < 0$;
(ii) the linear relation $\widetilde{\mathbf{S}}_\lambda$ is m-sectorial for all λ, $\operatorname{Re}\lambda < 0$;
(iii) $\operatorname{dom}(\mathbf{S}^*) \subseteq D[\mathbf{S}_N]$, where $D[\mathbf{S}_N]$ is the domain of the closure of the sesquilinear form associated with $\mathbf{S}_N$;
(iv) for all $x \in \operatorname{dom}(\mathbf{S})$ and some $\lambda, \operatorname{Re}\lambda < 0$ holds

$$\operatorname{Re}(\mathbf{S}(x), x) \geq k(\lambda)\big\|P_{\mathfrak{N}_\lambda} x\big\|^2,$$

where $k(\lambda) > 0$.

Remark 4.3. The resolvent formula for m-sectorial extensions of a densely defined closed sectorial operator S under one of the conditions (i)–(iv) and Theorem 4.1 have been established in [9].

References

[1] T. Ando *Topics on operator inequalities.* Division of Applied Mathematics, Research Institute of Applied Electricity, Hokkaido University, Sapporo, 1978. ii+44

[2] T. Ando and K. Nishio, *Positive selfadjoint extensions of positive symmetric operators.* Tohoku Math. J. **22** (1970), 251–270.

[3] Yu.M. Arlinskii, *Contractive extensions of a dual pair of contractions and their resolvents.* Ukrain. Math.Journ. **37**(1985), No. 2, 247–250.

[4] Yu.M. Arlinskiĭ, *On a class of contractions in a Hilbert space.* Ukrain. Math. Journ. **39** (1987), No. 6, 691–696 (Russian).

[5] Yu.M. Arlinskiĭ, *Characteristic functions of operators of the class $C(\alpha)$.* Izv.Vyssh. Uchebn. Zav., Matematika **No. 2** (1991), 13–21.

[6] Yu.M. Arlinskiĭ, *Extensions of sectorial operators and their fractional-linear transformations.* Preprint 92.01, Inst. Appl. Math. Mech. Akad. Nauk Ukraine, Donetsk (1992), 76p. (Russian).

[7] Yu.M. Arlinskiĭ, *On class of extensions of a $C(\alpha)$-suboperators.* Dokl. Akad. Nauk Ukraine **No. 8** (1992), 12–16 (Russian).

[8] Yu.M. Arlinskiĭ, *Extremal extensions of sectorial linear relations.* Math. Studii **7**, No. 1 (1997), 81–96.

[9] Yu.M. Arlinskiĭ, *On functions connected with sectorial operators and their extensions.* Int. Equat. Oper. Theory. **33**, No. 2 (1999), 125–152.

[10] Yu.M. Arlinskiĭ, *On a class of nondensely defined contractions on a Hilbert space and their extensions.* Journ. Math. Sci. **97** (1999), No. 5, 4390–4419.

[11] Yu.M. Arlinskiĭ and E.R. Tsekanovskiĭ, *Quasiselfadjoint contractive extensions of Hermitian contractions*, Teor. Funkts., Funkts. Anal. Prilozhen **50** (1988), 9–16.

[12] D.Z. Arov, *Passive linear stationary dynamical systems.* Sibirsk. Math. Journ. **20** (1979), No. 2, 211–228. (Russian).

[13] Gr. Arsene and A. Gheondea, *Completing matrix contractions*, J. Operator Theory **7** (1982), 179–189.

[14] E. Coddington and H.S.V. de Snoo, *Positive selfadjoint extensions of positive symmetric subspaces.* Math. Z. **159** (1978), 203–214.

[15] M.G. Crandall, *Norm preserving extensions of linear transformations in Hilbert space.* Proc. Amer. Math. Soc. **21** (1969), 335–340.

[16] Ch. Davis, W.M. Kahan, and H.F. Weinberger, *Norm-preserving dilations and their applications to optimal error bound.* Siam J. Numerical Anal. **19** (1982), No. 3, 445–469.

[17] R.G. Douglas, *On majorization, factorization and range inclusion of operators in Hilbert space*, Proc. Amer. Math. Soc. **17** (1966), 413–416.

[18] T. Kato, *Perturbation theory*, Springer-Verlag, Berlin-Heidelberg-New York, 1966.

[19] M.G. Kreĭn, *Theory of selfadjoint extensions of semibounded operators and its applications. I* Mat. Sb. **20** No. 3 (1947), 431–498.

[20] V.U. Kolmanovich and M.M. Malamud, *Extensions of sectorial operators and dual pairs of contractions.* Manuscript No. 4428-85, Deposited at VINITI,(1985), 1–57 (Russian).

[21] M.M. Malamud, *On extensions of Hermitian and sectorial operators and dual pairs of contractions.* Sov. Math. Dokl. **39** (1989), No. 2, 253–254 (Russian).

[22] M.M. Malamud, *On some classes of extensions of sectorial operators and dual pair of contractions.* Operator Theory: Advances and Appl. **124** (2001), 401–448.

[23] F.S. Rofe-Beketov, *Numerical range of linear relation and maximal linear relations.* Functions Theory, Functional Anal. and their Appl. **44** (1985), 103–111 (Russian).

[24] Yu.L. Shmul'yan and R.N. Yanovskaya, *Blocks of a contractive operator matrix*, Izv. Vyssh. Uchebn. Zaved., Mat. **7** (1981), 72–75.

[25] A.V. Shtraus, *On extensions of semibounded operators.* Dokl. Akad. Nauk SSSR **211** (1973), No. 3, 543–546 (Russian).

[26] B. Sz.-Nagy and C. Foias, *Harmonic analysis of operators on Hilbert space*, North-Holland, New York, 1970.

Yury Arlinskiĭ
Department of Mathematics
East Ukrainian National University
Kvartal Molodyozhny, 20-A
91034 Lugansk
Ukraine
e-mail: `yma@snu.edu.ua`

Operator Theory:
Advances and Applications, Vol. 162, 71–83

A Variational Principle for Linear Pencils of Forms

P. Binding and R. Hryniv

Abstract. Eigenvalues are characterized by a double extremum principle for pencils $a - \lambda b$ under rather weak assumptions on the symmetric sesquilinear forms a and b. For example, they can be indefinite and/or degenerate. The treatment unifies many cases previously studied separately, and also gives new eigenvalues previously uncharacterized.

Mathematics Subject Classification (2000). 49R50; 47B50; 47A75.

Keywords. Variational principle; Linear pencils; Quadratic forms.

1. Introduction

Double extremum principles for the generalized eigenvalue problem

$$Ax = \lambda Bx$$

(A and B being self-adjoint operators in a Hilbert space $\mathscr{H}$) can be considered classical in the "right-definite" case where B is positive definite. See, e.g., [WS] for a general result characterizing eigenvalues as double extrema of the generalized Rayleigh quotient

$$r(x) = \frac{(Ax, x)}{(Bx, x)}, \tag{1.1}$$

and for a bibliography of the subject. We note that the denominator in (1.1) does not vanish provided $x \neq 0$.

Various authors have produced analogous results in cases when (Bx, x) can vanish nontrivially. Already in 1904, Holmgren considered a case with B semidefinite, but the result is in general incorrect. See [Al, BEL, EL] for modern versions involving partial differential and abstract operator equations. The study of indefinite inner product spaces has provided impetus for results with indefinite (but invertible) B, and we refer to [KŠ, Kü] for an early special case involving a single

Research of P. B. partially supported by I.W. Killam Foundation and NSERC of Canada.
Research of R. H. partially supported by A. von Humboldt Foundation.

extremum, and to [BHN] for a recent general treatment. Cases where B is indefinite and noninvertible have also been treated (see [AM, BNY]), and linearisation of (say quadratic) pencils frequently leads to such indefinite situations (cf. [BN1, Section 1]).

We note that the methods used for many of the above results have been rather different, and some (e.g., [BHN, BNY]) are technically involved. One of our goals here is to simplify and unify the above situations. We achieve this in two ways. One is to consider a form pencil $p(\lambda) = a - \lambda b$ rather than an operator pencil $P(\lambda) = A - \lambda B$. We remark that $r(x)$ of (1.1), which is at the heart of the extremum principles, involves forms, and that eigenvalues, although traditionally expressed in terms of operators, can be defined via

$$a(x, y) = \lambda b(x, y)$$

for an appropriate set of y. Indeed this is standard in the context of partial differential equations in variational form.

We spend the greater part of Section 2 setting up rather weak sets of equivalent assumptions on a and b and in Section 3 we show that the situations cited earlier, and others, do satisfy our assumptions. Roughly, we require that $p(\lambda)$ be closable and have a finite maximal dimension of nonpositivity on some real λ interval. The main result (Theorem 2.6) is not of traditional form, but it is proved quite simply. It gives equality of the quantities σ_k (roughly, the supremum-infimum of $a(x)/b(x)$ over subspaces of codimension $k-1$) and τ_k (the supremum of those λ for which $p(\lambda)$ has maximal dimension of negativity less than k).

Armed with this, we proceed in Section 4 to characterize eigenvalues in terms of the σ_k. It should be pointed out that when $B \geq 0$, the kth eigenvalue λ_k is in general given not by σ_k as in the case $B > 0$, but by σ_{k+s} where the "index shift" s seems to have been observed first in [Al] in the partial differential equation context. When B is indefinite, only certain eigenvalues ν_k can be characterized, again as σ_{k+s}, and elaborate algorithms for "cancelling" the remaining eigenvalues have been proposed, cf. [BHN]. Here we shall use two-parameter eigencurves to specify the shift and cancellation quite simply. These curves are graphs of eigenvalues for operators A_λ associated with $p(\lambda)$, and ν_k is the maximal value of λ where the $(k+s)$th curve intersects the λ-axis. We shall also show how to carry out these calculations in terms of Jordan chains for $p(\lambda)$, e.g., Jordan bases of $B^{-1}A$ when this operator exists.

Our second goal is extension of known results. Our weak assumptions already guarantee this, and for example we know of no previous treatment of elliptic (even Sturm–Liouville) problems with unbounded indefinite weight functions vanishing on sets of positive Lebesgue measure (see Example 3.3). In Section 5 we extend our characterization to include eigenvalues previously cancelled, for a modified version of σ_k. We know of no previous results of this type.

Throughout, we consider eigenvalues of "generalised positive type" (Definition 5.1), for which $b(y) > 0$ for appropriate y. Generalised negative type eigenvalues can be characterized by dual principles which will be left to the reader.

2. An extremum principle for quadratic forms

We start with two symmetric sesquilinear forms a and b defined on a dense linear subset $\mathscr{D}$ of a Hilbert space $\mathscr{H}$ under the assumptions (I) and (II) below.

(I) There exists an interval (m, M) such that $a - \lambda b$ is closable for every $\lambda \in (m, M)$.

We write $n^-(t)$ for the maximal dimension of a subspace in $\mathscr{D}$ on which a given symmetric form t is negative definite, and we call t *quasi-uniformly positive* (QUP for short) if $n^-(t-\varepsilon)$ is finite for some $\varepsilon > 0$. Our second assumption is that

(II) For every $\lambda \in (m, M)$, the form $a - \lambda b$ is QUP.

Before proceeding, we shall discuss some properties related to our assumptions.

Remark 2.1. *Observe that a closable QUP form t is necessarily bounded below. Indeed, we may assume that t is already closed on $\mathscr{D}(t)$ and non-degenerate (just replace t with $t-\varepsilon$ for a suitable $\varepsilon > 0$ if necessary). Let $\mathscr{L}^- \subset \mathscr{D}(t)$ be a subspace of dimension $n^-(t) < \infty$, on which t is negative definite. We put*

$$\mathscr{L}^+ = \{x \in \mathscr{D}(t) \mid t(x, y) = 0 \text{ for all } y \in \mathscr{L}^-\};$$

then $t(x) \geq 0$ for all $x \in \mathscr{L}^+$ and $\mathscr{D}(t) = \mathscr{L}^+ \dotplus \mathscr{L}^-$ by [Bo, Lemma I.9.8]. *In fact, t is positive definite on $\mathscr{L}^+$* [La, Theorem 5.2] *and $\mathscr{D}(t)$ is a Hilbert space $\mathscr{H}_1$ with respect to the norm $\|x\|_1^2 := -t(x^-) + t(x^+)$, where $x = x^- + x^+$, $x^\pm \in \mathscr{L}^\pm$. It suffices to show that there is a constant $C > 0$ such that $\|x\|^2 \geq C\|x\|_1^2$ for all $x \in \mathscr{D}(t)$ with $t(x) \leq 0$. The result then follows from the inequalities $t(x) \geq -\|x\|_1^2 \geq -C^{-1}\|x\|^2$.*

Assume the contrary, i.e., that there is a sequence $(x_n) \subset \mathscr{D}(t)$ such that $\|x_n\|_1 = 1$, $t(x_n) \leq 0$, and $\|x_n\| \to 0$ as $n \to \infty$. Without loss of generality we may assume that x_n converge weakly to some x in $\mathscr{H}_1$; since $\mathscr{H}_1$ is continuously and densely embedded into $\mathscr{H}$, it follows that x_n converge weakly to x also in $\mathscr{H}$, whence $x = 0$. Write $x_n = x_n^- + x_n^+$ with $x_n^\perp \in \mathscr{L}^\pm$; then x_n^- converge weakly (and hence strongly, by finite dimensionality of $\mathscr{L}^-$) to 0 in the space $\mathscr{H}_1$. Thus $t(x_n^+) \leq -t(x_n^-) \to 0$ as $n \to \infty$, whence $\|x_n\|_1 \to 0$, contradiction.

Thus (I) and (II) imply that the following stronger form of (I) holds.

(I′) As for (I), but with $a - \lambda b$ also bounded below for every $\lambda \in (m, M)$.

We shall also consider a further strengthening of (I), viz.,

(I″) As for (I′), and the domain of the closure $(a-\lambda b)^\sim$ of $a - \lambda b$ does not depend on λ.

Lemma 2.2. (I′) *implies* (I″).

Proof. Take an arbitrary $\nu \in (m, M)$ and choose $\delta > 0$ such that $\nu \pm \delta \in (m, M)$. By assumption, there exist numbers $\gamma_\pm \in \mathbb{R}$ such that

$$(a - \nu b)(x) \pm \delta b(x) \geq -\gamma_\pm \|x\|^2$$

for all $x \in \mathscr{D}$. From this we conclude that

$$\mp b(x) \le \frac{1}{\delta}(a - \nu b)(x) + \frac{\gamma_\pm}{\delta}\|x\|^2,$$

whence

$$|b(x)| \le \frac{1}{\delta}|(a - \nu b)(x)| + \frac{|\gamma_+| + |\gamma_-|}{\delta}\|x\|^2,$$

so the form b is bounded relative to $a - \nu b$ with relative bound not greater than $1/\delta$. Theorem VI.1.33 of [Ka] now shows that the domains of $(a-\lambda b)^\sim$ and $(a-\nu b)^\sim$ coincide provided $|\lambda - \nu| < \delta$. Since ν was chosen arbitrarily in (m, M), the claim follows. □

Thus when assuming (I) and (II), we may also assume (I″). We remark that under these assumptions the forms a and b themselves need not be closable.

A simple but useful alternative criterion for (I) and (II) is given by the following. We start with a weaker version which holds at a single point:

(III) there exists $\nu \in \mathbb{R}$ such that $a - \nu b$ is closable and QUP,

and we add the extra assumption

(IV) b is bounded relative to $a - \nu b$, i.e., there exist $\alpha, \beta \ge 0$ such that

$$|b(x)| \le \beta|(a - \nu b)(x)| + \alpha\|x\|^2$$

for all $x \in \mathscr{D}$.

Lemma 2.3. (I) *and* (II) *hold if and only if* (III) *and* (IV) *hold.*

Proof. If (I) and (II) hold, then so does (I′), and the proof of Lemma 2.2 shows that b is bounded relative to $a - \nu b$, for every $\nu \in (m, M)$.

In the reverse direction, assume (III) and (IV). Then the form $a - \lambda b$ will also be closable for all real λ with $|\lambda - \nu| < 1/\beta$ by [Ka, Theorem VI.1.33]. Also, by definition, there exist $\varepsilon > 0$ and a subspace $\mathscr{H}_+$ of finite codimension such that

$$(a - \nu b)(x) > \varepsilon\|x\|^2$$

for all $x \in \mathscr{D} \cap \mathscr{H}_+$. Assuming that ε is sufficiently small (so that $2\beta\varepsilon < \alpha$), we conclude that for every λ with $|\lambda - \nu| < \varepsilon/(2\alpha)$ and every $x \in \mathscr{D} \cap \mathscr{H}_+$ the inequality

$$\begin{aligned}(a - \lambda b)(x) &= (a - \nu b)(x) + (\lambda - \nu)b(x)\\ &\ge (a - \nu b)(x) - \frac{\varepsilon}{2\alpha}\left[\beta(a - \nu b)(x) + \alpha\|x\|^2\right]\\ &= \left(1 - \frac{\beta\varepsilon}{2\alpha}\right)(a - \nu b)(x) - \frac{\varepsilon}{2}\|x\|^2 \ge \frac{\varepsilon}{4}\|x\|^2\end{aligned}$$

holds. Thus both (I) and (II) continue to hold in the open $\varepsilon/(2\alpha)$-neighborhood of ν, which may be taken as (m, M). □

Having fixed the properties of the forms a and b, we shall proceed now with the variational principle for the pencil $a - \lambda b$. We emphasize that no assumptions are made about definiteness or nondegeneracy of a and b.

Lemma 2.4. *The set* $\{(a(x), b(x)) \mid x \in \mathscr{D}\}$ *is a convex cone in* $\mathbb{R}^2$.

Proof. After identification of $(s_1, s_2) \in \mathbb{R}^2$ with $s_1 + is_2 \in \mathbb{C}$ the above set becomes the range of the quadratic form $a(x) + ib(x)$ on $\mathscr{D}$ and thus is convex by the Toeplitz–Hausdorff theorem [Ka, Theorem V.3.1]. A direct proof is given in by Atkinson [At, Theorem 2.8.1]. □

For any subspace $S \subseteq \mathscr{H}$, define $\imath(S)$ as

$$\inf\{a(x)/b(x) \mid x \in S \cap \mathscr{D},\ b(x) > 0\} \tag{2.1}$$

if the set $\{x \in S \cap \mathscr{D},\ b(x) > 0\}$ is nonempty and put $\imath(S) = -\infty$ otherwise.

Consider the following three conditions:

(i) b is strictly indefinite, and $b(x) = 0$ implies $a(x) \geq 0$;
(ii) $\imath(\mathscr{H}) > -\infty$;
(iii) $a - \imath(\mathscr{H})b \geq 0$ on $\mathscr{D}$.

It follows easily from Lemma 2.4 that (i) $\Longrightarrow$ (ii) $\Longrightarrow$ (iii). See also Kühne [Kü], Krein–Šmulian [KŠ], and Uhlig [Uh] for a review of similar two-form results. We shall need the second implication in the following form (see also [BY] in a finite dimensional case):

Corollary 2.5. *The inequality* $\imath(S) > -\infty$ *yields* $a(x) - \imath(S)b(x) \geq 0$ *for all* $x \in S \cap \mathscr{D}$.

The main result of this section concerns the constructions

$$\sigma_k = \sup\{\imath(S) \mid \operatorname{codim} S = k - 1\} \tag{2.2}$$

and

$$\tau_k = \sup\{\lambda \in (m, M) \mid n^-(a - \lambda b) < k\}. \tag{2.3}$$

Theorem 2.6. *If* σ_k *and* τ_k *both give numbers in* (m, M), *then* $\sigma_k = \tau_k$ *and both suprema in* (2.2) *and* (2.3) *are attained.*

Proof. Assume that $\tau_k \in (m, M)$ and put $d_k := n^-(a - \tau_k b)$. Then by (II) d_k is finite and there is a d_k-dimensional subspace $\mathscr{L}^-$ on which $(a - \tau_k b)(x) \leq -\varepsilon\|x\|^2$, for some $\varepsilon > 0$. Proceeding as in the proof of Lemma 2.3, we see that $a - \lambda b$ is negative definite on $\mathscr{L}^-$ for all λ sufficiently close to τ_k. Thus by the definition of τ_k, $d_k < k$ so the supremum in (2.3) is attained.

Second, as in the proof of Remark 2.1, there exists a subspace $\mathscr{L}^+$ of codimension at most $k - 1$ such that

$$(a - \tau_k b)(x) \geq 0 \tag{2.4}$$

for all $x \in \mathscr{L}^+ \cap \mathscr{D}$. Choosing ξ in (τ_k, M), we have $n^-(a - \xi b) \geq k$, so $(a - \xi b)(y) < 0$ for some $y \in \mathscr{L}^+ \cap \mathscr{D}$. Thus $\xi b(y) > a(y) \geq \tau_k b(y)$ whence $b(y) > 0$. Using (2.4) we see that $\imath(\mathscr{L}^+) \geq \tau_k$ and $\sigma_k \geq \tau_k$.

Third, assume that a subspace $S \subseteq \mathscr{H}$ of codimension $k - 1$ satisfies $\imath(S) = \mu \in (m, M)$; Corollary 2.5 then shows that $(a - \mu b)(x) \geq 0$ for all $x \in S \cap \mathscr{D}$. Thus $a - \mu b$ is nonnegative on S so $n^-(a - \mu b) \leq k - 1 < k$. It follows that $\tau_k \geq \mu$, whence $\tau_k \geq \sigma_k$ and $\sigma_k = \tau_k$. Moreover, in the notation of the second part of this proof, the supremum in (2.2) is attained on any subspace of codimension $k-1$ in $\mathscr{L}^+$ containing y. □

3. Operators

So far, we have dealt exclusively with forms, but there is an implicit connection with operators, since in view of (I), (II), and Remark 2.1 the form $a - \lambda b$ is closable and bounded below for $\lambda \in (m, M)$. Thus [Ka, Theorem VI.2.1] shows that a self-adjoint operator A_λ can be constructed from the closure of $a - \lambda b$. Defining n^- and QUP for an operator via its form (restricted to the domain of the operator) we see that A_λ is QUP.

We can formally write $A_\lambda = A - \lambda B$, where $A := A_0$ (if $m < 0 < M$ – otherwise we can translate the λ origin), but in general there may be no operator B. In this section we give three examples from the literature where an operator B does exist.

Example 3.1 (Right semidefinite case). We consider the situation of [EL] (cf. [Al, BEL, BV] for related cases). Let A and B be symmetric operators in $\mathscr{H}$ defined on $\mathscr{D}(A)$ and satisfying

(1a) A is self-adjoint and $B \geq 0$.

(1b) B has bound zero relative to A, i.e., for every $\beta > 0$ there is $\alpha(\beta)$ such that $\|Bu\| \leq \alpha(\beta)\|u\| + \beta\|Au\|$ for all $u \in \mathscr{D}(A)$.

(1c) For some real ν, $A - \nu B$ is boundedly invertible and $n^-(A - \nu B)$ is finite.

By [Ka, Theorem VI.1.38], (1b) implies that the sesquilinear form $b(x,y) := (Bx, y)$ is bounded relative to a with relative bound zero, so assumption (IV) holds. Now by (1a) and [Ka, Theorems VI.1.27, 33] the form $(a - \nu b)(x,y) := ((A - \nu B)x, y)$ is closable on $\mathscr{D} := \mathscr{D}(A)$.

Since $a - \nu b$ is QUP by (1c), (III) holds and, in view of Lemma 2.3, so do (I) and (II). In fact for all $\lambda \in \mathbb{R}$ the closure of $a - \lambda b$ has domain $\mathscr{D}(\tilde{a}) = \mathscr{D}(|A|^{1/2})$, i.e., the form domain of A. Moreover $n^-(a - \lambda b)$ is obviously nondecreasing in λ, so the largest possible interval (m, M) is given by $m = -\infty$ and $M = \sup\{\lambda \in \mathbb{R} \mid n^-(a - \lambda b) < \infty\}$.

Observe also that by (1c) the operator $A - \nu B$ is bounded below and thus by (1b) so is A. This was required explicitly in earlier references.

Example 3.2 (Krein space case). We consider the situation of [BHN] (cf. [BN1, ĆN, La] for related cases). Let T be a self-adjoint operator on a Krein space $\mathscr{K}$ (with inner product $[\cdot,\cdot]$) satisfying

(2) For some real $\nu \in \mathbb{R}$, $T - \nu I$ is QUP.

This is to be interpreted in terms of the form $[(T - \nu I)x, x]$. If $\mathscr{K}$ has a fundamental symmetry J, then $(x,y) := [Jx, y]$ makes $\mathscr{K}$ into a Hilbert space which we denote by $\mathscr{H}$. Further, $A := JT$ and $B := J$ are self-adjoint in $\mathscr{H}$ and $[(T - \nu I)x, x] = ((A - \nu B)x, x)$. Thus $A - \nu B$ is QUP and since B (although indefinite in general) is bounded (and boundedly invertible) we can argue as for Example 3.1 to see that (I)–(IV) hold.

To identify the maximal possible interval (m, M) on which (I) and (II) hold, we recall first that (2) implies that T possesses a spectral function E [La]. This function is defined on "admissible" intervals, i.e., with endpoints not in the finite set of critical points, and takes values in the set of orthoprojectors in $\mathscr{K}$. Moreover, for an admissible interval Δ, the subspace $E(\Delta)$ is invariant under T and the restriction of T to $E(\Delta)$ has no spectrum outside $\overline{\Delta}$.

For a subspace S of $\mathscr{K}$, we denote by $\varkappa^+(S)$ (resp. $\varkappa^-(S)$) the maximal dimension of nonnegative (resp. nonpositive) subspaces of S and we set

$$\varkappa^{\pm}(\lambda) = \inf\{\varkappa^{\pm}(E(\Delta)) \mid \Delta \ni \lambda\},$$

where Δ runs through all admissible intervals. We claim that

$$m := \sup\{\lambda \in \mathbb{R} \mid \varkappa^-(\lambda) = \infty\}, \qquad M := \inf\{\lambda \in \mathbb{R} \mid \varkappa^+(\lambda) = \infty\}$$

give the maximal choice of (m, M). Indeed, by [BHN, Theorem A.3] assumption (2) implies that $m < M$ and that $T - \lambda I$ is QUP for all $\lambda \in (m, M)$. If $\lambda < m$, we can take admissible $\Delta \ni m$ to the right of λ such that $\varkappa^+ := \varkappa^+(E(\Delta)) < \infty$. Then T (restricted to $\Pi := E(\Delta)$) is self-adjoint in the Pontryagin space Π and hence possesses a T-invariant nonpositive subspace $\mathscr{L}^-$ of codimension $\varkappa^+$. The spectral theorem for T [La] implies that $T - \lambda I$ is J-nonpositive on $\mathscr{L}^-$, so $T - \lambda I$ is not QUP if $\lambda < m$, and similarly not if $\lambda > M$.

Example 3.3. (Sturm–Liouville problems with L_1 coefficients). We consider the problem

$$-(py')' + qy = \lambda ry \tag{3.1}$$

on $(0, 1)$ under the regularity conditions

(3) $p > 0$ and $1/p, q, r \in L_1(0, 1)$.

This problem has an extensive literature developed particularly over the last 50 years. We write A for the self-adjoint operator in $\mathscr{H} = L_2(0, 1)$ corresponding to the left side of (3.1) with self-adjoint boundary conditions, and B for the operator of multiplication by r. Now $A - \lambda B$ is self-adjoint on $\mathscr{D}(A)$, is bounded below, and has discrete spectrum, and hence is QUP. Thus both (I) and (II) hold for all $\lambda \in \mathbb{R}$, and the maximal interval (m, M) is given by $m = -\infty$ and $M = \infty$. In fact the form domain of A can be taken as $\mathscr{D}$. (This is characterized explicitly in [BĆ], and consists of those $y \in W_2^1(0, 1)$ satisfying the essential boundary conditions from A when p is bounded).

Note that in general B is unbounded, noninvertible, indefinite and (unless $r \in L_2(0, 1)$) unbounded relative to A, so that Examples 3.1 and 3.2 do not apply. The analysis extends to more general regular or singular $2n$th order ordinary differential equations under the integrability conditions of [Na, Chapter V]. It also gives a significant improvement on previous variational principles for uniformly elliptic partial differential equations with indefinite weight functions, cf. [BN1, FL]. Here we may use the coefficient conditions of, say, [EE, Chapters VI, VII].

Remark 3.4. Variational principles have been derived in various further situations reducible to ours. For example, in the "uniformly left definite" case where A is positive definite and A^{-1} and B are bounded [We], one can consider the right definite operator pencil $B - \lambda^{-1}A$ which satisfies (III) and (IV) for negative $\nu := \lambda^{-1}$ of sufficiently large absolute value. In this case the "generalised Rayleigh quotient" is of the form $b(x)/a(x)$ and the denominator is uniformly positive for $x \neq 0$. A similar situation occurs where the quotient $a(x)/(Ax, Ax)$ is sometimes used to characterize eigenvalues of a boundedly invertible operators A. As shown in [EL], this can be treated as in Example 3.1 via the pencil $A - \lambda^{-1}A^2$.

Variational principles for various nonlinear eigenvalue problems can be derived via linearisation to the above examples, e.g., quadratic problems in [BN1, BEL, EL] and problems with "floating singularities" in [EL]. Finally, we note that much of the literature on variational principles for indefinite eigenvalue problems treats finite dimensional cases, and the most general result we know of this type is in [BNY]. Here A and B can both be indefinite and/or singular, but (III) and (IV) follow as for Example 3.3.

4. Noncancelled eigenvalues

We define λ as an eigenvalue of the pencil $p : \lambda \mapsto a - \lambda b$ if $A_\lambda x = 0$ for some nonzero $x \in \mathscr{H}$. If $b(x) > 0$ for all such x, λ has *positive type*. Although Theorem 2.6 is expressed in terms of the integer-valued function n^-, it is useful to have a continuous analogue provided by the eigencurves. We define $-\mu_k(\lambda)$ to be the kth eigenvalue of A_λ (counted by multiplicity, $\mu_1(\lambda)$ being maximal) whenever $n^-(A_\lambda)$ is finite.

Lemma 4.1. *The functions μ_k can be chosen holomorphic in a left (and also in a right) neighborhood of any point of (m, M) where they are nonnegative.*

Proof. The arguments of Section 2 (based on (I), (II), (IV) and Theorem VI.1.33 of [Ka]) show that $a - \lambda b$ is sectorial for λ in some complex neighborhood of (m, M). Hence $a - \lambda b$ is a holomorphic family of type (a) in Kato's sense, and so A_λ is holomorphic [Ka, Theorem VII.4.2]. Now the general machinery of [Ka, Chapter VII] can be applied to the finite system of eigenvalues $\mu_1(\lambda), \dots, \mu_k(\lambda)$. □

Evidently $n^-(a - \lambda b) = n^-(A_\lambda)$ is the number of indices k for which $\mu_k(\lambda) > 0$. Thus Theorem 2.6 and Lemma 4.1 give the following

Corollary 4.2. *$\lambda = \sigma_k$ is an eigenvalue of p satisfying $\mu_k(\sigma_k) = 0$.*

Remark 4.3. As noted in Theorem 2.6, the supremum in (2.2) is attained, say on a subspace S. In the above notation, S is orthogonal to the eigenvectors corresponding to the positive $\mu_j(\lambda)$ and is thus a nonnegative spectral subspace of A_λ.

The graph of μ_k is called the kth eigencurve. It is well known that for an eigenvalue λ, the expressions $\mu_k'(\lambda\pm)$ (for those k satisfying $\mu_k(\lambda) = 0$) equal

$b(u_j)$ for appropriate $u_j \in \mathscr{N}(A_\lambda)$. Thus λ is of positive type if and only if all eigencurves through $(\lambda, 0)$ have positive (left and right) slopes.

Example 4.4. Let us consider Example 3.1 again, and assume that $A_\lambda x = 0$ for some $x \neq 0$. If $b(x) = 0$, then $Bx = 0$ so $A_\mu x = 0$ for all $\mu \in (m, M)$ contradicting (1c). Thus $b(x) > 0$ and λ is an eigenvalue of positive type. If we index the eigenvalues of p, counted by multiplicity, as $\lambda_1 \leq \lambda_2 \leq \cdots$, and assume that (m, M) is maximal, then Corollary 4.2 gives

$$\lambda_k = \sigma_{k+s}.$$

Here the "index shift" s is given by

$$s = \#\{j \mid \mu_j(\lambda) > 0 \text{ for all } \lambda \in (m, M)\}, \tag{4.1}$$

i.e., the number of eigencurves entirely above the interval (m, M) of the λ-axis. It also follows that the integer valued function n^- can be calculated explicitly via

$$n^-(A_\lambda) = s + \#\{k \mid \lambda_k < \lambda\}.$$

One can also specify the index shift s of (4.1) directly in terms of the forms a and b which we shall regard as defined on $\mathscr{D}(a) := \mathscr{D}(|A|^{1/2})$.

Proposition 4.5. *The index shift s equals the maximal dimension d of a subspace of the set $\mathscr{D}^- := \{x \in \mathscr{D}(a) \mid b(x) = 0,\, a(x) \leq 0\}$.*

Proof. Since the number d does not change if we replace a by $a - \lambda b$, we may assume without loss of generality that $\lambda_1 > 0$ and thus that $s = n^-(a)$. As a is now nondegenerate, we have

$$s = \max\{\dim \mathscr{L} \mid \mathscr{L} \subseteq \mathscr{D}(a), a(x) \leq 0 \text{ for all } x \in \mathscr{L}\},$$

which shows that $s \geq d$. The quadratic form b is bounded and nonnegative in the Pontryagin space $\Pi := \mathscr{D}(a)$ with inner product $[x, y] := a(x, y)$ and thus $b(x) = [\tilde{B}x, x]$ for some bounded nonnegative operator $\tilde{B}$. Since $\lambda_1 > 0$ implies that $a - \lambda b$ is nondegenerate for $\lambda < 0$, $\tilde{B}$ has no spectrum on the negative semiaxis. As in [EL], we conclude now that there is an s-dimensional nonpositive subspace $\mathscr{L}^-$ in the kernel of $\tilde{B}$. But then $b(x) = [\tilde{B}x, x] = 0$ for every $x \in \mathscr{L}^-$, so $\mathscr{L}^- \subseteq \mathscr{D}^-$ and $d \geq s$. □

Observe that the operator $\tilde{B}$ in the above proof is equal to $(\overline{A})^{-1}\overline{B}$, where $\overline{A}$ and $\overline{B}$ are the closures of A and B respectively as operators from $\mathscr{D}(|A|^{1/2})$ into $\mathscr{D}(|A|^{-1/2})$ (such closures exist since A and B are bounded below, cf. [Ka, Ch. VI.1.5]). If $\mathscr{D}(B) \supseteq \mathscr{D}(|A|^{1/2})$, then $\overline{B}$ coincides with B restricted to $\mathscr{D}(|A|^{1/2})$, so the set $\{x \in \mathscr{D}(a) : b(x) = 0\}$ coincides with $\mathscr{N}(B) \cap \mathscr{D}(a)$, and thus we recover a result in [EL, Theorem 3.1].

Returning to the general case, we note that $\lambda = \sigma_k$ in Corollary 4.2 is in fact the *maximal* solution of the equation $\mu_k(\lambda) = 0$. If we "cancel" all eigenvalues of p which are not such maximal solutions, then the noncancelled ones, say, $\nu_1 \leq \nu_2 \leq \cdots$, are precisely the eigenvalues characterized by Corollary 4.2 and if (m, M) is maximal then $\nu_j = \sigma_{j+s}$ where s is given by (4.1). As we shall see, one can specify the index shift, and the cancellation, in terms of properties of p that do not involve all the eigencurve information.

Suppose initially that $\dim \mathscr{N}(A_{\lambda_k}) = 1$, where λ_k is an eigenvalue of p in (m, M), so $\mu_j(\lambda_k) = 0$ for just one j. The leading term (of sign ε_k and degree d_k) in the Taylor expansion of μ_j about λ_k (for example, in half neighborhoods of λ_k) determines the behavior of $n^-(a - \lambda b)$ for λ near λ_k. Indeed, if we write $n^-_\pm := n^-(A_{\lambda_k \pm})$ and $n^-_0 := n^-(A_{\lambda_k})$, then $n^-_+ - n^-_0 = 1$ if $\varepsilon_k > 0$ and $= 0$ otherwise; $n^-_- - n^-_0 = 1$ if $\varepsilon_k > 0$ and d_k is even, or $\varepsilon_k < 0$ and d_k is odd, and $= 0$ otherwise. Adding all the contributions together in the case of nonsimple λ_k, we have the following

Lemma 4.6. *Let $l^\pm_{\mathrm{o}}$ (resp. $l^\pm_{\mathrm{e}}$) denote the number of eigencurves through $(\lambda_k, 0)$ for which $\pm\varepsilon_k > 0$ and d_k is odd (resp. even). Then*

$$n^-_+ - n^-_0 = l^+_{\mathrm{o}} + l^+_{\mathrm{e}}, \quad n^-_- - n^-_0 = l^-_{\mathrm{o}} + l^+_{\mathrm{e}}, \quad n^-_+ - n^-_- = l^+_{\mathrm{o}} - l^-_{\mathrm{o}}.$$

It follows that knowledge of $n^-(m)$, $n^-(M)$ and the integers $l^\pm_{\mathrm{o}}$ and $l^\pm_{\mathrm{e}}$ at each eigenvalue of p is enough to specify $n^-(a - \lambda b)$ for each $\lambda \in (m, M)$, and hence to determine the shift and cancellation mentioned earlier.

Example 4.7. Let us return to Example 3.2. Then the d_k and ε_k (and hence the $l^\pm_{\mathrm{o}}$ and $l^\pm_{\mathrm{e}}$) may be determined from the Jordan decomposition of T restricted to the root subspace at a given eigenvalue (of T, or equivalently, of p). Specifically, d_k is the block size and ε_k its signature, i.e., the sign of $[x_1, x_{d_k}]$ for a Jordan basis $x_1, \dots, x_{d_k}$ of the given block. To see this, we may apply [GLR, Theorem I.3.19] and [Ma, Section 108] to the appropriate (finite dimensional) Riesz projection of $T - \lambda I$. For further study of the variation of $n^-(A - \lambda B)$ we refer to [BB, DG].

When T has only simple eigenvalues in (m, M), the rule for cancelling eigenvalues is to remove adjacent pairs $(\lambda^+_j, \lambda^-_{j+1})$, where the sign in the superscript denotes the type, and repeat until no more $(+, -)$ pairs remain. This leaves the ν_j defined above, preceded by the minimal solutions of $\mu_j(\lambda) = 0$, which are of (generalised, as in Section 5) negative type and are characterized by a dual principle. See [BHN] for details, and for the extension to nonsimple eigenvalues. The shift s can also be calculated as the total number of cancelled pairs and nonreal eigenvalue pairs of T. See [BN2] for a detailed study of this "minimal index".

Example 4.8. In the situation of Example 3.3 there is in general no operator T, but shift and cancellation can be carried out via the μ_j as in Lemma 4.6. For separated boundary conditions, the Prüfer angle $\theta(x, \lambda)$ is defined and $\theta(1, \lambda)$ is continuous (hence constant) along the eigencurves. Thus for this case we also see that σ_k is the maximal eigenvalue for which a corresponding eigenfunction has $k - 1$ zeros in $(0, 1)$. The index shift s is therefore the minimal number of zeros which can be achieved by an eigenfunction. Calculations of some of the quantities in Lemma 4.6 are given in [BLM] for a special case via the Prüfer angle and the Titchmarsh–Weyl m-function.

5. Generalised positive type eigenvalues

We start with the following

Definition 5.1. *An eigenvalue* λ *has* generalised positive type, *if* μ_j *is increasing in a right neighborhood of* λ *for at least one* j *satisfying* $\mu_j(\lambda) = 0$.

This means, in the notation of Section 4, that $\varepsilon_k > 0$ for at least one k with $\mu_k(\lambda) = 0$. The maximality in Theorem 2.6 shows that all eigenvalues characterized by Corollary 4.2 are of generalised positive type, but not conversely in general because of "cancellation". Suppose λ_+ is an arbitrary eigenvalue of this type, let λ_* be the next greatest eigenvalue of p (or M if there is none), and choose $\alpha \in (\lambda_+, \lambda_*)$. It will be convenient to choose (m, M) as a maximal interval satisfying (I) and (II).

Now we choose a new eigenparameter $\rho := (\alpha - \lambda)^{-1}$ and replace the form pair (a, b) by $(b, \alpha b - a)$. Note that

$$b - \rho(\alpha b - a) = \rho(a - \lambda b). \tag{5.1}$$

By (5.1) we see that $b - \rho(\alpha b - a)$ satisfies analogues of (I) and (II) on the maximal ρ interval $((\alpha - m)^{-1}, \infty) \subset \mathbb{R}^+$.

Let S be a subspace of $\mathscr{H}$. To emphasize the dependence on a and b, we shall replace the notation $\imath(S)$ from (2.1) et seq. by $\imath(S, a, b)$, and we shall write $\imath(S, a, b, c)$ if $c(x) > 0$ is imposed in addition to $b(x) > 0$, for some quadratic form c. We now apply Corollary 4.2 to obtain

$$(\alpha - \lambda_+)^{-1} = \max_{\operatorname{codim} S = k-1} \imath(S, b, \alpha b - a) \tag{5.2}$$

where k is such that $\lambda_+ = \sup\{\lambda \in (m, \alpha) \mid n^-(a - \lambda b) < k\}$.

Since $\alpha - \lambda_+ > 0$, we can take $b(x) > 0$ in (5.2) and then straightforward manipulations give

$$\alpha - \lambda_+ = \min_{\operatorname{codim} S = k-1} (\alpha - \imath(S, a, b, \alpha b - a)).$$

This leads us to our final

Corollary 5.2.

$$\lambda_+ = \max_{\operatorname{codim} S = k-1} \imath(S, a, b, \alpha b - a).$$

Thus λ_+ is now characterized by a principle similar to that of Corollary 4.2, but with $(\alpha b - a)(x) > 0$, i.e., $\frac{a(x)}{b(x)} < \alpha$, imposed in addition to $b(x) > 0$.

Remarks

1. We can express k in terms of "local" cancellation and shift, applied to the interval (m, α) instead of (m, M).
2. Further eigenvalues below λ_+ may be characterized via the restricted principle of Corollary 5.2. Thus in general an eigenvalue has many characterizations, depending on the choice of α.
3. If (m, M) is not maximal then the above principles are modified in general. For example, if m is increased to exceed the minimal eigenvalue ν_1, then ν_1

is no longer characterized, but the shift index s is increased (by one if ν_1 is simple, and in general according to Lemma 4.6). Similarly if M is decreased, then new eigenvalues may appear that were previously cancelled, as in Corollary 5.2, but the "inf" operation is over a more restricted cone. We remark that this result may be proved via a direct modification of Theorem 2.6, but we have preferred the presentation given for the sake of clarity.

Acknowledgment

The authors thank the referees for their helpful comments and suggestions. The research of R. H. was conducted while visiting the Institute for Applied Mathematics of the University of Bonn and supported in part by the Alexander von Humboldt foundation.

References

[Al] Allegretto, W., Second order elliptic equations with degenerate weight, *Proc. Amer. Math. Soc.* **107**(1989), 989–998.

[AM] Allegretto, W., and Mingarelli, A.B., Boundary problems of the second order with an indefinite weight-function, *J. Reine Angew. Math.* **398**(1989), 1–24.

[At] Atkinson, F.V., *Multiparameter Eigenvalue Problems*, Vol. 1, Academic Press, 1972.

[BB] Binding, P., and Browne, P.J., Applications of two parameter spectral theory to symmetric generalised eigenvalue problems. *Appl. Anal.* **29**(1988), 107–142.

[BĆ] Binding, P., and Ćurgus, B., Form domains and eigenfunction expansions for differential equations with eigenparameter dependent boundary conditions, *Canad. J. Math.* **54**(2002), 1142–1164.

[BEL] Binding, P., Eschwé, D., and Langer, H., Variational principles for real eigenvalues of self-adjoint operator pencils, *Integr. Equat. Oper. Theory* **38**(2000), 190–206.

[BHN] Binding, P., Hryniv, R., and Najman, B., A variational principle in Krein space II, *Integr. Equat. Oper. Theory* **51**(2005), 477–500.

[BLM] Binding, P., Langer, H., and Möller, M., Oscillation results for Sturm-Liouville problems with an indefinite weight function, *J. Comput. Appl. Math.* **171**(2004), 93–101.

[BN1] Binding, P., and Najman, B., A variational principle in Krein space, *Trans. Amer. Math. Soc.* **342**(1994), no. 4, 489–499.

[BN2] Binding, P., and Najman, B., The minimal index of a self-adjoint pencil. *Glas. Mat. Ser. III* **35**(2000), 25–44.

[BNY] Binding, P., Najman, B., and Ye, Q., A variational principle for eigenvalues of pencils of Hermitian matrices, *Integr. Equat. Oper. Theory,* **35**(1999), 398–422.

[BV] Binding, P., and Volkmer, H., Right semidefinite eigenvalue problems, *Proc. Edinb. Math. Soc. (2)* **46**(2003), 561–573.

[BY] Binding, P. and Ye, Q., Variational principles for indefinite eigenvalue problems, *Linear Algebra Appl.* **218**(1995), 251–262.

[Bo] Bognar, J., *Indefinite Inner Product Spaces*, Ergebnisse der Mathematik und ihrer Grenzgebiete, Band **78**, Springer-Verlag, New York, 1974.

[ĆN] Ćurgus, B. and Najman, B., Quasi-uniformly positive operators in Krein space, *Oper. Theory: Adv. Appl.* **80**(1995), 90–99.

[DG] Dijksma, A. and Gheondea, A., On the signatures of selfadjoint pencils, *Oper. Theory: Adv. Appl.* **80**(1995), 112-136.

[EE] Edmunds, D.E., and Evans, W.D., *Spectral Theory and Differential Operators*, Oxford University Press, New York, 1987.

[EL] Eschwé D., and Langer, M., Variational principles for eigenvalues of self-adjoint operator functions, *Integr. Equat. Oper. Theory* **49**(2004), no. 3, 287–321.

[FL] Fleckinger, J., and Lapidus, M., Eigenvalues of elliptic boundary value problems with an indefinite weight function, *Trans. Amer. Math. Soc.* **295**(1986), 305–324.

[GLR] Gohberg, I., Lancaster, P., and Rodman, L., *Matrices and Indefinite Scalar Products*, Oper. Theory: Adv. Appl. **8**, Birkhäuser Verlag, Basel, 1983.

[Ka] Kato, T., *Perturbation Theory for Linear Operators*, 2nd edition, Grundlehren der Mathematischen Wissenschaften, Band 132, Springer-Verlag, Berlin–New York, 1976.

[KŠ] Krein, M.G., and Šmul'jan, Ju.L., Plus-operators in a space with indefinite metric, *Amer. Math. Soc. Transl.* **85**(1969), 93–113.

[Kü] Kühne, R., Über eine Klasse J-selbstadjungierten Operatoren, *Math. Ann.* **154**(1964), 56–69.

[La] Langer, H., Spectral functions of definitizable operators in Krein spaces, *Lecture Notes in Mathematics* **984**, Springer-Verlag, Berlin, 1982, 1–46.

[Ma] Mal'cev, A.I., *Foundations of Linear Algebra*, Freeman, 1963.

[Na] Naimark, M.A., *Linear Differential Operators, Part II*, Ungar, New York, 1968.

[Uh] Uhlig, F., A recurring theorem about pairs of quadratic forms and extensions: a survey, *Linear Algebra Appl.* **25**(1979), 219–237.

[We] Weinberger, H.F., *Variational Methods for Eigenvalue Approximation*, Regional Conf. Ser. Appl. Math. **15**, SIAM, 1974.

[WS] Weinstein, A., and Stenger, W. *Methods of Intermediate Problems for Eigenvalues. Theory and Ramifications*, Academic Press, 1972.

P. Binding
Department of Mathematics and Statistics
University of Calgary
AB, Canada T2N 1N4
e-mail: `binding@ucalgary.ca`

R. Hryniv
Institute for Applied Problems
of Mechanics and Mathematics
3b Naukova str.
79601 Lviv
Ukraine
e-mail: `rhryniv@iapmm.lviv.ua`

Operator Theory:
Advances and Applications, Vol. 162, 85–101

Selfadjoint Extensions with Several Gaps: Finite Deficiency Indices

J.F. Brasche, M.M. Malamud and H. Neidhardt

Abstract. Let A be a closed symmetric operator on a separable Hilbert space with equal finite deficiency indices $n(A) < \infty$ and let J be an open subset of $\mathbb{R}$. It is shown that if there is a self-adjoint extension A_0 of A such that J is contained in the resolvent set of A_0 and the associated Weyl function of the pair $\{A, A_0\}$ is monotone with respect to J, then for any self-adjoint operator R on some separable Hilbert space $\mathfrak{R}$ obeying $\dim(E_R(J)\mathfrak{R}) \leq n(A)$ there exists a self-adjoint extension $\widetilde{A}$ such that the spectral parts $\widetilde{A}_J$ and R_J are unitarily equivalent. The result generalizes a corresponding result of M.G. Krein for a single gap.

Mathematics Subject Classification (2000). Primary 47A56; Secondary 47B25.

Keywords. symmetric operators, self-adjoint extensions, abstract boundary conditions, Weyl function.

1. Introduction

Let A be a densely defined symmetric operator in a separable Hilbert space $\mathfrak{H}$ with deficiency indices $n_+(A) = n_-(A) \leq \infty$. We recall that a bounded open interval $J = (\alpha, \beta)$ is called a gap for A if

$$\|2Af - (\alpha + \beta)f\| \geq \quad (\beta - \alpha)\|f\|, \quad f \in \mathrm{dom}(A). \tag{1}$$

If $\alpha \to -\infty$, then (1) turns into $(Af, f) \geq \beta\|f\|^2$, for all $f \in \mathrm{dom}(A)$, meaning that $(-\infty, \beta)$ is a gap for A if A is semi-bounded from below with the lower bound β. The problem whether there exist self-adjoint extensions $\widetilde{A}$ of A preserving the gap (α, β) has been extensively investigated in the middle of the thirties. It has been positively solved by M. Stone, K. Friedrichs and H. Freudental for operators semi-bounded from below ($\alpha = -\infty$) (see, [1, 19]) and by M.G. Krein [16] for the case of a finite gap. The problem to describe completely the set $\mathrm{Ext}_A(\alpha, \beta)$ of all self-adjoint extensions $\widetilde{A}$ of A preserving the gap has been solved by M.G. Krein

[16], [17](see also [1],[19]) in the case $J = (-\infty, \beta)$ and in [14] for a finite gap $J = (\alpha, \beta)$.

M.G. Krein [16] has investigated the spectrum of self-adjoint extensions $\widetilde{A}$ within a gap J of a densely defined symmetric operator A with finite deficiency indices. Namely, Krein has shown that if R is any self-adjoint operator on some auxiliary separable Hilbert space $\mathfrak{R}$ such that $\dim(E_R(J)\mathfrak{R}) \leq n$, then there exists a self-adjoint extension $\widetilde{A}$ such that the part $R_J := R \upharpoonright E_R(J)\mathfrak{R}$ of R is unitarily equivalent to $\widetilde{A}_J := \widetilde{A} \upharpoonright E_{\widetilde{A}}(J)\mathfrak{H}$, i.e., $\widetilde{A}_J \cong R_J$, where $E_R(\cdot)$ and $E_{\widetilde{A}}(\cdot)$ are the spectral measures of R and $\widetilde{A}$, respectively.

The result was generalized to the case of infinite deficiency indices in [7]. In this case it was shown that if R is any self-adjoint operator with pure point spectrum, then there exists a self-adjoint extension $\widetilde{A}$ such that $\widetilde{A}_J \cong R_J$. Naturally, the question arises whether we can put other kind of spectra into J, for instance, absolutely continuous or singular continuous spectrum. This problem has been investigated in a series of papers [2, 5, 7, 8, 9, 10]. For the class of (weakly) significant deficient symmetric operators (for the definition see [2, 8]) it was shown [2, Theorem 6.2] that for any auxiliary self-adjoint operator R and any open subset $J_0 \subseteq J$ there exists a self-adjoint extension $\widetilde{A}$ such that

$$\widetilde{A}^{pp} \cong R^{pp}_J, \tag{2}$$

$$\widetilde{A}^{ac}_J \cong R^{ac}_J, \tag{3}$$

$$\sigma_{sc}(\widetilde{A}) \cap J = \overline{J_0} \cap J \tag{4}$$

where R^{ac}, $\widetilde{A}^{ac}$ and R^{pp}, $\widetilde{A}^{pp}$ denote the absolutely continuous and pure point parts of R, $\widetilde{A}$, respectively. Notice that the deficiency indices of (weakly) significant deficient symmetric operators are always infinite. The assumption that A is a (weakly) significant deficient symmetric operator was essentially used in the first proof of (3) and (4). Later on this assumption was dropped for the third relation (4), see [9]. However, one has to mention that the singular continuous spectrum obtained in [9] belongs to a certain class of sets which excludes a wide class of possible sets, for instance, Cantor sets.

In [10] an attempt was made to remove all these restrictions assuming that the symmetric operator A has a special structure, namely,

$$A = \bigoplus_{k=1}^{\infty} S_k \quad \text{on} \quad \mathfrak{H} = \bigoplus_{k=1}^{\infty} \mathfrak{K}_k, \tag{5}$$

where each of the operators S_k is unitarily equivalent to a fixed (i.e., k-independent) densely defined closed symmetric operator S in a separable Hilbert space and S has positive deficiency indices. If J is a gap of S (and therefore of S_k for every k), then for any self-adjoint operator R on some separable Hilbert space $\mathfrak{R}$ there exists a self-adjoint extension $\widetilde{A}$ of A such that the relations (2) and (3) hold as well as $\sigma_{sc}(\widetilde{A}) \cap J = \sigma_{sc}(R) \cap J$, cf. [10, Theorem 10]. We note that if $n_\pm(S) < \infty$, then the operator A is not (weakly) significant deficient. Thus [10, Theorem 10] weakens considerable the property (4) for the special case (5). The proof relies on a technique which is quite different from that of [2, 7, 8, 9] and which is called the method of boundary triples and associated Weyl functions. We describe the method briefly below.

The previous results advise the assertion that for any densely defined closed symmetric operator A with infinite deficiency indices and gap J there is a self-adjoint extension $\widetilde{A}$ such that the conditions (2), (3) and $\widetilde{A}^{sc}_J \cong R^{sc}_J$ are satisfied for any auxiliary self-adjoint operator R. Indeed, this is true and was proved in [5, Theorem 1], see also [6, Theorem 27]. In particular, $\widetilde{A}$ has the same spectrum, the same absolutely continuous and singular continuous spectrum and the same eigenvalues inside J as R.

Since for one gap the problem on the spectral properties of self-adjoint extensions is completely solved, naturally the question arises whether it is possible to extend the results to the case of several gaps. It turns out that an analogous statement is wrong if J is the union of disjoint gaps. In general, there does not even exist a self-adjoint extension $\widetilde{A}$ of A such that $J \subset \rho(\widetilde{A})$. Indeed, let us consider the following example.

Example 1.1. Let $\mathfrak{H} = L^2((0,1))$. By A we denote the closed symmetric operator

$$\begin{aligned} (Af)(x) &:= -i\tfrac{d}{dx}f(x), \quad x \in (0,1), \\ f \in \mathrm{dom}(A) &:= \{f \in W^1_2((0,1)) : f(0) = f(1) = 0\}, \end{aligned}$$

which is simple and has deficiency indices $(1,1)$. We recall that a symmetric operator is simple if it is completely non-selfadjoint, that is, it does not exist a subspace which reduces the operator to a self-adjoint one. We note that A^* is given by $(A^*f)(x) := -i\frac{d}{dx}f(x)$, $f \in \mathrm{dom}(A^*) := W^1_2((0,1))$. If $\alpha \in [0, 2\pi)$, then one can associated a self-adjoint extension $A(\alpha)$ of A defined by

$$\mathrm{dom}(A(\alpha)) := \{f \in W^{1,2}((0,1)) : f(1) - e^{i\alpha}f(0)\}$$

and $A(\alpha) = A^* \restriction \mathrm{dom}(A(\alpha))$. It turns out that the family $\{A(\alpha)\}_{\alpha\in[0,2\pi)}$ of self-adjoint extension of A exhausts all of them. The spectrum of $A(\alpha)$ is discrete and consists of isolated simple eigenvalues $\lambda_n(\alpha) = \alpha + 2\pi n$, $n \in \mathbb{Z}$. Obviously, the intervals

$$\Delta_n(\alpha) := (\alpha + 2\pi n, \alpha + 2\pi(n+1)), \quad n \in \mathbb{N}, \quad \alpha \in [0, 2\pi),$$

are gaps of the symmetric operator A. Setting $J := \Delta_0(0) \cup \Delta_1(\pi) = (0, 2\pi) \cup (3\pi, 5\pi)$ we get a union of gaps of A. The intervals $\Delta_0(0) = (0, 2\pi)$ and $\Delta_1(\pi) = (3\pi, 5\pi)$ are disjoint. Moreover, one easily verifies that there is no self-adjoint extension $\widetilde{A}$ of A such that J is gap of $\widetilde{A}$. Moreover, if $\mu \in (0, \pi] \subseteq J$, then there is a self-adjoint extension $\widetilde{A}$ such that μ is an eigenvalue of $\widetilde{A}$, however, $\mu + 4\pi \in \Delta_1(\pi)$ is also in eigenvalue of $\widetilde{A}$. Similarly, if $\mu \in (\pi, 2\pi)$ is an eigenvalue, then $\mu + 2\pi \in \Delta_1(\pi)$ is an eigenvalue of $\widetilde{A}$, too. In other words, putting by extension an eigenvalue into $\Delta_0(0)$ it automatically appears an eigenvalue in $\Delta_1(\pi)$. However, choosing $J = (0, \pi) \cup (3\pi, 4\pi)$ it is not hard to see that for any $\mu \in J$ there is a self-adjoint extension $\widetilde{A}$ such that μ is the unique eigenvalue of $\widetilde{A}$ in J.

Taking into account Example 1.1 we always assume that for the open set $J \subseteq \mathbb{R}$ there exists a self-adjoint extension A_0 of A such that $J \subseteq \rho(A_0)$ where $\rho(A_0)$ denotes the resolvent set of A_0. Under this assumption we are interested in the following problem:

Problem 1.2. *Let A be a closed symmetric operator on a separable Hilbert space $\mathfrak{H}$ with (equal) deficiency indices $n_\pm(A)$ and let $J \subseteq \rho(A_0)$ be an open subset of $\mathbb{R}$. Further, let R be a self-adjoint operator on a separable Hilbert space $\mathfrak{R}$ satisfying $\dim(E_R(J)\mathfrak{R}) \le n(A) := n_\pm(A)$. Does there exist a self-adjoint extension $\widetilde{A}$ of A such that $\widetilde{A}_J \cong R_J$?*

Due to Example 1.1 the answer to Problem 1.2 is in general no which means, that the solution of this problem requires additional assumptions. To formulate these additional assumptions we rely on the theory of abstract boundary conditions:

Definition 1.3. *Let A be a densely defined closed symmetric operator on $\mathfrak{H}$ with equal deficiency indices $n_\pm(A)$. A triple $\Pi = \{\mathcal{H}, \Gamma_0, \Gamma_1\}$ consisting of an auxiliary Hilbert space $\mathcal{H}$ and linear mappings $\Gamma_i : \mathrm{dom}(A^*) \longrightarrow \mathcal{H}$, $i = 0, 1$, is called a boundary triple for the adjoint operator A^* if the following two conditions are satisfied:*

(i) *The second Green's formula takes place:*

$$(A^*f, g) - (f, A^*g) = (\Gamma_1 f, \Gamma_0 g) - (\Gamma_0 f, \Gamma_1 g), \quad f, g \in \mathrm{dom}(A^*).$$

(ii) *The mapping $\Gamma := \{\Gamma_0, \Gamma_1\} : \mathrm{dom}(A^*) \longrightarrow \mathcal{H} \oplus \mathcal{H}$, $\Gamma f := \{\Gamma_0 f, \Gamma_1 f\}$, is surjective.*

Example 1.4. Let A be the symmetric operator A of Example 1.1 and let

$$\Gamma_0 f := \frac{f(0) - f(1)}{\sqrt{2}}, \quad \Gamma_1 f := i\frac{f(0) + f(1)}{\sqrt{2}}, \quad f \in \mathrm{dom}(A^*) = W_2^1((0,1)). \tag{6}$$

A straightforward computation verifies that $\Pi = \{\mathcal{H}, \Gamma_0 \Gamma_1\}$, $\mathcal{H} = \mathbb{C}$, is a boundary triple for A^*.

If $\Pi := \{\mathcal{H}, \Gamma_0, \Gamma_1\}$ is a boundary triple for A^*, then $A_0 := A^* \restriction \ker(\Gamma_0)$ or $A_1 := A^* \restriction \ker(\Gamma_1)$ define self-adjoint extensions of the symmetric operator A. Moreover, it can be shown that there is bijective correspondence between the set of self-adjoint extensions Ext_A of A and the set of self-adjoint linear relations in $\mathcal{H}$ given by $\widetilde{A} \longleftrightarrow \Theta := \Gamma(\mathrm{dom}(\widetilde{A}))$. In particular, one has $A_i \longleftrightarrow \Theta_i$, $i = 0, 1$, where $\Theta_0 = \{0\} \times \mathcal{H}$ and $\Theta_1 = \mathcal{H} \times \{0\}$. In the following we use the notation $A_\Theta \longleftrightarrow \Theta$. In particular, if $\Theta = G(B)$, where $G(B)$ is the graph of a densely defined self-adjoint operator B on $\mathcal{H}$, then we write $A_B := A_{G(B)}$.

Example 1.5. Let $\Pi = \{\mathcal{H}, \Gamma_0, \Gamma_1\}$, $\mathcal{H} = \mathbb{C}$, the boundary triple of Example 1.4. We find that

$$\mathrm{dom}(A_0) = \{W_2^1((0,1)) : f(0) = f(1)\}$$

and

$$\mathrm{dom}(A_1) = \{W_2^1((0,1)) : f(0) = -f(1)\}.$$

Since any self-adjoint linear relation Θ in $\mathcal{H}$ either coincides with Θ_0 or is of the form $\Theta = G(B)$, $B\xi = b\xi$, $\xi \in \mathbb{C}$, $b \in \mathbb{R}$, we get that any extension $\widetilde{A}$ coincides either with A_0 or is given by the domain

$$\mathrm{dom}(\widetilde{A}) = \{W_2^1((0,1)) : f(1) = \frac{b-i}{b+i} f(0)\}, \quad b \in \mathbb{R},$$

which corresponds to Example 1.1. Notice that $A_0 = A(0)$ and $A_1 = A(\pi)$.

We note that for each pair $\{A, A_0\}$ consisting of a symmetric operator A and a self-adjoint extension A_0 there is boundary triple $\Pi = \{\mathcal{H}, \Gamma_0, \Gamma_1\}$ such that $A_0 = A^* \upharpoonright \ker(\Gamma_0)$. In particular, if B is a bounded self-adjoint operator and the self-adjoint extension $\widetilde{A}$ corresponds to the graph $G(B)$, then one verifies that $\Pi_B = \{\mathcal{H}, \Gamma_0^B, \Gamma_1^B\}$, $\Gamma_0^B := B\Gamma_0 - \Gamma_1$, $\Gamma_1^B := \Gamma_0$, is a boundary triple for A^* such that $\mathrm{dom}(\widetilde{A}) = \ker(\Gamma_0^B)$.

Having fixed a boundary triple Π for A^* one associates a so-called Weyl function $M(\cdot) : \rho(A_0) \longrightarrow [\mathcal{H}]$ with it where $[\mathcal{H}]$ denotes the set of bounded operators on $\mathcal{H}$.

Definition 1.6. ([13, 14]) *Let A be a densely defined closed symmetric operator on $\mathfrak{H}$ and let $\Pi = \{\mathcal{H}, \Gamma_0, \Gamma_1\}$ be a boundary triple for A^*. The unique mapping $M(\cdot) : \rho(A_0) \longrightarrow [\mathcal{H}]$ defined by*

$$\Gamma_1 f_z = M(z)\Gamma_0 f_z, \quad f_z \in \mathcal{N}_z = \ker(A^* - z), \quad z \in \rho(A_0),$$

is called the Weyl function corresponding to the boundary triple Π for A^*.

It is well known (cf. [13, 14]) that the above implicit definition of the Weyl function is correct and that $M(\cdot)$ is a Nevanlinna function.

We recall that an operator-valued function $F : \mathbb{C}_+ \longrightarrow [\mathcal{H}]$ is said to be a Nevanlinna function [1, 18, 20] if it is holomorphic and takes values in the set of dissipative operators on $\mathcal{H}$, i.e.,

$$\Im\mathrm{m}(F(z)) := \frac{F(z) - F(z)^*}{2i} \geq 0, \quad z \in \mathbb{C}_+.$$

If $F(\cdot)$ is a Nevanlinna function, then there exists an unbounded operator measure $\Sigma_F(\cdot) : \mathcal{B}_b(\mathbb{R}) \longrightarrow [\mathcal{H}]$ defined on the bounded Borel sets of $\mathbb{R}$, which obeys

$$T_F := \int_{-\infty}^{+\infty} \frac{1}{1+t^2}\, d\Sigma_F(t) \in [\mathcal{H}], \tag{7}$$

as well as operators $C_k = C_k^* \in [\mathcal{H}]$, $k = 0,1$, $C_1 \geq 0$, such that the representation

$$F(z) = C_0 + C_1 z + \int_{-\infty}^{+\infty} \left(\frac{1}{t-z} - \frac{t}{1+t^2} \right) d\Sigma_F(t), \quad z \in \mathbb{C}_+, \tag{8}$$

holds. The representation (8) is a generalization (see [12]) of a well-known result for scalar Nevanlinna functions (cf. [1, 4, 18, 20]). The integrals in (7) and (8) are understood in the strong sense. The measure $\Sigma_F(\cdot)$ is uniquely determined by the Nevanlinna function $F(\cdot)$. By $\mathrm{supp}(F)$ we denote the topological (minimal closed) support of Σ_F. Since $\mathrm{supp}(F)$ is closed the set $\mathcal{O}_F := \mathbb{R} \setminus \mathrm{supp}(F)$ is open. The Nevanlinna function $F(\cdot)$ admits an analytic continuation to $\mathcal{O}_F$ given by

$$F(\lambda) = C_0 + C_1 \lambda + \int_{-\infty}^{+\infty} \left(\frac{1}{t-\lambda} - \frac{t}{1+t^2} \right) d\Sigma_F(t), \quad \lambda \in \mathcal{O}_F. \tag{9}$$

If $F(\cdot)$ is a Weyl function, that is $F(\cdot) = M(\cdot)$, then one has in addition that $C_1 = 0$ and $0 \in \rho(\Im\mathrm{m}(M(i)))$. Since $T_M = \Im\mathrm{m}(M(i))$ we find that the operator T is boundedly invertible for Weyl functions.

Notice that if A is simple, then the Weyl function $M(\cdot)$ of the boundary triple Π determines the pair $\{A, A_0\}$ uniquely up to unitary equivalence (cf. [13, 14]).

Example 1.7. With respect to the boundary triple of Example 1.4 one easily verifies that the associated Weyl function is scalar and given by

$$m(z) = -\frac{\cos(z/2)}{\sin(z/2)} = -\cot(z/2), \quad z \in \mathbb{C}_+.$$

For the scalar Weyl function $m(\cdot)$ the measure Σ_m of the representation (8) is a scalar atomic Borel measure μ such that $\mathrm{supp}(\mu) = \cup_{n\in\mathbb{Z}}\{2n\pi\}$ and $\mu(\{2n\pi\}) = 2$. Indeed, from the representation

$$m(z) = \int_{-\infty}^{+\infty} \left(\frac{1}{t-z} - \frac{t}{1+t^2} \right) d\mu(t), \quad z \in \mathbb{C}_+,$$

we obtain

$$m(z) = 2\sum_{n\in\mathbb{Z}} \left(\frac{1}{2n\pi - z} - \frac{2n\pi}{1+4n^2\pi^2} \right).$$

Hence

$$m(z) = -\frac{2}{z} - 2\sum_{n=1}^{\infty} \left(\frac{1}{z-2n\pi} + \frac{1}{z+2n\pi} \right).$$

which yields

$$m(z) = -\frac{1}{z/2} - \sum_{n=1}^{\infty} \left(\frac{1}{\frac{1}{2}z - n\pi} + \frac{1}{\frac{1}{2}z + n\pi} \right).$$

By [15, XII.441.9] we immediately get that $m(z) = -\cot(z/2)$, $z \in \mathbb{C}_+$. The Weyl function $m(\cdot)$ admits an extension to $\mathcal{O}_m = \mathbb{R} \setminus \cup_{n\in\mathbb{Z}}\{2n\pi\}$ defined by $m(\lambda) = -\cot(\lambda/2)$, $\lambda \in \mathcal{O}_m$, which is periodic with period 2π.

In fact, from the Weyl function one can obtain all information on the self-adjoint extension A_0, in particular, on the spectrum $\sigma(A_0)$.

Proposition 1.8. *Let A be a simple closed symmetric operator and let $\Pi=\{\mathcal{H},\Gamma_0,\Gamma_1\}$ be a boundary triple for A^* with Weyl function $M(\lambda)$. Suppose that Θ is a self-adjoint linear relation in $\mathcal{H}$ and $\lambda \in \rho(A_0)$. Then*

(i) $\sigma(A_0) = \mathrm{supp}(M)$.
(ii) $\lambda \in \rho(A_\Theta)$ *if and only if* $0 \in \rho(\Theta - M(\lambda))$.
(iii) $\lambda \in \sigma_\tau(A_\Theta)$ *if and only if* $0 \in \sigma_\tau(\Theta - M(\lambda))$, $\tau \in \{p, c\}$.

Here we denote by $\sigma_p(\cdot)$ and $\sigma_c(\cdot)$ the point and the continuous spectrum.

Example 1.9. Let Π be the boundary triple of Example 1.4. From Proposition 1.8 (i) and Example 1.7 we restore that $\sigma(A_0) = \cup_{n\in\mathbb{Z}}\{2n\pi\}$.

Let Θ be a self-adjoint linear relation in $\mathcal{H}$ different from Θ_0. In this case we have $\Theta = G(B)$ where B is defined in Example 1.5. Taking into account Proposition 1.8 we find for the self-adjoint extension $\widetilde{A} \longleftrightarrow G(B)$ that $\lambda \in \rho(\widetilde{A}) \cap \rho(A_0)$ if and only if $m(\lambda) \neq b$ or $\cot(\lambda/2) \neq -b$.

In terms of the Weyl function it becomes clear why it can happen that putting by extension an eigenvalue into J one gets automatically further eigenvalues in J.

Example 1.10. Let J be an open set such that $J \subseteq \rho(A_0)$. Then for any $\mu \in J$ there is a self-adjoint extension $\widetilde{A}$ of A such that $\mu \in J$ is the unique eigenvalue of $\widetilde{A}$ in J if and only if the restricted Weyl function $m_J(\cdot) := m(\cdot \upharpoonright J) : J \longrightarrow \mathbb{R}$ is injective. One easily verifies that this is equivalent to the following monotonicity property: for any two component intervals J_1 and J_2 of the open set J one has either $m(\lambda_1) < m(\lambda_2)$ or $m(\lambda_1) > m(\lambda_2)$, $\lambda_1 \in J_1$, $\lambda_2 \in J_2$, which is a kind of monotonicity of the restricted Weyl function $m_J(\cdot)$. Notice that for the open set $J = (0, 2\pi) \cup (3\pi, 5\pi)$, cf. Example 1.1, this monotonicity property is not satisfied but for $J = (0, \pi) \cup (3\pi, 4\pi)$, cf. Example 1.1.

The last formulation admits an extension to the class of Nevanlinna functions.

Definition 1.11. A Nevanlinna function $F(\cdot)$ is monotone with respect to the open set $J \subseteq \mathcal{O}_F$ if for any two component intervals J_1 and J_2 of J one has either $F(\lambda_1) \leq F(\lambda_2)$ or $F(\lambda_1) \geq F(\lambda_2)$ for all $\lambda_1 \in J_1$ and $\lambda_2 \in J_2$.

We note that inside of a component interval one always has monotonicity, i.e., $F(\lambda_1) < F(\lambda_2)$ for $\lambda_1, \lambda_2 \in J_0$ and $\lambda_1 < \lambda_2$ where J_0 is a component interval of J.

Let $L \in \mathbb{N} \cup \infty$ be the number of component intervals of J. Obviously, if $F(\cdot)$ is monotone with respect to J and $L < \infty$, then there exists an enumeration $\{J_k\}_{k=1}^L$ of the components of J such that

$$F(\lambda_1) \leq F(\lambda_2) \leq \cdots \leq F(\lambda_L)$$

holds for $\{\lambda_1, \lambda_2, \ldots, \lambda_L\} \in J_1 \times J_2 \times \cdots \times J_L$. If $L = \infty$, then it can happen that such an enumeration does not exist.

In [3] Problem 1.2 was solved under the additional assumptions that A admits a boundary triple such that the corresponding Weyl function is monotone and of scalar-type. A Nevanlinna function $F : C_+ \longrightarrow [\mathcal{H}]$ is of scalar-type if there is scalar Nevanlinna function $m(\cdot)$ such that

$$F(z) = m(z)I_{\mathcal{H}}, \quad z \in \mathbb{C}_+,$$

where $I_{\mathcal{H}}$ is the identity on $\mathcal{H}$.

Theorem 1.12 (Theorem 4.4 of [3]). *Let A be a densely defined closed symmetric operator in a separable Hilbert space $\mathfrak{H}$ with equal deficiency indices $n_{\pm}(A) =: n(A)$. Further, let $\Pi = \{\mathcal{H}, \Gamma_0, \Gamma_1\}$ be a boundary triple for A^* with scalar-type Weyl function $M(\cdot) = m(\cdot)\, I_{\mathcal{H}}$. If the Weyl function $M(\cdot)$ is monotone with respect to the open set $J \subseteq \mathcal{O}_M(\subset \rho(A_0))$, then for any auxiliary self-adjoint operator R on some separable Hilbert space $\mathfrak{R}$ obeying $\dim(E_R(J)\mathfrak{R}) \leq n(A)$ there exists a self-adjoint extension $\widetilde{A}$ of A such that $\widetilde{A}_J \cong R_J$.*

The assumption that the Weyl function has to be of scalar-type is very restrictive. Indeed, in [3] it was shown that this implies a special structure of the symmetric operator A.

Proposition 1.13 (Proposition 4.8 of [3]). *Let A be a simple symmetric operator in $\mathfrak{H}$ with equal deficiency indices $n_{\pm}(A) =: n(A)$ and let $\Pi = \{\mathcal{H}, \Gamma_0, \Gamma_1\}$ be a boundary triple for A^*. The corresponding Weyl function $M(\cdot)$ is of scalar-type if and only if A and $A_0 := A^* \upharpoonright \ker(\Gamma_0)$ admit the decompositions*

$$A = \bigoplus_{k=1}^{n(A)} S_k \quad \textit{and} \quad A_0 = \bigoplus_{k=1}^{n(A)} S_{k,0} \tag{10}$$

such that

(i) *S_k, $k = 1, 2, \ldots, n(A)$, are closed symmetric operators with deficiency indices $n_{\pm}(S_k) = 1$ which are unitarily equivalent to each other,*

(ii) *$S_{k,0}$, $k = 1, 2, \ldots, n(A)$, are self-adjoint extensions of S_k which are unitarily equivalent to each other,*

(iii) *there is a boundary triple $\Pi_k = \{\mathcal{H}_k, \Gamma_0^k, \Gamma_1^k\}$ for S_k^* and each $k = 1,2,\ldots,n(A)$, such that $S_{k,0} = S_k^* \upharpoonright \ker(\Gamma_1^k)$ and the corresponding Weyl function coincides with $m(\cdot)$ for each $k = 1, 2, \ldots, n(A)$.*

The decomposition (10) is not unique.

We note that Theorem 1.12 and Proposition 1.13 in fact improve the results of [10], cf. (5). The assumption that A admits a scalar-type Weyl function has far going spectral implications beyond the gap.

Theorem 1.14 (Theorem 5.2 of [3]). *Let A be a simple symmetric operator in $\mathfrak{H}$ with infinite deficiency indices. Further, let $\Pi = \{\mathcal{H}, \Gamma_0, \Gamma_1\}$ be a boundary triple for A^* with scalar-type Weyl function $M(z) = m(z)I_{\mathcal{H}}$, $z \in \mathbb{C}_+$, and let B be a densely defined self-adjoint operator.*

(i) *Then $\sigma_{ac}(A_B) \supset \sigma_{ac}(A_0)$, $A_0 := A^* \upharpoonright \ker(\Gamma_0)$ where $\sigma_{ac}(\cdot)$ denotes the absolutely continuous spectrum of an operator.*

(ii) *If the operator B is purely absolutely continuous, then the self-adjoint extension A_B is purely absolutely continuous, too.*

Further one has

Theorem 1.15 (Theorem 5.6 of [3]). *Let A be a simple symmetric operator in $\mathfrak{H}$ with infinite deficiency indices. Further, let $\Pi = \{\mathcal{H}, \Gamma_0, \Gamma_1\}$ be a boundary triple for A^* with scalar-type Weyl function $M(z) = m(z)I_{\mathcal{H}}$, $z \in \mathbb{C}_+$, and let B be a densely defined self-adjoint operator.*

(i) *If B is singular, i.e., $B^s = B$, then the absolutely continuous parts A_B^{ac} and A_0^{ac} are unitarily equivalent, in particular, $\sigma_{ac}(A_B) = \sigma_{ac}(A_0)$.*
(ii) *If B and A_0 are singular, then A_B is singular.*
(iii) *If B is pure point and the spectrum of A_0 consists of isolated eigenvalues, then A_B is pure point.*

In [3] it was conjectured that already the monotonicity assumption is sufficient to solve the Problem 1.2. In the following we make a first step to verify this conjecture for the special case that the deficiency indices are finite.

2. Finite deficiency indices

Let us assume that A is a densely defined closed symmetric operator on $\mathfrak{H}$ with finite or infinite equal deficiency indices. Further, let $\Pi = \{\mathcal{H}, \Gamma_0, \Gamma_1\}$ be a boundary triple for A^*. Let B be a bounded self-adjoint operator on $\mathcal{H}$. Let A_B be the self-adjoint extension which corresponds to the linear self-adjoint relation $\Theta = G(B)$, i..e. $A_B \longleftrightarrow G(B)$. The boundary triple $\Pi_B := \{\mathcal{H}, \Gamma_0^B, \Gamma_1^B\}$, $\Gamma_0^B := B\Gamma_0 - \Gamma_1$, $\Gamma_1^B := \Gamma_0$, has the property that $\mathrm{dom}(A_B) = \ker(\Gamma_0^B)$. One easily verifies that the Weyl function to the boundary triple Π_B is given by

$$M_B(z) = (B - M(z))^{-1}, \quad z \in \mathbb{C}_+,$$

Let $\mu \in \rho(A_0)$ and $B := M(\mu)$. We consider the self-adjoint extension $A_\mu := A_{M(\mu)}$. The corresponding boundary triple is given by $\Pi_\mu := \Pi_{M(\mu)} = \{\mathcal{H}, \Gamma_0^\mu := \Gamma_0^{M(\mu)}, \Gamma_1^\mu := \Gamma_1^{M(\mu)}\}$ where $\Gamma_1^\mu := \Gamma_1$ and $\Gamma_0^\mu = M(\mu)\Gamma_0 - \Gamma_1$ and the corresponding Weyl function $M_\mu(\cdot) := M_{M(\mu)}(\cdot)$ by

$$M_\mu := (M(\mu) - M(z))^{-1}, \quad z \in \mathbb{C}_+. \tag{11}$$

Lemma 2.1. *Let A be a simple symmetric operator in $\mathfrak{H}$ with equal deficiency indices $n_\pm(A) \le \infty$. Further, let $\Pi = \{\mathcal{H}, \Gamma_0, \Gamma_1\}$ be a boundary triple for A^* with function $M(\cdot)$. If the Weyl function $M(\cdot)$ is monotone with respect to $J \subseteq \rho(A_0)$ and $\mu \in J$, then the Weyl function $M_\mu(\cdot)$ is monotone with respect to $J_\mu := J \backslash \{\mu\}$.*

Proof. Since J is open it admits a decomposition into component intervals J_k,

$$J = \bigcup_{k=1}^{L} J_k, \quad L \in \mathbb{N} \cup \infty.$$

If $M(\cdot)$ is monotone with respect to J, then for any two component intervals J and J' we have either $M(\lambda) \le M(\lambda')$ or $M(\lambda') \le M(\lambda)$ for $\lambda \in J$ and $\lambda' \in J'$.

Let $\mu \in J_m$, $1 \le m \le L$. We set

$$\widetilde{J}_k := \begin{cases} J_r, & r = 1, \ldots, m-1 \\ J_m \setminus [\mu, \infty), & r = m, \\ J_m \setminus (-\infty, \mu], & r = m+1, \\ J_{r-1}, & r = m+2, \ldots, L+1 \end{cases}$$

Further we note that $(M(\lambda) - M(\lambda'))^{-1}$ exist and is bounded provided $\lambda, \lambda' \in J$ and $\lambda \neq \lambda'$. If $\lambda_l \in \widetilde{J}_l$ and $\lambda_n \in \widetilde{J}_n$, $l, n \in \{1, \ldots, L+1\}$, then either

$$\begin{array}{ccccc} M(\mu) & \le & M(\lambda_l) & \le & M(\lambda_n), \\ M(\lambda_l) & \le & M(\mu) & \le & M(\lambda_n), \\ M(\lambda_l) & \le & M(\lambda_n) & \le & M(\mu) \end{array} \tag{12}$$

or

$$\begin{array}{ccccc} M(\mu) & \le & M(\lambda_n) & \le & M(\lambda_l), \\ M(\lambda_n) & \le & M(\mu) & \le & M(\lambda_l), \\ M(\lambda_n) & \le & M(\lambda_l) & \le & M(\mu) \end{array} \tag{13}$$

for $\lambda_l \in \widetilde{J}_l$ and $\lambda_n \in \widetilde{J}_n$. Obviously, one gets from (12) and (13) that either

$$(M(\mu) - M(\lambda_l))^{-1} \le (M(\mu) - M(\lambda_n))^{-1}$$

or

$$(M(\mu) - M(\lambda_n))^{-1} \le (M(\mu) - M(\lambda_l))^{-1}$$

for $\lambda_l \in \widetilde{J}_l$ and $\lambda_n \in \widetilde{J}_n$, $l, n \in \{1, \ldots, L+1\}$, which proves the monotonicity of $M_\mu(\cdot)$ with respect to J_μ. □

Remark 2.2. We note that either $M_\mu(\lambda) \ge 0$ or $M_\mu(\lambda) \le 0$ for $\lambda \in \widetilde{J}_k$ where $\widetilde{J}_k$ is a component interval of J_μ.

Lemma 2.3. *Let A be a simple symmetric operator in $\mathfrak{H}$ with equal deficiency indices $n_\pm(A) \le \infty$. Further, let $\Pi = \{\mathcal{H}, \Gamma_0, \Gamma_1\}$ be a boundary triple for A^* with Weyl function $M(\cdot)$. If $\mu \in J \subseteq \rho(A_0)$, then A_μ is given by*

$$A_\mu = A^* \upharpoonright (\mathrm{dom}(A) + \mathcal{N}_\mu) \tag{14}$$

and μ is an eigenvalue of A_μ with eigenspace $\mathcal{N}_\mu$.

If the Weyl function $M(\cdot)$ is monotone with respect to $J \subseteq \rho(A_0)$ and $\mu \in J$, then $\sigma(A_\mu) \cap J = \{\mu\}$.

Proof. Let $H := A^* \upharpoonright (\mathrm{dom}(A) \dot{+} \mathcal{N}_\mu)$. A straightforward computation shows that H is symmetric and closed. Since $\mathrm{dom}(A^*) = \mathrm{dom}(A) \dot{+} \mathcal{N}_\mu = \mathrm{dom}(H)$ we get that H is self-adjoint. Since $\ker(A^* - \mu) = \mathcal{N}_\mu$ we obtain that $\ker(H - \mu) = \mathcal{N}_\mu$ which shows that μ is an eigenvalue of H with eigenspace $\mathcal{N}_\mu$. Since $\mathrm{dom}(A_\mu) = \ker(\Gamma_0^\mu) = \ker(M(\mu)\Gamma_0 - \Gamma_1)$ we find that $M(\mu)\Gamma_0 h - \Gamma_1 h = (M(\mu) - M(\mu))\Gamma_0 h = 0$ for $h \in \mathcal{N}_\mu$. Hence $\mathrm{dom}(A_\mu) \supseteq \mathcal{N}_\mu$ which implies $\mathrm{dom}(A_\mu) \supseteq \mathrm{dom}(A) \dot{+} \mathcal{N}_\mu = \mathrm{dom}(H)$. It follows that $A_\mu \supseteq H$ which shows that $A_\mu = H$.

By Proposition 1.8 one has $\lambda \in \rho(A_\mu)$ for $\lambda \in \rho(A_0)$ if and only if $0 \in \rho(M(\mu) - M(\lambda))$. Since $(M(\mu) - M(\lambda))^{-1}$ exists and is bounded for $\mu \neq \lambda \in J$ one has $0 \in \rho(M(\mu) - M(\lambda))$ for $\lambda \in J \setminus \{\mu\}$. Hence $J \setminus \{\mu\} \subseteq \rho(A_\mu)$. □

The considerations below are based on a lemma which immediately follows from Lemma 4.3 of [3].

Lemma 2.4. *Let A be a densely defined closed symmetric operator on a separable Hilbert space $\mathfrak{H}$ with equal deficiency indices. Further, let $\Pi = \{\mathcal{H}, \Gamma_0, \Gamma_1\}$ be a boundary triple for A^* with Weyl function $M(\cdot)$. If $\widehat{A}$ is a closed symmetric extension of A obeying*

$$A \subseteq \widehat{A} \subseteq A_0, \quad (\mathrm{dom}(A_0) = \ker(\Gamma_0)),$$

then there is a boundary triple $\widehat{\Pi} = \{\widehat{\mathcal{H}}, \widehat{\Gamma}_0, \widehat{\Gamma}_1\}$ such that one has $A_0 = \widehat{A}_0 := \widehat{A}^* \upharpoonright \ker(\widehat{\Gamma}_0)$ and the associated Weyl function $\widehat{M}(\cdot)$ is monotone with respect to $J \subseteq \mathcal{O}_M$ provided $M(\cdot)$ is monotone with respect to J.

Let $\mu \in J_m$, $1 \le m \le L$, where J_m is a component interval of the open set $J \subseteq \rho(A_0)$. We consider the symmetric operator

$$\widehat{A}_\mu := A^* \upharpoonright \mathrm{dom}(\widehat{A}_\mu), \qquad \mathrm{dom}(\widehat{A}_\mu) := \mathrm{dom}(A) + \widehat{\mathcal{N}}_\mu \tag{15}$$

where $\widehat{\mathcal{N}}_\mu$ is a subspace of $\mathcal{N}_\mu = \ker(A^* - \mu)$. The operator $\widehat{A}_\mu$ is symmetric and closed. We note that $A^* \upharpoonright \widehat{\mathcal{N}}_\mu = \mu I_{\widehat{\mathcal{N}}_\mu}$ where $I_{\widehat{\mathcal{N}}_\mu}$ is the identity on the subspace $\widehat{\mathcal{N}}_\mu$.

Lemma 2.5. *Let $\widehat{A}_\mu$, $\mu \in J_m \subseteq J \subseteq \rho(A_0)$, be the closed symmetric operator defined by (15). Then with respect to the decomposition $\mathfrak{H} = \widehat{\mathcal{N}}_\mu \oplus \widehat{\mathcal{N}}_\mu^\perp$ the operator $\widehat{A}_\mu$ admits the representation*

$$\widehat{A}_\mu = \mu I_{\widehat{\mathcal{N}}_\mu} \oplus G_\mu \tag{16}$$

where G_μ is a closed symmetric operator with deficiency indices $n_\pm(G_\mu)=\mathrm{dim}(\mathcal{N}_\mu \ominus \widehat{\mathcal{N}}_\mu)$ and gap J_m. Moreover, μ is an eigenvalue of $\widehat{A}_\mu$ with eigenspace $\widehat{\mathcal{N}}_\mu$ and multiplicity $\mathrm{dim}(\widehat{\mathcal{N}}_\mu)$.

Proof. Since $\sigma(\mu I_{\widehat{\mathcal{N}}_\mu}) \subset J_m$ we obtain from Lemma 2.1 of [2] that the operator $\widehat{A}_\mu$ admits the decomposition (16). Since $\ker(\widehat{A}_\mu^* - \mu) = \mathcal{N}_\mu$ and

$$\widehat{A}_\mu^* = \mu I_{\widehat{\mathcal{N}}_\mu} \oplus G_\mu^* \tag{17}$$

we get that

$$\mathcal{N}_\mu = \widehat{\mathcal{N}}_\mu \oplus \ker(G_\mu^* - \mu)$$

which yields

$$\ker(G_\mu^* - \mu) = \mathcal{N}_\mu \ominus \widehat{\mathcal{N}}_\mu \tag{18}$$

and

$$n_\pm(G_\mu) = \mathrm{dim}(\ker(G_\mu^* - \mu)) = \mathrm{dim}(\mathcal{N}_\mu \ominus \widehat{\mathcal{N}}_\mu). \tag{19}$$

From (16) we see that μ is an eigenvalue of the extension $\widehat{A}_\mu$ with eigenspace $\widehat{\mathcal{N}}_\mu$ and multiplicity $\mathrm{dim}(\widehat{\mathcal{N}}_\mu)$. □

Lemma 2.6. *Let A be a simple symmetric operator in $\mathfrak{H}$ with equal deficiency indices $n_\pm(A)$. Further, let $\Pi = \{\mathcal{H}, \Gamma_0, \Gamma_1\}$ be a boundary triple for A^* with Weyl function $M(\cdot)$ and let $\mu \in J \subseteq \rho(A_0)$. If the Weyl function $M(\cdot)$ is monotone with respect to J, then the closed symmetric extension $\widehat{A}_\mu$ of A defined by* (15) *admits a boundary triple $\widehat{\Pi}_\mu = \{\widehat{\mathcal{H}}_\mu, \widehat{\Gamma}^\mu_0, \widehat{\Gamma}^\mu_1\}$ with a Weyl function $\widehat{M}_\mu(\cdot)$ which is monotone with respect to $J_\mu := J \setminus \{\mu\}$.*

Proof. The self-adjoint extension A_μ defined by Lemma 2.1 is also an extension of $\widehat{A}_\mu$ defined by (15). Indeed, from (14) we obtain that $\mathrm{dom}(A) + \mathcal{N}_\mu \subseteq \mathrm{dom}(A_\mu)$ which yields $\mathrm{dom}(\widehat{A}_\mu) \subseteq \mathrm{dom}(A_\mu)$. Applying Lemma 2.1 and Lemma 2.4 we complete the proof. □

Notice that the symmetric operator $\widehat{A}_\mu$ defined by (15) is not simple because it contains at least the self-adjoint part $\mu I_{\widehat{\mathcal{N}}_\mu}$, cf. decomposition (16). However, this does not mean that G_μ is simple because it can happen that G_μ contains further self-adjoint parts.

Next we are going to show that the boundary triple $\widehat{\Pi}_\mu = \{\widehat{\mathcal{H}}_\mu, \widehat{\Gamma}^\mu_0, \widehat{\Gamma}^\mu_1\}$ is in fact a boundary triple for G^*_μ

Corollary 2.7. *If $\widehat{\Pi}_\mu = \{\widehat{\mathcal{H}}_\mu, \widehat{\Gamma}^\mu_0, \widehat{\Gamma}^\mu_1\}$ is the boundary triple for $\widehat{A}^*_\mu$ of Lemma* 2.6, *then $\widehat{\Pi}'_\mu := \{\widehat{\mathcal{H}}_\mu, \widehat{\Gamma}'^\mu_0, \widehat{\Gamma}'^\mu_1\}$, $\widehat{\Gamma}'^\mu_0 := \widehat{\Gamma}^\mu_0 \upharpoonright \mathrm{dom}(G^*_\mu)$, $\widehat{\Gamma}'^\mu_1 := \widehat{\Gamma}^\mu_1 \upharpoonright \mathrm{dom}(G^*_\mu)$, is a boundary triple for G^*_μ with the same associated Weyl function $\widehat{M}_\mu(\cdot)$ (which is monotone with respect to $J_\mu := J \setminus \{\mu\}$).*

Proof. To show this it is sufficient to verify that $\widehat{\Gamma}^\mu_0 \mathrm{dom}(\widehat{\mathcal{N}}_\mu) = \widehat{\Gamma}^\mu_1 \mathrm{dom}(\widehat{\mathcal{N}}_\mu) = 0$. By Definition 1.3 the boundary triple $\widehat{\Pi}_\mu = \{\widehat{\mathcal{H}}_\mu, \widehat{\Gamma}^\mu_0, \widehat{\Gamma}^\mu_1\}$ satisfies Green's formula

$$(\widehat{A}^*_\mu f, g) - (f, \widehat{A}^*_\mu g) = (\widehat{\Gamma}^\mu_1 f, \widehat{\Gamma}^\mu_0 g) - (\widehat{\Gamma}^\mu_0 f, \widehat{\Gamma}^\mu_1 g)$$

and the mapping $\widehat{\Gamma}^\mu = \{\widehat{\Gamma}^\mu_0, \widehat{\Gamma}^\mu_1\}$ is surjective. Let $f \in \widehat{\mathcal{N}}_\mu$ and $g \in \mathrm{dom}(\widehat{A}^*_\mu)$. We find

$$(\widehat{A}^*_\mu f, g) - (f, \widehat{A}^*_\mu g) = \mu(f,g) - (f, \widehat{A}^*_\mu g) = \mu(f,g) - (\widehat{A}_\mu f, g) = 0$$

which implies

$$0 = (\widehat{\Gamma}^\mu_1 f, \widehat{\Gamma}^\mu_0 g) - (\widehat{\Gamma}^\mu_0 f, \widehat{\Gamma}^\mu_1 g), \quad f \in \widehat{\mathcal{N}}_\mu \quad g \in \mathrm{dom}(\widehat{A}^*_\mu).$$

Hence

$$0 = (\widehat{\Gamma}^\mu f, E_0 \widehat{\Gamma}^\mu g)$$

where $E_0 \widehat{\Gamma}^\mu g := \{-\widehat{\Gamma}^\mu_1 g, \widehat{\Gamma}^\mu_0 g\}$, $g \in \mathrm{dom}(\widehat{A}^*_\mu)$. Since the operator E_0 is unitary it follows that $\mathrm{ran}(E_0 \widehat{\Gamma}^\mu) = \widehat{\mathcal{H}}_\mu \times \widehat{\mathcal{H}}_\mu$. Therefore we find that $\widehat{\Gamma}^\mu f = 0$, $f \in \widehat{\mathcal{N}}_\mu$. □

Because $\widehat{M}_\mu(\cdot)$ is monotone with respect to J_μ the self-adjoint extension $G_{\mu,0} := G^*_\mu \upharpoonright \ker(\widehat{\Gamma}'^\mu_0)$ of G_μ contains J_μ in its resolvent set, i.e., $J_\mu \subseteq \rho(G_{\mu,0})$.

Finally, we note that the boundary triple $\widehat{\Pi}_\mu = \{\widehat{\mathcal{H}}_\mu, \widehat{\Gamma}^\mu_0, \widehat{\Gamma}^\mu_1\}$ with Weyl function $\widehat{M}_\mu(\cdot)$ of Lemma 2.6 can be calculated from the boundary triple $\Pi_\mu =$

$\{\mathcal{H}, \Gamma_1^\mu, \Gamma_1^\mu\}$ and the Weyl function $M_\mu(\cdot)$ defined by (11) explicitly. To see this we put $\mathcal{L} := \Gamma_1^\mu \mathrm{dom}(\widehat{A}_\mu)$. By π_μ we denote the orthogonal projection from $\mathcal{H}$ onto $\widehat{\mathcal{H}}_\mu = \mathcal{H} \ominus \mathcal{L}$. Further, we set $\widehat{\Gamma}_0^\mu := \Gamma_0^\mu \upharpoonright \mathrm{dom}(\widehat{A}_\mu^*)$ and $\widehat{\Gamma}_1^\mu := \pi_\mu \Gamma_1^\mu \upharpoonright \mathrm{dom}(\widehat{A}_\mu^*)$. One verifies that the so defined triple $\widehat{\Pi}_\mu = \{\widehat{\mathcal{H}}_\mu, \widehat{\Gamma}_0^\mu, \widehat{\Gamma}_1^\mu\}$ is a boundary triple for $\widehat{A}_\mu^*$. Its Weyl function $\widehat{M}(\cdot)$ is given by

$$\widehat{M}_\mu(z) = \pi_\mu M_\mu(z) \upharpoonright \widehat{\mathcal{H}}_\mu, \quad z \in \mathbb{C}_+. \tag{20}$$

Since the Weyl function $M_\mu(\cdot)$ is monotone with respect to J_μ by Lemma 2.1 one gets that $\widehat{M}_\mu(\cdot)$ is monotone with respect to J_μ, too.

Lemma 2.8. *Let A be a simple symmetric operator in $\mathfrak{H}$ with equal deficiency indices $n_\pm(A)$. Further, let $\Pi = \{\mathcal{H}, \Gamma_0, \Gamma_1\}$ be a boundary triple for A^* with Weyl function $M(\cdot)$ and let $\mu \in J \subseteq \rho(A_0)$. If the Weyl function $M(\cdot)$ is monotone with respect to J, then the closed symmetric extension $\widehat{A}_\mu$ of A defined by (15) admits a boundary triple $\widetilde{\Pi}_\mu = \{\widetilde{\mathcal{H}}_\mu, \widetilde{\Gamma}_0^\mu, \widetilde{\Gamma}_1^\mu\}$ with a Weyl function $\widetilde{M}_\mu(\cdot)$ which is monotone with respect to J.*

Proof. Let us introduce the triple $\widetilde{\Pi}_\mu := \{\widetilde{\mathcal{H}}_\mu, \widetilde{\Gamma}_0^\mu, \widetilde{\Gamma}_1^\mu\}$ defined by $\widetilde{\mathcal{H}}_\mu ;= \widehat{\mathcal{H}}_\mu$, $\widetilde{\Gamma}_0^\mu := -\widehat{\Gamma}_1^\mu$ and $\widetilde{\Gamma}_1^\mu := \widehat{\Gamma}_0^\mu$ which arises from $\widehat{\Pi}_{\mu,B} = \{\widehat{\mathcal{H}}_\mu, \widehat{\Gamma}_0^{\mu,B}, \widehat{\Gamma}_1^{\mu,B}\}$ by setting $B = 0$. The associated Weyl function $\widetilde{M}_\mu(\cdot)$ is given by

$$\widetilde{M}_\mu(z) = -(\widehat{M}_\mu(z))^{-1}, \quad z \in \mathbb{C}_+. \tag{21}$$

Let us show that $\widetilde{M}_\mu(\cdot)$ is monotone with respect to J_μ using the monotonicity of $\widehat{M}_\mu(\cdot)$ with respect to J_μ. By Remark 2.2 the open set J_μ can be divided into two disjoint open subsets J_μ^+ and J_μ^-, $J_\mu^+ \cup J_\mu^- = J_\mu$, such that $\widehat{M}_\mu(\lambda) \geq 0$ for $\lambda \in J_\mu^+$ and $\widehat{M}_\mu(\lambda) \leq 0$ for $\lambda \in J_\mu^-$. Let $\widetilde{J}_k$ and $\widetilde{J}_l$ are component intervals of J_μ. If $\widetilde{J}_k \subseteq J_\mu^+$ and $\widetilde{J}_l \subseteq J_\mu^-$, then

$$\widehat{M}_\mu(\lambda) \geq 0 \geq \widehat{M}_\mu(\lambda'), \quad \lambda \in \widetilde{J}_k, \quad \lambda' \in \widetilde{J}_l,$$

which proves

$$\widetilde{M}_\mu(\lambda) \leq 0 \leq \widetilde{M}_\mu(\lambda'), \quad \lambda \in \widetilde{J}_k, \quad \lambda' \in \widetilde{J}_l. \tag{22}$$

If $\widetilde{J}_k \subseteq J_\mu^-$ and $\widetilde{J}_l \subseteq J_\mu^+$, then we similarly prove that

$$\widetilde{M}_\mu(\lambda) \geq 0 \geq \widetilde{M}_\mu(\lambda'), \quad \lambda \in \widetilde{J}_k, \quad \lambda' \in \widetilde{J}_l. \tag{23}$$

If $\widetilde{J}_k \subseteq J_\mu^+$ and $\widetilde{J}_l \subseteq J_\mu^+$, then it follows from the monotonicity of $\widehat{M}_\mu(\cdot)$ with respect to J_μ that either

$$\widehat{M}_\mu(\lambda) \geq \widehat{M}_\mu(\lambda') \geq 0, \quad \text{or} \quad \widehat{M}_\mu(\lambda') \geq \widehat{M}_\mu(\lambda) \geq 0, \quad \lambda \in \widetilde{J}_k, \quad \lambda' \in \widetilde{J}_l,$$

which yields either

$$\widetilde{M}_\mu(\lambda') \leq \widetilde{M}_\mu(\lambda) \leq 0, \quad \text{or} \quad \widetilde{M}_\mu(\lambda) \leq \widetilde{M}_\mu(\lambda') \leq 0, \quad \lambda \in \widetilde{J}_k, \quad \lambda' \in \widetilde{J}_l. \tag{24}$$

Similarly we verify that $\widetilde{J}_k \subseteq J_\mu^-$ and $\widetilde{J}_l \subseteq J_\mu^-$ implies either

$$\widetilde{M}_\mu(\lambda') \geq \widetilde{M}_\mu(\lambda) \geq 0, \quad \text{or} \quad \widetilde{M}_\mu(\lambda) \geq \widetilde{M}_\mu(\lambda') \geq 0, \quad \lambda \in \widetilde{J}_k, \quad \lambda' \in \widetilde{J}_l. \tag{25}$$

From (22)–(25) we obtain that $\widetilde{M}_\mu(\cdot)$ is monotone with respect to J_μ. To show that in fact $\widetilde{M}_\mu(\cdot)$ is monotone with respect to J it is sufficient to verify that

$$s - \lim_{\lambda \uparrow \mu} \widetilde{M}_\mu(\lambda) = s - \lim_{\lambda \downarrow \mu} \widetilde{M}_\mu(\lambda) = 0. \tag{26}$$

To this end we note that

$$\lim_{\lambda \uparrow \mu} (\widehat{M}_\mu(\lambda) f, f) = \infty$$

for $f \in \widehat{\mathcal{H}}_\mu \setminus \{0\}$. Applying Lemma 1.1 of [17] we prove the first part of (26). The second part can be proved analogously. □

Lemma 2.8 shows that there is a boundary triple $\widetilde{\Pi}_\mu = \{\widetilde{\mathcal{H}}_\mu, \widetilde{\Gamma}_0^\mu, \widetilde{\Gamma}_1^\mu\}$ for G_μ^* with Weyl function $\widetilde{M}_\mu$ which is monotone with respect to J.

Theorem 2.9. *Let A be a simple symmetric operator in $\mathfrak{H}$ with finite deficiency indices $n := n_\pm(A)$. Further, let $\Pi = \{\mathcal{H}, \Gamma_0, \Gamma_1\}$ be a boundary triple for A^* with Weyl function $M(\cdot)$. If the Weyl function $M(\cdot)$ is monotone with respect to $J \subseteq \rho(A_0)$, then for any sequence of real numbers $\{\mu_j\}_{j=1}^s$, $\mu_j \in J$, and any sequence of integers $\{p_j\}_{j=1}^s$, $p_j \in \mathbb{N}$ obeying $\sum_{j=1}^s p_j \leq n$ there is a self-adjoint extension $\widetilde{A}$ of A such that*

$$\sigma(\widetilde{A}) \cap J = \sigma_p(\widetilde{A}) \cap J = \bigcup_{j=1}^{s} \{\mu_j\}$$

and the multiplicities of the eigenvalues μ_j coincides with p_j, $j = 1, 2, \ldots, s$.

Proof. We set $\mu = \mu_1$ and and choose a subspace $\widehat{\mathcal{N}}_{\mu_1} \subseteq \mathcal{N}_{\mu_1}(A)$ such that $\dim(\widehat{\mathcal{N}}_{\mu_1}) = p_1$. By the procedure above there is a closed symmetric extension $\widehat{A}_{\mu_1}$ of A admitting the orthogonal decomposition

$$\widehat{A}_{\mu_1} = \mu_1 I_{\widehat{\mathcal{N}}_{\mu_1}} \oplus G_{\mu_1}.$$

By (19) we get that $n_\pm(G_{\mu_1}) = n - p_1$. Moreover, by Lemma 2.8 the closed symmetric operator G_{μ_1} admits a boundary triple $\widetilde{\Pi}_{\mu_1} = \{\widetilde{\mathcal{H}}_{\mu_1}, \widetilde{\Gamma}_0^{\mu_1}, \widetilde{\Gamma}_0^{\mu_1}\}$ with Weyl function $\widetilde{M}_{\mu_1}(\,)$ which is monotone with respect to J.

In the next step we repeat the procedure for $A := G_{\mu_1}$, $\mu = \mu_2$ and $\widehat{\mathcal{N}}_{\mu_2} \subseteq \mathcal{N}_{\mu_2}(G_{\mu_1})$, $\dim(\widehat{\mathcal{N}}_{\mu_2}) = p_2$. Hence we find a closed symmetric extension $\widehat{A}_{\mu_2}$ of $A = G_{\mu_1}$ admitting the orthogonal decomposition

$$\widehat{A}_{\mu_2} = \mu_2 I_{\widehat{\mathcal{N}}_{\mu_2}} \oplus G_{\mu_2}$$

such that $n_\pm(G_{\mu_2}) = n - p_1 - p_2$. As in the previous step the closed symmetric operator G_{μ_2} admits a boundary triple $\widetilde{\Pi}_{\mu_2} = \{\widetilde{\mathcal{H}}_{\mu_2}, \widetilde{\Gamma}_0^{\mu_2}, \widetilde{\Gamma}_1^{\mu_2}\}$ with Weyl function $\widetilde{M}_{\mu_2}(\cdot)$ which is monotone with respect to J. We note that

$$A_{\mu_1\mu_2} := \mu_1 I_{\widehat{\mathcal{N}}_{\mu_1}} \oplus \mu_2 I_{\widehat{\mathcal{N}}_{\mu_2}} \oplus G_{\mu_2}$$

defines a closed symmetric extension of A such that $p_1 = \dim(\widehat{\mathcal{N}}_{\mu_1})$, $p_2 = \dim(\widehat{\mathcal{N}}_{\mu_2})$ and G_2 a closed symmetric operator with deficiency indices $n_\pm(G_2) = n - p_1 - p_2$ admitting a boundary triple $\widetilde{\Pi}_{\mu_2} = \{\widetilde{\mathcal{H}}_{\mu_2}, \widetilde{\Gamma}_0^{\mu_2}, \widetilde{\Gamma}_1^{\mu_2}\}$ such that the corresponding Weyl function $\widetilde{M}_{\mu_2}(\cdot)$ is monotone with respect to J.

We repeat this procedure s times and obtain a closed symmetric extension

$$A_{\mu_1\mu_2\ldots\mu_s} := \mu_1 I_{\widehat{\mathcal{N}}_{\mu_1}} \oplus \mu_2 I_{\widehat{\mathcal{N}}_{\mu_2}} \oplus \cdots \oplus \mu_s I_{\widehat{\mathcal{N}}_{\mu_s}} \oplus G_s$$

of A such that $p_j = \dim(\widehat{\mathcal{N}}_{\mu_j})$, $j = 1, 2, \ldots, s$, and G_s is a closed symmetric operator with deficiency indices $n_\pm(G_s) = n - \sum_{j=1}^s p_j$ admitting a boundary triple $\widetilde{\Pi}_{\mu_s} = \{\widetilde{\mathcal{H}}_{\mu_s}, \widetilde{\Gamma}_0^{\mu_s}, \widetilde{\Gamma}_1^{\mu_s}\}$ such that the Weyl function $\widetilde{M}_{\mu_s}(\cdot)$ is monotone with respect to J provided $n_\pm(G_s) \geq 1$.

Finally, if $n_\pm(G_s) = 0$, then G_s is a self-adjoint operator such that $J \subseteq \rho(G_s)$. Hence the operator $A_{\mu_1\mu_2\ldots\mu_s}$ is the desired self-adjoint extension of A. If $n_\pm(G_s) \geq 1$, then we choose the self-adjoint extension $\widetilde{G}_{s,0} = G_s^* \upharpoonright \ker(\widetilde{\Gamma}_0^{\mu_s})$ of G_s obeying $J \subseteq \rho(\widetilde{G}_{s,0})$. Setting

$$A'_{\mu_1\mu_2\ldots\mu_s} := \mu_1 I_{\widehat{\mathcal{N}}_{\mu_1}} \oplus \mu_2 I_{\widehat{\mathcal{N}}_{\mu_2}} \oplus \cdots \oplus \mu_s I_{\widehat{\mathcal{N}}_{\mu_s}} \oplus G_{s,0}.$$

we obtain the desired self-adjoint extension of A. □

Remark 2.10. Theorem 2.9 gives rise for the following problems:

(i) Can Theorem 2.9 be extended to the case of infinite deficiency indices of A? In other words, let A be a closed symmetric operator with infinite deficiency indices admitting a boundary triple $\Pi = \{\mathcal{H}, \Gamma_0, \Gamma_1\}$ with Weyl function $M(\cdot)$ which is monotone with respect to the open set $J \subseteq \rho(A_0)$. Is it true that for any pure point operator R on some separable Hilbert space $\mathfrak{R}$ there is a self-adjoint extension $\widetilde{A}$ such that $\widetilde{A}_J \cong R_J$?

(ii) Is the assumption on the monotonicity of the Weyl function $M(\cdot)$ not only sufficient but also necessary? In other words, let $J \subseteq \mathbb{R}$ be an open set and let A be a closed symmetric operator admitting a self-adjoint extension A_0 such that $J \subseteq \rho(A_0)$. Further, let $\Pi = \{\mathcal{H}, \Pi_0, \Pi_1\}$ be a boundary triple for A^* such that $A_0 = A^* \upharpoonright \ker(\Gamma_0)$ with Weyl function $M(\cdot)$. Is it true that the Weyl function $M(\cdot)$ is monotone with respect to J if and only if for any symmetric extension $\widehat{A}$ of A satisfying $A \subseteq \widehat{A} \subseteq A_0$ and for any pure point operator R on some separable Hilbert space $\mathfrak{R}$ obeying $\dim(E_R(J)\mathfrak{R}) \leq n_\pm(\widehat{A})$ there is a self-adjoint extension $\widetilde{A}$ of $\widehat{A}$ such that $\widetilde{A}_J \cong R_J$.

References

[1] N.I. Achieser and I.M. Glasmann: *Theorie der linearen Operatoren im Hilbertraum*, Akademie-Verlag, Berlin 1975.

[2] A. Albeverio, J.F. Brasche and H. Neidhardt: On inverse spectral theory for self-adjoint extensions. *J. Funct. Anal.* 154 (1998), 130–173.

[3] A. Albeverio, J.F. Brasche, M. Malamud, H. Neidhardt: Inverse spectral theory for symmetric operators with several gaps: scalar-type Weyl functions, *J. Funct. Anal.* 228 (2005), 144–188.

[4] W.O. Amrein and D.B. Pearson: Stability criteria for the Weyl m-function, *Math. Physics. Analysis and Geometry*, 1 (1998), 193–221.

[5] J.F. Brasche: The projection operator valued measures of the self-adjoint extensions of a symmetric operator A on a gap of A. Preprint. Available at http://www.math.chalmers.se/~brasche/gap1.pdf

[6] J.F. Brasche: Spectral theory for self-adjoint extensions, pp. 51–96 in R. del Rio, C. Villega-Blas (eds.): *Spectral Theory of Schrödinger operators*, Contemporary Mathematics 340, AMS, 2004.

[7] J.F. Brasche, H. Neidhardt and J. Weidmann: On the point spectrum of self-adjoint extensions, *Math.Zeitschr.* 214 (1993), 343–355.

[8] J.F. Brasche and H. Neidhardt: On the absolutely continuous spectrum of self-adjoint extensions, *J. Funct. Analysis* 131 (1995), 364–385.

[9] J.F. Brasche and H. Neidhardt: On the singular continuous spectrum of self-adjoint extensions, *Math. Zeitschr.* 222 (1996), 533–542.

[10] J.F. Brasche, M.M. Malamud and H. Neidhardt: Weyl functions and singular continuous spectra of self-adjoint extensions. Gesztesy, Fritz (ed.) et al., Stochastic processes, physics and geometry: New interplays. II. A volume in honor of Sergio Albeverio. *Proceedings of the conference on infinite-dimensional (stochastic) analysis and quantum physics.* Leipzig, Germany, January 18–22, 1999. Providence, RI: American Mathematical Society (AMS). CMS Conf. Proc. 29, 75–84 (2000).

[11] J.F. Brasche, M.M. Malamud and H. Neidhardt: Weyl functions and spectral properties of self-adjoint extensions. *Integral Equations and Operator Theory* 43 (2002), no. 3, 264–289.

[12] M.A. Brodski; *Triangular and Jordan representations of linear operators*, Nauka, Moscow, 1969; English translation in Transl. Math. Monogr. vol. 32, AMS, Providence, 1971.

[13] V.A. Derkach and M.M. Malamud, On the Weyl function and Hermitian operators with gaps, *Dokl. AN SSSR* 293, no. 5 (1987), 1041–1046.

[14] V.A. Derkach and M.M. Malamud, Generalized resolvents and the boundary value problems for Hermitian operators with gaps, *J. Funct. Anal.* 95 (1991), 1–95.

[15] Fichtenholz, G.M., *Differential- und Integralrechnung, Vol. II*, VEB Deutscher Verlag der Wissenschaften, Berlin 1964.

[16] M.G. Krein: The theory of self-adjoint extensions of semi-bounded Hermitian transformations and its applications. *Mat. Sb.*, N. Ser. 20(62), (1947) 431–495.

[17] M.G. Krein and I.E. Ovcharenko: On Q-functions and sc-extensions of Hermitian contraction with non-dense domain, *Sibirsk. Mat. Zh.* 18 (1977), no. 5, 1032–1056.

[18] M.G. Krein and A.A. Nudel'man, *The Markov moment problem and extremal problems*, Moscow 1973, Engl. transl. *Amer. Math. Soc.*, Providence, RI, 1977.

[19] Riesz, F.; Sz.-Nagy, B.: *Leçons d'analyse fonctionnelle*, Gauthier-Villars, Paris 1965; Akademiai Kiado, Budapest 1965.

[20] B. Simon: Spectral Analysis of rank one perturbations and applications, *CRM Proceedings and Lecture Notes* 8 (1995), 109–149.

J.F. Brasche
Institut für Mathematik
TU Clausthal
Erzstr. 1
D-38678 Clausthal-Zellerfeld
Germany
e-mail: brasche@math.tu-clausthal.de

M.M. Malamud
Department of Mathematics
Donetsk National University
Universitetskaja 24
340055 Donetsk
Ukraine
e-mail: mdm@dc.donetsk.ua

H. Neidhardt
Weierstrass Institute for
Applied Analysis and Stochastics (WIAS)
Mohrenstr. 39
D-10117 Berlin
Germany
e-mail: neidhard@wias-berlin.de

Operator Theory:
Advances and Applications, Vol. 162, 103–116

The Spectrum of the Multiplication Operator Associated with a Family of Operators in a Banach Space

R. Denk, M. Möller and C. Tretter

Abstract. An operator family of densely defined closed linear operators and the multiplication operator associated with it are considered. The spectrum of this multiplication operator is expressed in terms of the spectra of the operators in the given family.

Mathematics Subject Classification (2000). Primary 47A10; Secondary 47A56, 47B40, 47E05, 76B99.

Keywords. Multiplication operator, operator family, vector measure, spectrum.

1. Introduction

When considering problems from mathematical physics modelled by linear differential operators, separation of variables (often with respect to the time variable on the one hand and the space variables on the other hand) leads to spectral problems where the spectrum gives information about stability and discrete states. However, a further separation of variables in the space variables is often useful; for example, if differentiation does not occur with respect to all space variables. Therefore, the original spectral problem is split into a family of spectral problems. Here we investigate the question how the spectrum of the original problem can be described by the spectra of the operators in this family. This allows, for example, to describe the spectrum of certain PDE problems in terms of spectra of a family of associated ODE problems.

More precisely, we consider an operator family $(A(\rho))_{\rho\in X}$ of closed densely defined operators on a Banach space E, where X is a locally compact space. With

The second author has been partially supported through a grant by the NRF of South Africa, GUN 2053746, and by the John Knopfmacher Centre for Applicable Analysis and Number Theory at the University of the Witwatersrand, Johannesburg.

this family we associate an operator $\mathfrak{A}$ on $L^p(X,\mu,E)$, $1 \le p < \infty$, for a given Radon measure μ on X such that

$$(\mathfrak{A}f)(\rho) = A(\rho)f(\rho), \quad \rho \in X,$$

which we call a multiplication operator. Our main result describes the spectrum of $\mathfrak{A}$ in terms of the spectra of the operators $A(\rho)$.

There are two main assumptions on the operator family $A(\rho)$, namely that the domains $\mathcal{D}(A(\rho))$ are independent of ρ and that the operator family depends continuously on ρ on a compactification of X, where the common domain is equipped with a graph norm. For self-adjoint operators on Hilbert spaces, this concept was developed, e.g., in Reed and Simon, [9, Section XIII.16]. There it is only assumed that $A(\rho)$ depends measurably on ρ; however, the characterization of the spectrum is more complicated, see [9, Theorem XIII.85]. For not necessarily self-adjoint operators in Hilbert space one can use the theory of direct integrals, see, e.g., Azoff, [1], and Dixmier, [5, Chapter II, §2]. For usual multiplication operators, i.e., multiplication by matrix functions, the spectrum has been investigated, e.g., by Hardt and Wagenführer in [7].

As was pointed out in [1], in general there is little resemblance between the spectra of the family $A(\rho)$ and the spectrum of $\mathfrak{A}$. Therefore we will require continuity of the family $A(\rho)$; see below for a precise definition. Although our assumptions on $A(\rho)$ seem quite restrictive, many problems in mathematical physics lead to operator functions of this type. We note that a particular example has been studied by Binding and Volkmer in [2] in the setting of two-parameter problems.

In [3] we have considered a particular example from magnetohydrodynamics in L^2. In this paper we give a more general theoretical background and extend the example into a more general setting. In a forthcoming paper we will consider the more general case that the assumptions on $A(\rho)$ are replaced by the assumption that $A(\rho)$ depends continuously on ρ with respect to the gap topology on the space of closed operators in H. This allows the domains of $A(\rho)$ to depend on ρ.

The paper is organized as follows. In Section 2 we define the multiplication operator $\mathfrak{A}$ associated with $(A(\rho))_{\rho\in X}$ and prove that $\mathfrak{A}$ is closed. In Section 3 it is shown that the spectrum of $\mathfrak{A}$ is the union of the spectra of $A(\rho)$ over ρ in the compactification of X. Results on the point spectrum and the essential spectrum are established in Section 4. In Section 5 results are obtained for cases where $A(\rho)$ is only continuous on X. In Section 6 the general results are applied to some classes of examples to illustrate the reduction process.

2. The multiplication operator associated with an operator family

Throughout this paper, X denotes a nonempty locally compact Hausdorff space, μ a Radon measure on X with $\operatorname{supp}\mu = X$, E a Banach space with norm $\|\cdot\|$, $1 \le p < \infty$, and

$$\mathfrak{H} := L^p(X,\mu,E)$$

the space of L^p-functions on X with respect to μ with values in E. It is well known that $\mathfrak{H}$ is a Banach space with dual $\mathfrak{H}^* \supset L^{p'}(X, \mu, E^*)$, see, e.g., [6, Theorem III.6.6] and [4, p. 97], where $1/p + 1/p' = 1$. Note that [4] only deals with finite measure spaces, but it is easily seen that finiteness is not needed here. Let $\mathcal{C}(E)$ denote the set of closed operators on E. For a subspace $D \subset E$, we denote by $\mathcal{C}_D(E)$ the subset of $\mathcal{C}(E)$ consisting of those closed operators T with domain $\mathcal{D}(T)$ being exactly D.

On $\mathcal{C}_D(E)$ we define a topology as follows. For an operator $G \in \mathcal{C}_D(E)$, we denote by $\|\cdot\|_G$ the graph norm of G on D given by

$$\|x\|_G := (\|x\|^p + \|Gx\|^p)^{\frac{1}{p}}, \quad x \in D,$$

and set

$$\rho_G(S, T) := \|S - T\|_G := \sup_{\substack{x \in D \\ \|x\|_G = 1}} \|(S - T)x\|, \quad S, T \in \mathcal{C}_D(E).$$

Then ρ_G is a metric on $\mathcal{C}_D(E)$. We note that the topology induced by the metric ρ_G does not depend on the choice of the operator G since all the graph norms on D are equivalent by the closed graph theorem, and that $(\mathcal{C}_D(E), \|\cdot\|_G)$ can be identified with a subset of the space $\mathcal{B}((D, \|\cdot\|_G), E)$ of bounded linear operators from $(D, \|\cdot\|_G)$ to E.

Let Y be a compactification of X. We consider an operator function

$$A : Y \to \mathcal{C}(E)$$

with the following properties:

(a) $D = \mathcal{D}(A(\rho))$, $\rho \in Y$, is independent of ρ and a dense subspace of E,

(b) $A : Y \to \mathcal{C}_D(E)$ is continuous,

where $\mathcal{C}_D(E)$ is equipped with the above mentioned topology.

Proposition 2.1. *Let assumptions* (a) *and* (b) *be satisfied. Then there are positive constants* c_G, m_G *such that*

$$\|A(\rho)\|_G \le c_G, \quad \rho \in Y, \tag{2.1}$$

and

$$m_G\|x\|_G \le \|x\|_{A(\rho)} \le M_G\|x\|_G, \quad x \in D, \ \rho \in Y, \tag{2.2}$$

where $M_G := (1 + c_G^p)^{\frac{1}{p}}$.

Proof. The existence of c_G such that (2.1) holds is an immediate consequence of assumption (b) and the compactness of Y.

Now let $x \in D$ and $\rho \in Y$. Then

$$\|x\|_{A(\rho)}^p = \|x\|^p + \|A(\rho)x\|^p \le \|x\|^p + c_G^p (\|x\|^p + \|Gx\|^p) \le (1 + c_G^p) \|x\|_G^p,$$

which proves the right inequality in (2.2).

Assume that the left inequality in (2.2) is false for any positive constant m_G. Then there are a sequence $(x_n)_1^\infty$ in D and a sequence $(\rho_n)_1^\infty$ in Y such that $\|x_n\|_G = 1$ for all $n \in \mathbb{N}$ and $\|x_n\|_{A(\rho_n)} \to 0$ as $n \to \infty$. Since Y is compact, there is a limit point $\rho \in Y$ of $(\rho_n)_1^\infty$. The continuity assumption (b) implies that for every $\varepsilon > 0$ there is a positive integer n_ε such that $\|A(\rho_{n_\varepsilon}) - A(\rho)\|_G < \varepsilon$ and $\|x_{n_\varepsilon}\|_{A(\rho_{n_\varepsilon})} < \varepsilon$. This leads to

$$\begin{aligned}\|x_{n_\varepsilon}\|_{A(\rho)}^p &= \|x_{n_\varepsilon}\|^p + \|A(\rho)x_{n_\varepsilon}\|^p \\ &\le \|x_{n_\varepsilon}\|^p + 2^{p-1}\|A(\rho)x_{n_\varepsilon} - A(\rho_{n_\varepsilon})x_{n_\varepsilon}\|^p + 2^{p-1}\|A(\rho_{n_\varepsilon})x_{n_\varepsilon}\|^p \\ &\le \|x_{n_\varepsilon}\|^p + 2^{p-1}\|A(\rho) - A(\rho_{n_\varepsilon})\|_G^p\|x_{n_\varepsilon}\|_G^p + 2^{p-1}\|A(\rho_{n_\varepsilon})x_{n_\varepsilon}\|^p \\ &\le 2^p\varepsilon^p.\end{aligned}$$

But this contradicts the equivalence of the two graph norms $\|\cdot\|_{A(\rho)}$ and $\|\cdot\|_G$. □

Proposition 2.2. *Let assumptions* (a) *and* (b) *be satisfied. Then there is a unique bounded linear operator* $\widetilde{A}$ *from* $L^p(X,\mu,D)$ *into* $L^p(X,\mu,E)$ *such that* $(\widetilde{A}f)(\rho) = A(\rho)f(\rho)$ *for all* $f \in L^p(X,\mu,D)$ *and almost all* $\rho \in X$.

Proof. Let $f \in L^p(X,\mu,D)$ be a simple function, that is, $f = \sum_{i=1}^n \chi_{A_i} f_i$ with measurable $A_i \subset X$ and $f_i \in D$, $i = 1,2,\dots,n$, where χ_{A_i} denotes the characteristic function of A_i. Then

$$A(\rho)f(\rho) = \sum_{i=1}^n \chi_{A_i}(\rho)A(\rho)f_i,$$

and hence $\rho \mapsto A(\rho)f(\rho)$ is measurable by assumption (b). Further, by (2.1),

$$\int_X \|A(\rho)f(\rho)\|^p\,d\mu(\rho) \le \int_X \|A(\rho)\|_G^p\|f(\rho)\|_G^p\,d\mu(\rho) \le c_G^p \int_X \|f(\rho)\|_G^p\,d\mu(\rho).$$

Hence there is a unique bounded linear operator defined on the subset of simple functions of $L^p(X,\mu,D)$ with the desired property. Since this subset is dense in $L^p(X,\mu,D)$, see [6, Corollary III.3.8], the proof is complete. □

Theorem 2.3. *Let assumptions* (a) *and* (b) *be satisfied. Then the operator* $\mathfrak{A}$ *in* $\mathfrak{H} = L^p(X,\mu,E)$ *defined on* $\mathcal{D}(\mathfrak{A}) := L^p(X,\mu,D)$ *by* $\mathfrak{A}f = \widetilde{A}f$, *i.e.,*

$$(\mathfrak{A}f)(\rho) = A(\rho)f(\rho) \tag{2.3}$$

for $f \in L^p(X,\mu,D)$ *and almost all* $\rho \in X$, *is closed.*

Proof. Let $(\mathfrak{f}_n)_1^\infty$ be a sequence in $L^p(X,\mu,D)$ such that $\mathfrak{f}_n \to \mathfrak{f}$ and $\mathfrak{A}\mathfrak{f}_n \to \mathfrak{g}$ in $L^p(X,\mu,E)$ for some $\mathfrak{f},\mathfrak{g} \in L^p(X,\mu,E)$. As every L^p-convergent sequence contains a subsequence converging almost everywhere, see [6, Theorem III.3.6 and Corollary III.6.13], we may assume that $\mathfrak{f}_n(\rho) \to \mathfrak{f}(\rho)$, $(\mathfrak{A}\mathfrak{f}_n)(\rho) \to \mathfrak{g}(\rho)$ for almost all $\rho \in X$. Hence, since $(\mathfrak{A}\mathfrak{f}_n)(\rho) = A(\rho)\mathfrak{f}_n(\rho)$ by definition of $\mathfrak{A}$ and since $A(\rho)$ is closed, it follows that $\mathfrak{f}(\rho) \in D$ and $A(\rho)\mathfrak{f}(\rho) = \mathfrak{g}(\rho)$ for almost all $\rho \in X$.

To finish the proof we have to show that $\mathfrak{f} \in L^p(X,\mu,D)$, for which it suffices to prove that $(\mathfrak{f}_n)_1^\infty$ is a Cauchy sequence there. In view of Proposition 2.1 we conclude that

$$\begin{aligned}\|\mathfrak{f}_n - \mathfrak{f}_m\|^p_{L^p(X,\mu,D)} &= \int_X \|\mathfrak{f}_n(\rho) - \mathfrak{f}_m(\rho)\|^p_G \, d\mu(\rho) \\ &\le m_G^{-p} \int_X \|\mathfrak{f}_n(\rho) - \mathfrak{f}_m(\rho)\|^p_{A(\rho)} \, d\mu(\rho) \\ &= m_G^{-p} \int_X \left(\|\mathfrak{f}_n(\rho) - \mathfrak{f}_m(\rho)\|^p + \|A(\rho)(\mathfrak{f}_n(\rho) - \mathfrak{f}_m(\rho))\|^p\right) d\mu(\rho) \\ &= m_G^{-p} \left[\|\mathfrak{f}_n - \mathfrak{f}_m\|^p_{\mathfrak{H}} + \|\mathfrak{A}\mathfrak{f}_n - \mathfrak{A}\mathfrak{f}_m\|^p_{\mathfrak{H}}\right] \to 0\end{aligned}$$

as n, m tend to ∞, which completes the proof. □

Because of (2.3) we call $\mathfrak{A}$ the multiplication operator associated with the operator family A. The simplest examples of multiplication operators associated with a family of operators are operators of multiplication by scalar functions or, more generally, by matrix functions. But multiplication operators may also arise from differential operators the highest derivatives of which do not contain derivatives in all directions, see Section 6.

3. The spectrum of the multiplication operator

In the following for an operator T we denote its spectrum by $\sigma(T)$, its point spectrum, i.e., the set of its eigenvalues, by $\sigma_p(T)$, and its essential spectrum, i.e., the set of all points $\lambda \in \sigma(T)$ where $T - \lambda$ is not a Fredholm operator, by $\sigma_{\text{ess}}(T)$.

Theorem 3.1. *Let assumptions* (a) *and* (b) *be satisfied. Then the spectra of the operator* $\mathfrak{A}$ *and the operator family* A *are related as follows:*

$$\sigma(\mathfrak{A}) = \bigcup_{\rho \in Y} \sigma(A(\rho)).$$

Proof. Let $\lambda \in \bigcup_{\rho \in Y} \sigma(A(\rho))$. Choose $\rho_0 \in Y$ such that $\lambda \in \sigma(A(\rho_0))$. In view of assumption (b) we can choose a sequence of open neighborhoods Y_n of ρ_0 in Y such that $\|A(\rho) - A(\rho_0)\|_G < \frac{1}{n}$ for all $\rho \in Y_n$. For $\lambda \in \sigma(A(\rho_0))$ there are two cases: either $A(\rho_0) - \lambda$ has a closed range which is a proper subspace of E, or there is a sequence $(f_n)_1^\infty$ in D with $\|f_n\| = 1$ and $\|(A(\rho_0) - \lambda) f_n\| < \frac{1}{n}$. We first consider the second case. Since X is dense in Y, $X_n := Y_n \cap X$ is a nonempty open subset of X. Since μ is a Radon measure on the locally compact space X with support X, we can find a measurable subset M_n of X_n such that $\mu(M_n)$ is a finite positive number. Let $\alpha_n := (\mu(M_n))^{-\frac{1}{p}}$ and set

$$\mathfrak{f}_n(\rho) := \alpha_n \chi_{M_n}(\rho) \frac{f_n}{\|f_n\|_G}, \qquad \rho \in X.$$

Obviously, $\mathfrak{f}_n \in L^p(X, \mu, D)$ and $\|\mathfrak{f}_n\|_{L^p(X,\mu,D)} = 1$.

Furthermore,

$$\begin{aligned}\|(\mathfrak{A}-\lambda)\mathfrak{f}_n\|_{\mathfrak{H}}^p &= \alpha_n^p \int_{M_n} \Big\|(A(\rho)-\lambda)\frac{f_n}{\|f_n\|_G}\Big\|^p \, d\mu(\rho) \\ &\le \alpha_n^p\,\mu(M_n) \sup_{\rho\in M_n} \Big(\|A(\rho)-A(\rho_0)\|_G + \frac{1}{\|f_n\|_G}\|(A(\rho_0)-\lambda)f_n\|\Big)^p < \frac{2^p}{n^p}\end{aligned}$$

since $\|f_n\|_G \ge \|f_n\| = 1$, which proves $\lambda \in \sigma(\mathfrak{A})$.

Now assume that $A(\rho_0)-\lambda$ has a closed range which is a proper subspace of E. Then there is $h \in E^* \setminus \{0\}$ such that $\langle (A(\rho_0)-\lambda)f, h\rangle = 0$ for all $f \in D$. Suppose that $\lambda \in \rho(\mathfrak{A})$. Choose M_n and α_n as above, let $\beta_n := (\mu(M_n))^{-\frac{1}{p'}}$, and set

$$\mathfrak{h}_n(\rho) := \beta_n \chi_{M_n}(\rho) h, \quad \rho \in X.$$

Then $\mathfrak{h}_n \in L^{p'}(X,\mu,E^*)$ and $\|\mathfrak{h}_n\|_{L^{p'}(X,\mu,E^*)} = \|h\|$. Choose $g \in E$ such that $\langle g, h\rangle = 1$ and set

$$\mathfrak{g}_n(\rho) := \alpha_n \chi_{M_n}(\rho) g, \quad \rho \in X.$$

Then $\mathfrak{g}_n \in L^p(X,\mu,E)$, $\|\mathfrak{g}_n\|_{L^p(X,\mu,E)} = \|g\|$, and $\mathfrak{f}_n := (\mathfrak{A}-\lambda)^{-1}\mathfrak{g}_n \in L^p(X,\mu,D)$ because $\mathcal{D}(\mathfrak{A}) = L^p(X,\mu,D)$ and the closed graph theorem imply that $(\mathfrak{A}-\lambda)^{-1} \in \mathcal{B}(\mathfrak{H}, L^p(X,\mu,D))$. Thus we would obtain

$$\begin{aligned}1 &= |\langle g,h\rangle| \\ &= \beta_n \int_{M_n} |\langle \alpha_n g, h\rangle| \, d\mu(\rho) \\ &= \beta_n \int_{M_n} |\langle (A(\rho)-\lambda)\mathfrak{f}_n(\rho), h\rangle| \, d\mu(\rho) \\ &= \beta_n \int_{M_n} |\langle (A(\rho)-A(\rho_0))\mathfrak{f}_n(\rho), h\rangle| \, d\mu(\rho) \\ &= \int_{M_n} |\langle (A(\rho)-A(\rho_0))\mathfrak{f}_n(\rho), \mathfrak{h}_n(\rho)\rangle| \, d\mu(\rho) \\ &\le \int_{M_n} \|A(\rho)-A(\rho_0)\|_G \, \|\mathfrak{f}_n(\rho)\|_G \, \|\mathfrak{h}_n(\rho)\| \, d\mu(\rho) \\ &\le \frac{1}{n} \|\mathfrak{f}_n\|_{L^p(X,\mu,D)} \, \|\mathfrak{h}_n\|_{L^{p'}(X,\mu,E^*)} \\ &= \frac{1}{n} \|\mathfrak{f}_n\|_{L^p(X,\mu,D)} \, \|h\|,\end{aligned}$$

and hence $\|\mathfrak{f}_n\|_{L^p(X,\mu,D)} = \|(\mathfrak{A}-\lambda)^{-1}\mathfrak{g}_n\|_{L^p(X,\mu,D)} \to \infty$ as $n \to \infty$. This contradiction shows that also in this case $\lambda \in \sigma(\mathfrak{A})$.

Conversely, let $\lambda \in \sigma(\mathfrak{A})$. If λ is an eigenvalue of $\mathfrak{A}$, then there exists a non-zero $f \in \mathcal{D}(\mathfrak{A}) = L^p(X,\mu,D)$ such that $\mathfrak{A}f = \lambda f$. Hence, by definition of $\mathfrak{A}$, $A(\rho)f(\rho) = \lambda f(\rho)$ for almost all $\rho \in X$. Since $f \ne 0$, $f(\rho) \ne 0$ for all ρ in some set of positive measure. Hence there is $\rho_0 \in X$ such that $f(\rho_0) \ne 0$ and

$A(\rho_0)f(\rho_0) = \lambda f(\rho_0)$. This proves

$$\sigma_p(\mathfrak{A}) \subset \bigcup_{\rho \in X} \sigma_p(A(\rho)). \tag{3.1}$$

Now assume that $\lambda \in \sigma(\mathfrak{A})$ is not an eigenvalue of $\mathfrak{A}$. Then $\mathfrak{A} - \lambda$ is injective, but not surjective, and we can find an element $\mathfrak{g} \in \mathfrak{H}$ such that $(\mathfrak{A} - \lambda)\mathfrak{f} \neq \mathfrak{g}$ for all $\mathfrak{f} \in L^p(X, \mu, D)$. Assume $\lambda \notin \bigcup_{\rho \in Y} \sigma(A(\rho))$. For (almost all) $\rho \in X$ we define

$$\mathfrak{h}(\rho) := (A(\rho) - \lambda)^{-1}\mathfrak{g}(\rho).$$

From assumption (b) and the continuity of the inversion, see [8, Theorem IV.1.16], it follows that the mapping $\rho \mapsto (A(\rho) - \lambda)^{-1}$ from Y into $\mathcal{B}(E, D)$ is continuous. Hence $\mathfrak{h} : X \to D$ is measurable and

$$\left(\int_X \|\mathfrak{h}(\rho)\|_G^p \, d\mu(\rho) \right)^{\frac{1}{p}} \leq \sup_{\rho \in Y} \|(A(\rho) - \lambda)^{-1}\|_G \, \|\mathfrak{g}\|_{\mathfrak{H}} < \infty \tag{3.2}$$

since Y is compact. Thus, by (3.2), $\mathfrak{h} \in L^p(X, \mu, D) = \mathcal{D}(\mathfrak{A})$ and $(\mathfrak{A} - \lambda)\mathfrak{h} = \mathfrak{g}$, a contradiction. □

Remark 3.2. i) It is a remarkable fact that the spectrum of $\mathfrak{A}$ is independent of the chosen measure μ as long as $\operatorname{supp} \mu = X$.
ii) Also, the spectrum of $\mathfrak{A}$ is independent of p. This is a property which often holds for differential operators in L^p spaces.
iii) The assumption that $\operatorname{supp} \mu = X$ is not essential in that one can replace X with $\operatorname{supp} \mu$ and Y with the closure of $\operatorname{supp} \mu$ in Y. Assumptions (a) and (b) clearly remain true for these smaller sets.

Example. Let $n \in \mathbb{N}$, $-\infty < a < b < \infty$, let $A \in M_n(C[a, b])$ be an $n \times n$ matrix the entries of which are continuous functions on $[a, b]$, and consider the Lebesgue measure λ on $[a, b]$. Then the family $A : [a, b] \to M_n(\mathbb{C})$ of matrices satisfies conditions (a) and (b), and the multiplication operator $\mathfrak{A}$ acting in the space $L^p([a, b], \lambda, \mathbb{C}^n) = L^p([a, b])^n$ defined in Theorem 2.3 by

$$(\mathfrak{A}f)(x) = A(x)f(x)$$

for $f \in L^p([a, b])^n$ and almost all $x \in [a, b]$ is the usual operator of multiplication by the matrix function A. By Theorem 3.1 it follows that (compare [7]),

$$\sigma(\mathfrak{A}) = \bigcup_{x \in [a,b]} \sigma_p(A(x)) = \bigcup_{x \in [a,b]} \{\lambda \in \mathbb{C} : \det(A(x) - \lambda) = 0\}.$$

In particular, if $n = 1$ and $u \in C[a, b]$, we obtain the well-known result that the spectrum of the operator $\mathfrak{A}$ of multiplication by the function u is given by

$$\sigma(\mathfrak{A}) = \bigcup_{x \in [a,b]} u(x) = u([a, b]).$$

Proposition 3.3. *Let assumptions* (a) *and* (b) *be satisfied and suppose in addition that for all* $\lambda \notin \overline{\bigcup_{\rho \in X} \sigma(A(\rho))}$ *there exists a constant* $M_\lambda > 0$ *such that*

$$\sup_{\rho \in X} \|(A(\rho) - \lambda)^{-1}\|_G \le M_\lambda. \tag{3.3}$$

Then

$$\sigma(\mathfrak{A}) = \overline{\bigcup_{\rho \in X} \sigma(A(\rho))}.$$

Proof. The inclusion

$$\sigma(\mathfrak{A}) \supset \overline{\bigcup_{\rho \in X} \sigma(A(\rho))}$$

follows from Theorem 3.1 since $\sigma(\mathfrak{A})$ is closed. For the point spectrum the converse inclusion (even without the closure) has been proved in (3.1). The proof of the inclusion for the whole spectrum follows if we modify the last paragraph of the proof of Theorem 3.1 using assumption (3.3) in order to show in (3.2) that the function $\mathfrak{h}$ therein belongs to $L^p(X, \mu, D)$. □

In the next theorem we will see that assumption (3.3) is fulfilled if all operators $A(\rho)$, $\rho \in X$, are self-adjoint. However, the following example shows that, even if conditions (a) and (b) hold, it may happen that

$$\bigcup_{\rho \in Y} \sigma(A(\rho)) \not\subset \overline{\bigcup_{\rho \in X} \sigma(A(\rho))}.$$

Example. Consider the family of operators in $\ell^2(\mathbb{Z})$ given by

$$A(\rho) = A_0 + \rho A_1, \quad \rho \in (0, 1],$$

where A_0 is a modified left shift operator in $\ell^2(\mathbb{Z})$ defined by

$$A_0 x_0 = 0, \quad A_0 x_n = x_{n-1}, \quad n \in \mathbb{Z},\ n \neq 0,$$

and the operator A_1 in $\ell^2(\mathbb{Z})$ is given by

$$A_1 x_0 = x_{-1}, \quad A_1 x_n = 0, \quad n \in \mathbb{Z},\ n \neq 0.$$

It is not difficult to show, see [8, Chapter IV, Example 3.8], that

$$\sigma(A(\rho)) \subset \{\lambda \in \mathbb{C} : |\lambda| = 1\}, \quad \rho \in (0, 1],$$

but in the limit $\rho \to 0$ one has

$$\sigma(A_0) = \{\lambda \in \mathbb{C} : |\lambda| \le 1\}.$$

Theorem 3.4. *Let* E *be a Hilbert space and let assumptions* (a) *and* (b) *be satisfied. Assume that* A *is self-adjoint, i.e.,* $A(\rho)$ *is self-adjoint for all* $\rho \in X$. *Then*

$$\sigma(\mathfrak{A}) = \overline{\bigcup_{\rho \in X} \sigma(A(\rho))},$$

and $\mathfrak{A}$ *is self-adjoint if* $p = 2$.

Proof. We first note that we can take $p = 2$, see Remark 3.2 i). Also, the self-adjointness of each $A(\rho)$ implies that for any $\rho \in X$ and $\lambda \notin \overline{\bigcup_{\rho' \in X} \sigma(A(\rho'))}$ we have the estimate

$$\|(A(\rho) - \lambda)^{-1}\| = (\operatorname{dist}(\lambda, \sigma(A(\rho))))^{-1} \leq \Big(\operatorname{dist}\Big(\lambda, \bigcup_{\rho' \in X} \sigma(A(\rho'))\Big)\Big)^{-1} =: \eta < \infty,$$

where η is independent of ρ. Then, with the aid of (2.2), it follows that

$$\begin{aligned} m_G^2 \,\|(A(\rho) - \lambda)^{-1}x\|_G^2 &\leq \|(A(\rho) - \lambda)^{-1}x\|_{A(\rho)}^2 \\ &= \|(A(\rho) - \lambda)^{-1}x\|^2 + \|A(\rho)(A(\rho) - \lambda)^{-1}x\|^2 \\ &= \|(A(\rho) - \lambda)^{-1}x\|^2 + \|(I + \lambda(A(\rho) - \lambda)^{-1})x\|^2 \\ &\leq \left(2 + (1 + 2\,|\lambda|^2)\,\|(A(\rho) - \lambda)^{-1}\|^2\right)\|x\|^2 \leq \left(2 + (1 + 2\,|\lambda|^2)\,\eta^2\right)\|x\|^2 \end{aligned}$$

for $\rho \in X$ and $x \in E$ and hence condition (3.3) of Corollary 3.3 is satisfied, which proves the assertion about the spectrum of $\mathfrak{A}$.

It remains to be shown that $\mathfrak{A}$ is self-adjoint. For $\mathfrak{f}, \mathfrak{g} \in L^2(X, \mu, D)$ we have

$$(\mathfrak{A}\mathfrak{f}, \mathfrak{g}) = \int_X (A(\rho)\mathfrak{f}(\rho), \mathfrak{g}(\rho))\, d\mu(\rho) = \int_X (\mathfrak{f}(\rho), A(\rho)\mathfrak{g}(\rho))\, d\mu(\rho) = (\mathfrak{f}, \mathfrak{A}\mathfrak{g}),$$

and hence $\mathfrak{A}$ is symmetric. For all $\rho \in X$ we have $\sigma(A(\rho)) \subset \mathbb{R}$ since $A(\rho)$ is self-adjoint. From what has already been proved it follows that $\sigma(\mathfrak{A}) \subset \mathbb{R}$. Since the operator $\mathfrak{A}$ is symmetric and closed, this implies that $\mathfrak{A}$ is self-adjoint. □

4. The point spectrum of the multiplication operator

In this section we assume that E is a Hilbert space and that $p = 2$.

Theorem 4.1. *Let assumptions* (a) *and* (b) *be satisfied, where E is a separable Hilbert space, and suppose $p = 2$. Then $\lambda \in \sigma_p(\mathfrak{A})$ if and only if there exists a measurable subset M of X such that $\mu(M) > 0$ and*

$$\lambda \in \sigma_p(A(\rho)) \qquad \textit{for all} \;\; \rho \in M\,. \tag{4.1}$$

Proof. If $\lambda \in \sigma_p(\mathfrak{A})$, we have already seen in the proof of Theorem 3.1 that there exists a set E of positive measure such that (4.1) holds.

Conversely, let $\lambda \in \mathbb{C}$ be given for which a measurable set M with $\mu(M) > 0$ and (4.1) exists. Due to the fact that μ is a Radon measure, every measurable set of infinite measure contains a measurable subset of positive finite measure. Therefore we may assume $\mu(M) < \infty$. We want to show that $\lambda \in \sigma_p(\mathfrak{A})$.

The main part of the proof consists in showing that the orthogonal projection in D onto the null space $N(A(\rho) - \lambda)$ is measurable. To see this, for every $\rho \in X$ we consider the operator $A(\rho) - \lambda$ as a bounded operator from D to E and define its adjoint operator $(A(\rho) - \lambda)^* \in \mathcal{B}(E, D)$. Due to assumption (b) and the definition of the adjoint operator, for every fixed $f \in D$ and $g \in E$ the scalar-valued function

$$\rho \mapsto \left(f, (A(\rho) - \lambda)^* g\right)_G$$

is a continuous (and thus measurable) function on X. Here $(\cdot,\cdot)_G$ denotes the scalar product in the Hilbert space D induced by $\|\cdot\|_G$.

Now we fix an orthonormal basis $\{e_n\}_1^\infty$ of E. Then $\mathfrak{f}_n(\rho) := (A(\rho)-\lambda)^* e_n$ is a measurable function of ρ, and for every fixed $\rho \in X$ the set $\{\mathfrak{f}_n(\rho) : n \in \mathbb{N}\}$ is complete in the range $R(A(\rho)-\lambda)^* \subset D$ in the sense that the closure of all finite linear combinations of $\mathfrak{f}_n(\rho)$ contains this range. Applying the Gram–Schmidt orthogonalization to $\{\mathfrak{f}_n(\rho)\}_1^\infty$, it is possible to construct an orthonormal basis $\{\mathfrak{f}_n'(\rho)\}_1^\infty$ of $\overline{R(A(\rho)-\lambda)^*}$ (orthonormal with respect to $(\cdot,\cdot)_G$) which depends measurably on ρ, see [5, Chapter II, §1, Lemma 1]. This implies the measurability of the orthogonal projection in D onto $\overline{R(A(\rho)-\lambda)^*}$, i.e., the measurability of

$$\rho \mapsto P'(\rho)f := \sum_{n=1}^{\infty} \big(f, \mathfrak{f}_n'(\rho)\big)_G \mathfrak{f}_n'(\rho) \ \in D$$

for every fixed $f \in D$. Therefore, for the projection $P(\rho) := I - P'(\rho)$, $\rho \mapsto P(\rho)f$ is also measurable for all $f \in D$. But $P(\rho)$ is the orthogonal projection in D onto

$$\overline{R(A(\rho)-\lambda)^*}^{\perp_D} = N(A(\rho)-\lambda)\,.$$

As D endowed with the norm $\|\cdot\|_G$ is isomorphic to the graph of G, which is a closed subspace of the separable Hilbert space $E \times E$, D is separable, too. We fix an orthonormal basis $\{h_n\}_1^\infty$ of D and define $\mathfrak{f}(\rho) := P(\rho)h_{N(\rho)}$ for $\rho \in M$ where

$$N(\rho) := \min\{n \in \mathbb{N} : P(\rho)h_n \neq 0\}\,,$$

adapting an idea from [1], proof of Lemma 5.7. Note that for every $\rho \in M$ at least one $n \in \mathbb{N}$ exists with $P(\rho)h_n \neq 0$ because $N(A(\rho)-\lambda) \neq \{0\}$. For $\rho \in X \setminus M$, we define $\mathfrak{f}(\rho) := 0$.

As $\rho \mapsto P(\rho)h_n$ is measurable for every $n \in \mathbb{N}$, the same is true for $\rho \mapsto \mathfrak{f}(\rho)$. Moreover, we have

$$\|\mathfrak{f}\|_{L^2(X,\mu,D)} \le \mu(M)^{\frac{1}{2}} \sup_{\substack{\rho\in M\\ n\in\mathbb{N}}} \|P(\rho)h_n\|_G \le \mu(M)^{\frac{1}{2}} < \infty,$$

and therefore the function $\mathfrak{f}$ belongs to the domain of $\mathfrak{A}$. By definition of $\mathfrak{f}$, we have $(A(\rho)-\lambda)\mathfrak{f}(\rho) = 0$ for all $\rho \in X$ and $\mathfrak{f}(\rho) \neq 0$ for all $\rho \in M$ which shows that λ is an eigenvalue of $\mathfrak{A}$. □

If one takes a singleton $\{\rho\}$ for X, then obviously $\mathfrak{A}$ is isomorphic to $A(\rho)$, and $\sigma_{\mathrm{ess}}(A(\rho)) \neq \sigma(A(\rho))$ implies $\sigma_{\mathrm{ess}}(\mathfrak{A}) \neq \sigma(\mathfrak{A})$. Below we shall see that this latter property cannot happen if μ is non-atomic, i.e., if for every measurable subset M of X with $\mu(M) > 0$ there is a measurable subset $M_0 \subset M$ such that $0 < \mu(M_0) < \mu(M)$.

Theorem 4.2. *Let assumptions* (a) *and* (b) *be satisfied, where E is a Hilbert space, and suppose $p = 2$. Assume that μ is non-atomic. Then $\sigma_{\mathrm{ess}}(\mathfrak{A}) = \sigma(\mathfrak{A})$.*

Proof. First we show that every eigenvalue has an infinite-dimensional eigenspace. If $(\mathfrak{A}-\lambda)\mathfrak{f} = 0$, $\mathfrak{f} \in \mathfrak{H}$, $\mathfrak{f} \neq 0$, then choose $M_1 \supset M_2 \supset M_3 \supset \dots$, measurable subsets of X, such that $\mathfrak{f}(\rho) \neq 0$ for all $\rho \in M_1$ and $\mu(M_1) > \mu(M_2) > \mu(M_3) > \dots$. Then $(\mathfrak{A} - \lambda)\mathfrak{f}\chi_{M_n} = 0 \cdot \chi_{M_n} = 0$, i.e., $\{\mathfrak{f}\chi_{M_n} : n \in \mathbb{N}\}$ belongs to the null space of $\mathfrak{A} - \lambda$ and is obviously a set of linearly independent functions. Hence λ is an eigenvalue with infinite-dimensional eigenspace.

Now assume $\mathfrak{f} \in \mathfrak{H} \setminus \{0\}$ is orthogonal to the range of $\mathfrak{A} - \lambda$. Choosing sets $M_1, M_2, \dots$ as above we obtain for all $\mathfrak{g} \in \mathcal{D}(\mathfrak{A})$ that

$$((\mathfrak{A} - \lambda)\mathfrak{g}, \mathfrak{f}\chi_{M_n}) = ((\mathfrak{A} - \lambda)\mathfrak{g}\chi_{M_n}, \mathfrak{f}) = 0,$$

i.e., $\{\mathfrak{f}\chi_{M_n} : n \in \mathbb{N}\}$ is orthogonal to the range of $\mathfrak{A} - \lambda$, and thus the range of $\mathfrak{A} - \lambda$ cannot be a proper subspace of $\mathfrak{H}$ with a finite-dimensional complement. □

Remark 4.3. The statement of Theorem 4.2 remains true for all $1 < p < \infty$, all σ-finite measures μ and all Banach spaces E with the Radon-Nikodým property since then

$$(L^p(X, \mu, E))^* = L^{p'}(X, \mu, E^*). \tag{4.2}$$

Note that Hilbert spaces have the Radon-Nikodým property, see [4, Corollary IV.1.4]. In case μ is a finite measure, (4.2) can be found in [4, Theorem IV.1.1]; this easily extends to σ-finite measures.

5. The spectrum of $\mathfrak{A}$ under weakened assumptions

In the previous sections we assumed that A is defined on Y. But the operators $A(\rho)$ are naturally defined only on X, and even though continuous dependence on ρ might be a reasonable assumption, requiring that we have a continuous extension to Y could be too restrictive. However, the estimates (2.2) are essential to define $\mathfrak{A}$ and to show that $\mathfrak{A}$ is closed. Hence we shall consider the conditions

(a′) $D = \mathcal{D}(A(\rho))$, $\rho \in X$, is independent of ρ and a dense subspace of E,
(b′) $A : X \to \mathcal{C}_D(E)$ is continuous,
(c′) There are positive constants m_G and M_G such that

$$m_G\|x\|_G \leq \|x\|_{A(\rho)} \leq M_G\|x\|_G, \quad x \in D,\ \rho \in X.$$

It is clear from Proposition 2.1 that the assumptions (a′), (b′), (c′) are weaker than the assumptions (a), (b). It is now easy to see that

Remark 5.1. The statements of Proposition 2.1 and Theorem 2.3 remain true if the assumptions (a) and (b) are replaced by (a′), (b′), and (c′).

Revisiting the proofs of Theorem 3.1 and Proposition 3.3, we obtain

Corollary 5.2. *Let assumptions* (a′), (b′), *and* (c′) *be satisfied. Then*

$$\overline{\bigcup_{\rho \in X} \sigma(A(\rho))} \subset \sigma(\mathfrak{A}) \quad \text{and} \quad \sigma_p(\mathfrak{A}) \subset \bigcup_{\rho \in X} \sigma_p(A(\rho)).$$

If, additionally, for all $\lambda \notin \overline{\bigcup_{\rho\in X}\sigma(A(\rho))}$ there exists a constant $M_\lambda > 0$ such that

$$\sup_{\rho\in X}\|(A(\rho)-\lambda)^{-1}\|_G \le M_\lambda, \qquad \textit{then} \qquad \sigma(\mathfrak{A}) = \overline{\bigcup_{\rho\in X}\sigma(A(\rho))}.$$

Finally, we note

Remark 5.3. In Theorems 3.4, 4.1 and 4.2 assumptions (a) and (b) can be replaced by (a′), (b′), and (c′).

6. Examples

6.1. Let $I=[a,b]$ and X be intervals, $-\infty<a<b<\infty$, μ a Radon measure on X with support X, $1\le p<\infty$, $n\in\mathbb{N}$, and set $\mathfrak{H}=(L^p(X\times I,\mu))^n$ (actually, the measure should be the product measure $\mu\times$ Lebesgue measure, but the notation for Lebesgue measure will always be suppressed). We note that, by Fubini's theorem, $\mathfrak{H}=(L^p(X,\mu,L^p(I)))^n$. Let Y be the standard compactification of X in $\overline{\mathbb{R}}$. Consider a continuous function $B: Y\times I\to M_n(\mathbb{C})$, where $M_n(\mathbb{C})$ denotes the set of $n\times n$ matrices with entries in $\mathbb{C}$, and define the operator $\mathfrak{A}$ on $\mathfrak{H}$ by

$$\begin{aligned}\mathcal{D}(\mathfrak{A}) &= \big\{f\in(L^p(X\times I,\mu))^n : \partial_2 f\in(L^p(X\times I,\mu))^n,\ f(\cdot,a)=f(\cdot,b)\big\},\\ \mathfrak{A}f &= \partial_2 f - Bf, \quad f\in\mathcal{D}(\mathfrak{A}),\end{aligned}$$

where ∂_2 denotes differentiation with respect to the second variable.

For each $\rho\in Y$ define $A(\rho)$ by

$$\begin{aligned}\mathcal{D}(A(\rho)) &= \{g\in(W^{1,p}(I))^n : g(a)=g(b)\},\\ A(\rho)g &= g' - B(\rho,\cdot)g, \quad g\in\mathcal{D}(A(\rho)),\end{aligned}$$

where $W^{1,p}(I)$ is the usual Sobolev space.

Theorem 6.1. *The operator $\mathfrak{A}$ is closed, $\sigma(\mathfrak{A})=\sigma_{\mathrm{ess}}(\mathfrak{A})$, and*

$$\sigma(\mathfrak{A}) = \bigcup_{\rho\in Y}\sigma(A(\rho)) = \overline{\bigcup_{\rho\in X}\sigma(A(\rho))}. \tag{6.1}$$

Proof. Each of the operators $A(\rho)$ is a closed operator since it is a relatively compact perturbation of the system of differential operators with B replaced by zero. From the continuity of B on $Y\times I$ it follows immediately that A depends continuously on ρ as a mapping from Y into $\mathcal{C}_{W^{1,p}(I)}(L^p(I))$. Observe that $f\in\mathcal{D}(\mathfrak{A})$ if and only if $f,\partial_2 f\in(L^p(X\times I,\mu))^n$, i.e., $f\in L^p(X,\rho,W^{1,p}(I))$. Hence $\mathfrak{A}$ is the multiplication operator associated with the family $A(\rho)_{\rho\in X}$ and therefore closed by Theorem 2.3. The assertion on the essential spectrum and the left identity above follow from Theorems 3.1, 4.2, and Remark 4.3.

To prove the right identity we first note that neither side of this equation depends on p. Therefore it is sufficient to show that (c′) holds for $p=2$. To this end let $g \in (W^{1,2}(I))^n$. Then

$$\begin{aligned}\|g\|^2_{(W^{1,2}(I))^n} &= \|g\|^2_{(L^2(I))^n} + \|g'\|^2_{(L^2(I))^n} \\ &\le \|g\|^2_{(L^2(I))^n} + 2\|g' - B(\rho,\cdot)g\|^2_{(L^2(I))^n} + 2\|B(\rho,\cdot)g\|^2_{(L^2(I))^n} \\ &\le \|g\|^2_{(L^2(I))^n} + 2\|A(\rho)g\|^2_{(L^2(I))^n} + 2\sup_{\substack{\rho\in X\\ t\in I}} \|B(\rho,t)\|^2\,\|g\|^2_{(L^2(I))^n} \\ &\le 2(1+\sup_{\substack{\rho\in X\\ t\in I}}\|B(\rho,t)\|^2)\|g\|^2_{A(\rho)},\end{aligned}$$

where $\|B(\rho,t)\|$ denotes the matrix operator norm associated with the Euclidean norm on $\mathbb{C}^n$. Similarly,

$$\|g\|^2_{A(\rho)} \le 2(1+\sup_{\substack{\rho\in X\\ t\in I}}\|B(\rho,t)\|^2)\|g\|^2_{(W^{1,2}(I))^n}. \qquad \square$$

6.2. Let $I = [a,b]$ and X be intervals, $-\infty < a < b < \infty$, μ a Radon measure on X with support X, $n \in \mathbb{N}$, and set $\mathfrak{H} = L^2(X \times I, \mu)$. Consider continuous and bounded functions $a_j : X \times I \to \mathbb{C}$, $j = 0,\dots,n$, such that a_n is never zero and a_n^{-1} is bounded, and define the operator $\mathfrak{A}$ on $\mathfrak{H}$ by

$$\begin{aligned}\mathcal{D}(\mathfrak{A}) =& \{f \in L^2(X\times I,\mu) : \partial_2^{(j)} f \in L^2(X\times I,\mu),\ f^{(j-1)}(\cdot,a) = f^{(j-1)}(\cdot,b), \\ & \qquad j=1,\dots,n\},\\ \mathfrak{A}f =& \sum_{j=0}^{n} a_j \partial_2^{(j)} f, \quad f \in \mathcal{D}(\mathfrak{A}),\end{aligned}$$

where ∂_2 denotes differentiation with respect to the second variable.

For each $\rho \in X$ define $A(\rho)$ by

$$\begin{aligned}\mathcal{D}(A(\rho)) &= \{g \in (H^n(I))^n : g^{(j-1)}(a) = g^{(j-1)}(b), j = 1,\dots,n\},\\ A(\rho)g &= \sum_{j=0}^{n} a_j(\rho,\cdot)g^{(j)}, \quad g \in \mathcal{D}(A(\rho)).\end{aligned}$$

As above we obtain

Theorem 6.2. *The operator $\mathfrak{A}$ is closed, $\sigma(\mathfrak{A}) = \sigma_{\mathrm{ess}}(\mathfrak{A})$, and*

$$\sigma(\mathfrak{A}) = \overline{\bigcup_{\rho\in X} \sigma(A(\rho))}.$$

The picture, however, changes if a_n is allowed to have zeros. In this case, the domains of the operators $A(\rho)$ are no longer independent of ρ. This problem will be considered in a forthcoming paper.

References

[1] Azoff, E.A., Spectrum and direct integral, Trans. Amer. Math. Soc. 197 (1974), 211–223.

[2] Binding, P., Volkmer, H., A spectrum determined by eigencurves, Math. Nachr. 202 (1999), 1–15.

[3] Denk, R., Möller, M., Tretter, C., The spectrum of the multiplication operator associated with a family of operators, J. London Math. Soc. 65:2 (2002), 483–492.

[4] Diestel, J., Uhl, J.J., Vector measures, American Math. Soc., Providence, Rhode Island, 1977.

[5] Dixmier, J., Les algèbres d'opérateurs dans l'espace hilbertien (algèbres de von Neumann), Deuxième édition, Gauthier-Villars Éditeur, Paris, 1969.

[6] Dunford, N., Schwartz, J., Linear operators, Part I, Wiley

[7] Hardt, V., Wagenführer, E., Spectral properties of a multiplication operator, Math. Nachr. 178 (1996), 135–156.

[8] Kato, T., Perturbation theory for linear operators, 2nd ed., Springer-Verlag, Berlin, 1980.

[9] Reed, M., Simon, B., Methods of modern mathematical physics. IV. Analysis of operators, Academic Press, New York, 1978.

R. Denk
Fachbereich Mathematik und Statistik
Universität Konstanz
78457 Konstanz
Germany
e-mail: robert.denk@uni-konstanz.de

M. Möller
School of Mathematics
University of the Witwatersrand
2050 WITS
South Africa
e-mail: manfred@maths.wits.ac.za

C. Tretter
FB 3 – Mathematik
Universität Bremen
Bibliothekstr. 1
28359 Bremen
Germany
e-mail: ctretter@math.uni-bremen.de

Operator Theory:
Advances and Applications, Vol. 162, 117–133

A Factorization Model for the Generalized Friedrichs Extension in a Pontryagin Space

Vladimir Derkach, Seppo Hassi and Henk de Snoo

Abstract. An operator model for the generalized Friedrichs extension in the Pontryagin space setting is presented. The model is based on a factorization of the associated Weyl function (or Q-function) and it carries the information on the asymptotic behavior of the Weyl function at $z = \infty$.

Mathematics Subject Classification (2000). Primary 47A06, 47B25, 47B50; Secondary 47A10, 47A55.

Keywords. Generalized Nevanlinna function, Pontryagin space, symmetric operator, selfadjoint extension, generalized Friedrichs extension, operator model, factorization.

1. Introduction

Let $\mathbf{N}_\kappa$ be the class of generalized Nevanlinna functions, i.e., meromorphic functions on $\mathbb{C} \setminus \mathbb{R}$ with $Q(\bar{z}) = \overline{Q(z)}$ and such that the kernel

$$\mathsf{N}_Q(z,\lambda) = \frac{Q(z) - \overline{Q(\lambda)}}{z - \bar{\lambda}}, \quad z, \lambda \in \rho(Q), \quad z \neq \bar{\lambda},$$
$$\mathsf{N}_Q(z,\bar{z}) = Q'(z), \quad z \in \rho(Q),$$

has κ negative squares on the domain of holomorphy $\rho(Q)$ of $Q(z)$, see [14]. Every function $Q \in \mathbf{N}_\kappa$ can be interpreted as a Q-function of a symmetric operator in a Pontryagin space $\mathfrak{H}$, see [14], and in particular with $\kappa = 0$ in a Hilbert space, or as a Weyl function of an abstract boundary triplet, see [9, 1]. A function $Q \in \mathbf{N}_\kappa$ is said to belong to the subclass $\mathbf{N}_{\kappa,0}$ if it admits the following asymptotic property

$$Q(z) = \gamma + O(1/z), \quad \gamma \in \mathbb{R}, \quad \text{as } z \widehat{\rightarrow} \infty,$$

where $z \widehat{\rightarrow} \infty$ means that z tends to ∞ nontangentially ($0 < \varepsilon < \arg z < \pi - \varepsilon$), see [4]. The subclass $\mathbf{N}_{\kappa,0}$ can be characterized in terms of operator representations

The research was supported by the Academy of Finland (project 208055), the Research Institute for Technology at the University of Vaasa, and by the Dutch Organization for Scientific Research NWO (B 61-552).

as follows: $Q \in \mathbf{N}_{\kappa,0}$ if and only if Q admits the operator representation

$$Q(z) = \gamma + [(A - z)^{-1}\omega, \omega], \tag{1.1}$$

with a selfadjoint operator A in a Pontryagin space $\mathfrak{H}$ and $\omega \in \mathfrak{H}$, cf. [15, 4]. Moreover, the representation (1.1) can be taken minimal in the sense that ω is a cyclic vector for A:

$$\mathfrak{H} = \overline{\operatorname{span}}\, \{\, (A - z)^{-1}\omega : z \in \rho(A) \,\}. \tag{1.2}$$

In this case the negative index $\mathrm{sq}_-\mathfrak{H}$ of $\mathfrak{H}$ is equal to κ. The operator S which is defined as a restriction of A to the domain

$$\operatorname{dom} S = \{\, f \in \operatorname{dom} A : [f, \omega] = 0 \,\}, \tag{1.3}$$

is a symmetric operator in $\mathfrak{H}$ with defect numbers $(1, 1)$. The selfadjoint extensions of S which are operators are one-dimensional perturbations of A given by

$$A + \tau[\cdot, \omega]\omega, \quad \tau \in \mathbb{R}.$$

One further selfadjoint extension of S is given by

$$S_F = S \mathbin{\widehat{+}} (\{0\} \times \operatorname{span}\{\omega\}), \tag{1.4}$$

which admits a natural interpretation as the generalized Friedrichs extension of S; see [4] and the references therein. The selfadjoint extension S_F in (1.4) is a linear relation with a nontrivial multivalued part $\operatorname{mul} S_F = \operatorname{span}\{\omega\}$. The vector ω can be positive, negative, or neutral. In the last case the root subspace

$$\mathsf{R}_\infty(S_F) = \{\, h \in \mathfrak{H} : \{0, h\} \in S_F^n \text{ for some } n \in \mathbb{N} \,\}$$

of S_F at ∞ can be larger than $\operatorname{mul} S_F$. If ω is neutral and the root subspace is nondegenerate (degenerate), then $z = \infty$ is called a regular (singular) critical eigenvalue of S_F. According to [5] the root subspace $\mathsf{R}_\infty(S_F)$ is nondegenerate of dimension $n + 1$ if and only if in the operator representation (1.1) ω satisfies the following conditions:

$$\omega \in \operatorname{dom} A^n, \quad [A^i\omega, \omega] = 0 \quad \text{for } i = 0, 1, \dots, n - 1, \quad [A^n\omega, \omega] \neq 0.$$

This condition can be equivalently stated by means of the function $Q_\infty = 1/(\gamma - Q)$, which is the Weyl function associated to S_F (and the Q-function of the pair $\{S, S_F\}$): the root subspace $\mathsf{R}_\infty(S_F)$ is nondegenerate of dimension $n + 1$ if and only if $Q_\infty(z) = P(z) + o(z)$ as $z \widehat{\to} \infty$ for some polynomial P of degree $n + 1$. This asymptotic behavior of Q_∞ at ∞ is also equivalent to the following asymptotic expansion of Q

$$Q(z) = \gamma - \sum_{j=n}^{2n} \frac{s_j}{z^{j+1}} + o\left(\frac{1}{z^{2n+1}}\right), \quad s_n \neq 0, \quad z \widehat{\to} \infty, \tag{1.5}$$

where $s_n, \dots, s_{2n} \in \mathbb{R}$, see [5, Theorem 5.2]. In the terminology of [5] the condition (1.5) means that $Q \in \mathbf{N}_{\kappa,-2n}$, where n is the least integer such that $s_n \neq 0$. In general the asymptotic behavior of the Weyl functions Q and $Q_\infty = 1/(\gamma - Q)$

at ∞ is governed by the properties of the root subspace $\mathsf{R}_\infty(S_F)$; see [7] for the case that $\kappa = 1$. In the present paper an operator model is presented for the function $Q_\infty = 1/(\gamma - Q)$. This model leads to a full characterization of asymptotic expansions for the function Q (cf. (1.5)) which will appear elsewhere, see [8].

An operator model for a selfadjoint operator in a Pontryagin space was constructed in [13]. Another operator model for generalized Nevanlinna functions was given in [2] using a recent factorization result which states that every function $Q_\infty \in \mathbf{N}_\kappa$ admits an essentially unique representation

$$Q_\infty(z) = r^\sharp(z)Q_0(z)r(z), \quad Q_0 \in \mathbf{N}_0, \tag{1.6}$$

where $r = \widetilde{q}/\widetilde{p}$ is a rational function with relatively prime polynomials $\widetilde{p}$ and $\widetilde{q}$; see [10], cf. also [4]. If $\deg \widetilde{q} > \deg \widetilde{p}$ such a model necessarily involves a selfadjoint relation in a Pontryagin space, which is not an operator. When $r = q$ is a (matrix) polynomial a similar, but simpler operator model for factorized (matrix) Nevanlinna functions of the form (1.6) was constructed in [6]; however, in this case $\kappa = \mathrm{sq\_}\mathsf{R}_\infty(S_F)$. In the present paper this model is adapted to the case where Q_∞ is factorized as follows

$$Q_\infty(z) = q(z)q^\sharp(z)Q_0(z), \quad Q_0 \in \mathbf{N}_{\kappa'} \text{ for some } \kappa' \in \mathbb{N}. \tag{1.7}$$

Hence, in this factorization the function Q_0 is allowed to be a generalized Nevanlinna function, too. Such models typically arise when S is nondensely defined, i.e., S is of the form (1.3) with a vector ω which is either negative or neutral. This model is analyzed in detail in the case where q is a "proper" divisor of the numerator $\widetilde{q}$ of r in the canonical factorization (1.6) such that $\deg q = \deg \widetilde{q} - \deg \widetilde{p}$. In this case the underlying model operator S either is a simple symmetric operator in a Pontryagin space, or differs from it by a finite rank selfadjoint operator in a Hilbert space.

Operator models which are based on factorizations of the form (1.7) with proper divisors can be used in particular for separating in the operator model of Q_∞ the part which corresponds to the generalized pole of Q_∞ of nonpositive type at $z = \infty$ (if $\kappa_\infty(Q_\infty) > 0$), or the part which corresponds to the generalized zero of nonpositive type of a generalized Nevanlinna function Q at $z = \infty$ (if $\pi_\infty(Q) > 0$). The case where $z = \infty$ is a generalized pole of nonpositive type is associated with the generalized Friedrichs extension S_F of S in (1.4) with $[\omega, \omega] \leq 0$, and the case where $z = \infty$ is a generalized zero of nonpositive type is associated with the selfadjoint extension A of S whose Weyl function (Q-function) is of the form $Q(z) = [(A - z)^{-1}\omega, \omega]$ with $[\omega, \omega] \leq 0$.

The organization of the sections is as follows. In Section 2 some preparatory results are given involving boundary triplets and their Weyl functions in a Pontryagin space. Moreover, some basic notions related to the canonical factorization of generalized Nevanlinna functions are recalled. In Section 3 factorizations of generalized Nevanlinna functions of the form (1.7) are treated and in particular an operator model for such functions is given. The results motivate the notion of proper factorization for functions of the form (1.7) which is introduced in Section 4. In this section also the corresponding minimal factorization model is produced.

2. Preliminaries

2.1. Boundary triplets and Weyl functions

Let S be a closed symmetric relation in a Pontryagin space $\mathfrak{H}$ with defect numbers (n,n) and let S^* be the adjoint of S. A triplet $\Pi = \{\mathbb{C}^n, \Gamma_0, \Gamma_1\}$ is said to be a boundary triplet for S^*, if the following two conditions are satisfied:

(i) the mapping $\Gamma : \widehat{f} \to \{\Gamma_0 \widehat{f}, \Gamma_1 \widehat{f}\}$ from S^* to $\mathbb{C}^n \oplus \mathbb{C}^n$ is surjective;
(ii) the abstract Green's identity

$$[f', g] - [f, g'] = (\Gamma_1 \widehat{f}, \Gamma_0 \widehat{g}) - (\Gamma_0 \widehat{f}, \Gamma_1 \widehat{g}) \tag{2.1}$$

holds for all $\widehat{f} = \{f, f'\}$, $\widehat{g} = \{g, g'\} \in S^*$.

It is easily seen that $A_0 = \ker \Gamma_0$ and $A_1 = \ker \Gamma_1$ are selfadjoint extensions of S. Associated to every boundary triplet there are the Weyl function Q and the γ-field γ defined by

$$\gamma(z) = p_1 \left(\Gamma_0 \upharpoonright \widehat{\mathfrak{N}}_z\right)^{-1}, \quad Q(z)\Gamma_0 \widehat{f}_z = \Gamma_1 \widehat{f}_z, \quad z \in \rho(A_0),$$

where p_1 is the orthogonal projection onto the first component of $\mathbb{C}^n \oplus \mathbb{C}^n$, $\widehat{f}_z := \{f_z, z f_z\} \in \widehat{\mathfrak{N}}_z$, and $\mathfrak{N}_z = \ker(S^* - z)$ denotes the defect subspace of S at $z \in \mathbb{C}$. The vector function γ is holomorphic on $\rho(A_0)$ and satisfies the identity

$$\gamma(z) = (I + (z - z_0)(A_0 - z)^{-1})\gamma(z_0). \tag{2.2}$$

It follows from (2.1) and (2.2) that the Weyl function Q satisfies the identity

$$\frac{Q(z) - Q(z_0)^*}{z - \bar{z}_0} = \gamma(z_0)^* \gamma(z). \tag{2.3}$$

This means that Q is a Q-function of the pair $\{S, A_0\}$, see [14]. If S is simple, so that $\mathfrak{H} = \overline{\text{span}}\,\{\mathfrak{N}_z : z \in \rho(A_0)\}$, then the Weyl function Q belongs to the class $\mathbf{N}_\kappa$, otherwise $Q \in \mathbf{N}_{\kappa'}$ with $\kappa' \leq \kappa$.

In the case where S is given by (1.3) one can define a boundary triplet for S^* via the following result.

Proposition 2.1. (cf. [4]) *Let A be a selfadjoint operator in a Pontryagin space $\mathfrak{H}$ and let the restriction S of A be defined by (1.3) with $\omega \in \mathfrak{H}$. Then the adjoint S^* of S in $\mathfrak{H}$ is of the form*

$$S^* = \{\{f, Af + c\omega\} : f \in \operatorname{dom} A,\ c \in \mathbb{C}\} \tag{2.4}$$

and a boundary triplet $\Pi^\infty = \{\mathbb{C}, \Gamma_0^\infty, \Gamma_1^\infty\}$ for S^* is determined by

$$\Gamma_0^\infty \widehat{f} = [f, \omega], \quad \Gamma_1^\infty \widehat{f} = c, \quad \widehat{f} = \{f, Af + c\omega\} \in S^*. \tag{2.5}$$

The corresponding γ-field γ_∞ and the Weyl function Q_∞ are given by

$$\gamma_\infty(z) = \frac{(A - z)^{-1}\omega}{[(A - z)^{-1}\omega, \omega]}, \quad Q_\infty(z) = -\frac{1}{[(A - z)^{-1}\omega, \omega]}, \quad z \in \rho(A_0). \tag{2.6}$$

The function Q_∞ is the Q-function of S and its generalized Friedrichs extension $S_F = \ker \Gamma_0^\infty$ in (1.4). Likewise, the function $Q = -1/Q_\infty$ is the Q-function of S and $A = \ker \Gamma_1^\infty$. Observe, that if $\Pi_1 = \{\,\mathbb{C}, \Gamma_0^{(1)}, \Gamma_1^{(1)}\,\}$ and $\Pi_2 = \{\,\mathbb{C}, \Gamma_0^{(2)}, \Gamma_1^{(2)}\,\}$ are two boundary triplets for S^* such that $\ker \Gamma_0^{(1)} = \ker \Gamma_0^{(2)}$, then

$$\Gamma_0^{(2)} = \frac{1}{k}\Gamma_0^{(1)}, \quad \Gamma_1^{(2)} = b\Gamma_0^{(1)} + \bar{k}\Gamma_1^{(1)} \tag{2.7}$$

and the corresponding γ-fields and Weyl functions are connected by

$$\gamma_2(z) = k\gamma_1(z), \quad Q_2(z) = \gamma + |k|^2 Q_1(z), \tag{2.8}$$

where $\gamma = bk \in \mathbb{R}$ and $k \neq 0$ $(b, k \in \mathbb{C})$; see [3, Proposition 3.13].

The following statement is contained in [1, Theorem 3.1], but the proof given below is simpler, and this makes the presentation also self-contained.

Lemma 2.2. *Let S be a closed symmetric operator in the Pontryagin space $\mathfrak{H}$ with defect numbers* $(1,1)$, *let* $\Pi = \{\,\mathbb{C}, \Gamma_0, \Gamma_1\,\}$ *be a boundary triplet for* S^*, *and let Q be the corresponding Weyl function. Then the multivalued part* $\operatorname{mul} A_0$ *of* $A_0 = \ker \Gamma_0$ *is trivial if and only if*

$$\lim_{z \widehat{\to} \infty} \frac{Q(z)}{z} = 0. \tag{2.9}$$

The subspace $\operatorname{mul} A_0$ *is nondegenerate if and only if the limit in* (2.9) *is finite. In this case there is a vector* $g \in \operatorname{mul} A_0$ *such that*

$$\lim_{z \widehat{\to} \infty} \gamma(z) = g, \quad \lim_{z \widehat{\to} \infty} \frac{Q(z)}{z} = \Gamma_1\{0, g\} = [g, g]. \tag{2.10}$$

Proof. The first statement is well known (see [14], [1]); it is immediate from the equality

$$\frac{Q(z) - \overline{Q(z_0)}}{z - \bar{z}_0} = [(I + (z - z_0)(A_0 - z)^{-1})\gamma(z_0), \gamma(z_0)].$$

Assume that $\operatorname{mul} A_0 = \operatorname{span}\{\omega\}$ is nontrivial. Then $A_0 = S \,\widehat{+}\, (\{0\} \times \operatorname{span}\{\omega\})$ and the operators S and $A = A_1$ are connected via (1.3). Here $\operatorname{mul} A = \operatorname{span}\{\omega\}$ is nondegenerate if and only if $[\omega, \omega] \neq 0$. Consider the boundary triplet for S^* given by (2.5) in Proposition 2.1. It follows from (2.6) that

$$\lim_{z \widehat{\to} \infty} \gamma_\infty(z) = \frac{\omega}{[\omega, \omega]}, \quad \lim_{z \widehat{\to} \infty} \frac{Q_\infty(z)}{z} = \frac{1}{[\omega, \omega]}. \tag{2.11}$$

The equalities (2.10) are obtained by setting $g = \frac{\omega}{[\omega,\omega]}$. In particular, the equality $\Gamma_1^\infty\{0, g\} = [g, g]$ is implied by the equalities (2.5). Since every other boundary triplet Π with $\ker \Gamma_0 = S \,\widehat{+}\, (\{0\} \times \operatorname{span}\{\omega\})$ is connected to Π_∞ via (2.7) the corresponding properties remain true when Π_∞ is replaced by Π, see (2.7), (2.8). □

Lemma 2.3. *Let S be a closed symmetric relation in a Pontryagin space $\mathfrak{H}$ with defect numbers* $(1,1)$, *let* $\Pi = \{\,\mathbb{C}, \Gamma_0, \Gamma_1\,\}$ *be a boundary triplet for* S^*, *and let Q be the corresponding Weyl function. Moreover, let* $\alpha \in \mathbb{R}$ *and let* $\ker (S - \alpha)$ *be* *trivial. Then the linear relation* $A_0 - \alpha$ *has a trivial kernel if and only if*

$$\lim_{z \widehat{\to} \alpha} (z-\alpha)Q(z) = 0. \tag{2.12}$$

The subspace $\ker(A_0 - \alpha)$ *is nondegenerate if and only if the limit in* (2.12) *is finite. In this case there is a vector* $g_\alpha \in \ker(A_0-\alpha)$ *such that*

$$\lim_{z \widehat{\to} \alpha} (z-\alpha)\gamma(z) = g_\alpha, \quad \lim_{z \widehat{\to} \alpha} (z-\alpha)Q(z) = \Gamma_1\{g_\alpha, \alpha g_\alpha\} = -[g_\alpha, g_\alpha]. \tag{2.13}$$

Proof. Consider a new symmetric relation $\widetilde{S} := (S-\alpha)^{-1}$. Since $\ker(S-\alpha)=\{0\}$ the transform $\widetilde{S}$ is an operator. Clearly, $\widetilde{S}^* = (S^*-\alpha)^{-1}$, i.e.,

$$\widetilde{S}^* = \{\, \{f'-\alpha f, f\} : \{f,f'\} \in S^*\,\}.$$

Define a boundary triplet $\widetilde{\Pi} = \{\mathbb{C}, \widetilde{\Gamma}_0, \widetilde{\Gamma}_1\}$ for $\widetilde{S}^*$ by

$$\widetilde{\Gamma}_0\{f'-\alpha f, f\} = \Gamma_0\{f,f'\}, \quad \widetilde{\Gamma}_1\{f'-\alpha f, f\} = -\Gamma_1\{f,f'\}$$

and let $\widetilde{A}_j = \ker \widetilde{\Gamma}_j$, $j=1,2$. Then $\ker(A_0-\alpha) = \operatorname{mul}\widetilde{A}_0$. Moreover, the γ-field $\widetilde{\gamma}(\lambda)$ and the Weyl function $\widetilde{Q}(\lambda)$ corresponding to $\widetilde{\Pi}$ are given by

$$\widetilde{\gamma}(1/(z-\alpha)) = (z-\alpha)\gamma(z), \quad \widetilde{Q}(1/(z-\alpha)) = -Q(z). \tag{2.14}$$

If $\ker(A_0-\alpha)$ is nondegenerate Lemma 2.2 applied to $\widetilde{S}$ shows that there is a vector $g_\alpha \in \operatorname{mul}\widetilde{A}_0 = \ker(A_0-\alpha)$, such that

$$\lim_{z \widehat{\to} \alpha} (z-\alpha)\gamma(z) = \lim_{z \widehat{\to} \alpha} \widetilde{\gamma}(1/(z-\alpha)) = g_\alpha,$$

and

$$\lim_{z \widehat{\to} \alpha} (z-\alpha)Q(z) = -\lim_{z \widehat{\to} \alpha} \frac{\widetilde{Q}(\frac{1}{z-\alpha})}{\frac{1}{z-\alpha}} = -\widetilde{\Gamma}_1\{0, g_\alpha\} = \Gamma_1\{g_\alpha, \alpha g_\alpha\}. \tag{2.15}$$

The converse statements are obvious from (2.15) by Lemma 2.2. □

2.2. Canonical factorization of generalized Nevanlinna functions

According to Lemma 2.3 the eigenspace $\ker(A_0-\alpha)$, $\alpha \in \mathbb{R}$, is positive (negative) if and only if the second limit in (2.13) is negative (positive). This motivates the following definitions (cf. [17]).

A point $\alpha \in \mathbb{R}$ is called a generalized pole of nonpositive type of the function $Q \in \mathbf{N}_\kappa$ with multiplicity $\kappa_\alpha(Q)$ if

$$-\infty < \lim_{z \widehat{\to} \alpha} (z-\alpha)^{2\kappa_\alpha+1}Q(z) \le 0, \quad 0 < \lim_{z \widehat{\to} \alpha} (z-\alpha)^{2\kappa_\alpha-1}Q(z) \le \infty. \tag{2.16}$$

Similarly, the point ∞ is called a generalized pole of nonpositive type of Q with multiplicity $\kappa_\infty(Q)$ if

$$0 \le \lim_{z \widehat{\to} \infty} \frac{Q(z)}{z^{2\kappa_\infty+1}} < \infty, \quad -\infty \le \lim_{z \widehat{\to} \infty} \frac{Q(z)}{z^{2\kappa_\infty-1}} < 0. \tag{2.17}$$

A point $\beta \in \mathbb{R}$ is called a generalized zero of nonpositive type of the function $Q \in \mathbf{N}_\kappa$ if β is a generalized pole of nonpositive type of the function $-1/Q$.

The multiplicity $\pi_\beta(Q)$ of the generalized zero of nonpositive type β of Q can be characterized by the inequalities:

$$0 < \lim_{z \widehat{\to} \beta} \frac{Q(z)}{(z-\beta)^{2\pi_\beta+1}} \le \infty, \quad -\infty < \lim_{z \widehat{\to} \beta} \frac{Q(z)}{(z-\beta)^{2\pi_\beta-1}} \le 0. \tag{2.18}$$

Similarly, the point ∞ is called a generalized zero of nonpositive type of Q with multiplicity $\pi_\infty(Q)$ if

$$-\infty \le \lim_{z \widehat{\to} \infty} z^{2\pi_\infty+1} Q(z) < 0, \quad 0 \le \lim_{z \widehat{\to} \infty} z^{2\pi_\infty-1} Q(z) < \infty. \tag{2.19}$$

It was shown in [16] for $Q \in \mathbf{N}_\kappa$ that the total number of poles (zeros) in $\mathbb{C}_+$ and generalized poles (zeros) of nonpositive type in $\mathbb{R} \cup \{\infty\}$ is equal to κ. The generalized poles and zeros of nonpositive type of a generalized Nevanlinna function give rise to the following factorization result ([10], see also [4]).

Theorem 2.4. *Let $Q \in \mathbf{N}_\kappa$ and let $\alpha_1, \dots, \alpha_l$ ($\beta_1, \dots, \beta_m$) be all the generalized poles (zeros) of nonpositive type in $\mathbb{R}$ and the poles (zeros) in $\mathbb{C}_+$ with multiplicities $\kappa_1, \dots, \kappa_l$ ($\pi_1, \dots, \pi_m$). Then the function Q admits a canonical factorization of the form*

$$Q(z) = r(z) r^\sharp(z) Q_0(z), \tag{2.20}$$

where $Q_0 \in \mathbf{N}_0$ and $r = \widetilde{q}/\widetilde{p}$ with relatively prime polynomials

$$\widetilde{p}(z) = \prod_{j=1}^{l} (z - \alpha_j)^{\kappa_j}, \quad \widetilde{q}(z) = \prod_{j=1}^{m} (z - \beta_j)^{\pi_j},$$

of degree $\kappa - \kappa_\infty(Q)$ and $\kappa - \pi_\infty(Q)$, respectively.

Since $Q_0 \in \mathbf{N}_0$, it follows from (2.20) that the function $Q \in \mathbf{N}_\kappa$ admits the integral representation

$$Q(z) = r(z) r^\sharp(z) \left(a + bz + \int_{\mathbb{R}} \left(\frac{1}{t-z} - \frac{t}{1+t^2} \right) d\rho(t) \right), \quad r = \frac{\widetilde{q}}{\widetilde{p}}, \tag{2.21}$$

where $a \in \mathbb{R}$, $b \ge 0$, and $\rho(t)$ is a nondecreasing function satisfying the integrability condition

$$\int_{\mathbb{R}} \frac{d\rho(t)}{t^2+1} < \infty. \tag{2.22}$$

3. Operator models for factorized $\mathbf{N}_\kappa$-functions

Operator models for generalized Nevanlinna functions with the only generalized pole of nonpositive type at ∞ have been constructed in [5] and [12]. Such functions admit the canonical factorization of the form

$$Q_\infty(z) = q(z) q^\sharp(z) Q_0(z), \tag{3.1}$$

where $Q_0 \in \mathbf{N}_0$, $q^\sharp(z) = \overline{q(\bar{z})}$, and $q(z) = z^k + q_{k-1} z^{k-1} + \dots + q_0$ is a polynomial. The model in [5] was constructed as a coupling of a symmetric operator S_0 in a

Hilbert space $\mathfrak{H}_0$ and a symmetric operator S_q in a finite-dimensional Pontryagin space $\mathfrak{H}_q$ determined by the polynomial q via the 2×2-matrix function

$$\begin{pmatrix} 0 & q(z) \\ q^\sharp(z) & 0 \end{pmatrix}, \tag{3.2}$$

which belongs to the class $\mathbf{N}_\kappa$ with $\kappa = \deg q$. The basic idea in the coupling is to specify a symmetric extension S of the orthogonal sum $S_0 \oplus S_q$ in $\mathfrak{H}_0 \oplus \mathfrak{H}_q$, such that $Q_\infty = q(z)q^\sharp(z)Q_0(z)$ becomes a Weyl function of S. Using boundary triplets the symmetric extension S can be easily described by means of abstract boundary conditions on the adjoint $S_0^* \oplus S_q^*$. In [2] a similar procedure was used to constructed a model for an arbitrary scalar $\mathbf{N}_\kappa$-function Q, which relies on the canonical factorization $Q(z) = r(z)r^\sharp(z)Q_0(z)$ of Q in (2.20), cf. [10], [4]. The model from [2] has been recently analyzed with reproducing kernel space methods in [11].

In this section the model from [5] and [2] is adapted to the case where the function Q_0 is allowed to be a generalized Nevanlinna function, too. More precisely, let the function $Q \in \mathbf{N}_\kappa$ be factorized as follows

$$Q(z) = r(z)r^\sharp(z)Q_0(z), \quad Q_0 \in \mathbf{N}_{\kappa'}, \quad \kappa' \in \mathbb{N}, \tag{3.3}$$

where $r(z) = q(z)/p(z)$ and p, q are monic and relatively prime polynomials. The simplest case here occurs when $r\,(= q)$ is a monic polynomial. To construct a factorization model in this case the following notations will be needed. Let the $k\times k$ matrices B_q and C_q be defined by

$$B_q = \begin{pmatrix} q_1 & \dots & q_{k-1} & 1 \\ \vdots & & 1 & 0 \\ & & & \vdots \\ q_{k-1} & & & \\ 1 & 0 & \dots & 0 \end{pmatrix}, \quad C_q = \begin{pmatrix} 0 & 1 & \dots & 0 \\ \vdots & \ddots & \ddots & 0 \\ 0 & 0 & \dots & 1 \\ -q_0 & -q_1 & \dots & -q_{k-1} \end{pmatrix}, \tag{3.4}$$

and let the operators $\mathcal{B}$ and $\mathcal{C}$ in $\mathfrak{H}_q = \mathbb{C}^k \oplus \mathbb{C}^k$ be given by the block matrices

$$\mathcal{B} = \begin{pmatrix} 0 & B_q \\ B_{q^\sharp} & 0 \end{pmatrix}, \quad \mathcal{C} = \begin{pmatrix} C_{q^\sharp} & 0 \\ 0 & C_q \end{pmatrix}. \tag{3.5}$$

Moreover, let $\Lambda = (1, \lambda, \dots, \lambda^{k-1})$, $\lambda \in \mathbb{C}$. Then $\sigma(C_q) = \sigma(q)$ and $B_qC_q = C_q^\top B_q$, which implies that $\mathcal{C}$ is selfadjoint in the Pontryagin space $\mathfrak{H}_q := (\mathbb{C}^k \oplus \mathbb{C}^k, \langle \mathcal{B}\cdot, \cdot \rangle)$. The next result is an extension of [6, Theorem 4.2]; here a short direct proof is presented.

Theorem 3.1. *Let S_0 be a closed symmetric relation in the Pontryagin space $\mathfrak{H}_0$ with the boundary triplet $\Pi^0 = \{\mathcal{H}, \Gamma_0^0, \Gamma_1^0\}$ and the Weyl function Q_0 and let q be a monic polynomial of degree $k = \deg q \geq 1$. Then:*

(i) *the linear relation*

$$S = \left\{ \left\{ \begin{pmatrix} f_0 \\ f \\ \widetilde{f} \end{pmatrix}, \begin{pmatrix} f_0' \\ C_{q^\sharp} f \\ C_q \widetilde{f} + \Gamma_0^0 \widehat{f}_0 e_k \end{pmatrix} \right\} : \begin{array}{c} \widehat{f}_0 = \{f_0, f_0'\} \in S_0^*, \\ f_1 = \Gamma_1^0 \widehat{f}_0, \\ \widetilde{f}_1 = 0 \end{array} \right\} \tag{3.6}$$

is closed and symmetric in $\mathfrak{H}_0 \oplus \mathfrak{H}_q$ *and has defect numbers* $(1,1)$;

(ii) *the adjoint* S^* *of* S *is given by*

$$S^* = \left\{ \left\{ \begin{pmatrix} f_0 \\ f \\ \widetilde{f} \end{pmatrix}, \begin{pmatrix} f_0' \\ C_{q^\sharp} f + \widetilde{\varphi} e_k \\ C_q \widetilde{f} + \Gamma_0^0 \widehat{f}_0 e_k \end{pmatrix} \right\} : \begin{array}{c} \widehat{f}_0 = \{f_0, f_0'\} \in S_0^*, \\ f_1 = \Gamma_1^0 \widehat{f}_0, \\ \widetilde{\varphi} \in \mathbb{C} \end{array} \right\}; \tag{3.7}$$

(iii) *a boundary triplet* $\Pi = \{\mathcal{H}, \Gamma_0, \Gamma_1\}$ *for* S^* *is determined by*

$$\Gamma_0(\widehat{f}_0 \oplus \widehat{F}) = \widetilde{f}_1, \quad \Gamma_1(\widehat{f}_0 \oplus \widehat{F}) = \widetilde{\varphi}, \quad \widehat{f}_0 \oplus \widehat{F} \in S^*;$$

(iv) *the corresponding* γ*-field* γ_∞ *and the Weyl function* Q_∞ *are of the form*

$$\gamma_\infty(\lambda) = (q(\lambda)\gamma_0(\lambda), q(\lambda)Q_0(\lambda)\Lambda, \Lambda)^\top, \tag{3.8}$$

$$Q_\infty(\lambda) = q(\lambda)q^\sharp(\lambda)Q_0(\lambda). \tag{3.9}$$

Proof. (i) & (ii) Denote the linear relation on the right side of (3.7) by T. Clearly S and T in (3.6), (3.7) are closed as finite-dimensional extensions of S_0^*. Moreover, S is obtained from T as a 2-dimensional restriction. Let $\{F, F'\}, \{G, G'\} \in T$ with $F = \{f_0, f, \widetilde{f}\}$, $G = \{g_0, g, \widetilde{g}\}$ and $\widetilde{\varphi}, \widetilde{\psi} \in \mathbb{C}$ as in (3.7). Then a straightforward calculation shows that

$$[F', G] - [F, G'] = \widetilde{\varphi}\overline{\widetilde{g}_1} - \widetilde{f}_1\overline{\widetilde{\psi}}, \tag{3.10}$$

which implies that $T \subset S^*$ and that S is symmetric. Moreover, by means of the identity $C_{q^\sharp}\Lambda^\top = \lambda\Lambda^\top - q^\sharp(\lambda)e_k$ it is easy to see that

$$\{\gamma_\infty(\lambda), \lambda\gamma_\infty(\lambda)\} \in T \quad \text{with } \widetilde{\varphi} = q(\lambda)Q_0(\lambda)q^\sharp(\lambda), \tag{3.11}$$

$\lambda \in \rho(Q_0)$. Therefore, by a dimension argument $T = S^*$ and S has defect numbers $(1,1)$. Thus (i) and (ii) are proven.

(iii) This statement is now clear from (3.10).

(iv) According to (3.11), $\Gamma_0\{\gamma_\infty(\lambda), \lambda\gamma_\infty(\lambda)\} = 1$ and $\Gamma_1\{\gamma_\infty(\lambda), \lambda\gamma_\infty(\lambda)\} = q(\lambda)Q_0(\lambda)q^\sharp(\lambda)$, and this proves (iv). □

Remark 3.2. (i) The extension of the model that was constructed in [2] for the canonical factorization of generalized Nevanlinna functions (see (2.20)) to the case of factorizations of the form (3.3) with a rational function $r = q/p$ with some relatively prime polynomials p and q can be carried out in a completely analogous manner as was done in Theorem 3.1 above.

(ii) The formulas in Theorem 3.1 are consequences of the coupling method used in [2], [6] (cf. also [3]), where the starting point is the orthogonal sum of S_0 and S_q, where S_q is the restriction of $\mathcal{C}$ in (3.5) to the subspace

$$\operatorname{dom} S_q = \left\{ f \oplus \widetilde{f} \in \mathfrak{H}_q : f_1 = \widetilde{f}_1 = 0 \right\}$$

and has the defect numbers $(2,2)$ and the Weyl function (3.2). The orthogonal sum of the corresponding γ-fields and Weyl functions, in the case of Theorem 3.1 are given by

$$\gamma_M(\lambda) = \gamma_0(\lambda) \oplus \begin{pmatrix} \Lambda^\top & 0 \\ 0 & \Lambda^\top \end{pmatrix}, \quad M(\lambda) = Q_0(\lambda) \oplus \begin{pmatrix} 0 & q(\lambda) \\ q^\sharp(\lambda) & 0 \end{pmatrix}.$$

The boundary conditions in Theorem 3.1 implicitly express the transform of γ_M determined by the vector function $K(\lambda) = (q(\lambda), q(\lambda)Q_0(\lambda), 1)^\top$ via

$$\gamma_\infty(\lambda) = \gamma_M(\lambda)K(\lambda),$$

so that

$$\begin{aligned}(\lambda - \bar{\mu})[\gamma_\infty(\lambda), \gamma_\infty(\mu)] &= Q_\infty(\lambda) - Q_\infty(\mu)^* \\ &= K(\mu)^*(M(\lambda) - M(\mu)^*)K(\lambda),\end{aligned}$$

cf. [6, p. 15]. These formulas are typically needed in the construction of Q-functions.

In general the symmetric extension S of $S_0 \oplus S_q$ in Theorem 3.1 need not be simple, even if S_0 is a simple symmetric operator in $\mathfrak{H}_0$.

Example 1. Consider the function $Q_\infty(z) = -z$. Then for every simple symmetric operator $\widetilde{S}$ the model space associated to $Q_\infty(z) = -z$ is one-dimensional. If $Q_\infty(z) = q(z)q^\sharp(z)Q_0(z)$, where q is a polynomial of degree $\deg q \geq 1$, then $\dim \mathfrak{H}_q \geq 2$, $\dim \mathfrak{H}_0 \geq 1$, and S as constructed in Theorem 3.1 acts on the space $\mathfrak{H}_0 \oplus \mathfrak{H}_q$ with $\dim(\mathfrak{H}_0 \oplus \mathfrak{H}_q) \geq 3$, and therefore S cannot be simple. Here the canonical factorization $Q(z) = z^2 Q_0(z)$ with $Q_0(z) = -1/z$ gives rise to S which acts on a 3-dimensional space. Observe, that the Weyl function (or Q-function) associated to the pair $\{S_0, A_1^0\}$ is $-Q_0(z)^{-1} = z$, so that the selfadjoint extension A_1^0 of S_0 in $\mathfrak{H}_0$ is not an operator.

Lemma 3.3. *Let the linear relation S be given by (3.6), where S_0 is a symmetric operator in the Pontryagin space $\mathfrak{H}_0$, and let $A_i^0 = \ker \Gamma_i^0 (\supset S_0)$, $i = 0, 1$. Then:*

(i) $\operatorname{mul} S$ *is nontrivial if and only if* $\operatorname{mul} A_1^0$ *is nontrivial and in this case*

$$\operatorname{mul} S = \{ (g, 0, \Gamma_0^0 \widehat{g} e_k)^\top : \widehat{g} = \{0, g\} \in A_1^0 \}; \tag{3.12}$$

(ii) *if* $\operatorname{mul} S = \{0\}$ *then* $\kappa_\infty(Q_\infty) \geq k$;

(iii) *if* $\operatorname{mul} S$ *is nontrivial, then it is spanned by a positive vector if and only if* $\kappa_\infty(Q_\infty) = k$, *if* $\operatorname{mul} S$ *is negative, then* $\kappa_\infty(Q_\infty) = k - 1$, *and if* $\operatorname{mul} S$ *is neutral, then* $\kappa_\infty(Q_\infty) \leq k - 1$;

(iv) *if* $\operatorname{mul} A_0^0$ *is nontrivial, then it is spanned by a positive vector if and only if* $\kappa_\infty(Q_\infty) = k$, *if* $\operatorname{mul} S$ *is negative, then* $\kappa_\infty(Q_\infty) = k + 1$, *and if* $\operatorname{mul} S$ *is neutral, then* $\kappa_\infty(Q_\infty) \geq k + 1$.

Proof. (i) The description (3.12) is obtained from (3.6). Clearly, $\operatorname{mul} S = \{0\}$ if and only if $\operatorname{mul} A_1^0 = \{0\}$.

(ii) If $\operatorname{mul} S = \{0\}$ or equivalently $\operatorname{mul} A_1^0 = \{0\}$ then $\lim_{z \widehat{\to} \infty} zQ_0(z) = \infty$ by Lemma 2.2, so that $\pi_\infty(Q_0) = 0$ and consequently $\kappa_\infty(Q_\infty) \geq k$.

(iii) Assume that $\operatorname{mul} S$ is nontrivial. Then there is a vector $g \in \operatorname{mul} A_1^0$, $g \neq 0$, and the vector $G = (g, 0, \Gamma_0^0 \widehat{g} e_k)^\top$ spans $\operatorname{mul} S$. It follows from $[G, G] = [g, g]_{\mathfrak{H}_0}$ and Lemma 2.2 that the vector G is positive if and only if

$$0 < \lim_{z \widehat{\to} \infty} \frac{-1/Q_0(z)}{z} < \infty. \tag{3.13}$$

This implies that

$$-\infty < \lim_{z \widehat{\to} \infty} zQ_0(z) < 0, \quad \lim_{z \widehat{\to} \infty} \frac{Q_0(z)}{z} = 0, \tag{3.14}$$

and

$$-\infty < \lim_{z \widehat{\to} \infty} \frac{Q_\infty(z)}{z^{2k-1}} < 0, \quad \lim_{z \widehat{\to} \infty} \frac{Q_\infty(z)}{z^{2k+1}} = 0. \tag{3.15}$$

Thus, ∞ is a generalized pole of nonpositive type of Q_∞ of multiplicity k. If the vector G is negative, then by Lemma 2.2 the limit in (3.13) is also negative and one obtains

$$0 < \lim_{z \widehat{\to} \infty} \frac{Q_\infty(z)}{z^{2k-1}} < \infty,$$

which implies that $\kappa_\infty(Q_\infty) = k - 1$. Finally, if the vector g is neutral, then by Lemma 2.2 the limit in (3.13) cannot be finite and this leads to the estimate $\kappa_\infty(Q_\infty) < k$.

(iv) Assume that $\operatorname{mul} A_0^0$ is nontrivial and is spanned by the vector $g (\neq 0)$. If the vector g is positive then again by Lemma 2.2

$$0 < \lim_{z \widehat{\to} \infty} \frac{Q_0(z)}{z} < \infty, \tag{3.16}$$

and hence

$$0 < \lim_{z \widehat{\to} \infty} \frac{Q_\infty(z)}{z^{2k+1}} < \infty. \tag{3.17}$$

This implies that $\kappa_\infty(Q_\infty) = k$. If g is negative, then the limits in (3.16) and (3.17) are negative and $\kappa_\infty(Q_\infty) = k + 1$ or the limits in (3.16) and (3.17) are infinite and then $\kappa_\infty(Q_\infty) > k$. □

Lemma 3.4. *Let the linear relation S be given by (3.6), where S_0 is a symmetric operator in the Pontryagin space $\mathfrak{H}_0$ such that $\sigma_p(S_0) = \emptyset$. Moreover, let $A_i^0 = \ker \Gamma_i^0 (\supset S_0)$, $i = 0, 1$, and let k_α be the multiplicity of $\alpha \in \mathbb{C}$ as a zero of the polynomial q. Then:*

(i) *$\sigma_p(S) = \sigma_p(A_0^0) \cap \sigma(q^\sharp)$ and for $\alpha \in \sigma_p(A_0^0) \cap \sigma(q^\sharp)$ one has*

$$\ker (S - \alpha) = \{ (g_0, \Gamma_1^0 \widehat{g}_0 \Lambda|_{\lambda=\alpha}, 0)^\top : g_0 \in \ker (A_0^0 - \alpha) \}; \tag{3.18}$$

(ii) *if $\ker (S - \alpha) = \{0\}$, then $\pi_\alpha(Q_\infty) \geq k_\alpha$;*

(iii) *if* $\ker(S-\alpha)$ *or equivalently* $\ker(A_0^0-\alpha)$ *is nontrivial, then it is spanned by a positive vector if and only if* $\alpha \in \mathbb{R}$ *and* $\pi_\alpha(Q_\infty) = k_\alpha$, *if* $\ker(S-\alpha)$ *is negative, then* $\pi_\alpha(Q_\infty) = k_\alpha - 1$, *and if* $\ker(S-\alpha)$ *is neutral, then* $\pi_\alpha(Q_\infty) \le k_\alpha - 1$;

(iv) *if* $\ker(A_1^0-\alpha)$ *is nontrivial, then it is spanned by a positive vector if and only if* $\alpha \in \mathbb{R}$ *and* $\pi_\alpha(Q_\infty) = k_\alpha$, *if* $\ker(A_1^0-\alpha)$ *is negative, then* $\pi_\alpha(Q_\infty) = k_\alpha + 1$, *and if* $\ker(A_1^0-\alpha)$ *is neutral, then* $\pi_\alpha(Q_\infty) \ge k_\alpha + 1$.

Proof. (i) Let $G = (g_0, g, \widetilde{g})^\top \in \ker(S-\alpha)$ for some $\widehat{g}_0 = \{g_0, g_0'\} \in S_0^*$, $g, \widetilde{g} \in \mathbb{C}^k$. Then by (3.6) the equalities

$$C_q\widetilde{g} + \Gamma_0^0\widehat{g}_0 e_k = \alpha\widetilde{g}, \quad \widetilde{g}_1 = 0, \quad g_0' = \alpha g_0, \tag{3.19}$$

and

$$C_{q^\sharp} g = \alpha g, \quad g_1 = \Gamma_1^0\widehat{g}_0, \tag{3.20}$$

hold. It follows from the equalities (3.19) that $\widetilde{g} = 0$ and $\Gamma_0^0\widehat{g}_0 = 0$, i.e., $\widehat{g}_0 \in A_0^0$. The equalities (3.20) show that $\Gamma_1^0\widehat{g}_0 \neq 0$ if and only if $g \neq 0$. In particular, $g_0 \neq 0$ if and only if $G \neq 0$. Hence, if $\alpha \in \sigma_p(S)$, then $\alpha \in \sigma_p(A_0^0)$. Here $\Gamma_1^0\widehat{g}_0 \neq 0$, since otherwise $\alpha \in \sigma_p(S_0)$. Hence, (3.20) shows that $\alpha \in \sigma(q^\sharp)$. Thus, $\sigma_p(S) \subset \sigma_p(A_0^0) \cap \sigma(q^\sharp)$ and the vector G has the representation (3.18). The reverse inclusion in (3.18) is easily checked and hence $\sigma_p(A_0^0) \cap \sigma(q^\sharp) \subset \sigma_p(S)$ holds, too.

(iii) Assume that $\alpha \in \sigma_p(A_0^0) \cap \sigma(q^\sharp)$, and let $G \in \ker(S-\alpha)$. Then $[G, G] = [g_0, g_0]$. If the vector $g_0 \in \ker(A_0^0 - \alpha)$ is positive, then $\alpha \in \mathbb{R}$ and

$$-\infty < \lim_{z \widehat{\to} \alpha} (z-\alpha) Q_0(z) < 0, \tag{3.21}$$

and hence

$$-\infty < \lim_{z \widehat{\to} \alpha} \frac{Q_\infty(z)}{(z-\alpha)^{2k_\alpha - 1}} < 0. \tag{3.22}$$

Therefore, $\pi_\alpha(Q_\infty) = k_\alpha$. If the vector g_0 is negative or neutral, then the first limit in (3.21) is either positive or infinite and hence

$$0 < \lim_{z \widehat{\to} \alpha} \frac{Q_\infty(z)}{(z-\alpha)^{2k_\alpha - 1}} < \infty \ (\le \infty).$$

Due to (2.18) this implies that $\pi_\alpha(Q_\infty) = k_\alpha - 1$ or $\pi_\alpha(Q_\infty) \le k_\alpha - 1$, respectively.

The proofs of (ii) and (iv) are analogous, cf. the proof of Lemma 3.3. □

4. Proper factorizations and a minimal model for the generalized Friedrichs extension

The next definition is motivated by Lemma 3.3 and Lemma 3.4.

Definition 4.1. The factorization $Q_\infty(z) = q(z)q^\sharp(z)Q_0(z)$ will be called *proper* if q is a divisor of degree $\kappa_\infty(Q_\infty) > 0$ of the polynomial $\widetilde{q}$ in the canonical factorization (2.20).

Clearly, proper factorizations of $Q_\infty \in \mathbf{N}_\kappa$ always exist, but are not unique if $\widetilde{q}$ has more than one zero and $\kappa_\infty(Q_\infty) < \kappa$. Proper factorizations of $Q_\infty \in \mathbf{N}_\kappa$ can be characterized also without using the canonical factorization of Q_∞ by means of the multiplicities $\kappa_\infty(Q_\infty)$ and $\pi_\alpha(Q_\infty)$ as follows:

$$\kappa_\infty(Q_\infty) = k \text{ and } \pi_\alpha(Q_\infty) \geq k_\alpha \text{ for all the zeros } \alpha \text{ of } q,$$

where $k = \deg q$ and k_α is the multiplicity of α as the zero of q.

The symmetric relation S constructed in Theorem 3.1 need not be simple even if the corresponding factorization of Q_∞ in (3.1) is proper, cf. Example 1 in Section 3. However, when the factorization is proper, the simple part of S can be described easily and a minimal model from the model constructed in Theorem 3.1 can be produced by means of a reducing subspace which is positive.

Observe, that the selfadjoint extensions $A_0 = \ker \Gamma_0$ and $A_1 = \ker \Gamma_1$ of the symmetric relation S in Theorem 3.1 are given by

$$A_0 = \left\{ \left\{ \begin{pmatrix} f_0 \\ f \\ \widetilde{f} \end{pmatrix}, \begin{pmatrix} f_0' \\ C_{q^\sharp} f + \widetilde{\varphi} e_k \\ C_q \widetilde{f} + \Gamma_0^0 \widehat{f}_0 e_k \end{pmatrix} \right\} : \begin{array}{c} \widehat{f}_0 = \{f_0, f_0'\} \in S_0^*, \\ f_1 = \Gamma_1^0 \widehat{f}_0, \\ \widetilde{f}_1 = 0, \ \widetilde{\varphi} \in \mathbb{C} \end{array} \right\}, \tag{4.1}$$

and

$$A_1 = \left\{ \left\{ \begin{pmatrix} f_0 \\ f \\ \widetilde{f} \end{pmatrix}, \begin{pmatrix} f_0' \\ C_{q^\sharp} f \\ C_q \widetilde{f} + \Gamma_0^0 \widehat{f}_0 e_k \end{pmatrix} \right\} : \begin{array}{c} \widehat{f}_0 = \{f_0, f_0'\} \in S_0^*, \\ f_1 = \Gamma_1^0 \widehat{f}_0 \end{array} \right\}. \tag{4.2}$$

The above formulas for A_0 and A_1 show that $\operatorname{mul} A_1 = \operatorname{mul} S$ and

$$\begin{aligned} \operatorname{mul} A_0 &= \left\{ \begin{pmatrix} f_0' \\ \widetilde{\varphi} e_k \\ \Gamma_0^0 \widehat{f}_0 e_k \end{pmatrix} : \widehat{f}_0 = \{0, f_0'\} \in A_1^0, \ \widetilde{\varphi} \in \mathbb{C} \right\} \\ &= \operatorname{mul} S \oplus \operatorname{span}\{\omega_0\}, \end{aligned} \tag{4.3}$$

where $\omega_0 = (0, e_k, 0)^\top$.

Theorem 4.2. *Let $Q_\infty \in \mathbf{N}_\kappa$, let the factorization of Q_∞ in (3.1) be proper, let S_0 be a simple symmetric operator in the Pontryagin space $\mathfrak{H}_0$ with the boundary triplet $\Pi^0 = \{\mathcal{H}, \Gamma_0^0, \Gamma_1^0\}$ and the Weyl function Q_0. Moreover, let S, A_1, and A_0 be given by (3.6), (4.1), and (4.2), respectively. Then:*

(i) *S is simple if and only if $\sigma_p(S) = \emptyset$. In this case the linear relations S, $A = A_1$, and $S_F = A_0$ satisfy the equalities (1.3) and (1.4) with $\omega = \omega_0$ and the operator representation*

$$-1/Q_\infty(\lambda) = [(A - \lambda)^{-1} \omega, \omega] \tag{4.4}$$

is minimal.

(ii) *If S is not simple, then the subspace*

$$\mathfrak{H}'' = \operatorname{span}\{ \operatorname{mul} S, \ker(S - \alpha) : \alpha \in \sigma_p(A_0^0) \cap \sigma(q) \}$$

is positive and reducing for S.

The simple part of S coincides with the restriction S' of S to $\mathfrak{H}' := \mathfrak{H} \ominus \mathfrak{H}''$. The compressions S', A', S'_F of S, A_1, and A_0 to the subspace $\mathfrak{H}'$ satisfy the equalities (1.3) *and* (1.4), *with $\omega' \in \mathfrak{H}'$ given by*

$$\omega' = \begin{cases} \omega_0, & \text{if } k > 1, \\ (g, -1/\overline{\Gamma_0^0 \widehat{g}}, \Gamma_0^0 \widehat{g})^\top, & \text{if } k = 1, \end{cases} \tag{4.5}$$

where $\widehat{g} = \{0, g\} \in A_1^0$, $[g, g] = 1$, and the function $-1/Q_\infty$ admits the minimal representation

$$-1/Q_\infty(\lambda) = [(A' - \lambda)^{-1}\omega', \omega']. \tag{4.6}$$

Proof. (i) In the case when $\mathfrak{H}_0$ is a Hilbert space the first statement in (i) was proved in [6, Theorem 5.4]. In the case when $\mathfrak{H}_0$ is a Pontryagin space this result follows from part (ii). Namely, S is simple if and only if $S = S'$, or equivalently $\mathfrak{H}'' = \{0\}$, which by Lemmas 3.3, 3.4 means that $\sigma_p(S) = \emptyset$.

As to the other statements in (i) observe that the equalities (1.3) and (1.4) for S, A, and S_F follow immediately from the descriptions given in (3.6), (4.2), and (4.1). The cyclicity of ω is implied by the equality

$$(A - \lambda)^{-1}\omega = -\frac{\gamma(\lambda)}{Q_\infty(\lambda)}, \quad \lambda \in \rho(A_1^0)\backslash(\sigma(q) \cup \sigma(q^\sharp)), \tag{4.7}$$

where $\gamma(\lambda)(\in \mathfrak{N}_\lambda)$ is given by (3.8). In this case $\operatorname{mul} S = \{0\}$ and (4.3) shows that $\operatorname{mul} S_F = \operatorname{span}\{\omega_0\}$. The operator representation (4.4) is an immediate consequence of (4.7), since in view of (3.5) and (3.8) one has $[\gamma(\lambda), \omega] = 1$.

(ii) If $\operatorname{mul} S$ or, equivalently, $\operatorname{mul} A_1^0$ is nontrivial and $\kappa_\infty(Q_\infty) = k$ then according to Lemma 3.3 $\operatorname{mul} S$ is positive. Since the eigenspaces $\ker(S - \alpha_j)$, $\alpha_j \in \sigma_p(A_0^0) \cap \sigma(q)$, are also positive due to Lemma 3.4 and mutually orthogonal it follows that $\mathfrak{H}'$ is a positive subspace in $\mathfrak{H}$. Clearly, $\mathfrak{H}'$ is a reducing subspace for S. To prove the simplicity of the restriction S' of S it is enough to prove that

$$\mathfrak{H} \subset \widetilde{\mathfrak{H}} \oplus \mathfrak{H}'', \text{ where } \widetilde{\mathfrak{H}} = \overline{\operatorname{span}}\{\, \mathfrak{N}_\lambda(S^*) : \lambda \in \rho(A_0^0)\,\}.$$

Step 1. First it is shown that

$$\{0\} \oplus \mathbb{C}^k \oplus \{0\} \subset \widetilde{\mathfrak{H}} \oplus \mathfrak{H}''. \tag{4.8}$$

It follows from (3.8) that for every $\lambda \in \rho(A_1^0)\backslash\sigma(q)$:

$$\gamma(\lambda)q(\lambda)^{-1}Q_0(\lambda)^{-1} = \begin{pmatrix} \gamma_0(\lambda)Q_0(\lambda)^{-1} \\ \Lambda^\top \\ q(\lambda)^{-1}Q_0(\lambda)^{-1}\Lambda^\top \end{pmatrix}. \tag{4.9}$$

Since $\gamma_0(\lambda)Q_0(\lambda)^{-1}$ and $-Q_0(\lambda)^{-1}$ are the γ-field and the Weyl function of A_1^0, it follows from Lemma 2.2 that there is a vector $g \in \operatorname{mul} A_1^0$ such that

$$\lim_{\lambda \widehat{\to} \infty} \gamma_0(\lambda)Q_0(\lambda)^{-1} = g, \quad \lim_{\lambda \widehat{\to} \infty} \frac{1}{\lambda} Q_0(\lambda)^{-1} = \Gamma_0^0 \widehat{g}. \tag{4.10}$$

In the case when $\operatorname{mul} A_1^0$ is trivial one can take $g = 0$ and both these limits are equal to 0. It follows from (4.9) and (4.10) that $\gamma(\lambda)q(\lambda)^{-1}Q_0(\lambda)^{-1}$ has the following asymptotic expansion at ∞

$$\frac{\gamma(\lambda)}{q(\lambda)Q_0(\lambda)} = \lambda^{k-1}\begin{pmatrix}0\\ e_k\\ 0\end{pmatrix} + \cdots + \lambda\begin{pmatrix}0\\ e_2\\ 0\end{pmatrix} + \begin{pmatrix}g\\ e_1\\ \Gamma_0^0\widehat{g}e_k\end{pmatrix} + o(1).$$

Since $\operatorname{mul} S$ is spanned by the vector $(g, 0, \Gamma_0^0\widehat{g}e_k)^\top$ it follows that the vectors $(0, e_j, 0)^\top$, $j = 1, \ldots, k$, belong to $\widetilde{\mathfrak{H}} \oplus \mathfrak{H}''$.

Step 2. Next it is shown that

$$\{0\} \oplus \{0\} \oplus \mathbb{C}^k \subset \widetilde{\mathfrak{H}} \oplus \mathfrak{H}''. \tag{4.11}$$

It follows from Lemma 2.3 that for every $\alpha \in \sigma(q)$ there is a vector $g_\alpha \in \ker(A_0^0 - \alpha)$ such that

$$\lim_{\lambda \widehat{\to} \alpha}(\lambda - \alpha)\gamma_0(\lambda) = g_\alpha, \quad \lim_{\lambda \widehat{\to} \alpha}(\lambda - \alpha)Q_0(\lambda) = \Gamma_1^0\widehat{g}_\alpha, \quad \widehat{g}_\alpha = \{g_\alpha, \alpha g_\alpha\}. \tag{4.12}$$

In the case when $\ker(A_0^0 - \alpha)$ is trivial one can take $g_\alpha = 0$ in these equalities. It follows from (4.9) and (4.12) that $\gamma(\lambda)$ has the following asymptotic expansion at every $\alpha(\in \sigma(q))$

$$\gamma(\lambda) = \sum_{j=0}^{k_\alpha - 2}(\lambda - \alpha)^j\begin{pmatrix}0\\ 0\\ V_j(\alpha)\end{pmatrix} + (\lambda - \alpha)^{k_\alpha - 1}\begin{pmatrix}g_\alpha\\ \Gamma_1^0\widehat{g}_\alpha\Lambda|_\alpha\\ V_{k_\alpha - 1}(\alpha)\end{pmatrix} + (\lambda - \alpha)^{k_\alpha - 1}o(1),$$

where k_α is a multiplicity of the zero α of q and the Vandermonde vector $V_j(\alpha)$ is given by

$$V_j(\alpha) = \frac{1}{j!}\left(\frac{d}{d\lambda}\right)^j \Lambda^\top\big|_{\lambda=\alpha}, \quad j = 0, \ldots, k_\alpha - 1.$$

Let $\alpha_1, \alpha_2, \ldots, \alpha_l$ be all the zeros of q with the multiplicities $k_{\alpha_1}, \ldots, k_{\alpha_l}$, respectively. Using the formula (3.18) for $\ker(S - \alpha_j)$ one obtains

$$\operatorname{span}\{(0, 0, V_j(\alpha_j))^\top : i = 0, 1, \ldots, l;\ j = 0, 1, \ldots, k_{\alpha_i} - 1; \alpha_i \in \sigma(q)\} \subset \widetilde{\mathfrak{H}} \oplus \mathfrak{H}''.$$

This proves (4.11), since

$$\operatorname{span}\{V_j(\alpha_i) : i = 0, 1, \ldots, l;\ j = 0, 1, \ldots, k_{\alpha_i} - 1; \alpha_i \in \sigma(q)\} = \mathbb{C}^k.$$

Step 3. The inclusion $\mathfrak{H}_0 \oplus \{0\} \oplus \{0\} \subset \widetilde{\mathfrak{H}} \oplus \mathfrak{H}''$ is implied by the simplicity of S_0.

The representation (4.6) is implied by (4.7) and the equality

$$(A - \lambda)^{-1}\omega = (A' - \lambda)^{-1}\omega', \quad \lambda \in \rho(A).$$

This completes the proof. □

Observe that in part (i) of Theorem 4.2 the vector ω is neutral, i.e., $[\omega, \omega] = 0$, and that in part (ii) of Theorem 4.2 one has $[\omega', \omega'] \le 0$. Moreover,

$$[\omega', \omega'] < 0 \text{ if and only if } \operatorname{mul} S \neq \{0\} \text{ and } k = \deg q = 1,$$

in which case $\kappa_\infty(Q_\infty) = 1$.

In the special case $\kappa = k = 1$ the above model and the operator representations (4.4) and (4.6) for $-1/Q_\infty$ reduce to those constructed in [7].

References

[1] V.A. Derkach, "On generalized resolvents of Hermitian relations", J. Math. Sciences, 97 (1999), 4420–4460.

[2] V.A. Derkach and S. Hassi, "A reproducing kernel space model for $\mathbf{N}_\kappa$-functions", Proc. Amer. Math. Soc., 131 (2003), 3795–3806.

[3] V.A. Derkach, S. Hassi, M.M. Malamud, and H.S.V. de Snoo, "Generalized resolvents of symmetric operators and admissibility", Methods of Functional Analysis and Topology, 6 (2000), 24–55.

[4] V.A. Derkach, S. Hassi, and H.S.V. de Snoo, "Operator models associated with Kac subclasses of generalized Nevanlinna functions", Methods of Functional Analysis and Topology, 5 (1999), 65–87.

[5] V.A. Derkach, S. Hassi, and H.S.V. de Snoo, "Generalized Nevanlinna functions with polynomial asymptotic behaviour and regular perturbations", Oper. Theory Adv. Appl., 122 (2001), 169–189.

[6] V.A. Derkach, S. Hassi, and H.S.V. de Snoo, "Operator models associated with singular perturbations", Methods of Functional Analysis and Topology, 7 (2001), 1–21.

[7] V.A. Derkach, S. Hassi, and H.S.V. de Snoo, "Rank one perturbations in a Pontryagin space with one negative square", J. Funct. Anal., 188 (2002), 317–349.

[8] V.A. Derkach, S. Hassi, and H.S.V. de Snoo, "Asymptotic expansions of generalized Nevanlinna functions and their spectral properties", submitted.

[9] V.A. Derkach and M.M. Malamud, "The extension theory of hermitian operators and the moment problem", J. Math. Sciences, 73 (1995), 141–242.

[10] A. Dijksma, H. Langer, A. Luger, and Yu. Shondin, "A factorization result for generalized Nevanlinna functions of the class $\mathbf{N}_\kappa$", Integral Equations and Operator Theory, 36 (2000), 121–125.

[11] A. Dijksma, H. Langer, A. Luger, and Yu. Shondin, "Minimal realizations of scalar generalized Nevanlinna functions related to their basic factorization", Oper. Theory Adv. Appl., 154 (2004), 69–90.

[12] A. Dijksma, H. Langer, Yu.G. Shondin, and C. Zeinstra, "Self-adjoint operators with inner singularities and Pontryagin spaces", Oper. Theory Adv. Appl., 118 (2000), 105–175.

[13] P. Jonas, H. Langer, and B. Textorius, "Models and unitary equivalence of cyclic selfadjoint operators in Pontryagin spaces", Oper. Theory Adv. Appl., 59 (1992), 252–284.

[14] M.G. Kreĭn and H. Langer, "Über die Q-Funktion eines π-hermiteschen Operators im Raume Π_κ", Acta. Sci. Math. (Szeged), 34 (1973), 191–230.

[15] M.G. Kreĭn and H. Langer, "Über einige Fortsetzungsprobleme, die eng mit der Theorie hermitescher Operatoren im Raume Π_κ zusammenhängen. I. Einige Funktionenklassen und ihre Darstellungen", Math. Nachr., 77 (1977), 187–236.

[16] M.G. Kreĭn and H. Langer, "Some propositions on analytic matrix functions related to the theory of operators in the space Π_κ", Acta Sci. Math. (Szeged), 43 (1981), 181–205.

[17] H. Langer, "A characterization of generalized zeros of negative type of functions of the class $\mathbf{N}_\kappa$", Oper. Theory Adv. Appl., 17 (1986), 201–212.

Vladimir Derkach
Department of Mathematics
Donetsk National University
Universitetskaya str. 24
83055 Donetsk
Ukraine
e-mail: derkach@univ.donetsk.ua

Seppo Hassi
Department of Mathematics and Statistics
University of Vaasa
P.O. Box 700
65101 Vaasa
Finland
e-mail: sha@uwasa.fi

Henk de Snoo
Department of Mathematics and Computing Science
University of Groningen
P.O. Box 800
9700 AV Groningen
Nederland
e-mail: desnoo@math.rug.nl

Operator Theory:
Advances and Applications, Vol. 162, 135–144

Generalized Schur Functions and Augmented Schur Parameters

Aad Dijksma and Gerald Wanjala

Abstract. Every Schur function $s(z)$ is the uniform limit of a sequence of finite Blaschke products on compact subsets of the open unit disk. The Blaschke products in the sequence are defined inductively via the Schur parameters of $s(z)$. In this note we prove a similar result for generalized Schur functions.

Mathematics Subject Classification (2000). Primary 30D30, 30D50, 30C80.

Keywords. Generalized Schur function, Schur parameter, Blaschke product.

1. Introduction

A Schur function is a holomorphic function defined on the open unit disk $\mathbb{D}$, which is bounded by 1 there. We denote the class of such functions by $\mathbf{S}$. If $s(z) \in \mathbf{S}$ is not identically equal to a unimodular constant, then by Schwarz' Lemma (see, for example, [9, Theorem 6.1]) the function

$$\widehat{s}(z) = \frac{1}{z}\frac{s(z) - s(0)}{1 - s(0)^* s(z)}$$

is again in the class $\mathbf{S}$. The map $s(z) \mapsto \widehat{s}(z)$ is called the Schur transformation on $\mathbf{S}$ and $\widehat{s}(z)$ is called the Schur transform of $s(z)$. To a Schur function $s(z)$ which is not equal to a unimodular constant we can associate a sequence of Schur functions $(s_j(z))_{j \geq 0}$ by repeatedly applying the Schur transformation:

$$s_0(z) := s(z), \ s_1(z) = \widehat{s}_0(z), \ \ldots, \ s_j(z) = \widehat{s}_{j-1}(z), \ldots.$$

This repeated application of the Schur transformation is called the Schur algorithm. The sequence $(s_j(z))_{j \geq 0}$ is finite and terminates at the nth step of the algorithm with $s_n(z)$ if and only if $|s_n(0)| = 1$. For then, by the maximum modulus principle, $s_n(z) \equiv s_n(0)$ and the Schur transformation is not defined for $s_n(z)$.

This occurs if and only if $s(z)$ is a Blaschke product of order n, that is, of the form

$$s(z) = c\prod_{j=0}^{n} \frac{z-\alpha_j}{1-\alpha_j^* z}, \quad \alpha_j \in \mathbb{D},\ c \in \mathbb{T},$$

where $\mathbb{T}$ stands for the unit circle. The numbers $\gamma_j = s_j(0)$, $j = 0, 1, \ldots$, are called the Schur parameters associated with $s(z)$. If the sequence $(\gamma_j)_{j\geq 0}$ is infinite then $|\gamma_j| < 1$ for all $j = 0, 1, \ldots$; if it stops with γ_n then $|\gamma_j| < 1$ for $j = 0, 1, \ldots, n-1$ and $|\gamma_n| = 1$. A sequence of complex numbers with these properties will be called a Schur sequence.

The sequence of Schur parameters determines the function. To see this, let m be an integer ≥ 0 and define the rational functions

$$B_{m,0}(s;z) = \begin{cases} \dfrac{z+\gamma_m}{1+\gamma_m^* z} & \text{if } |\gamma_m| < 1, \\ \gamma_m & \text{if } |\gamma_m| = 1, \end{cases}$$

and

$$B_{m,j}(s;z) = \frac{zB_{m,j-1}(s;z) + \gamma_{m-j}}{1+\gamma_{m-j}^* zB_{m,j-1}(s;z)}, \quad j = 1, 2, \ldots, m.$$

Hence $B_{m,j-1}(s;z) = \widehat{B}_{m,j}(s;z)$. If $|\gamma_m| < 1$, then $B_{m,j}(s;z)$ is a Blaschke product of order $j+1$ for $j = 0, 1, \ldots, m$; if $|\gamma_m| = 1$, then $B_{m,j}(s;z)$ is a Blaschke product of order j for $j = 0, 1, \ldots, m$. Moreover, the sequence of Schur parameters associated with $B_{m,m}(s;z)$ is finite and given by $\gamma_0, \gamma_1, \ldots, \gamma_m, 1$ if $|\gamma_m| < 1$ and $\gamma_0, \gamma_1, \ldots, \gamma_m$ if $|\gamma_m| = 1$. Thus the first $m+1$ Schur parameters of $s(z)$ coincide with the first $m+1$ Schur parameters of $B_{m,m}(s;z)$. I. Schur showed that this implies that the difference $s(z) - B_{m,m}(s;z)$ has a zero at $z = 0$ of order $\geq m+1$. This can also be seen by proving by induction that for $j = 0, 1, \ldots, m$, the difference $s_{m-j}(z) - B_{m,j}(s;z)$ has a zero at $z = 0$ of order at least $j+1$; see [14, Theorem I.2.1]. Since $|s(z)| \leq 1$ and $|B_{m,m}(s;z)| \leq 1$ on $\mathbb{D}$, Schwarz' Lemma implies that

$$|s(z) - B_{m,m}(s;z)| \leq 2|z|^{m+1}, \quad z \in \mathbb{D}, \tag{1.1}$$

and hence Carathéodory's theorem holds: If the sequence of Schur parameters breaks up at γ_n with $|\gamma_n| = 1$ then $s(z) = B_{n,n}(s;z)$; otherwise

$$s(z) = \lim_{m\to\infty} B_{m,m}(s;z), \tag{1.2}$$

where the limit is uniform in z on compact subsets of $\mathbb{D}$.

If we introduce the Möbius transform

$$\tau_n(w) = \frac{zw+\gamma_n}{1+\gamma_n^* zw} = \gamma_n + \frac{(1-|\gamma_n|^2)z}{\gamma_n^* z + 1/w},$$

then the composition formulas

$$\begin{aligned} s(z) &= \tau_0 \circ \tau_1 \circ \cdots \circ \tau_n(s_{n+1}(z)), \\ B_{m,m}(s;z) &= \begin{cases} \tau_0 \circ \tau_1 \circ \cdots \circ \tau_m(1) & \text{if } |\gamma_m| < 1, \\ \tau_0 \circ \tau_1 \circ \cdots \circ \tau_{m-1}(\gamma_m) & \text{if } |\gamma_m| = 1, \end{cases} \end{aligned} \tag{1.3}$$

hold and they show the close relation between the Schur algorithm, the Schur parameters, and continued fractions. In this note we do not pursue this connection but refer to the recent paper [16]. We only recall that if $(\gamma_j)_{j\geq 0}$ is a Schur sequence of complex numbers, then $(B_m(z))_{m\geq 0}$ with $B_m(z)$ as in (1.3) is a Cauchy sequence of finite Blaschke products, which converges to a function $s(z) \in \mathbf{S}$ whose sequence of Schur parameters coincides with $(\gamma_j)_{j\geq 0}$. This follows from the arguments leading up to (1.1) which also imply that

$$|B_m(z) - B_n(z)| \leq 2|z|^{\min\{m,n\}+1}, \quad z \in \mathbb{D}.$$

An excellent account of Schur's work on analysis, including the Schur algorithm, can be found in [13]; we refer to this paper for the complete list of works of I. Schur in this area.

If $s(z)$ is a generalized Schur function which is holomorphic in $z = 0$ (see Section 2 below), then the Krein–Langer factorization $s(z) = B(z)^{-1}s_0(z)$, where $s_0(z)$ is a Schur function and $B(z)$ is a finite Blaschke product with $B(0) \neq 0$, implies

$$s(z) = \lim_{m\to\infty} B(z)^{-1}B_{m,m}(s_0; z), \tag{1.4}$$

where the convergence is uniform in z on compact subsets of $\mathbb{D} \setminus \{$ poles of $s(z)\}$. In fact, because of (1.1), for every compact set $K \in \mathbb{D} \setminus \{$ poles of $s(z)\}$, there is a real number $M \geq 0$ such that

$$|s(z) - B(z)^{-1}B_{m,m}(s_0; z)| \leq M|z|^{m+1}, \quad z \in K. \tag{1.5}$$

However, the functions on the right-hand side of (1.4) are not related to the generalized Schur algorithm for $s(z)$ nor to any form of continued fractions similar to what we described in the foregoing paragraphs. In this note we prove a result for generalized Schur functions which is analogous to (1.2) and is related to the generalized Schur algorithm. For that we introduce the sequence of augmented Schur parameters (see Section 3), which plays the same role for generalized Schur functions as the sequence of Schur parameters does for Schur functions. In Section 2 we present the preliminaries: generalized Schur functions, the generalized Schur transformation, and related notions and results needed in the sequel. Section 3 contains the two theorems of this note.

2. Generalized Schur functions and the generalized Schur transformation

For any integer $\kappa \geq 0$, by $\mathbf{S}_\kappa$ we denote the set of complex-valued functions $s(z)$ which are meromorphic on $\mathbb{D}$ and have the following equivalent properties:

1. $s(z)$ has κ poles (counted according to their multiplicities) and

$$\limsup_{r\uparrow 1}|s(re^{it})| \leq 1, \text{ for almost all } t \in [0, 2\pi).$$

2. The kernel

$$K_s(z,w) = \frac{1 - s(z)s(w)^*}{1 - zw^*}, \qquad z, w \in \Omega(s),$$

has κ negative squares, where $\Omega(s)$ is the domain of holomorphy of $s(z)$.
3. The 2×2 matrix kernel

$$D_s(z,w) = \begin{pmatrix} \dfrac{1 - s(z)s(w)^*}{1 - zw^*} & \dfrac{s(z) - s(w^*)}{z - w^*} \\ \dfrac{\widetilde{s}(z) - \widetilde{s}(w^*)}{z - w^*} & \dfrac{1 - \widetilde{s}(z)\widetilde{s}(w)^*}{1 - zw^*} \end{pmatrix}$$

has κ negative squares on $\Omega(s) \cap \Omega(\widetilde{s})$, where $\widetilde{s}(z) = s(z^*)^*$.
4. The function $s(z)$ admits the Krein–Langer factorization

$$s(z) = B(z)^{-1} s_0(z), \quad B(z) = \prod_{j=1}^{\kappa} \frac{z - \alpha_j}{1 - \alpha_j^* z},$$

where $s_0(z) \in \mathbf{S}$, $\alpha_j \in \mathbb{D}$ and $s_0(\alpha_j) \neq 0$, $j = 1, 2, \ldots, \kappa$.

Evidently, $\mathbf{S}_0 = \mathbf{S}$. The functions of the class $\mathbf{S}_\kappa$ are called generalized Schur functions with κ negative squares. They were introduced and studied in [15]. For the equivalence of these properties, see [15], [7], and [8, Section 3.4]. By $\mathbf{S}_\kappa^0$ we denote the set of functions $s(z) \in \mathbf{S}_\kappa$ which are holomorphic at $z = 0$ and we set $\mathbf{S}^0 = \cup_{\kappa \geq 0} \mathbf{S}_\kappa^0$. Consider a function $s(z) \in \mathbf{S}^0$ which is not identically equal to a unimodular constant and assume it has the Taylor expansion

$$s(z) = \sigma_0 + \sigma_1 z + \sigma_2 z^2 + \cdots + \sigma_k z^k + \sigma_{k+1} z^{k+1} + \cdots .$$

Then the generalized Schur transform $\widehat{s}(z)$ of $s(z)$ is defined as follows.

(1) If $|\sigma_0| < 1$, then

$$\widehat{s}(z) = \frac{1}{z} \frac{s(z) - \sigma_0}{1 - \sigma_0^* s(z)}.$$

This formula coincides with the "classical" formula in the Introduction.

(2) If $|\sigma_0| > 1$ then the case $s(z) \equiv \sigma_0$ does not arise since this implies that $s(z) \notin \mathbf{S}^0$. This means there exists an integer $k \geq 1$ such that $\sigma_1 = \sigma_2 = \cdots = \sigma_{k-1} = 0$ and $\sigma_k \neq 0$. In this case,

$$\widehat{s}(z) = z^k \frac{1 - \sigma_0^* s(z)}{s(z) - \sigma_0}.$$

Note: k is the order of the pole of the quotient on the right-hand side.

(3) If $|\sigma_0| = 1$ then there exists an integer $k \geq 1$, such that $\sigma_1 = \sigma_2 = \cdots = \sigma_{k-1} = 0$ and $\sigma_k \neq 0$ since we assume that $s(z)$ is not a unimodular constant. With this k, determine complex numbers c_j, $j = 0, 1, \ldots, k-1$, such that

$$(s(z) - \sigma_0)(c_0 + c_1 z + \cdots + c_j z^j + \cdots) = \sigma_0 z^k$$

(so that $c_0 \neq 0$), define the polynomial $p(z) = c_0 + c_1 z + \cdots + c_{k-1} z^{k-1}$, and, finally, set $Q(z) = p(z) - z^{2k} p(1/z^*)^*$. In this case,

$$\widehat{s}(z) = z^q \frac{(Q(z) - z^k)s(z) - \sigma_0 Q(z)}{\sigma_0^* Q(z)s(z) - (Q(z) + z^k)},$$

where $q \geq 0$ is the order of the pole of the quotient on the right-hand side. Note: (i) q is finite because $\sigma_0^* Q(z)s(z) - (Q(z) + z^k) \not\equiv 0$ (see [1, page 5]). (ii) For some complex number t_{2k}, we have

$$\begin{aligned} \sigma_0^* Q(z)s(z) - (Q(z) + z^k) &= t_{2k} z^{2k} + \cdots, \\ (Q(z) - z^k)s(z) - \sigma_0 Q(z) &= (\sigma_0 t_{2k} - \sigma_k) z^{2k} + \cdots. \end{aligned}$$

(see [8, Lemma 3.3.1 and its proof]), and hence if $q = 0$, then $t_{2k} \neq 0$ and so $\widehat{s}(0) \neq \sigma_0$.

In all these cases $\widehat{s}(z)$ belongs to the class $\mathbf{S}^0$; see [8, Lemma 3.4.4] and also [1, Theorem 3.1]. In fact, the following result holds.

Lemma 2.1. *If* $s(z) \in \mathbf{S}^0_\kappa$, *then* $\widehat{s}(z) \in \mathbf{S}^0_{\widehat{\kappa}}$, *where* $\widehat{\kappa} = \kappa$, $\kappa - k$, *and* $\kappa - k - q$ *in cases* (1), (2), *and* (3), *respectively.*

For a proof we refer to [1, Theorems 5.1, 6.1, and 8.1]. The definition of the generalized Schur transformation goes back to [10], [12], [11], and [8, Definition 3.3.1]. In [8] it is applied to solve the problem: When is a formal power series around $z = 0$ the Taylor expansion of a generalized Schur function. In [1, 2, 4, 6, 17] it is studied for its effect on the coisometric and unitary operator realizations of a generalized Schur function, including those whose state spaces are the reproducing kernel Pontryagin spaces with kernels $K_s(z,w)$ and $D_s(z,w)$; in [3] it is shown to provide an algorithm for the unique factorization of a 2×2 matrix polynomial which is J-unitary on $\mathbb{T}$ (for the definition, see below) in normalized elementary factors; and, finally, in [5] (see also [12]) it is used in solving a basic interpolation problem for generalized Schur functions.

The inverse of the generalized Schur transformation in each of these three cases can be written as

$$s(z) = \frac{a(z)\widehat{s}(z) + b(z)}{c(z)\widehat{s}(z) + d(z)},$$

where the coefficient matrix

$$\Theta(z) = \begin{pmatrix} a(z) & b(z) \\ c(z) & d(z) \end{pmatrix}$$

can be chosen as

$$\Theta_1(z) = \frac{1}{\sqrt{1 - |\sigma_0|^2}} \begin{pmatrix} 1 & \sigma_0 \\ \sigma_0^* & 1 \end{pmatrix} \begin{pmatrix} z & 0 \\ 0 & 1 \end{pmatrix}, \text{ if } |\sigma_0| < 1, \tag{2.1}$$

$$\Theta_2(z) = \frac{1}{\sqrt{|\sigma_0|^2 - 1}} \begin{pmatrix} \sigma_0 & 1 \\ 1 & \sigma_0^* \end{pmatrix} \begin{pmatrix} 1 & 0 \\ 0 & z^k \end{pmatrix}, \text{ if } |\sigma_0| > 1, \tag{2.2}$$

$$\Theta_3(z) = \begin{pmatrix} Q(z) + z^k & -\sigma_0 z^q Q(z) \\ \sigma_0^* Q(z) & -z^q(Q(z) - z^k) \end{pmatrix}, \text{ if } |\sigma_0| = 1. \tag{2.3}$$

These 2×2 matrix polynomials are J-unitary on $\mathbb{T}$, that is, satisfy $\Theta(z)^* J \Theta(z) = J$, $|z| = 1$, where

$$J = \begin{pmatrix} 1 & 0 \\ 0 & -1 \end{pmatrix}.$$

3. The augmented Schur parameters

To $s(z) \in \mathbf{S^0}$ which is not identically equal to a unimodular constant we can apply the generalized Schur algorithm:

$$s_0(z) := s(z),\ s_1(z) = \widehat{s}_0(z), \ldots,\ s_j(z) = \widehat{s}_{j-1}(z), \ldots,$$

where now $\widehat{s}_{j-1}(z)$ denotes the generalized Schur transform of $s_{j-1}(z)$ for $j = 1, 2, \ldots$. We set $\gamma_j = s_j(0)$, $j = 0, 1, \ldots$. The sequence of functions $(s_j(z))_{j\geq 0}$ terminates at $s_n(z)$ if $s_n(z) \equiv \gamma_n$ with $|\gamma_n| = 1$, because in that case the generalized Schur transform of $s_n(z)$ is not defined. The number γ_j if $|\gamma_j| < 1$, the pair (γ_j, k_j) if $|\gamma_j| > 1$, and the quadruple $(\gamma_j, k_j, q_j, Q_j(z))$ if $|\gamma_j| = 1$, which are defined in accordance with the definitions of the generalized Schur transformation (see Section 2), will be called the augmented Schur parameter and briefly denoted by $\widehat{\gamma}_j$. The sequence $(\widehat{\gamma}_j)_{j\geq 0}$ will be called the sequence of augmented Schur parameters. It is finite and stops at $\widehat{\gamma}_n$, when the sequence $(s_j(z))_{j\geq 0}$ terminates at $s_n(z)$; in this case, $\widehat{\gamma}_n$ carries no further information, that is, $\widehat{\gamma}_n = \gamma_n$ and $|\gamma_n| = 1$. From the definition of the generalized Schur transformation we see that for $j = 0, 1, \ldots$ (and up to $n - 1$ if the sequence $(\widehat{\gamma}_j)_{j\geq 0}$ stops at $\widehat{\gamma}_n$ with $n \geq 1$) the following implications hold:

$$\begin{cases} |\gamma_j| > 1 & \Longrightarrow \quad \gamma_{j+1} \neq 0, \\ |\gamma_j| = 1, q_j > 0 & \Longrightarrow \quad \gamma_{j+1} \neq 0, \\ |\gamma_j| = 1, q_j = 0 & \Longrightarrow \quad \gamma_{j+1} \neq \gamma_j. \end{cases} \tag{3.1}$$

Moreover, by [8, Lemma 3.4.5], see also [1, Corollary 9], there is an integer $j_0 \geq 0$ such that $s_j(z) \in \mathbf{S}$ for all $j \geq j_0$ and hence $\widehat{\gamma}_j = \gamma_j$ with $|\gamma_j| \leq 1$, $j \geq j_0$.

With the sequence $(\widehat{\gamma}_j)_{j\geq 0}$ we define for $m \geq 0$,

$$B_{m,0}(s; z) = \gamma_m \quad \text{if } |\gamma_m| = 1 \text{ and } (\widehat{\gamma}_j)_{j\geq 0} \text{ stops with } \widehat{\gamma}_m, \tag{3.2}$$

otherwise,

$$B_{m,0}(s; z) = \begin{cases} \dfrac{z + \gamma_m}{1 + \gamma_m^* z} & \text{if } |\gamma_m| < 1, \\[2ex] \dfrac{z^{k_m} + \gamma_m}{\gamma_m^* z^{k_m} + 1} & \text{if } |\gamma_m| > 1, \\[2ex] \dfrac{(Q_m(z) + z^{k_m}) - \gamma_m z^{q_m} Q_m(z)}{\gamma_m^* Q_m(z) - z^{q_m}(Q_m(z) - z^{k_m})} & \text{if } |\gamma_m| = 1, \end{cases}$$

and for $j = 1, 2, \ldots, m$,

$$B_{m,j}(s;z) = \begin{cases} \dfrac{\gamma_{m-j} + zB_{m,j-1}(s;z)}{1 + \gamma^*_{m-j} zB_{m,j-1}(s;z)} & \text{if } |\gamma_{m-j}| < 1, \\ \dfrac{z^{k_{m-j}} + \gamma_{m-j} B_{m,j-1}(s;z)}{\gamma^*_{m-j} z^{k_{m-j}} + B_{m,j-1}(s;z)} & \text{if } |\gamma_{m-j}| > 1, \\ \dfrac{(Q_{m-j}(z) + z^{k_{m-j}})B_{m,j-1}(s;z) - \gamma_{m-j} z^{q_{m-j}} Q_{m-j}(z)}{\gamma^*_{m-j} Q_{m-j}(z) B_{m,j-1}(s;z) - z^{q_{m-j}}(Q_{m-j}(z) - z^{k_{m-j}})} & \text{if } |\gamma_{m-j}| = 1. \end{cases}$$

Each $B_{m,j}(s;z)$ is of the form $B_1(z)^{-1}B_2(z)$ where $B_1(z)$ and $B_2(z)$ are finite Blaschke products with $B_1(0) \neq 0$ that is, a rational generalized Schur function holomorphic at $z = 0$ and having unimodular values on $\mathbb{T}$ (in particular, it has no poles in $\mathbb{T}$). Clearly, $B_{m,j-1}(s;z) = \widehat{B}_{m,j}(s;z)$, $j = 1, 2, \ldots, m$. Similarly as above, these formulas can be expressed in terms of Möbius transformations and hence are related to continued fractions. The sequence of augmented Schur parameters for $B_{m,m}(s;z)$ is $\widehat{\gamma}_0, \widehat{\gamma}_1, \ldots, \widehat{\gamma}_m$ in case $B_{m,0}(s;z)$ is given by (3.2), otherwise it is $\widehat{\gamma}_0, \widehat{\gamma}_1, \ldots, \widehat{\gamma}_m, 1$.

Let $j_0 \geq 0$ be an integer such that $s_j(z) \in \mathbf{S}$ for all $j \geq j_0$. From the definition of the generalized Schur transformation in Subsection 2 we have

$$s(z) = \frac{\theta_{11}(z)s_{j_0}(z) + \theta_{12}(z)}{\theta_{21}(z)s_{j_0}(z) + \theta_{22}(z)}, \tag{3.3}$$

where the coefficient matrix can be written as the product

$$\Theta_s(z) := \begin{pmatrix} \theta_{11}(z) & \theta_{12}(z) \\ \theta_{21}(z) & \theta_{22}(z) \end{pmatrix} = \Theta_{(0)}(z)\Theta_{(1)}(z)\cdots\Theta_{(j_0-1)}(z),$$

in which each factor $\Theta_{(i)}(z)$ is of one of the forms $\Theta_1(z)$, $\Theta_2(z)$, and $\Theta_3(z)$ given by (2.1), (2.2), and (2.3), respectively.

Thus $\Theta_s(z)$ is a 2×2 matrix polynomial which is J-unitary on $\mathbb{T}$. It follows from (2.1), (2.2), and (2.3) (see also [8, Lemma 3.4.2 v)]) that for some integer $\ell_0 \geq j_0$,

$$\det \Theta_s(z) = z^{\ell_0}. \tag{3.4}$$

It is easy to see that

$$\rho(s) := \ell_0 - j_0 = \sum_{j:\,\widehat{\gamma}_j \text{ is a pair}} (k_j - 1) + \sum_{j:\,\widehat{\gamma}_j \text{ is a quadruple}} (2k_j + q_j - 1).$$

In particular, if j_1 and ℓ_1 are defined in the same way as j_0 and ℓ_0, then $\ell_1 - j_1 = \ell_0 - j_0$.

Theorem 3.1. *Let $s(z) \in \mathbf{S}^0$ and let $\rho(s)$ be as defined above. Then for each compact subset $K \subset \mathbb{D} \setminus \{\text{poles of } s(z)\}$, there exist a real number $M > 0$ and an integer $m_0 \geq j_0$ such that for all $z \in K$ and all $m \geq m_0$,*

$$|s(z) - B_{m,m}(s;z)| \leq M|z|^{\rho(s)+m+1}.$$

The estimate in the theorem is an improvement of (1.5) by a factor $|z|^{\rho(s)}$, but only holds for sufficiently large m. The theorem implies that

$$s(z) = \lim_{m\to\infty} B_{m,m}(s;z)$$

uniformly in z on compact subsets of $\mathbb{D} \setminus \{\text{poles of } s(z)\}$.

Proof. If the sequence $(\widehat{\gamma}_j)_{j\geq 0}$ of augmented Schur parameters corresponding to $s(z)$ is finite and terminates with $|\gamma_n| = 1$, then $s(z) = B_{n,n}(s;z)$ and the theorem holds true. We now assume that the sequence $(\widehat{\gamma}_j)_{j\geq 0}$ is infinite. From (3.3),

$$B_{j_0+m,j_0+m}(s;z) = \frac{\theta_{11}(z)B_{j_0+m,m}(s;z) + \theta_{12}(z)}{\theta_{21}(z)B_{j_0+m,m}(s;z) + \theta_{22}(z)},$$

and

$$B_{j_0+m,m}(s;z) = B_{m,m}(s_{j_0};z)$$

we obtain

$$s(z) - B_{j_0+m,j_0+m}(s;z) = \frac{\det\Theta_s(z)(s_{j_0}(z) - B_{m,m}(s_{j_0};z))}{(\theta_{21}(z)s_{j_0}(z) + \theta_{22}(z))(\theta_{21}(z)B_{m,m}(s_{j_0};z) + \theta_{22}(z))}. \tag{3.5}$$

By (1.1) and (3.4), the numerator of the quotient on the right-hand side satisfies the inequality

$$|\det\Theta_s(z)(s_{j_0}(z) - B_{m,m}(s_{j_0};z))| \leq 2|z|^{\ell_0+m+1}, \quad z \in \mathbb{D}. \tag{3.6}$$

We claim that the factor $\theta_{21}(z)s_{j_0}(z) + \theta_{22}(z)$ in the denominator does not vanish in $\mathbb{D} \setminus \{\text{poles of } s(z)\}$. To see this, assume that for some $z_0 \in \mathbb{D} \setminus \{\text{poles of } s(z)\}$ we do have that

$$\theta_{21}(z_0)s_{j_0}(z_0) + \theta_{22}(z_0) = 0. \tag{3.7}$$

Then, by (3.3) and since $z = z_0$ is not a pole of $s(z)$, we also have

$$\theta_{11}(z_0)s_{j_0}(z_0) + \theta_{12}(z_0) = 0.$$

The last two equations can be written in matrix form:

$$\Theta_s(z_0)\begin{pmatrix} s_j(z_0) \\ 1 \end{pmatrix} = 0.$$

This implies $\det\Theta_s(z_0) = 0$ and so, on account of (3.4), $z_0 = 0$. However, from [8, Lemma 3.4.2 iii) and v)] (or from [3, Theorem 6.6]) it follows that there are complex numbers $k_0 \neq 0$, k_1, …, such that

$$\theta_{21}(z)s_{j_0}(z) + \theta_{22}(z) = \frac{\det\Theta_s(z)}{\theta_{11}(z) - \theta_{21}(z)s(z)} = k_0 + k_1 z + \cdots.$$

This contradicts (3.7) with $z_0 = 0$ and proves the claim. Let K be a compact subset of $\mathbb{D} \setminus \{\text{poles of } s(z)\}$ and let

$$\varepsilon = \min_{z\in K}|\theta_{21}(z)s_{j_0}(z) + \theta_{22}(z)|. \tag{3.8}$$

Because of the claim just proved, $\varepsilon > 0$. Applying (1.1), we find that for all $z \in \mathbb{D}$,

$$|(\theta_{21}(z)s_{j_0}(z) + \theta_{22}(z)) - (\theta_{21}(z)B_{m,m}(s_{j_0};z) + \theta_{22}(z))| \leq 2|z|^{m+1}\max_{z\in\mathbb{D}}|\theta_{21}(z)|$$

and hence for some integer $m_1 \geq 0$ we have that for all $m \geq m_1$,

$$|\theta_{21}(z)B_{m,m}(s_{j_0};z) + \theta_{22}(z)| \geq \frac{1}{2}\varepsilon, \quad z \in K. \tag{3.9}$$

Combining (3.5), (3.6), (3.8), and (3.9), we see that for $m \geq m_1$,

$$|s(z) - B_{j_0+m,j_0+m}(s;z)| \leq \frac{4}{\varepsilon^2}\,|z|^{\ell_0+m+1}, \quad z \in K.$$

This readily implies the theorem with $m_0 = m_1 + j_0$ and $M = 4/\varepsilon^2$. □

A sequence $(\widehat{\gamma}_j)_{j\geq 0}$ will be called an augmented Schur sequence if:

(a) except for at most finitely many values of j, $\widehat{\gamma}_j$ is a complex number γ_j with $|\gamma_j| < 1$;

(b) in the exceptional cases, $\widehat{\gamma}_j$ is either a pair (γ_j, k_j) consisting of a complex number γ_j with $|\gamma_j| > 1$ and an integer $k_j \geq 1$ or a quadruple $(\gamma_j, k_j, q_j, Q_j(z))$ consisting of a unimodular complex number γ_j, integers $k_j \geq 1$ and $q_j \geq 0$, and a polynomial $Q_j(z) = p_j(z) - z^{2k_j}p_j(1/z^*)^*$, where $p_j(z)$ is a polynomial of degree $< k_j$ and $p_j(0) \neq 0$;

(c) in case the sequence is finite and ends with $\widehat{\gamma}_n$, also $\widehat{\gamma}_n$ is exceptional: $\widehat{\gamma}_n = \gamma_n$ with $|\gamma_n| = 1$; and

(d) the implications (3.1) hold.

Theorem 3.2. *Let $(\widehat{\gamma}_j)_{j\geq 0}$ be an augmented Schur sequence. Then there is a unique $s(z) \in \mathbf{S}^0$ such that $(\widehat{\gamma}_j)_{j\geq 0}$ is the corresponding sequence of augmented Schur parameters. The number κ of negative squares of $s(z)$ is given by*

$$\kappa = \sum_{j:\,\widehat{\gamma}_j \text{ is a pair}} k_j + \sum_{j:\,\widehat{\gamma}_j \text{ is a quadruple}} k_j + q_j.$$

Proof. Let m be an integer ≥ 0 such that for all $j \geq m$, $\widehat{\gamma}_j$ is a complex number γ_j with $|\gamma_j| < 1$. Let $s_m(z)$ be the Schur function whose sequence of Schur parameters coincides with $(\gamma_{m+j})_{j\geq 0}$. Define $B_{m,0}(z) = s_m(z)$ and define $B_{m,j}(z)$ in the same inductive way as $B_{m,j}(s;z)$ is defined starting with $B_{m,0}(s;z)$, $j = 1, 2, \ldots, m$. Then $s(z) = B_{m,m}(z)$ has the desired properties. The formula for κ follows from Lemma 2.1. □

References

[1] D. Alpay, T.Ya. Azizov, A. Dijksma, and H. Langer, *The Schur algorithm for generalized Schur functions* I*: Coisometric realizations*, Operator Theory: Adv., Appl., vol. 129, Birkhäuser Verlag, Basel, 2001, 1–36.

[2] D. Alpay, T.Ya. Azizov, A. Dijksma, and H. Langer, *The Schur algorithm for generalized Schur functions* II*: Jordan chains and transformation of characteristic functions*, Monatsh. Math. **138** (2003), 1–29.

[3] D. Alpay, T.Ya. Azizov, A. Dijksma, and H. Langer, *The Schur algorithm for generalized Schur functions* III*: J-unitary polynomials on the circle*, Linear Algebra Appl. **369** (2003), 113–144.

[4] D. Alpay, T.Ya. Azizov, A. Dijksma, H. Langer, and G. Wanjala *Characteristic functions of coisometric colligations and fractional linear transformations arising from a basic interpolation problem*, Department of Mathematics, University of Groningen, The Netherlands, Research report IWI 2002.

[5] D. Alpay, T.Ya. Azizov, A. Dijksma, H. Langer, and G. Wanjala, *A basic interpolation problem for generalized Schur functions and coisometric realizations*, Operator Theory: Adv. Appl., vol 143, Birkhäuser Verlag, Basel, 2003, 39–76.

[6] D. Alpay, T.Ya. Azizov, A. Dijksma, H. Langer, and G. Wanjala, *The Schur algorithm for generalized Schur functions* IV*: unitary realizations*, Operator Theory: Adv., Appl., vol. 149, Birkhäuser Verlag, Basel, 2004, 23–45.

[7] D. Alpay, A. Dijksma, J. Rovnyak, and H. de Snoo, *Schur functions, operator colligations, and reproducing kernel Pontryagin spaces*, Operator Theory: Adv. Appl., vol. 96, Birkhäuser Verlag, Basel, 1997.

[8] M.J. Bertin, A. Decomps-Guilloux, M. Grandet-Hugot, M. Pathiaux-Delfosse, and J.P. Schreiber, *Pisot and Salem numbers*, Birkhäuser Verlag, Basel, 1992.

[9] R.B. Burckel, An introduction to classical complex analysis, volume 1, Academic press, New York, San Francisco, 1979.

[10] C. Chamfy, *Fonctions méromorphes sur le cercle unité et leurs séries de Taylor*, Ann. Inst. Fourier **8** (1958), 211–251.

[11] P. Delsarte, Y. Genin, and Y. Kamp, *Pseudo-Carathéodory functions and hermitian Toeplitz matrices*, Philips J. Res. **41**(1) (1986), 1–54.

[12] J. Dufresnoy, *Le problème des coefficients pour certaines fonctions méromorphes dans le cercle unité*, Ann. Acad. Sc. Fenn. Ser. A.I, **250**,9 (1958), 1–7.

[13] H. Dym and V. Katsnelson, *Contributions of Issai Schur to analysis*, Studies in memory of Issai Schur (Chevaleret/Rehovot, 2000), xci–clxxxviii, Progr. Math., 210, Birkhäuser, Boston, MA, 2003.

[14] J.B. Garnett, *Bounded analytic functions*, Academic press, New York, 1981.

[15] M.G. Krein and H. Langer, *Über einige Fortsetzungsprobleme, die eng mit der Theorie hermitescher Operatoren im Raume* π_κ *zusammenhängen, Teil I: Einige Funktionenklassen und ihre Darstellungen*, Math. Nachr. **77** (1977), 187–236.

[16] S. Khrushchev, *Schur's algorithm, orthogonal polynomials and convergence of Wall's continued fractions in* $L^2(\mathbb{T})$, J. of Approximation Theory, vol. **108** (2001), 161–248.

[17] G. Wanjala, *Closely connected unitary realizations of the solutions to the basic interpolation problem for generalized Schur functions*, Operator Theory: Adv. Appl., vol. 160, Birkhäuser Verlag, Basel, 2005, 439–466.

Aad Dijksma
Department of Mathematics, University of Groningen
P.O. Box 800, 9700 AV Groningen, The Netherlands
e-mail: dijksma@math.rug.nl

Gerald Wanjala
Department of Mathematics, Mbarara University of Science and Technology
P.O. Box 1410, Mbarara, Uganda
e-mail: wanjalag@yahoo.com

Operator Theory:
Advances and Applications, Vol. 162, 145–163

On Nonmonic Quadratic Matrix Polynomials with Nonnegative Coefficients

K.-H. Förster and B. Nagy

Abstract. The matrix polynomial $Q(\lambda) = \lambda I - S(\lambda)$, where $S(\cdot)$ is a nonmonic quadratic matrix polynomial with (entrywise) nonnegative square matrix coefficients, will be studied. We describe the distribution of the eigenvalues of $Q(\cdot)$, depending on the sign of function $r \longmapsto r - \varrho(S(r))$ (here $\varrho(\cdot)$ denotes the spectral radius). The existence of a nonnegative (spectral) matrix root of $Q(\cdot)$ will be related to the existence of a positive $r > \varrho(S(r))$. Assuming that $S(t)$ is irreducible for one positive t, we describe the spectrum of $Q(\cdot)$ on the circles with radius r for any $r = \varrho(S(r)) > 0$, and describe the possibilities for the existence of a nonnegative matrix root of $Q(\cdot)$, for the properties of a corresponding M-matrix and the spectral properties of $Q(\cdot)$, depending on the function $\varrho(S(\cdot))$ and on its derivatives.

Mathematics Subject Classification (2000). Primary 15A28; Secondary 15A48, 47A56, 15A18.

Keywords. Matrix polynomials, (entrywise) nonnegative matrix coefficients, irreducible matrix polynomials, nonnegative matrix roots, eigenvalues, spectral radius.

1. Introduction

We consider polynomials $Q(\cdot)$ of the form

$$Q(\lambda) = \lambda I_{n\times n} - (\lambda^2 A_2 + \lambda A_1 + A_0),$$

where the coefficients A_2, A_1 and A_0 are nonnegative square matrices of size n, $A_2 \neq 0$ and $I_{n\times n}$ denotes the identity matrix of size n.

This is an important class of matrix polynomials; see [19, §2], [15, §23.2], [4, §4], [10] and others.

This work was completed with partial support of the Hungarian National Science Grants OTKA Nos T-030042 and T- 047276 and partial support of the DAAD and the Technical University of Berlin.

For a motivation of this paper and also for possible applications in the theory of Markov chains see, for example, [6], [7] and [8]. Note that we do *not* assume that the matrices A_0, A_1, A_2 are embedded in the (infinite) transition matrix P of a (particular type of a) Markov chain which is assumed to be irreducible in [6], [7] and [8]. In this sense our results are extensions of theirs to this more general situation. We emphasize that our method differs from theirs mainly by the consequent application of the theory of nonnegative matrices and of the properties of the spectral radius of the matrix polynomial $S(\cdot)$ for nonnegative values of the variable (notation see below).

In Section 2 we study the behavior of the spectral radius of the values of the function $r \mapsto S(r) = r^2A_2 + rA_1 + A_0$ for nonnegative values of r. The results here are partly of a preparatory character, partly significant in themselves. Propositions 2.3 and 2.4 describe the distribution of the spectral points (i.e., eigenvalues counted with algebraic multiplicities) in dependence on the relation of the spectral radius of $S(r)$ to r. In Section 3 we study conditions for the existence of a (right) matrix root W of the polynomial $Q(\cdot)$ or, equivalently, for the factorizability of $Q(\cdot)$ into linear factors. The main results here are Theorem 3.4 and Proposition 3.5, in which we relate the existence of a spectral root W to the existence of a positive r strictly majorizing the spectral radius of $S(r)$.

In Section 4 we assume that the matrix $S(r)$ is irreducible for one (and hence for every) positive r. One of the main results here is Proposition 4.5, connecting the value of the index of the phase imprimitivity of the graph of the matrix polynomial S to the spectral behavior of Q on the circle $|z| = r$ for every positive r equaling the spectral radius of $S(r)$. The final result (Theorem 4.10) establishes 8 pairwise exclusive cases for the existence of a nonnegative root of $Q(\cdot)$, for the properties of a corresponding M-matrix B and the spectral properties of $Q(\cdot)$ in dependence on the behavior of the spectral radius of $S(r)$ and of its derivative. We illustrate the obtained results throughout with instructive examples.

Our terminology is mostly traditional. For standard facts in the theory of matrix polynomials we refer to [9], concerning nonnegative matrices to [12] or [3], for some terminology in connection with Markov chains to [6]. We recall here only that the spectrum of a matrix polynomial $P(\cdot)$ is defined as the set of all complex numbers (*eigenvalues*) λ for which the matrix $P(\lambda)$ is not invertible. The positive integer $\dim\ker P(\lambda)$ is the *geometric*, and the multiplicity of λ as a zero of $\det P(\lambda)$ is the *algebraic* multiplicity of λ. Note that *multiplicity* without qualification will always mean algebraic multiplicity. Finally, we want to fix the following

Notation

- By 0_n and $\mathbf{1}_n$ we denote the vectors in $\mathbb{C}^n$ with all components zero or one, respectively.

Let u be a vector in $\mathbb{C}^n$, we write:

- $u > 0_n$ all components of u are nonnegative and $u \neq 0$.
- $u \gg 0_n$ if all components are positive. In this case we call u strictly positive.

- $|u|$ for the absolute value of u; we use the same notations for matrices.
- $\langle u, v\rangle$ for the standard inner product in $\mathbb{C}^n$.
- $\|u\|$ the (Euclidian) norm of u.
- By $I_{n\times n}$ we denote the identity matrix of size $n \times n$,
- By $0_{m\times n}$ and $\mathbf{1}_{m\times n}$ denote the matrices of size $m \times n$ with all entries zero or one, respectively.
- For a matrix A we denote by A^T, $\operatorname{adj}(A)$, $\ker(A)$, $\operatorname{ran}(A)$, $\det(A)$, $\operatorname{rank}(A)$, $\|A\|$, $\operatorname{trace}(A)$, $\sigma(A)$, $\varrho(A)$ and $A(r,s)$ its transpose, adjoint (adjugate in the terminology of [12]), kernel (= nullspace), range, determinant, rank, (spectral) norm, trace, spectrum (= set of its eigenvalues), spectral radius and its entry in the rth row and sth column, respectively.

For a matrix polynomial $P(\cdot)$ its spectrum is denoted by

- $\sigma(P(\cdot)) = \{\lambda \in \mathbb{C} \mid P(\lambda)\}$ is singular, and by
- $\hat{\sigma}(P(\cdot)) = \sigma(P(\cdot)) \cup \{\infty\}$ when the leading coefficient is singular.
- For a positive number r we set
$$\mathbb{D}_r = \{\lambda \in: \mathbb{C} : |\lambda| < r\}, \quad \mathbb{T}_r = \{\lambda \in \mathbb{C} : |\lambda| = r\}.$$
- For a linear space V we denote by $\dim(V)$ its dimension.
- For a positive integer n we set $\langle n\rangle = \{1, 2, \ldots, n\}$.
- We shall use the type of notation $Q(\cdot)$ for a function $z \mapsto Q(z)$ throughout. In particular, $B - \cdot A$ will denote the map $z \mapsto B - zA$.

2. The spectral radius of $S(\lambda) = \lambda^2 A_2 + \lambda A_1 + A_0$

Let A_0, A_1 and A_2 be $n \times n$ nonnegative matrices. We are interested in positive numbers r such that
$$r = \varrho(r^2 A_2 + rA_1 + A_0).$$
We set
$$S(\lambda) = \lambda^2 A_2 + \lambda A_1 + A_0.$$

By a result of E. Vesentini (see [1, Theorem 3.2.7], [8, Proposition 7, p. 545]), the function
$$\log \varrho_S : \mathbb{C} \to \mathbb{R} \text{ with } \lambda \longmapsto \log \varrho(S(\lambda)),$$
hence also the function ϱ_S itself are subharmonic.

Since the coefficients of $S(\cdot)$ are nonnegative, the function
$$\varrho_S : [0, \infty[\to \mathbb{R}_+ \quad \text{with} \quad r \longmapsto \varrho(S(r))$$
is continuous and nondecreasing, and $\varrho_S(r) = \max\{\varrho(S(\lambda)) : |\lambda| = r\}$, since $|S(\lambda)| \leq S(|\lambda|)$ and the spectral radius of matrices is monotone on the cone of the nonnegative matrices. From the theory of subharmonic functions (see [11, Theorem 2.13] or [8, Proposition 7, p. 545]) we obtain that the function $\log(\varrho_S(\cdot))$ is convex in $\log(r)$ on $]0, \infty[$, i.e.,
$$\eta_S :]0, \infty[\to \mathbb{R} \quad \text{with} \quad t \longmapsto \log(\varrho_S(e^t)) \tag{2.1}$$

is convex or, equivalently,

$$\varrho_S({r_1}^\tau {r_2}^{1-\tau}) \le \varrho_S(r_1)^\tau \varrho_S(r_2)^{1-\tau} \quad \text{for} \qquad r_1, r_2 \in]0,\infty[, \qquad \tau \in [0,1].$$

We call a function satisfying the last functional inequality log-log-convex.

Proposition 2.1. *Let $S(\cdot)$ be as above (for the last assertion: with nonnegative coefficients). Then the following assertions are equivalent:*

(I) *$\varrho_S(\cdot)$ is bounded.*

(II) *$\varrho_S(\cdot)$ is constant.*

(III) *$\sigma(S(\lambda))$ is independent of λ, therefore $\sigma(S(\lambda)) = \sigma(A_0)$ for all $\lambda \in \mathbb{C}$.*

If one of the conditions above holds, then A_1 and A_2 are nilpotent.

Proof. See [1, Theorem 3.4.14]. Note that the spectrum of an operator in a finite dimensional space is polynomially convex (has no holes). Since the coefficients are nonnegative matrices, $r^j A_j \le S(r)$ for positive r implies $\varrho(A_j) \le r^{-j}\varrho(S(r)) = r^{-j}\varrho_S(r)$, $j = 0,1,2$. Therefore A_1 and A_2 are nilpotent, if $\varrho_S(\cdot)$ is bounded. □

Example 2.2. Let

$$S(\lambda) = \begin{pmatrix} 0 & \lambda \\ \lambda^2 & 0 \end{pmatrix},$$

then A_1 and A_2 are nilpotent, and $\varrho_S(r) = r^{3/2}$ is not bounded.

Proposition 2.3. *Let $Q(\lambda) = \lambda I_{n\times n} - (\lambda^2 A_2 + \lambda A_1 + A_0)$ with nonnegative $n \times n$ matrices $A_j, j = 0,1,2$. Then*

(I) *Let r be such that $\varrho_S(r) < r$. Then $Q(\cdot)$ has exactly n eigenvalues (counting multiplicities) in the open disc around zero with radius r.*

(II) *Let $r > 0$ and $\delta > 0$ such that $\varrho_S(t) < t$ when $r < t < r + \delta$. Then $Q(\cdot)$ has exactly n eigenvalues (counting multiplicities) in the closed disc around zero with radius r.*

(III) *Let $0 \le r_1 < r_2$ such that $\varrho_S(r) < r$ for all $r \in]r_1, r_2[$. Then*

$$\sigma(Q(\cdot)) \cap \{\lambda \in \mathbb{C} \mid r_1 < |\lambda| < r_2\} = \emptyset.$$

Proof. (I): (see [6, Theorem 2]) For $\tau \in [0,1]$ the matrix $\nu I_{n\times n} - \tau S(\nu)$ is invertible for all $\nu \in \mathbb{T}_r$; indeed, $\varrho(\tau S(\nu)) \le \tau\varrho(|S(\nu)|) \le \tau\varrho(S(|\nu|)) < r = |\nu|$. Therefore the number of eigenvalues of $Q_\tau(\cdot)$ with $Q_\tau(\lambda) = \lambda I_{n\times n} - \tau S(\lambda)$ in $\mathbb{D}_r$ is independent of τ. For $\tau = 0$ this number is n, for $\tau = 1$ this number is the number of eigenvalues of $Q(\cdot)$ in $\mathbb{D}_r$.

(II): By (I), $Q(\cdot)$ has n eigenvalues in $\mathbb{D}_t$ for $r < t < r + \delta$. Let t go to r to obtain (II).

(III): This follows readily from (I). □

Proposition 2.4. *Let $S(\cdot)$ and $Q(\cdot)$ be as above, and let $0 < r_1 < r_2$ be such that $\varrho_S(r_j) = r_j$ for $j = 1,2$. Then we have exactly one of the following two possibilities:*

(I) *$\varrho_S(r) = r$ for all $r \in [r_1, r_2]$.*

(II) *$\varrho_S(r) < r$ for all $r \in]r_1, r_2[$.*

The first case is equivalent to $\sigma(Q(\cdot)) = \mathbb{C}$; the second case is equivalent to

$$\sigma(Q(\cdot)) \cap \{\lambda \in \mathbb{C} \mid r_1 < |\lambda| < r_2\} = \emptyset.$$

Proof. The log-log-convexity of $\varrho_S(\cdot)$ or the convexity of $\eta_S(\cdot)$ show that either (I) or (II) holds.

In case (I) we have $\det(Q(r)) = \det(\varrho(S(r))I_{n\times n} - S(r)) = 0$ for all $r \in [r_1, r_2]$, since $S(r)$ is a nonnegative matrix. Then the polynomial $\det(Q(\cdot))$ vanishes identically, which is equivalent to $\sigma(Q(\cdot)) = \mathbb{C}$. Conversely, $\{\lambda \in \mathbb{C} : \lambda \in \sigma(S(\lambda)\} = \sigma(Q(\cdot)) = \mathbb{C}$ implies that $\varrho_S(r) \geq r$ for all $r \in [r_1, r_2]$. Therefore (I) holds.

Assume that the second case holds. By Proposition 2.3 (III) we obtain $\sigma(Q(\cdot)) \cap \{\lambda \in \mathbb{C} \mid r_1 < |\lambda| < r_2\} = \emptyset$. The converse is true, since $\sigma(Q(\cdot)) \cap \{\mu \in \mathbb{C} \mid r_1 < |\mu| < r_2\} = \emptyset$ implies that $\sigma(Q(\cdot)) \neq \mathbb{C}$. Therefore (I) cannot hold. □

The following example shows that in the second case of the last proposition $\varrho_S(0) = 0$ is possible.

Example 2.5. Let

$$S(\lambda) = \begin{pmatrix} \lambda^2 & 1 \\ \lambda p & \lambda^2 \end{pmatrix},$$

where p is a positive real. Then $\varrho_S(r) = r^2 + (pr)^{1/2}$. It is not difficult to see that for $0 < p < 4/27$ we are in the second case of Proposition 2.4, and of course $\varrho_S(0) = 0$.

Example 2.6. Let $n \in \mathbb{N}$ and $k_j \in \mathbb{N}_0 = \mathbb{N} \cup \{0\}$ for $j = 1, 2, \ldots, n$. Consider the $n \times n$ (weighted cyclic) matrix

$$S(\lambda) = \begin{pmatrix} 0 & \lambda^{k_1} & 0 & \cdots & 0 \\ 0 & 0 & \lambda^{k_2} & & \vdots \\ \vdots & \vdots & \ddots & \ddots & \vdots \\ 0 & 0 & \cdots & 0 & \lambda^{k_{n-1}} \\ \lambda^{k_n} & 0 & \cdots & 0 & 0 \end{pmatrix}, \quad \lambda \in \mathbb{C}.$$

We assume that $2 = \max_{1\leq j\leq n} k_j$. Then $S(\cdot)$ is a quadratic matrix polynomial with entrywise nonnegative matrices as coefficients.

Let $\lambda \neq 0$. Then $\sigma(S(\lambda)) = \{z \in \mathbb{C} : z^n = \lambda^k\}$ where $k = k_1 + \cdots + k_n$. The eigenvalues of $S(\lambda)$ have (geometric and) algebraic multiplicities equal to 1, and the corresponding eigenvectors of $S(\lambda)$ to $z \in \sigma(S(\lambda))$ are scalar multiples of

$$(1, \lambda^{-k_1}z, \lambda^{-k_1-k_2}z^2, \ldots, \lambda^{-k_1-\cdots-k_{n-1}}z^{n-1}).$$

For $r > 0$ the matrix $S(r)$ is nonnegative and irreducible. $S(1)$ is row-stochastic; i.e., $S(1)\mathbf{1}_n = \mathbf{1}_n$.

We have $\varrho_S(r) = r^{\frac{k}{n}}$.

If $k < n$ then $\varrho_S(\cdot)$ is concave and $\varrho_S(r) = r$ for $r = 0, 1$. This shows that Proposition 2.4 is not true for $r_1 = 0$.

For $k = n$ we obtain $\varrho_S(r) = r$ for $r > 0$, therefore $\varrho_S(1) = \varrho_S'(1) = 1$ and $\varrho_S''(1) = 0$ in this case.

For $Q(\lambda) = \lambda I_{n\times n} - S(\lambda)$ we have $\sigma(Q(\cdot)) = \{z \in \mathbb{C} : z^n = z^k\}$. In particular, $\sigma(Q(\cdot)) = \mathbb{C}$ if $n = k$. Note that, due to more stringent initial conditions, this case becomes impossible in [6] (see Proposition 15, p. 135).

Proposition 2.7. *Under the first assumption of Proposition* 2.3 *the following statements hold:*

(I) *Let $\varrho_S(\cdot)$ be differentiable in $r > 0$. Then $\varrho_S(r) = r$ implies*

$$(\det Q)'(r) = (1 - \varrho_S'(r))\,\mathrm{trace}(\mathrm{adj}(\varrho_S(r)I_{n\times n} - S(r))).$$

(II) *Let $\varrho_S(\cdot)$ be twice differentiable in $r > 0$. Then $\varrho_S(r) = r$ and $\varrho_S'(r) = 1$ imply*

$$(\det Q)''(r) = -\varrho_S''(r)\,\mathrm{trace}(\mathrm{adj}(\varrho_S(r)I_{n\times n} - S(r))).$$

Proof. (I): For the derivative of the function $\det Q : \lambda \longmapsto \det(Q(\lambda))$ we have $(\det Q)'(\lambda) = \sum_{j=1}^{n} \det(Q_{(j)}(\lambda))$, where $Q_{(j)}(\lambda)$ is the matrix that coincides with the matrix $Q(\lambda)$ except that the entry in the jth column is differentiated with respect to λ; see [13, p. 491]. Using the corresponding formula for the derivative of $\lambda \longmapsto \det(\varrho_S(\lambda)I_{n\times n} - S(\lambda))$ and the fact that $\varrho_S(r)I_{n\times n} - S(r)$ is singular for $r > 0$ (note that $S(r)$ is nonnegative), we obtain from $\varrho_S(r) = r$ that $(\det Q)'(r) = (1 - \varrho_S'(r)) \sum_{j=1}^{n} \det(Q_{[j]}(r))$, where $Q_{[j]}(r)$ is the matrix that coincides with the matrix $Q(r)$ except that the entry in the jth column is the jth vector of the canonical basis of $\mathbb{C}^n$. Therefore the sum in the last equality is the trace of the adjugate $\mathrm{adj}(\varrho_S(r) - S(r)) = \mathrm{adj}(Q(r))$.

(II): Using again the formula for the derivative of the determinant of a differentiable matrix-valued function and $\varrho_S(r) = r$, we obtain

$$(\det Q)''(r) = (1 - \varrho_S'(r)^2) \sum_{j,k=1}^{n} \det(Q_{[j,k]}(r)) - \varrho_S''(r) \sum_{j=1}^{n} \det(Q_{[j]}(r)),$$

where $Q_{[j,k]}(r)$ is the matrix that coincides with $Q(r)$ except that the entries in the jth and the kth column are the jth and the kth vector of the canonical basis of $\mathbb{C}^n$, respectively. Now the formula in (II) follows immediately. □

3. Nonnegative roots of the polynomial $Q(\cdot)$

In this section we study the problem: under what conditions does there exist a nonnegative $n \times n$ matrix W that is a right root of $Q(\cdot)$, i.e., satisfies the equation

$$W - (A_2W^2 + A_1W + A_0) = 0_{n\times n}, \tag{3.1}$$

where A_j are entrywise nonnegative matrices. The following results have corresponding versions for left roots of $Q(\cdot)$, whose proofs require only minor changes. Here we concentrate on right roots of $Q(\cdot)$.

If W is a right root of $Q(\cdot)$, then there exists a unique $n\times n$ matrix B such that

$$Q(\lambda) = (B-\lambda A_2)(\lambda I_{n\times n} - W) \quad \text{for all} \quad \lambda\in\mathbb{C}. \tag{3.2}$$

This holds if and only if

$$B = I_{n\times n} - A_1 - A_2W, \quad BW = A_0. \tag{3.3}$$

If W is a nonnegative root of $Q(\cdot)$ then B is a Z-matrix. We consider the fixed point iteration

$$W_{k+1} = A_2W_k^2 + A_1W_k + A_0 \quad \text{with} \quad 0\le W_0\le A_0 \tag{3.4}$$

for $k=0,1,2,\ldots$.
The nonnegativity of the matrices A_j ($j=1,2,3$) implies

$$0\le W_k\le W_{k+1}(\text{entrywise}) \text{ for } \quad k=0,1,2,\ldots. \tag{3.5}$$

Indeed, $0\le W_0\le A_0\le W_1$, and $W_{k+1}-W_k = A_2(W_k^2-W_{k-1}^2)+A_1(W_k-W_{k-1})\ge 0$ if $W_k\ge W_{k-1}\ge 0_{n\times n}$.
If W is a nonnegative right root of $Q(\cdot)$, then

$$W_k\le W \quad \text{for} \quad k=0,1,2,\ldots. \tag{3.6}$$

Indeed, $0\le W_0\le A_0\le W$ and $W_k\le W$ imply $W_{k+1} = A_2W_k^2 + A_1W_k + A_0\le A_2W^2+A_1W+A_0 = W$.

The next proposition follows simply from the preceding discussion.

Proposition 3.1. *Let $Q(\lambda) = \lambda I_{n\times n} - (\lambda^2A_2+\lambda A_1+A_0)$ with nonnegative matrices $A_j, j=0,1,2$. Then $Q(\cdot)$ has a nonnegative right root if and only if the fixed point iteration* (3.4) *converges. If* (3.4) *converges, its limit is the smallest nonnegative right root of $Q(\cdot)$.*

If the fixed point iteration (3.4) converges, then we denote its limit, i.e., the smallest nonnegative root of $Q(\cdot)$, by $\hat{W}$, and we denote by $\hat{B}$ the corresponding matrix in the factorization (3.2).

Proposition 3.2. *Let $Q(r)u\ge 0_n$ for some $r>0$ and some vector $u\gg 0_n$. Then $Q(\cdot)$ has a nonnegative root, $\hat{W}u\le ru$ and $\varrho(\hat{W})\le r$.*

Proof. We will show that the sequence $(W_k)_{k=0}^\infty$ of the fixed point iteration (3.4) converges. From $S(r)u\le ru$ it follows easily that $0_n\le W_ku\le ru$ for all k. Let $W_k(p,q)$ denote the element of the matrix W_k in its pth row and qth column. Then we obtain $0\le W_k(p,q)\le \frac{u_p}{u_q}$ for all $p,q\in\langle n\rangle$. Since $(W_k(p,q))_{k=0}^\infty$ is nondecreasing, it converges. Therefore $(W_k)_{k=0}^\infty$ converges.
Above we saw that $W_ku\le ru$; this implies $\hat{W}u\le ru$. Since $u\gg 0_n$, the last inequality implies $\varrho(\hat{W})\le r$, see [3, (1.11), p. 28]. □

Example 3.3. Let a and b be positive numbers and define

$$Q(\lambda) = \lambda I_{2\times 2} - S(\lambda) = \left(I_{2\times 2} - \lambda\begin{pmatrix} a & 0\\ 0 & 0\end{pmatrix}\right)\left(\lambda I_{2\times 2} - \begin{pmatrix} 0 & 0\\ 0 & b\end{pmatrix}\right),$$

Let $ab > 1$. Then there do not exist a vector $u \gg 0_2$ and a positive r such that $Q(r)u \geq 0_2$. This shows that the assumptions in Proposition 2.3 are not necessary for the existence of a nonnegative right root of $Q(\cdot)$. Also $\varrho_S(r) > r$ for all positive r.

For the next theorem we need the notion of a *spectral root* for which we refer to [15, § 22.3, p.115]

Theorem 3.4. *Let $Q(\lambda) = \lambda I_{n\times n} - (\lambda^2 A_2 + \lambda A_1 + A_0)$ with nonnegative matrices $A_j (j = 0, 1, 2)$ and $\varrho_S(r) < r$ for at least one positive r. Then $Q(\cdot)$ has a nonnegative right root; in this case $\varrho(\hat{W}) < r$, $\hat{B}$ is a nonsingular M-matrix (i.e., is invertible with nonnegative inverse) and $\varrho(\hat{B}^{-1}A_2) < 1/r$. Moreover, $\hat{W}$ is a spectral root of $Q(\cdot)$,*

$$\sigma(\hat{W}) = \sigma(Q(\cdot)) \cap \mathbb{D}_r \quad \textit{and} \quad \{\lambda \in \mathbb{C} : 1/\lambda \in \sigma(\hat{B}^{-1}A_2)\} = \sigma(Q(\cdot)) \cap (\mathbb{C}\backslash\overline{\mathbb{D}_r}).$$

Proof. $Q(r)$ is invertible and $Q(r)^{-1} = (rI_{n\times n} - S(r))^{-1} > 0_{n\times n}$. Take a vector $v \gg 0_n$ and define $u = Q(r)^{-1}v$. Since $Q(r)^{-1}$ has at least one positive element in each row, we get $u \gg 0_n$ and $Q(r)u \gg 0$. Apply Proposition 2.3.

Further, $\hat{W}u = (A_2\hat{W}^2 + A_1\hat{W} + A_0)u \leq S(r)u \ll ru$. Thus $S(r)u \leq tu$ for some $t < r$, but then $u \gg 0_n$ implies $\varrho(\hat{W}) \leq t < r$. Similarly $(A_2\hat{W} + A_1)u \leq (1/r)S(r)u \ll u$, and this implies $\varrho(A_2\hat{W} + A_1) < 1$. Therefore $\hat{B} = I_{n\times n} - A_1 - A_2\hat{W}$ is invertible and has a nonnegative inverse. From (3.2) it follows that $(I_{n\times n} - r\hat{B}^{-1}A_2)u = \hat{B}^{-1}Q(r)(rI_{n\times n} - \hat{W})^{-1}u \geq (1/r)\hat{B}^{-1}Q(r)u \gg 0_n$, and we get $\varrho(\hat{B}^{-1}A_2) < 1/r$ (see [3, (1.11), p. 28]). Since the spectrum of the linear polynomial $\hat{B} - \cdot A_2$ is equal to the set $\{\lambda \in \mathbb{C} : 1/\lambda \in \sigma(\hat{B}^{-1}A_2)\}$, and the factorization of $Q(\cdot)$ implies

$$\sigma[Q(\cdot)] = \sigma(\hat{W})\sigma(\hat{B} - \cdot A_2),$$

the last assertions follow immediately. □

Proposition 3.5. *Let $Q(\cdot)$ be as in Theorem 3.4 and let $r > 0$. Then there exist $n\times n$ matrices W and B such that $Q(\cdot)$ admits a factorization (3.2), W is nonnegative, $\varrho(W) < r$, B is a nonsingular M-matrix and $\varrho(B^{-1}A_2) < 1/r$ if and only if $\varrho_S(r) < r$.*

Moreover, if matrices W and B with the properties above exist, then $W = \hat{W}$ and $B = \hat{B}$.

Proof. For the "if"-part see Theorem 3.4. Further, if W and B with these properties exist, then from (3.2) we obtain that the Z-matrix $rI_{n\times n} - S(r)$ is invertible and its inverse is nonnegative. This implies $\varrho_S(r) < r$, see [3, (N_{38}), p. 137]. Since $\sigma(W) = \sigma(Q(\cdot)) \cap \mathbb{D}_r = \sigma(\hat{W})$, and W and $\hat{W}$ are spectral roots of $Q(\cdot)$, they are equal by [15, Lemma 22.8]. □

The following example shows that matrix polynomials of the type above can have different nonnegative roots:

Example 3.6. The polynomial

$$Q(\lambda) = \begin{pmatrix} \lambda & -p-(1-p)\lambda^2 \\ -q-(1-q)\lambda^2 & \lambda \end{pmatrix},$$

where $p, q \in]0,1[$, has the nonnegative roots

$$\begin{pmatrix} 0 & 1 \\ 1 & 0 \end{pmatrix} \quad \text{and} \quad \begin{pmatrix} 0 & p(1-q)^{-1} \\ q(1-p)^{-1} & 0 \end{pmatrix}.$$

The first one is the minimal nonnegative root if $(1-p)(1-q) < pq$, and the second one is the minimal nonnegative root if $(1-p)(1-q) > pq$.

Let $p = q \neq 1/2$. Then $\varrho_S(r) = p + (1-p)r^2$. Therefore $\varrho_S(r) = r$ if and only if $r = 1$ or $r = p(1-p)^{-1}$, and $\varrho_S(r) < r$ if r lies between 1 and $p(1-p)^{-1}$.

4. The irreducible case

For the notion of an irreducible matrix we refer to [12, §6.2]. For $0 < r \leq t$ we have

$$0_{n\times n} \leq \frac{r^2}{t^2} S(t) \leq S(r) \leq S(t) \leq \frac{t^2}{r^2} S(r).$$

Therefore, $S(r)$ is irreducible for one positive r if and only if $S(r)$ is irreducible for all positive r. In this section of the paper we always assume that $S(r)$ is irreducible for all $r > 0$. In this case we say "$S(\cdot)$ is irreducible".

From the Perron-Frobenius theory it follows that $\varrho_S(r)$ is a simple eigenvalue of $S(r)$, i.e., the algebraic multiplicity of $\varrho_S(r)$ as an eigenvalue of $S(r)$ is one, and there is a strictly positive eigenvector u_r of $S(r)$ to $\varrho_S(r)$.

Denoting (not necessarily orthogonal) direct sum by $\oplus$, we have

$$\ker(\varrho_S(r)I_{n\times n} - S(r)) \oplus \operatorname{ran}(\varrho_S(r)I_{n\times n} - S(r)) = \mathbb{C}^n$$

and the spectral projection $P(r)$ of $S(r)$ to $\varrho(S(r))$ [which is the projection of $\mathbb{C}^n$ onto $\ker(\varrho_S(r)I_{n\times n} - S(r))$ along $\operatorname{ran}(\varrho_S(r)I_{n\times n} - S(r))$] is strictly positive (i.e., all elements of $P(r)$ are positive), and has rank 1.

The analytic perturbation theory of eigenvalues (see [2, p. 93, 113, 144], [14, II-1, 2]) shows:

The maps

$$\varrho_S :]0,\infty[\to \mathbb{R}_+ \quad \text{with} \quad r \longmapsto \varrho(S(r)) \quad \text{and}$$
$$P :]0,\infty[\to \mathbb{C}^{n\times n} \quad \text{with} \quad r \longmapsto P(r)$$

are real analytic. Therefore, keeping in mind the log-log-convexity of $\varrho_S(\cdot)$, Proposition 2.4 implies immediately

Proposition 4.1. *Let $S(\cdot)$ be irreducible. Then either $\varrho_S(r) = r$ for all nonnegative r or there are at most two positive r with $\varrho_S(r) = r$.*

For $u > 0_n$ and $r > 0$ the vector $P(r)u$ is a strictly positive eigenvector of $S(r)$ corresponding to its spectral radius $\varrho_S(r)$, and $P^T(r)u$ is a strictly positive eigenvector of $S^T(r)$ to $\varrho_S(r)$; here $\cdot^T$ denotes matrix transposition.

A simple calculation shows that the first derivative of $\varrho_S(\cdot)$ is given by

$$\varrho_S'(r) = \frac{\langle S'(r)P(r)u, P^T(r)v\rangle}{\langle P(r)u, P^T(r)v\rangle} \quad \text{for} \quad r > 0, \quad u > 0_n \quad \text{and} \quad v > 0_n. \tag{4.1}$$

For a proof consider the identity

$$\langle \varrho_S(r)P(r)u, P^T(r)v\rangle = \langle S(r)P(r)u, P^T(r)v\rangle \quad (r > 0),$$

and take the derivatives with respect to r.

Proposition 4.2. *Let $Q(\lambda) = \lambda I_{n\times n} - (\lambda^2 A_2 + \lambda A_1 + A_0)$ with nonnegative matrices A_j $(j = 0, 1, 2)$ such that $S(\cdot)$ is irreducible.*

(I) *Let $Q(r)u = 0_n$ for some $r > 0$ and $u > 0_n$. Then $r = \varrho_S(r), u \gg 0_n$ and $\ker(Q(r)) = \operatorname{span}\{u\}$.*

(II) *Let $Q(r\xi)x = 0_n$ for some r with $\varrho_S(r) = r > 0, |\xi| = 1$. Then $Q(r)|x| = 0_n$; therefore $|x| \in \ker(Q(r))$ and $x \neq 0_n$ implies $|x| \gg 0_n$.*

(III) $\dim\ker(Q(r\xi)) \leq 1$ *for $r = \varrho_S(r) > 0$ and $|\xi| = 1$.*

Proof. (I) $Q(r)u = 0_n$ means that r is an eigenvalue of $S(r)$ and u is a corresponding eigenvector. $S(r)$ irreducible and $u > 0_n$ imply $r = \varrho_S(r)$ and $u \gg 0_n$, [16, I-1.3].

(II) $Q(r\xi)x = 0_n$ implies $r|x| = |S(r\xi)x| \leq S(r)|x|$. Since $S(r)$ is irreducible and $r = \varrho_S(r)$, we have $S(r)|x| = r|x|$, see [16, p. 12].

(III) Let x and y be nonzero vectors in $\ker(Q(r\xi))$. By (II), they have only nonzero components. Denote by x_1 and y_1 their first components, respectively. Then $y_1 x - x_1 y$ is a vector in $\ker(Q(r\xi))$ with 0 as its first component. By (II), this implies $y_1 x - x_1 y$ is zero. Therefore two or more vectors in $\ker(Q(r\xi))$ are linearly dependent. □

We will characterize the number of eigenvalues of $Q(\cdot)$ on $\mathbb{T}_r$ for $r = \varrho_S(r) > 0$ in a similar way as it is known for the number of peripheral eigenvalues of an irreducible matrix. We use some graph theoretical concepts used in [6, §4] to study the spectral properties of M/G/1 Markov chains.

We consider the infinite graph $G(A_0]A_1[A_2) = G = (V, E)$ with the set of vertices and edges

$$V = \{(j, p) \mid 1 \leq j \leq n, p \in \mathbb{Z}\} \quad \text{and}$$
$$E = \{((j, p), (k, q)) \mid A_t(j, k) > 0 \text{ where } t = 1 + q - p\}, \text{ respectively.}$$

Here $A_t(j, k)$ denotes the entry in the jth row and the kth column of the $n \times n$ matrix A_t. In our case we set $A_t = 0_{n\times n}$ for $t \in \mathbb{Z}\backslash\{0, 1, 2\}$. This graph has as adjacency matrix a doubly infinite block Toeplitz matrix with A_1 on its main diagonal.

According to [6] (and others) we call j the *phase* and p the *level* of $(j,p) \in V$. For a path in G with vertices (j_r, p_r), $r = 1, \ldots, s+1$

$$\sum_{r=1}^{s}(p_{r+1} - p_r) = \sum_{r=1}^{s}(t_r - 1) \quad \text{where } A_{t_r}(j_r, j_{r+1}) > 0$$

is called the *level displacement* of the path. Such a path is called a *phase cycle* if $j_1 = j_{s+1}$. Let now $S(1) = A_0 + A_1 + A_2$ be irreducible. Then for $j, k \in \langle n\rangle = \{1, \ldots, n\}$ with $j \neq k$ there exist $s, j_1, \ldots, j_{s+1} \in \langle n\rangle$ and $t_1, \ldots, t_{s+1} \in \{0, 1, 2\}$ such that

$$j_1 = j, \quad j_{s+1} = k \quad \text{and} \quad A_{t_r}(j_r, j_{r+1}) > 0 \quad \text{for } r = 1, \ldots, s.$$

Thus for (j,p) and (k,q) in V exists a path in G from (j,p) to (k,q) with level displacement $\sum\limits_{r=1}^{s}(t_r - 1)$.

Example 4.3. Let

$$A_0 = \begin{pmatrix} 0 & 0 \\ 1 & 0 \end{pmatrix}, \quad A_1 = 0_{n\times n} \quad , \quad A_2 = \begin{pmatrix} 0 & 1 \\ 0 & 0 \end{pmatrix}.$$

Then the level displacements of all phase cycles in $G(A_0]A_1[A_2)$ are zero.

The *index of phase imprimitivity* of G is defined as the g.c.d. (greatest common divisor) of the level displacements of all phase cycles in G; if the level displacements of all phase cycles in G are zero, its index of phase imprimitivity is 0, by definition. All these concepts are in a natural correspondence to those for the standard case $Q(\lambda) = \lambda I_{n\times n} - A_0$; for example, the level displacement of a path in $G(A_0]0_{n\times n}[0_{n\times n})$ is (up to its sign) equal to the length of the corresponding path in the adjacency graph of A_0.

Lemma 4.4. *Let $A_0 + A_1 + A_2$ be irreducible, and d be the index of phase imprimitivity of the graph $G(A_0]A_1[A_2)$. For $j \in \langle n\rangle$ let d_j be the g.c.d. of the level displacements of all cycles of G through j. Then $d_j = d$.*

The proof of this lemma (cf. [6, Proposition 14]) goes in exactly the same way as in the standard case, see [16, Lemma IV-3.1].

Proposition 4.5. *Let $Q(\cdot)$ satisfy the main assumption of Proposition 4.2, and let d be the index of phase imprimitivity of the graph $G(A_0]A_1[A_2)$. Then for all $r > 0$ with $\varrho_S(r) = r$ the following statements hold.*

(I) *$d = 0$ is equivalent to $\mathbb{T}_r \subset \sigma(Q(\cdot))$ (which is equivalent to $\sigma(Q(\cdot)) = \mathbb{C}$).*
(II) *Let $d \neq 0$. Then for $\theta \in [0, 2\pi[$ the complex number $re^{i\theta}$ is an eigenvalue of $Q(\cdot)$ if and only if $\theta \in \{0, \frac{2\pi}{d}, 2\cdot\frac{2\pi}{d}, \ldots, (d-1)\frac{2\pi}{d}\}$.*

Proof. Let $\varrho_S(r) = r > 0$ and let $u > 0_n$ such that $Q(r)u = 0_n$. For $x \in \mathbb{C}^n, x \neq 0_n$, and ξ with $|\xi| = 1$ and $Q(r\xi)x = 0_n$ it follows from Proposition 4.2 that $|x| = \tau u$ for some $\tau > 0$.

Therefore all components x_j of x are nonzero, $j = 1, \ldots, n$. It follows that

$$\frac{x_j}{x_k} = \xi^{t-1}\frac{u_j}{u_k} \quad \text{if} \quad A_t(j,k) \neq 0; \tag{4.2}$$

indeed, from $r\xi x = S(r\xi)x$ it follows that

$$\frac{x_j}{u_j} = \sum_{k=1}^{n}\sum_{t=0}^{2}\left(A_t(j,k)r^{t-1}\frac{u_k}{u_j}\right)\left(\xi^{t-1}\frac{x_k}{u_k}\right).$$

Now $ru = S(r)u$ implies for $j = 1, 2, \ldots, n$

$$1 = \sum_{k=1}^{n}\sum_{t=0}^{2} A_t(j,k)r^{t-1}\frac{u_k}{u_j}.$$

Therefore $\frac{x_j}{u_j}$ is a convex combination of $\xi^{t-1}\frac{x_k}{u_k}, k = 1, \ldots n$. Then $\left|\xi^{t-1}\frac{x_k}{u_k}\right| = \left|\frac{x_k}{u_k}\right| = \tau > 0$ for $k = 1, \ldots, n$ imply $\frac{x_j}{u_j} = \xi^{t-1}\frac{x_k}{u_k}$ if $A_t(j,k) \neq 0$, and (4.2) is proved.

Let $((j_l, p_l))_{l=1}^{s+1}$ be a phase cycle in the graph $G(A_0]A_1[A_2)$ with level displacement $\widetilde{d} = \sum_{l=1}^{s}(t_l - 1)$ where $A_{t_l}(j_l, j_{l+1}) > 0$ for $l = 1, \ldots s$. Then, by (4.2),

$$\xi^{\widetilde{d}} = \frac{x_{j_1}}{x_{j_{s+1}}}\prod_{l=1}^{s}\xi^{t_l-1}\frac{x_{j_{l+1}}}{x_{j_l}} = \frac{u_{j_1}}{u_{j_{s+1}}}\prod_{l=1}^{s}\frac{u_{j_{l+1}}}{u_{j_l}} = 1,$$

where $j_1 = j_{s+1}$.

If $\mathbb{T}_r \subset \sigma(Q(\cdot))$, the last equation implies that $\xi^{\widetilde{d}} = 1$ for all $\xi \in \mathbb{T}_1$, and we obtain $\widetilde{d} = 0$. Therefore the level displacements of all phase cycles in G are zero, i.e., $d = 0$.

If $d > 0$ the equation implies that $\xi^d = 1$, i.e., $\xi = e^{i\theta}$ with $\theta \in \{0, \frac{2\pi}{d}, \ldots, (d-1)\frac{2\pi}{d}\}$.

Now let $\xi \in \mathbb{T}_1$. We set $x_1 = u_1$ and

$$x_k = \xi^{-t+1}\frac{u_k}{u_j}x_j \text{ if } A_t(j,k) \neq 0$$

where $1 \leq k, j \leq n$.

In the first case all x_k for $k = 1, \ldots, n$ are well defined (i.e., they do not depend on t); indeed the irreducibility of $S(1)$ and $d = 0$ imply that $t = \tilde{t}$ if $A_t(j,k) > 0$ and $A_{\tilde{t}}(j,k) > 0$. In the second case they are well defined if $\xi^d = 1$, since $S(1)$ is irreducible and d divides $t - \tilde{t}$ if $A_t(j,k) > 0$ and $A_{\tilde{t}}(j,k) > 0$.

Now $\sum_{k=1}^{n}\sum_{t=0}^{2} A_t(j,k)(r\xi)^t\frac{x_k}{x_j} = \sum_{k=1}^{n}\sum_{t=0}^{2} A_t(j,k)r^t\xi\frac{u_k}{u_j} = r\xi$ for $j = 1, \ldots n$, i.e., $r\xi x = S(r\xi)x$. Therefore $r\xi \in \sigma(Q(\cdot))$. □

Proposition 4.6. *Under the main assumptions of Proposition 4.2 the following statements hold:*

(I) $Q(\cdot)$ *has a nonnegative root with positive spectral radius if and only if* $\varrho_S(r) = r$ *for some positive* r.

(II) *Let* W *be a nonnegative (right) root of* $Q(\cdot)$ *with* $r = \varrho(W) > 0$. *Then* $r = \varrho_S(r)$, $\ker(rI_{n\times n} - W) = \ker(Q(r)) = \operatorname{span}\{u\}$ *for some* $u \gg 0_n$ *and* W *has (algebraically) simple peripheral eigenvalues. If* B *(for* B *see equation (3.2)) is a nonsingular* M*-matrix, then* $B^{-1}A_2$ *has also simple peripheral eigenvalues.*

(III) *Let* W *be a nonnegative root of* $Q(\cdot)$. *Then* W *is irreducible or it has zero columns.*

(IV) *Let* $Q(\cdot)$ *have a nonnegative root. Then*

$$\varrho(\hat{W}) = \min\{r > 0 \mid \varrho_S(r) = r\} =: \hat{r},$$

$$\sigma(\hat{W}) = \sigma(Q(\cdot)) \cap \overline{\mathbb{D}}_{\hat{r}},$$

where $\hat{W}$ *denotes the minimal nonnegative root of* $Q(\cdot)$.

Proof. (I) If W is a nonnegative root of $Q(\cdot)$ and $r = \varrho(W) > 0$, then there exists an eigenvector $u > 0_{n\times n}$ of W corresponding to r. This implies $Q(r)u = 0_n$, what is equivalent to $ru = S(r)u$, i.e., r is a distinguished eigenvalue of $S(r)$. Now $S(r)$ is irreducible therefore $\varrho_S(r) = r$. If $\varrho_S(r) = r > 0$ the irreducibility of $S(r)$ implies that there exists a strictly positive u with $Q(r)u = ru - S(r)u = 0_n$, and we can apply Proposition 3.2.

(II) Clearly $\{0_n\} \neq \ker(rI_{n\times n} - W) \subset \ker(Q(r)) = \ker(rI_{n\times n} - S(r))$. Since $r = \varrho(W)$ and W is nonnegative, W has a nonnegative eigenvector u corresponding to r. The last inclusion shows that r is a distinguished eigenvalue of $S(r)$. The matrix $S(r)$ is irreducible, therefore $r = \varrho_S(r)$, $u \gg 0_n$ and $\ker(rI_{n\times n} - W) = \ker(rI_{n\times n} - S(r)) = \operatorname{span}\{u\}$. Now W has a strictly positive Perron vector from a 1-dimensional eigenspace, and this implies that its peripheral eigenvalues are simple, see [17, Ex.8(a), p. 43] , [18, Cor. 3.5] . If B is a nonsingular M-matrix, its inverse is nonnegative, and

$$Q^T(\lambda) = (\lambda I_{n\times n} - W^T)\lambda B^T(\lambda^{-1} - (A_2B^{-1})^T) \quad \text{for all} \quad \lambda \neq 0. \tag{4.3}$$

We can apply similar arguments as above to prove that the peripheral eigenvalue of $(A_2B^{-1})^T$ (and then of $B^{-1}A_2$ and A_2B^{-1}) are simple.

(III) Assume that W is a nonnegative root of $Q(\cdot)$ which has no zero column and is reducible. Then there exists a nonzero and irreducible $n_2 \times n_2$ matrix W_{22} with $1 \leq n_2 < n$ such that W is cogredient to the block matrix

$$\begin{bmatrix} W_{11} & 0 \\ W_{21} & W_{22} \end{bmatrix},$$

i.e., there exists an $n \times n$ permutation matrix P such that PWP^T is this block matrix. Let for $j = 0, 1, 2$

$$PA_jP^T = \begin{bmatrix} A_{j\,11} & A_{j\,12} \\ A_{j\,21} & A_{j\,22} \end{bmatrix},$$

where $A_{j\,22}$ is an $n_2 \times n_2$ matrix. We will show that $A_{j\,12} = 0$ for $j = 0, 1, 2$. Then $S(1)_{12} = A_{2\,12} + A_{1\,12} + A_{0\,12} = 0$, thus $S(1)$ is reducible; we got a contradiction. Note that $W = S(W)$ implies $(A_2W^2)_{12} + (A_1W)_{12} + A_{0\,12} = 0$. In this sum all summands are nonnegative matrices, thus $A_{0\,12} = 0, (A_1W)_{12} = 0$ and $(A_2W^2)_{12} = 0$. Now $0 = (A_1W)_{12} = A_{1\,12}W_{22}$ and W_{22} irreducible imply $A_{1\,12} = 0$. $(A_2W^2)_{12} = 0$ implies $(A_2)_{12}W_{22}^{h+1} = 0$ for $h = 1, 2, \ldots$. By [16, Problem 8, p. 66] there exists a h such that Z_{22}^{1+h} is irreducible, therefore we obtain finally $A_{2\,12} = 0$.

(IV) The first equality follows from the first part of this proposition, the second follows from Proposition 2.3. □

Example 4.7.

(a) If $A_0 = 0_{n\times n}$, then necessarily $\hat{W} = 0_{n\times n}$.

(b) Let

$$Q(\lambda) = \begin{pmatrix} \lambda & -\lambda^2 \\ -1 & \lambda \end{pmatrix}.$$

Then $S(\cdot)$ is irreducible, and

$$\begin{pmatrix} \alpha & \beta \\ 1 & 0 \end{pmatrix}$$

is a root of $Q(\cdot)$ for all α and β. Clearly, its minimal nonnegative root has a zero column, but is not the zero matrix.

Proposition 4.8. *Under the main assumption of Proposition 4.2 the following statements hold:*

(I) *Let* $\varrho_S(r) = r > 0$. *Then* $(\det Q)'(r) = 0$ *is equivalent to* $\varrho_S'(r) = 1$, $(\det Q)'(r) > 0$ *is equivalent to* $\varrho_S'(r) < 1$, *and* $(\det Q)'(r) < 0$ *is equivalent to* $\varrho_S'(r) > 1$.

(II) *Let* $\varrho_S(r) = r > 0$ *and* $\varrho_S'(r) = 1$. *Then* $(\det Q)''(r) \neq 0$ *is equivalent to* $\varrho_S''(r) \neq 0$.

Proof. Since $S(\cdot)$ is irreducible, the function ϱ_S is differentiable for all positive r and $\mathrm{adj}(\varrho_S(r) - S(r)) \gg 0$, see [16, Corollary 4.1, p. 16]. Therefore $\mathrm{trace}(\mathrm{adj}(\varrho_S(r) - S(r))) > 0$, and both statements follow from Proposition 2.7 (see also [6] and [8, p. 544]). □

Proposition 4.9. *Let* $Q(\cdot)$ *satisfy the main assumption of Proposition* 4.2, *and let* $\varrho_S(r) = r$.

(I) *If* $\varrho_S'(r) \neq 1$, *then the eigenvalues of* $Q(\cdot)$ *on* $\mathbb{T}_r$ *are simple; i.e., their geometric and algebraic multiplicities are 1.*

(II) *If* $\varrho_S'(r) = 1$, *then either* $\varrho_S(t) = t$ *for all* $t \geq 0$ *or* $\varrho_S(t) > t$ *for all positive* $t \neq r$; *in the second case* $\varrho_S''(r) > 0$, *and the eigenvalues of* $Q(\cdot)$ *on* $\mathbb{T}_r$ *have geometric multiplicity* 1 *and algebraic multiplicity* 2.

Proof. In case (I) there are positive t near r such that $\varrho_S(t) < t$. Therefore, by Theorem 3.4, we obtain that $Q(\cdot)$ has a minimal nonnegative root $\hat{W}$, and

$$Q(\lambda) = (\hat{B} - \lambda A_2)(\lambda I_{n\times n} - \hat{W}) \quad \text{for all} \quad \lambda \in \mathbb{C}, \tag{4.4}$$

where $\hat{B}$ has a nonnegative inverse.

Let $\varrho_S'(r) < 1$. Then, by Proposition 2.3 (II), $\varrho(\hat{W}) = r$ and $\varrho(\hat{B}^{-1}A_2) < 1/r$. By Proposition 4.6 (II), the peripheral eigenvalues of $\hat{W}$ are simple. This, equation (4.4) and $\varrho(\hat{B}^{-1}A_2) < 1/r$ prove the assertion.

Let $\varrho_S'(r) > 1$. Then $\varrho(\hat{W}) < r$ and $\varrho(\hat{B}^{-1}A_2) = 1/r$. A similar argument as in the case above gives the assertion. The proof of case (I) is complete.

Let $\varrho_S'(r) = 1$ and not $\varrho_S(t) = t$ for all $t \geq 0$. Set $u = \log(r)$. For the convex function $\eta_S(\cdot)$ from (2.1) we obtain $\eta_S(u) = u$ and $(\eta_S)'(u) = 1$. Since $\eta_S(\cdot)$ is convex and not $\eta_S(v) = v$ for all real v, we obtain $\eta_S(v) > v$ for all real $v \neq u$, that is equivalent to the first assertion in (II).

For $0 < \tau < 1$ we define $Q_\tau(\lambda) := \lambda I_{n\times n} - \tau S(\lambda)$. For $Q_\tau(\cdot)$ we are in the situation of Theorem 3.4. Therefore

$$\begin{aligned} Q_\tau(\lambda) &= (\hat{B}_\tau - \lambda\tau A_2)(\lambda I_{n\times n} - \hat{W}_\tau) \quad \text{for all} \quad \lambda \in \mathbb{C} \quad \text{and} \\ \hat{B}_\tau &= I_{n\times n} - \tau A_1 - \tau A_2 \hat{W}_\tau, \quad \hat{B}_\tau \hat{W}_\tau = \tau A_0, \end{aligned}$$

where $\hat{W}_\tau$ denotes the smallest nonnegative root of $Q_\tau(\cdot)$. From Proposition 4.6 we know that $Q(\cdot)$ has a (smallest) nonnegative root $\hat{W}$. By applying the fixed point iteration for $\hat{W}_\tau$ with initial matrix $0_{n\times n}$, it follows easily that $\hat{W}_{\tau_1} \leq \hat{W}_{\tau_2} \leq \hat{W}$ when $0 < \tau_1 \leq \tau_2 < 1$. Therefore the limit of $\hat{W}_\tau$ exists, when τ goes to 1. Using equation (3.1), it follows that this limit is a nonnegative root of $Q(\cdot)$ which is less than or equal to $\hat{W}$, but then it coincides with $\hat{W}$. From equation (3.3) we obtain that $\hat{B}$ is the limit of $\hat{B}_\tau$, when τ goes to 1.

We will show that $\hat{B}$ is a nonsingular M-matrix. First we show that $\hat{B}$ is nonsingular. $\sigma(Q(\cdot)) \neq \mathbb{C}$, by Proposition 2.4. Therefore $\sigma(\hat{B} - \cdot\ A_2) \neq \mathbb{C}$. Now $\sigma(\hat{B}_\tau - \cdot\ \tau A_2) \cap \overline{\mathbb{D}_r} = \emptyset$ for all $\tau \in]0,1[$, therefore $\sigma(\hat{B} - \cdot\ A_2) \cap \mathbb{D}_r = \emptyset$. Thus $0 \notin \sigma(\hat{B}_\tau - \cdot\ \tau A_2)$ and $\hat{B}$ is invertible. $\hat{B}_\tau$ has a nonnegative inverse, therefore the inverse of $\hat{B}$ is nonnegative and it is a nonsingular M-matrix, since it is a Z-matrix.

The log-log-convexity of $\varrho_S(\cdot)$ and the assumptions that $\varrho_S'(r) = 1$ and $\varrho_S(t) > t$ for $r \neq t > 0$ imply that there exists a positive $\tau_1 < 1$ such that for every τ satisfying $\tau_1 < \tau < 1$ there are exactly two positive numbers $r_{\tau,1}, r_{\tau,2}$ with the property that $\varrho_S(r_{\tau,j}) = r_{\tau,j}$ for $j = 1, 2$ and $r_{\tau,1} < r < r_{\tau,2}$. It is not hard to see that both $r_{\tau,j}$ converge to r when τ goes to 1.

Let d denote the index of phase imprimitivity of $G(A_0]A_1[A_2)$, which is nonzero by our assumptions, see Proposition 4.5. Since we have $G(A_0]A_1[A_2) = G(\tau A_0]\tau A_1[\tau A_2)$ by the proof of the second part of this proposition, the polynomials $\cdot I_{n\times n} - \hat{W}_\tau$ and $\hat{B}_\tau - \cdot\ \tau A_2$ have their eigenvalues for $\theta \in \{0, \frac{2\pi}{d}, 2\cdot\frac{2\pi}{d}, \ldots, (d-1)\frac{2\pi}{d}\}$ exactly at $r_{\tau,1}e^{i\theta}$ and $r_{\tau,2}e^{i\theta}$ respectively. When τ goes to 1, this implies that the polynomials $Q(\cdot)$, $\cdot I_{n\times n} - \hat{W}$ and $\hat{B} - \cdot\ A_2$ have their eigenvalues exactly at $re^{i\theta}$ for $\theta \in \{0, \frac{2\pi}{d}, 2\cdot\frac{2\pi}{d}, \dots, (d-1)\frac{2\pi}{d}\}$. By Proposition 4.6 (II), they are simple for the last two polynomials. Therefore $(\det Q)(re^{i\theta}) = (\det Q)'(re^{i\theta}) = 0$ and $(\det Q)''(re^{i\theta}) = 2 \times \det(\cdot I_{n\times n} - W)'(re^{i\theta}) \times \det(\hat{B} - \cdot\, A_2)'(re^{i\theta}) \neq 0$, i.e., these eigenvalues are of algebraic multiplicity 2 for $Q(\cdot)$. By Proposition 4.9 (I), they are of geometric multiplicity 1. From Proposition 4.8 (II) we obtain $\varrho_S''(r) \neq 0$ and the log-log-convexity of $\varrho_S(\cdot)$ implies $\varrho_S''(r) > 0$. □

We have made all the needed preparations in order to prove the main result of Section 4.

Theorem 4.10. *Let $Q(\lambda) = \lambda I_{n\times n} - (\lambda^2 A_2 + \lambda A_1 + A_0)$ with nonnegative matrices $A_j, j = 0,1,2$, such that $S(\cdot)$ is irreducible, and let d be the index of phase imprimitivity of the graph $G(A_0]A_1[A_2)$. Then, recalling the minimal nonnegative root $\hat{W}$ of $Q(\cdot)$ and the matrix $\hat{B}$ in (3.2), exactly one of the following eight cases holds.*

(I) *$\varrho_S(r) > r$ for all $r \geq 0$. Then $Q(\cdot)$ has no nonnegative root.*

(II) *$\varrho_S(r) > r$ for all $r > 0$ and $\varrho_S(0) = 0$. Then either $Q(\cdot)$ has no nonnegative root or $\hat{B}$ is not a regular M-matrix.*

(III) *There exists exactly one $r > 0$ with $\varrho_S(r) = r$ and $\varrho_S'(r) < 1$. Then $Q(\cdot)$ has a nonnegative root, $\varrho(\hat{W}) = r$, $\hat{B}$ is a nonsingular M-matrix and $\varrho(\hat{B}^{-1}A_2) = 0$; $Q(\cdot)$ has $n-d$ eigenvalues (counting multiplicities) in $\mathbb{D}_r$, d eigenvalues of algebraic multiplicity 1 on $\mathbb{T}_r$ at the dth roots of r^d and n eigenvalues at ∞ (counting multiplicities).*

(IV) *There exists exactly one $r > 0$ with $\varrho_S(r) = r$ and $\varrho_S'(r) > 1$. Then $Q(\cdot)$ has a nonnegative root, $\varrho(\hat{W}) = 0$, $\hat{B}$ is a nonsingular M-matrix and $\varrho(\hat{B}^{-1}A_2) = 1/r$; $Q(\cdot)$ has 0 as eigenvalue of multiplicity n, d eigenvalues of algebraic multiplicity 1 on $\mathbb{T}_r$ at the dth roots of r^d and $n-d$ eigenvalues (including ∞ and counting multiplicities) outside $\overline{\mathbb{D}_r}$.*

(V) *There exist exactly two numbers $r_2 > r_1 > 0$ with $\varrho_S(r_j) = r_j$ for $j = 1,2$. Then $Q(\cdot)$ has a nonnegative root, $\varrho(\hat{W}) = r_1$, $\hat{B}$ is a nonsingular M-matrix and $\varrho(\hat{B}^{-1}A_2) = 1/r_2$; $Q(\cdot)$ has $n-d$ eigenvalues (counting multiplicities) in $\mathbb{D}_{r_1}$, d simple eigenvalues on $\mathbb{T}_{r_1}$ and on $\mathbb{T}_{r_2}$ at the dth roots of r_1^d and r_2^d, respectively, and $n-d$ eigenvalues (including ∞ and counting multiplicities) outside $\overline{\mathbb{D}_{r_2}}$.*

(VI) *There exists exactly one $r > 0$ with $\varrho_S(r) = r$ and $\varrho_S'(r) = 1$. Then $Q(\cdot)$ has a nonnegative root, $\varrho(\hat{W}) = r$, $\hat{B}$ is a nonsingular M-matrix, $\varrho(\hat{B}^{-1}A_2) = 1/r$ and $\varrho_S''(r) > 0$; $Q(\cdot)$ has $n-d$ eigenvalues (counting multiplicities) in $\mathbb{D}_r$, and d eigenvalues of geometric multiplicity 1 and algebraic multiplicity 2 on $\mathbb{T}_r$ at the dth roots of r^d and $n-d$ eigenvalues (including ∞ and counting multiplicities) outside $\overline{\mathbb{D}_r}$.*

(VII) $\varrho_S(r) < r$ *for all* $r > 0$*. Then* $Q(\cdot)$ *has a nonnegative root,* $\varrho(\hat{W}) = 0$*,* $\hat{B}$ *is a nonsingular* M*-matrix and* $\varrho(\hat{B}^{-1}A_2) = 0$*;* $Q(\cdot)$ *has* 0 *and* ∞ *as eigenvalues both of multiplicity* n*.*

(VIII) $\varrho_S(r) = r$ *for all* $r \geq 0$*. Then* $Q(\cdot)$ *has a nonnegative root and* $\hat{B} - \lambda A_2$ *is not invertible for all* $\lambda \in \mathbb{C}$*;* $\sigma(Q(\cdot)) = \mathbb{C}$*.*

Proof. (I) From $\varrho_S(r) > r$ for all $r \geq 0$ and Proposition 4.6 (I) it follows that $Q(\cdot)$ can have only nilpotent nonnegative roots. If we assume that such a root W exists, then $A_0 \leq W$ by equation (3.1). Therefore we obtain the contradiction $0 = \varrho(W) \geq \varrho(A_0) = \varrho_S(0) > 0$.

(II) If we assume that $Q(\cdot)$ has a nonnegative root then, by Proposition 4.6 (I), it is nilpotent, therefore $\hat{W}$ is nilpotent. Assume that $\hat{B}$ is a nonsingular M-matrix. Then $\varrho(\hat{B}^{-1}A_2) = 0$, otherwise $\varrho_S(r) = \varrho_{S^T}(r) = r$ for $r = 1/\varrho(\hat{B}^{-1}A_2)$. Therefore $Q(r)$ is invertible for all positive r; now $\varrho_S(r) > r$ implies that its inverse is not nonnegative, since it is a Z-matrix. From $Q(r)^{-1} = (rI_{n\times n} - \hat{W})^{-1}(I_{n\times n} - r\hat{B}^{-1}A_2)^{-1}\hat{B}^{-1}$ it follows that $Q(r)^{-1}$ is nonnegative for small positive r. This is a contradiction.

(III) From Proposition 4.6 (I) it follows that $Q(\cdot)$ has a nonnegative root, $\varrho(\hat{W}) = r$, $\hat{B}$ is a nonsingular M-matrix and $\varrho(\hat{B}^{-1}A_2) < 1/r$. Therefore $\hat{B} - tA_2$ is invertible for all $t \in [0, r]$. For $t > r$ we have $\varrho_S(t) < t$, since r is the only positive number with $\varrho_S(t) = t$ and $\varrho_S(\cdot)$ is continuous. Then $Q(t)$ and therefore $\hat{B} - tA_2$ are invertible for $t > r$. Now $\hat{B} - tA_2$ is invertible for all nonnegative t, this is equivalent to $\varrho(\hat{B}^{-1}A_2) = 0$. The assertions on the eigenvalues of $Q(\cdot)$ follow from Propositions 2.3, 4.5 and 4.9 (I).

(IV) We have $\varrho_S(t) < t$ for all $t \in]0, r[$ since r is the only positive t with $\varrho_S(t) = t$ and $\varrho_S(\cdot)$ is continuous. Therefore $\varrho(\hat{W}) = 0$, by Theorem 3.4. $\hat{B}$ is a nonsingular M-matrix (see the proof of Proposition 4.9 (I)), and necessarily $\varrho(\hat{B}^{-1}A_2) = 1/r$. The assertions on the eigenvalues of $Q(\cdot)$ follow from Propositions 2.3, 4.5 and 4.9 (I).

(V) We have $\varrho_S(t) < t$ for $t \in]r_1, r_2[$, by Proposition 2.4 (II). From Theorem 3.4 it follows that $\hat{W}$ and $\hat{B}$ exist, further $\varrho(\hat{W}) \leq r_1$ and $\varrho(\hat{B}^{-1}A_2) \leq 1/r_2$. Since r_1 and r_2 are eigenvalues of $Q(\cdot)$ we have equality in the last two inequalities. The further assertions on the eigenvalues of $Q(\cdot)$ follow from Propositions 2.3, 4.5 and 4.9 (I).

(VI) The assertions follow immediately from the proof of Proposition 4.9 (II) and from Propositions 4.5 and 2.3.

(VII) The assertions follow immediately from Theorem 3.4.

(VIII) For a proof see Proposition 4.1 and Proposition 2.4. □

Example 4.11. We will show by examples that in case (II) both possibilities can occur.

(a) Let $S(\lambda) = \lambda^2 A_2 + \lambda I_{n\times n}$, where $A_2 > 0_{n\times n}$ is irreducible. Then

$$Q(\lambda) = -\lambda^2 A_2 = (0_{n\times n} - \lambda A_2)(\lambda I_{n\times n} - 0_{n\times n}),$$

$\varrho_S(r) = r + r^2 \varrho(A_2) > r$ for $r > 0$, and $Q(\cdot)$ has the minimal nonnegative root $0_{n\times n}$, but $\hat{B} = 0_{n\times n}$ is not invertible.

(b) Consider Example 2.5. It is not difficult to see that for $p > 4/27$ we have $\varrho_S(r) > r$ for $r > 0$. We will show that $Q(\cdot)$ has no nonnegative root. Assume that W is a nonnegative root of $Q(\cdot)$. Then it is nonzero and nilpotent. Therefore

$$W = \begin{pmatrix} 0 & 0 \\ w & 0 \end{pmatrix}, \quad \text{or} \quad W = \begin{pmatrix} 0 & w \\ 0 & 0 \end{pmatrix}$$

with $w > 0$. In both cases, from

$$BW = A_0 = \begin{pmatrix} 0 & 1 \\ 0 & 0 \end{pmatrix}$$

we obtain a contradiction.

Acknowledgment

The authors wish to thank a referee for his/her extremely careful work and suggestions which improved the presentation of the paper. We thank Dr. J. Behrndt for his help in the preparation of the final version of this paper.

References

[1] Aupetit, B.: A Primer on Spectral Theory, Springer-Verlag: New York, 1991.

[2] Baumgärtel, H.: Analytic Perturbation Theory for Matrices and Operators, Operator Theory: Advances and Applications, Vol. 15, Birkhäuser: Basel-Boston-Stuttgart, 1985.

[3] Berman, A., Plemmons, R.: Nonnegative Matrices in the Mathematical Sciences. SIAM: Philadelphia, 1994.

[4] Bini, D.A., Latouche, G., Meini, B.: Solving Matrix Polynomial Equations Arising in Queueing Problems. Linear Algebra Appl. **340**, 225–244 (2002).

[5] Butler, G.J., Johnson, C.R., Wolkowicz, H.: Nonnegative Solutions of a Quadratic Matrix Equation Arising from Comparison Theorems in Ordinary Differential Equations. SIAM J. Alg. Disc. Meth. **6**, 47–53 (1985).

[6] Gail, H.R., Hantler, S.L., Taylor, B.A.: Spectral Analysis of $M/G/1$ and $G/M/1$ Type Markov Chains, Adv. Appl. Prob. **28**, 114–165 (1996).

[7] Gail, H.R., Hantler, S.L., Taylor, B.A.: Non-skip-free $M/G/1$ and $G/M/1$ Type Markov Chains. Adv. Appl. Prob. **29**, 733–758 (1997).

[8] Gail, H.R., Hantler, S.L., Taylor, B.A.: Matrix-geometric Invariant Measures for G/M/1 Type Markov Chains, Commun. Statist.- Stochastic Models **14(3)**, 537–569 (1998).

[9] Gohberg, I., Lancaster, P., Rodman, L.: Matrix Polynomials, Academic Press: New York, 1982.

[10] Grassmann, W.K.: Real Eigenvalues of Certain Tridiagonal Matrix Polynomials with Queueing Applications. Linear Algebra App. **342**, 93–106 (2002).

[11] Hayman, W.K., Kennedy, P.B.: Subharmonic Functions, Vol. I, Academic Press: New York, 1976.

[12] Horn, R.A., Johnson, C.R.: Matrix Analysis, Cambridge University Press: Cambridge, 1985.

[13] Horn, R.A., Johnson, C.R.: Topics in Matrix Analysis, Cambridge University Press: Cambridge, 1991.

[14] Kato, T.: Perturbation Theory for Linear Operators, Springer-Verlag: Berlin-Heidelberg-New York, 1966.

[15] Marcus, A.S.: Introduction to the Spectral Theory of Polynomial Operator Pencils. Translation of Mathematical Monograph, Vol. 71, Amer. Math. Soc., Providence, 1988.

[16] Minc, H.: Nonnegative Matrices, Wiley: New York, 1988.

[17] Schaefer, H.H.: Banach Lattices and Positive Operators, Springer-Verlag: New York, 1974.

[18] Schneider, H.: The Influence of the Marked Reduced Graph of a Nonnegative Matrix on the Jordan Form and on Related Properties: A Survey. Linear Algebra Appl. **84**, 169–189 (1986)

[19] Tisseur, F., Meerbergen, K.: The Quadratic Eigenvalue Problem. SIAM Review **43**, 235–286 (2001).

K.-H. Förster
Technische Universität Berlin
Institut für Mathematik, MA 6-4
D-10623 Berlin
Germany
e-mail: `foerster@math.tu-berlin.de`

B. Nagy
University of Technology and Economics
Department of Analysis
Institute of Mathematics
H-1521 Budapest
Hungary
e-mail: `bnagy@math.bme.hu`

Operator Theory:
Advances and Applications, Vol. 162, 165–190

On Operator Representations of Locally Definitizable Functions

Peter Jonas

Abstract. Let Ω be some domain in $\overline{\mathbf{C}}$ symmetric with respect to the real axis and such that $\Omega \cap \overline{\mathbf{R}} \neq \emptyset$ and the intersections of Ω with the upper and lower open half-planes are simply connected. We study the class of piecewise meromorphic $\mathbf{R}$-symmetric operator functions G in $\Omega \setminus \overline{\mathbf{R}}$ such that for any subdomain Ω' of Ω with $\overline{\Omega'} \subset \Omega$, G restricted to Ω' can be written as a sum of a definitizable and a (in Ω') holomorphic operator function. As in the case of a definitizable operator function, for such a function G we define intervals $\Delta \subset \mathbf{R} \cap \Omega$ of positive and negative type as well as some "local" inner products associated with intervals $\Delta \subset \mathbf{R} \cap \Omega$.

Representations of G with the help of linear operators and relations are studied, and it is proved that there is a representing locally definitizable selfadjoint relation A in a Krein space which locally exactly reflects the sign properties of G: The ranks of positivity and negativity of the spectral subspaces of A coincide with the numbers of positive and negative squares of the "local" inner products corresponding to G.

Mathematics Subject Classification (2000). 47 B 50, 47 A 56, 47 A 60.

Keywords. definitizable operator functions, generalized Nevanlinna functions, selfadjoint and unitary operators in Krein spaces, locally definitizable operators, spectral points of positive and negative type.

1. Introduction

Let $(\mathcal{H}, [\cdot,\cdot])$ be a separable Krein space and let $\mathcal{L}(\mathcal{H})$ denote the algebra of bounded linear operators in $\mathcal{H}$. Recall that a piecewise meromorphic $\mathcal{L}(\mathcal{H})$-valued function G in $\mathbf{C} \setminus \mathbf{R}$ symmetric with respect to $\mathbf{R}$ (that is, $G(\overline{z}) = G(z)^+$ for all points z of holomorphy of G; " + " denotes the Krein space adjoint) is called *definitizable* if there exists an $\mathbf{R}$-symmetric scalar rational function r such that the

This work was supported by the Hochschul- und Wissenschaftsprogramm des Bundes und der Länder of Germany.

product rG is the sum of a Nevanlinna function N and an $\mathcal{L}(\mathcal{H})$-valued rational function P with the poles of P being points of holomorphy of G:

$$r(z)G(z) = N(z) + P(z)$$

for all points $z \in \mathbf{C} \setminus \mathbf{R}$ of holomorphy of rG. A rational operator function is by definition a meromorphic operator function in $\overline{\mathbf{C}}$ ([5]). The classes $N_k(\mathcal{L}(\mathcal{H}))$, $k = 0, 1, \ldots$, of generalized Nevanlinna operator functions, introduced and first studied by M.G. Krein and H. Langer, are contained in the set of the definitizable operator functions ([4], [5]).

By [5, Proposition 3.3] an $\mathbf{R}$-symmetric piecewise meromorphic $\mathcal{L}(\mathcal{H})$-valued function G in $\mathbf{C} \setminus \mathbf{R}$ is definitizable if and only if it has no more than a finite number of nonreal poles, the order of growth of G near $\overline{\mathbf{R}}$ is finite (see Section 2.1) and there is a finite (possibly empty) subset e of $\overline{\mathbf{R}}$ such that every connected component of $\overline{\mathbf{R}} \setminus e$ is of definite type with respect to G (see Definition 2.5). We can use this characterization of definitizability of operator functions to introduce a local variant of this notion ([6, Definition 4.1], see Definition 2.9 below), that is, we define, in a natural way, operator functions definitizable in some domain Ω. In the same way as for definitizable operator functions open subsets of $\overline{\mathbf{R}}$ of type π_+ and π_- can be defined, which gives a localization of the characteristic properties of the generalized Nevanlinna functions (Section 2.3).

Let, in the following, Ω be a domain in $\overline{\mathbf{C}}$ which is symmetric with respect to $\mathbf{R}$, such that $\Omega \cap \overline{\mathbf{R}} \neq \emptyset$, and $\Omega \cap \mathbf{C}^+$ and $\Omega \cap \mathbf{C}^-$ are simply connected. Here $\mathbf{C}^+$ and $\mathbf{C}^-$ denote the open upper and the open lower half planes, respectively. An operator function G is definitizable in Ω if and only if for every domain Ω' with the same properties as Ω, and with $\overline{\Omega}' \subset \Omega$, the restriction of G to Ω' can be written as a sum of a definitizable operator function and an operator function holomorphic in Ω' (see Proposition 2.10).

The main objective of the present paper are representations of operator functions definitizable in Ω with the help of selfadjoint operators or selfadjoint relations definitizable in Ω (Section 3). We consider representations of the form studied in [3] for generalized Nevanlinna functions and in [5] for definitizable functions. A local variant of the notion of minimality of a representation is introduced (Definition 3.2). If a representation of an operator function G is locally minimal, then the local "sign properties" of G (including multiplicities) are exactly reflected by the local "sign properties" of the representing relation. Moreover, if A_1 and A_2 are two locally minimal locally definitizable representing relations for G, results from [5] on the "local unitary equivalence" of A_1 and A_2 for the case of a definitizable G remain true in our more general situation.

In Section 3.2 we shall show that for every domain Ω' with the same properties as Ω, and with $\overline{\Omega'} \subset \Omega$, there exists a locally minimal representation of the restriction of G to Ω' with the help of some selfadjoint relation A in a Krein space which is definitizable over Ω'. This will be proved with the help of a variant of T. Ya. Azizov's theorem on the representation of operator functions (Theorem 3.7): there exists a minimal representing selfadjoint relation with spectrum outside of an arbitrarily chosen compact subset of the set of holomorphy of the operator function.

By a linear fractional transformation of the independent variable and by making use of the corresponding Cayley transformation all definitions and results mentioned above can be carried over to similar definitions and equivalent results for operator functions skew-symmetric with respect to the unit circle $\mathbf{T}$. It is often convenient to give the proofs in the $\mathbf{T}$-skew-symmetric situation. Therefore, we shall formulate all definitions and most of the results for both situations.

2. Locally definitizable operator functions

2.1. Preliminaries on R-symmetric and T-skew-symmetric operator functions

For every subset M of $\overline{\mathbf{C}}$ we set $M^* := \{\bar{\lambda} : \lambda \in M\}$ and $\hat{M} := \{\bar{\lambda}^{-1} : \lambda \in M\}$. For a scalar function f defined on a set $M \subset \overline{\mathbf{C}}$ with $M = M^*$ ($M = \hat{M}$) we set $f^*(\lambda) := \overline{f(\bar{\lambda})}$ (resp. $\hat{f}(\lambda) := \overline{f(\bar{\lambda}^{-1})}$). If the values of f are bounded linear operators in a Krein space $\mathcal{H}$ we set $f^*(\lambda) := f(\bar{\lambda})^+$ (resp. $\hat{f}(\lambda) := f(\bar{\lambda}^{-1})^+$).

Let, in this and the following sections, Ω be a domain in $\overline{\mathbf{C}}$ with the properties mentioned in the introduction. Let $\lambda_0 \in \Omega \cap \mathbf{C}^+$,

$$\psi(\lambda) := -(\lambda - \lambda_0)(\lambda - \overline{\lambda_0})^{-1} \quad \phi(z) := (\overline{\lambda_0} z + \lambda_0)(z + 1)^{-1}.$$

Then $\phi \circ \psi = \mathrm{id}$ and $\psi(\overline{\mathbf{R}}) = \mathbf{T}$. The domain $\psi(\Omega)$ is symmetric with respect to $\mathbf{T}$, $\psi(\Omega) \cap \mathbf{T}$ is not empty, $0, \infty \in \psi(\Omega)$, and $\psi(\Omega) \cap \mathbf{D}$ and $\psi(\Omega) \cap \hat{\mathbf{D}}$ are simply connected domains of $\overline{\mathbf{C}}$. Here $\mathbf{D}$ denotes the open unit disc.

Let G be an $\mathcal{L}(\mathcal{H})$-valued meromorphic function in $\Omega \setminus \overline{\mathbf{R}}$, $G = G^*$, such that no point of $\Omega \cap \overline{\mathbf{R}}$ is an accumulation point of nonreal poles of G. Let μ be a point of $\Omega \cap \overline{\mathbf{R}}$ such that G can be continued analytically in μ from $\Omega \cap \mathbf{C}^+$ and (hence, also) from $\Omega \cap \mathbf{C}^-$ and these analytic continuations coincide. In the following we will tacitly assume that G is defined also in these points μ, and by a "point of holomorphy" of G we will understand either a point of holomorphy of G in $\Omega \setminus \overline{\mathbf{R}}$ or a point $\mu \in \Omega \cap \overline{\mathbf{R}}$ with the property just mentioned.

If M is a closed subset of $\overline{\mathbf{C}}$ and $\mathcal{X}$ is a Banach space, the linear space of all locally holomorphic functions on M with values in $\mathcal{X}$ equipped with the usual topology (see [7, Section 27.4]) will be denoted by $H(M, \mathcal{X})$. We set $H(M) := H(M, \mathbf{C})$.

Assume that $\lambda_0 \in \Omega \cap \mathbf{C}^+$ is a point of holomorphy of G. Let $\mathcal{O}^+$ be a bounded C^∞-domain (not necessarily simply connected) with $\overline{\mathcal{O}^+} \subset \Omega \cap \mathbf{C}^+$ and $\lambda_0 \in \mathcal{O}^+$ such that G is locally holomorphic on $\overline{\mathcal{O}^+}$. Then by $G = G^*$, G is also locally holomorphic on $\overline{\mathcal{O}^-}$, $\mathcal{O}^- := (\mathcal{O}^+)^*$. For every $g \in H(\overline{\mathbf{C}} \setminus (\mathcal{O}^+ \cup \mathcal{O}^-))$ we define

$$S_G.g := -2i\,(\operatorname{Im}\lambda_0) \int_{\mathcal{C}} G(\lambda) g(\lambda)(\lambda - \lambda_0)^{-1}(\lambda - \overline{\lambda_0})^{-1}\, d\lambda, \qquad (2.1)$$

where $\mathcal{C} = \partial\mathcal{O}^+ \cup \partial\mathcal{O}^-$ †. Evidently, for every function g locally holomorphic on

$$(\overline{\mathbf{C}} \setminus \Omega) \cup \overline{\mathbf{R}} \cup \{\text{poles of } G\}$$

we may find some domain $\mathcal{O}^+$ as above and such that $g \in H(\overline{\mathbf{C}} \setminus (\mathcal{O}^+ \cup \mathcal{O}^-))$. Then the operator $S_G.g$ is defined, and it does not depend on the choice of $\mathcal{O}^+$. S_G is a continuous linear mapping of $H((\overline{\mathbf{C}} \setminus \Omega) \cup \overline{\mathbf{R}} \cup \{\text{poles of } G\})$ into $\mathcal{L}(\mathcal{H})$. It is easy to see that $S_G.g^* = (S_G.g)^+$.

It is not difficult to find some right inverse of the mapping $G \mapsto S_G$: Let $\sigma_0 = \sigma_0^*$ be a countable subset of $\Omega \setminus \overline{\mathbf{R}}$ which has no accumulation points in Ω, $\lambda_0 \notin \sigma_0$, and let S be a continuous linear mapping of $H((\overline{\mathbf{C}} \setminus \Omega) \cup \overline{\mathbf{R}} \cup \sigma_0)$ in $\mathcal{L}(\mathcal{H})$ such that

$$S\,.\,g^* = (S\,.\,g)^+ \quad \text{for all} \quad g \in H((\overline{\mathbf{C}} \setminus \Omega) \cup \overline{\mathbf{R}} \cup \sigma_0)$$

(or, equivalently, $S.g$ is selfadjoint for $g = g^* \in H((\overline{\mathbf{C}} \setminus \Omega) \cup \overline{\mathbf{R}} \cup \sigma_0))$ and S is of finite order at every point μ_0 of σ_0. That is, the restriction of S to the subspace of all functions $g \in H((\overline{\mathbf{C}} \setminus \Omega) \cup \overline{\mathbf{R}} \cup \sigma_0)$ which are zero in some neighborhood of $((\overline{\mathbf{C}} \setminus \Omega) \cup \overline{\mathbf{R}} \cup \sigma_0) \setminus \{\mu_0\}$ has the form $g \longmapsto \sum_{\nu=0}^{k} A_\nu g^{(\nu)}(\mu_0)$ where $A_\nu \in \mathcal{L}(\mathcal{H})$, $\nu = 0, \cdots, k$, for some $k \in \mathbf{N}$. We denote the linear space of these mappings by $\Phi(\Omega, \overline{\mathbf{R}} \cup \sigma_0; \mathcal{L}(\mathcal{H}))$. If G is as above then S_G belongs to this space where σ_0 is the set of poles of G in $\Omega \setminus \overline{\mathbf{R}}$. For $S \in \Phi(\Omega, \overline{\mathbf{R}} \cup \sigma_0; \mathcal{L}(\mathcal{H}))$ we define

$$G_S(\lambda) := S\,.\,g_\lambda \quad \text{where}$$

$$g_\lambda(w) := (4\pi)^{-1}(\operatorname{Im}\lambda_0)^{-1}(\lambda - \operatorname{Re}\lambda_0 + (\lambda - \lambda_0)(\lambda - \overline{\lambda_0})(w - \lambda)^{-1}).$$

The function G_S fulfils the general assumptions on the operator functions G considered in this section. It is not difficult to verify that

$$S_{G_S} = S.$$

If G is as at the beginning of this section, then

$$G(\lambda) - \frac{1}{2}(G(\lambda_0) + G(\lambda_0)^+) = S_G\,.\,g_\lambda (= G_{S_G}(\lambda))$$

for all points λ of holomorphy of G in $\Omega \setminus \mathbf{R}$.

As in [5, Section 3] besides the operator-valued functional S_G we consider a form-valued functional $S_G(\cdot,\cdot)$. Let $\mathcal{O}^+$, $\mathcal{O}^-$, $\mathcal{C}$ and g be as in the definition of S_G, and let u, v be $\mathcal{H}$-valued functions locally holomorphic on $\overline{\mathbf{C}} \setminus (\mathcal{O}^+ \cup \mathcal{O}^-)$, that is $u, v \in H(\overline{\mathbf{C}} \setminus (\mathcal{O}^+ \cup \mathcal{O}^-), \mathcal{H})$. Then we set

$$S_G(u,v).g := -2i\,(\operatorname{Im}\lambda_0) \int_{\mathcal{C}} [G(\lambda)u(\lambda), v(\overline{\lambda})] g(\lambda)(\lambda - \lambda_0)^{-1}(\lambda - \overline{\lambda_0})^{-1}\, d\lambda. \quad (2.2)$$

This defines $S_G(\cdot,\cdot).(\cdot)$ for all ($\mathcal{H}$-valued and scalar, respectively) functions locally holomorphic on $H((\overline{\mathbf{C}} \setminus \Omega) \cup \overline{\mathbf{R}} \cup \{\text{poles of } G\})$. If $g = g^*$, the sesquilinear form $(u, v) \mapsto S_G(u,v).g$ is hermitian.

Let F be a $\mathcal{L}(\mathcal{H})$-valued meromorphic function in $\psi(\Omega \setminus \overline{\mathbf{R}}) = \psi(\Omega) \setminus \mathbf{T}$ which is skew-symmetric with respect to the unit circle $\mathbf{T}$: $\hat{F} = -F$. Assume that no

† In the definition of S_G in [6], relation (3.8), a minus sign is missing.

point of $\psi(\Omega) \cap \mathbf{T}$ is an accumulation point of non-unimodular poles of F and that F is holomorphic at 0 and ∞. Then

$$G := iF \circ \psi \tag{2.3}$$

satisfies the assumptions mentioned at the beginning of this section. If $\mathcal{O}^+$ and $\mathcal{O}^-$ are as above then for every $f \in H(\overline{\mathbf{C}} \setminus \psi(\mathcal{O}^+ \cup \mathcal{O}^-))$ we define

$$T_F.\, f := \int_{\psi(\mathcal{C})} F(z) f(z) (iz)^{-1} dz, \tag{2.4}$$

where $\psi(\mathcal{C}) = \partial\psi(\mathcal{O}^+) \cup \partial\psi(\mathcal{O}^-)$. Similarly to the definition of S_G, in this way the operator $T_F.f$ is defined for every function f which is locally holomorphic on $(\overline{\mathbf{C}} \setminus \psi(\Omega)) \cup \mathbf{T} \cup \{\text{poles of } F\}$.

Below we will make use of some right inverse of the mapping $F \mapsto T_F$: Let $\tau_0 = \hat{\tau}_0$ be a countable bounded subset of $\psi(\Omega) \setminus \mathbf{T}$ which has no accumulation points in Ω, and let T be a continuous linear mapping of $H((\overline{\mathbf{C}} \setminus \psi(\Omega)) \cup \mathbf{T} \cup \tau_0)$ in $\mathcal{L}(\mathcal{H})$ such that

$$T \,.\, \hat{f} = (T \,.\, f)^+ \quad \text{for all} \quad f \in H((\overline{\mathbf{C}} \setminus \psi(\Omega)) \cup \mathbf{T} \cup \tau_0)$$

(or, equivalently, $T.f$ is selfadjoint for all $f = \hat{f} \in H((\overline{\mathbf{C}} \setminus \psi(\Omega)) \cup \mathbf{T} \cup \tau_0))$ and T is of finite order at every point of τ_0. The linear space of these mappings is denoted by $\Phi(\psi(\Omega), \mathbf{T} \cup \tau_0; \mathcal{L}(\mathcal{H}))$. If F is as above then T_F belongs to this space where τ_0 is the set of all poles of F in $\psi(\Omega) \setminus \mathbf{T}$. If $T \in \Phi(\psi(\Omega), \mathbf{T} \cup \tau_0; \mathcal{L}(\mathcal{H}))$ and $\zeta \in \psi(\Omega) \setminus (\mathbf{T} \cup \tau_0)$ we define an operator function F_T by

$$F_T(\zeta) := T \,.\, h_\zeta \quad \text{where} \quad h_\zeta(z) := (4\pi)^{-1}(z + \zeta)(z - \zeta)^{-1}.$$

Then F_T is meromorphic in $\psi(\Omega) \setminus \mathbf{T}$, the poles of F_T in $\psi(\Omega) \setminus \mathbf{T}$ are contained in τ_0 and we have $\widehat{F_T} = -F_T$. Moreover,

$$T_{F_T} = T. \tag{2.5}$$

If F is as above then

$$F(z) - \frac{1}{2}(F(0) + F(0)^+) = T_F.h_z (= F_{T_F}(z)) \tag{2.6}$$

for all points of holomorphy of F in $\psi(\Omega) \setminus \mathbf{T}$.

If, again, the operator function F is as above, $f \in H(\overline{\mathbf{C}} \setminus \psi(\mathcal{O}^+ \cup \mathcal{O}^-))$ and $p, q \in H(\overline{\mathbf{C}} \setminus \psi(\mathcal{O}^+ \cup \mathcal{O}^-), \mathcal{H})$, we define

$$T_F(p,q).f := \int_{\psi(\mathcal{C})} [F(z)p(z), q(\bar{z}^{-1})] f(z)(iz)^{-1} dz.$$

If (2.3) holds then for all functions f, p, q which are locally holomorphic on $(\overline{\mathbf{C}} \setminus \psi(\Omega)) \cup \mathbf{T} \cup \{\text{poles of } F\}$ (scalar and with values in $\mathcal{H}$, respectively) we have

$$T_F.f = S_G.(f \circ \psi), \tag{2.7}$$

$$T_F(u \circ \phi, v \circ \phi).(g \circ \phi) = S_G(u, v).g. \tag{2.8}$$

Let Δ be an open subset of $\Omega \cap \overline{\mathbf{R}}$, and let $m \geq 1$. We shall say that the *order of growth of* G *near* Δ *is* $\leq m$, if for every closed subset Δ' of Δ there exists a constant M and an open neighborhood $\mathcal{U}$ of Δ' in $\overline{\mathbf{C}}$ such that

$$\|G(\lambda)\| \leq M(1+|\lambda|)^{2m}|\mathrm{Im}\,\lambda|^{-m}$$

for all $\lambda \in \mathcal{U} \setminus \overline{\mathbf{R}}$. We do not exclude the case when $\Omega = \overline{\mathbf{C}}$ and $\Delta = \overline{\mathbf{R}}$.

Analogously, if Γ is an open subset of $\psi(\Omega) \cap \mathbf{T}$ we shall say that the *order of growth of* F *near* Γ *is* $\leq m$, if for every closed subset Γ' of Γ there exists a constant M and an $r_0 \in (0,1)$ such that

$$\|F(re^{i\Theta})\| \leq M|1-|r||^{-m}$$

for all $e^{i\Theta} \in \Gamma'$ and $r \in [r_0, 1) \cup (1, r_0^{-1}]$.

2.2. Extension of the functionals associated with G and F and its consequences

Assume that the order of growth of G near $\Delta \subset \Omega \cap \overline{\mathbf{R}}$ is $\leq m$. It is easy to verify that this is equivalent to the fact that the order of growth of F near $\psi(\Delta)$ is $\leq m$.

Let Γ_0 be the union of a finite number of pairwise disjoint open arcs of $\mathbf{T}$, $\Gamma_0 \neq \mathbf{T}$, and let $\delta_0 \in (0,1)$ be such that for

$$Q_0 := \{re^{i\Theta} : e^{i\Theta} \in \Gamma_0,\ r \in (\delta_0, 1) \cup (1, \delta_0^{-1})\} \tag{2.9}$$

the function F is locally holomorphic on $\overline{Q_0} \setminus \overline{\Gamma_0}$.

We denote by $D^{(p)}(\overline{\mathbf{C}} \setminus Q_0, \mathcal{H})$, p nonnegative integer, the linear space of all continuous $\mathcal{H}$-valued functions f on $\overline{\mathbf{C}} \setminus Q_0$ such that f is locally holomorphic on $\overline{\mathbf{C}} \setminus (Q_0 \cup \Gamma_0)$ and the restriction $f|\mathbf{T}$ is a C^p function. For $D^{(p)}(\overline{\mathbf{C}} \setminus Q_0, \mathbf{C})$ we simply write $D^{(p)}(\overline{\mathbf{C}} \setminus Q_0)$. We introduce a locally convex topology on $D^{(p)}(\overline{\mathbf{C}} \setminus Q_0, \mathcal{H})$: Let ϵ_0, $0 < \epsilon_0 < 1 - \delta_0$, be such that for $0 < \epsilon < \epsilon_0$ every component of Γ_0 contains a point of

$$\Gamma_\epsilon := \{e^{i\Theta} \in \Gamma_0 : \mathrm{dist}\,(e^{i\Theta}, \mathbf{T} \setminus \Gamma_0) > \epsilon\} \subset \Gamma_0,$$

and set

$$Q_\epsilon := \{re^{i\Theta} : e^{i\Theta} \in \Gamma_\epsilon,\ r \in (\delta_0 + \epsilon, 1) \cup (1, (\delta_0 + \epsilon)^{-1})\}.$$

Let $(\epsilon_n) \subset (0, \epsilon_0)$ be a decreasing null sequence and let $D_n^{(p)}$ be the subspace of $D^{(p)}(\overline{\mathbf{C}} \setminus Q_0, \mathcal{H})$ of all $f \in D^{(p)}(\overline{\mathbf{C}} \setminus Q_0, \mathcal{H})$ which can analytically be continued to $\overline{\mathbf{C}} \setminus \overline{(Q_{\epsilon_n} \cup \Gamma_{\epsilon_n})}$ such that f is continuous on $\overline{\mathbf{C}} \setminus (Q_{\epsilon_n} \cup \Gamma_{\epsilon_n})$. Evidently, we have $D^{(p)}(\overline{\mathbf{C}} \setminus Q_0, \mathcal{H}) = \bigcup_{n=1}^{\infty} D_n^{(p)}$. On the space $D_n^{(p)}$ we consider the norm

$$\begin{aligned} \|f\|_n^{(p)} := \sup &\left\{ \|f(z)\| : z \in \overline{\mathbf{C}} \setminus \overline{(Q_{\epsilon_n} \cup \Gamma_{\epsilon_n})} \right\} \\ &+ \sup \left\{ \|\tfrac{d^\nu}{d\Theta^\nu} f(e^{i\Theta})\| : e^{i\Theta} \in \Gamma_0,\ 0 \leq \nu \leq p \right\},\ f \in D_n^{(p)}. \end{aligned}$$

$(D_n^{(p)}, \|f\|_n^{(p)})$ is a Banach space. On the space $D^{(p)}(\overline{\mathbf{C}} \setminus Q_0, \mathcal{H})$ we consider the topology of the inductive limit of the spaces $D_n^{(p)}$, $n = 1, 2, \ldots$. One verifies as in [7, §27, 4.(2)] that this topology is separated. By well-known properties of the Abel-Poisson integral, $H(\overline{\mathbf{C}} \setminus Q_0)$ is dense in $D^{(p)}(\overline{\mathbf{C}} \setminus Q_0)$.

It was proved in [6, Theorem 3.1] that for $\overline{\Gamma}_0 \subset \psi(\Delta)$ and under the above growth assumption on F, T_F is continuous on $H(\overline{\mathbf{C}} \setminus Q_0)$ with respect to the topology of $D^{(m+1)}(\overline{\mathbf{C}} \setminus Q_0)$. Therefore T_F can be extended by continuity to $D^{(m+1)}(\overline{\mathbf{C}} \setminus Q_0)$. By (2.7) S_G is continuous on $H(\overline{\mathbf{C}} \setminus \phi(Q_0))$ with respect to the topology in $H(\overline{\mathbf{C}} \setminus \phi(Q_0))$ induced by the topology defined above and the mapping

$$H(\overline{\mathbf{C}} \setminus Q_0) \ni f \longmapsto f \circ \psi \in H(\overline{\mathbf{C}} \setminus \phi(Q_0)).$$

We extend S_G to all functions defined on $\overline{\mathbf{C}} \setminus \phi(Q_0)$ and belonging to the space $D^{(m+1)}(\overline{\mathbf{C}} \setminus Q_0) \circ \psi = \{f \circ \psi : \ f \in D^{(m+1)}(\overline{\mathbf{C}} \setminus Q_0)\}$:

$$S_G.(f \circ \phi) := T_F.f, \qquad f \in D^{(m+1)}(\overline{\mathbf{C}} \setminus Q_0).$$

In particular, the extended functionals T_F and S_G are defined on all functions $f \in D^{(m+1)}(\overline{\mathbf{C}} \setminus Q_0)$ and $g \in D^{(m+1)}(\overline{\mathbf{C}} \setminus Q_0) \circ \psi$, respectively, such that f and g are zero outside compact subsets of $\psi(\Delta)$ and Δ, respectively. In these cases, for brevity, we shall write $f \in C_0^{m+1}(\psi(\Delta))$ and $g \in C_0^{m+1}(\Delta)$. If we regard $\overline{\mathbf{R}}$ as a real-analytic manifold in the usual way, then the restriction of ψ to $\overline{\mathbf{R}}$ is a real-analytic diffeomorphism of $\overline{\mathbf{R}}$ onto $\mathbf{T}$, and therefore the linear space of the restrictions of the functions of $C_0^{m+1}(\Delta)$ to $\overline{\mathbf{R}}$ coincides with the linear space of the C^{m+1}-functions g on $\overline{\mathbf{R}}$ with supp $g \in \Delta$. If $f \in C_0^{m+1}(\psi(\Delta))$ is a real function, then it can be approximated in $D^{(m+1)}(\overline{\mathbf{C}} \setminus Q_0)$ by a sequence of functions $f_n \in H(\overline{\mathbf{C}} \setminus Q_0)$ with $f_n = \hat{f}_n$, hence, $T_F.f$ is selfadjoint. Similarly, $S_G.g$ is selfadjoint for real functions $g \in C_0^{m+1}(\Delta)$.

We will make use of the following proposition (cf. [4, Section 1.3]).

Proposition 2.1. *Assume that the order of growth of G near to the open subset Δ of $\Omega \cap \overline{\mathbf{R}}$ is $\leq m$, and let Δ_0 be the union of a finite number of pairwise disjoint connected open subsets of Δ such that $\overline{\Delta_0} \subset \Delta$. Then the following holds.*

(i) *G can be written as a sum*

$$G = G_0 + G_{(0)},$$

where G_0 and $G_{(0)}$ are $\mathcal{L}(\mathcal{H})$-valued meromorphic functions in $\Omega \setminus \overline{\mathbf{R}}$, $G_0 = G_0^*$ is locally holomorphic on $\overline{\mathbf{C}} \setminus \Delta$, has growth of order $\leq m+2$ near $\overline{\mathbf{R}}$, and $G_{(0)} = G_{(0)}^*$ is locally holomorphic on $\overline{\Delta_0}$.

(ii) *F can be written as a sum*

$$F = F_0 + F_{(0)},$$

where F_0 and $F_{(0)}$ are $\mathcal{L}(\mathcal{H})$-valued meromorphic functions in $\psi(\Omega) \setminus \mathbf{T}$ such that $F_0 = -\hat{F}_0$ is locally holomorphic on $\overline{\mathbf{C}} \setminus \psi(\Delta)$, has growth of order $\leq m+2$ near $\mathbf{T}$, and $F_{(0)} = -\hat{F}_{(0)}$ is locally holomorphic on $\overline{\psi(\Delta_0)}$.

Proof. It is sufficient to prove assertion (ii). For every point z of holomorphy of F we have (see (2.6))

$$F(z) = T_F.h_z + \frac{1}{2}(F(0) - F(0)^+).$$

Let $\alpha \in C_0^{m+1}(\psi(\Delta))$ be real on $\mathbf{T}$ and equal to 1 on some neighborhood of $\overline{\psi(\Delta_0)}$. We set

$$F_0(z) := T_F.\alpha h_z + \frac{1}{2}(F(0) - F(0)^+) \tag{2.10}$$

Let τ_0 denote the set of all poles of F in $\psi(\Omega)\setminus\mathbf{T}$. The operator $T_F.\alpha f$ is selfadjoint for every $f = \hat{f} \in H((\overline{\mathbf{C}}\setminus\psi(\Omega))\cup\mathbf{T}\cup\tau_0)$. Then it is easy to see that the functional

$$\alpha T_F : \ f \longmapsto T_F.\alpha f$$

belongs to $\Phi(\psi(\Omega), \mathbf{T}\cup\tau_0; \mathcal{L}(\mathcal{H}))$. Therefore, $\hat{F}_0 = -F_0$. By the continuity properties of T_F, F_0 is complex differentiable outside of the support of α and, hence, locally holomorphic on $\overline{\mathbf{C}} \setminus \psi(\Delta)$. We define

$$F_{(0)}(z) := T_F.(1-\alpha)h_z.$$

Then $F = F_0 + F_{(0)}$. Since $1-\alpha$ is zero in some neighborhood of $\overline{\psi(\Delta_0)}$ we conclude that $F_{(0)}$ is complex differentiable in some neighborhood (in $\overline{\mathbf{C}}$) of any point of $\overline{\psi(\Delta_0)}$.

Let K be a compact subset of $\mathbf{C}\setminus\{0\}$. Then by the definition of F_0 and the local C^{m+1}-continuity of T_F there exist constants M and M' such that $z \in K\setminus\mathbf{T}$ implies

$$\begin{aligned}\|F_0(z)\| &\le M \sup\left\{\left|\frac{d^k}{d\Theta^k}h_z(e^{i\Theta})\right| : \ \Theta\in[0,2\pi],\ k=0,\dots,m+1\right\}\\ &\le M'|1-|z||^{m+2}.\end{aligned}$$

That is, the growth of F_0 near $\mathbf{T}$ is of order $\le m+2$. □

Lemma 2.2. *Assume that G, G_0, F and F_0 are as in Proposition 2.1. Then*

$$S_G.g = S_{G_0}.g \quad \textit{for all} \quad g \in C_0^{m+3}(\Delta_0), \tag{2.11}$$

$$T_F.f = T_{F_0}.f \quad \textit{for all} \quad f \in C_0^{m+3}(\psi(\Delta_0)). \tag{2.12}$$

Proof. If $\Gamma_0 := \psi(\Delta_0)$ and Q_0 is as in (2.9), then every $f \in C_0^{m+3}(\Gamma_0)$ can be approximated in $D_n^{(m+3)}$ for some n by a sequence (f_k) of functions belonging to $H(\overline{\mathbf{C}}\setminus Q_0)$. Then, if $F_{(0)}$ is as in Proposition 2.1, by the definition of $T_{F_{(0)}}$, $T_{F_{(0)}}.f = \lim_{k\to\infty} T_{F_{(0)}}.f_k = 0$, which implies the lemma. □

Local growth properties of G and F imply also local continuity properties of the functionals $S_G(\cdot,\cdot)$ and $T_F(\cdot,\cdot)$ similar to those of S_G and T_F.

Proposition 2.3. *Assume that the order of growth of F near to the open subset Γ of $\psi(\Omega)\cap\mathbf{T}$ is $\le m$. Let Γ_0 be the union of a finite number of pairwise disjoint open subarcs of Γ such that $\overline{\Gamma}_0 \subset \Gamma$ and let Q_0 be as in (2.9). Then*

$$H(\overline{\mathbf{C}}\setminus Q_0,\mathcal{H})^2 \times H(\overline{\mathbf{C}}\setminus Q_0) \ni (p,q,f) \longmapsto T_F(p,q).f$$

is continuous with respect to the topology of $(D^{(m+3)}(\overline{\mathbf{C}}\setminus Q_0,\mathcal{H}))^2 \times D^{(m+3)}(\overline{\mathbf{C}}\setminus Q_0)$.

Proof. Let O_1^+ be a simply connected C^∞-subdomain of $\mathbf{D}$ with the following properties:

(i) $0 \in O_1^+$, $Q_0 \cap \mathbf{D} \subset O_1^+$, $\overline{O_1^+} \cap \mathbf{T} = \overline{\Gamma}_0$.

(ii) F is holomorphic in O_1^+ and in all points of $\overline{O_1^+} \setminus \mathbf{T}$.

Then there exists an $r_0 \in (0,1)$ such that for all $r \in [r_0, 1)$, F is holomorphic on the closure of $rO_1^+ := \{rz :\ z \in O_1^+\}$. We define $\mathcal{O}_r := rO_1^+ \cup (rO_1^+)\hat{}$, $r \in [r_0, 1]$.

Let F_0 and $F_{(0)}$ be as in Proposition 2.1, (ii), with $\psi(\Delta) = \Gamma$ and $\psi(\Delta_0) = \Gamma_0$. If $p, q \in H(\overline{\mathbf{C}} \setminus Q_0, \mathcal{H})$, $f \in H(\overline{\mathbf{C}} \setminus Q_0)$, then for sufficiently small $1 - r > 0$ we have

$$\begin{aligned} T_F(p,q).f &= \int_{\partial\mathcal{O}_r} [F(z)p(z), q(\bar{z}^{-1})] f(z)(iz)^{-1} dz \\ &= \int_{\partial\mathcal{O}_r} [F_0(z)p(z), q(\bar{z}^{-1})] f(z)(iz)^{-1} dz \\ &\quad + \int_{\partial\mathcal{O}_r} [F_{(0)}(z)p(z), q(\bar{z}^{-1})] f(z)(iz)^{-1} dz \\ &= \int_{\partial(r\mathbf{D} \cup r^{-1}\hat{\mathbf{D}})} [F_0(z)p(z), q(\bar{z}^{-1})] f(z)(iz)^{-1} dz \\ &\quad + \int_{\partial(\mathcal{O}_1 \setminus \bar{\Gamma}_0)} [F_{(0)}(z)p(z), q(\bar{z}^{-1})] f(z)(iz)^{-1} dz. \end{aligned} \tag{2.13}$$

F_0 has growth of order $\leq m+2$ near $\mathbf{T}$. Then by [4, Proposition 1.2] the first term on the right-hand side of (2.13) is continuous on $(C^{m+3}(\mathbf{T}, \mathcal{H}))^2 \times C^{m+3}(\mathbf{T})$. Since the topologies of $D^{(m+3)}(\overline{\mathbf{C}} \setminus Q_0, \mathcal{H})$ and $D^{(m+3)}(\overline{\mathbf{C}} \setminus Q_0)$ are stronger than those of $(C^{m+3}(\mathbf{T}, \mathcal{H}))$ and $C^{m+3}(\mathbf{T})$, respectively, the first term on the right-hand side of (2.13) is continuous with respect to the topology mentioned in the proposition. As to the second term on the right-hand side of (2.13), there is a constant M such that the absolute value of the second term can be estimated from above by

$$M \ \sup\{\|p(z)\| \ ; \ z \in \overline{\mathbf{C}} \setminus \bar{Q}_0\} \ \sup\{\|q(z)\| : \ z \in \overline{\mathbf{C}} \setminus \bar{Q}_0\} \\ \times \sup\{|f(z)| : \ z \in \overline{\mathbf{C}} \setminus \bar{Q}_0\}.$$

This implies Proposition 2.3. □

If the assumptions of Proposition 2.3 are fulfilled, by (2.8) a similar continuity statement holds for $S_G(\cdot,\cdot)$ and the topologies induced by the mapping $f \longmapsto f \circ \psi$. For the extended functional $S_G(\cdot,\cdot)$ we have

$$S_G(p \circ \phi, q \circ \phi).(f \circ \phi) := T_F(p,q).f, \\ p, q \in D^{(m+3)}(\overline{\mathbf{C}} \setminus Q_0, \mathcal{H}), \quad f \in D^{(m+3)}(\overline{\mathbf{C}} \setminus Q_0). \tag{2.14}$$

In the same way as in Lemma 2.2 and making use of (2.14) we verify the following.

Lemma 2.4. *If F, F_0, G and G_0 are as in Proposition 2.1, then*

$$T_F(p,q).f = T_{F_0}(p,q).f$$

for all $p, q \in C_0^{m+3}(\psi(\Delta_0), \mathcal{H})$, $f \in C_0^{m+3}(\psi(\Delta_0))$, *and*

$$S_G(u,v).g = S_{G_0}(u,v).g$$

for all $u, v \in C_0^{m+3}(\Delta_0, \mathcal{H})$, $g \in C_0^{m+3}(\Delta_0)$.

2.3. Open sets of positive and negative type with respect to an operator function

Let G and F be as in Sections 2.1 and 2.2. The following definitions of open sets of positive and negative type with respect to the operator functions G and F are equivalent to those in [6, Definitions 3.7 and 3.9]. For these and further equivalent descriptions of these sets see [6, Lemma 3.8 and Lemma 3.10].

Definition 2.5. An open subset $\Delta \subset \Omega \cap \overline{\mathbf{R}}$ is said to be of *positive type with respect to* G if for every $x \in \mathcal{H}$ and every sequence (λ_n) of points of holomorphy of G in $\Omega \cap \mathbf{C}^+$ which converges in $\overline{\mathbf{C}}$ to a point of Δ we have

$$\liminf_{n\to\infty} \operatorname{Im}\,[G(\lambda_n)x, x] \geq 0.$$

An open subset $\Delta \subset \Omega \cap \overline{\mathbf{R}}$ is said to be of *negative type with respect to* G if Δ is of positive type with respect to $-G$. Δ is said to be of *definite type with respect to* G if Δ is of positive type or of negative type with respect to G. A point $\lambda \in \Omega \cap \overline{\mathbf{R}}$ which is not contained in an open set of definite type with respect to G is called a *critical point of* G *in* Ω, we write $\lambda \in \mathrm{K}(G, \Omega)$.

Definition 2.5′. An open set $\Gamma \subset \psi(\Omega) \cap \mathbf{T}$ is said to be of *positive type with respect to* F if for every $x \in \mathcal{H}$ and every convergent sequence $(z_n) \subset \psi(\Omega) \cap \mathbf{D}$ of points of holomorphy of F with $\lim_{n\to\infty} z_n \in \Gamma$ we have

$$\liminf_{n\to\infty} \operatorname{Re}\,[F(z_n)x, x] \geq 0.$$

An open set $\Gamma \subset \psi(\Omega) \cap \mathbf{T}$ is said to be of *negative type with respect to* F if Γ is of positive type with respect to $-F$. Γ is said to be of *definite type with respect to* F if Γ is of positive type or of negative type with respect to F. A point $z \in \psi(\Omega) \cap \mathbf{T}$ which is not contained in an open set of definite type with respect to F is called a *critical point of* F *in* $\psi(\Omega)$, we write $z \in \mathrm{K}(F, \psi(\Omega))$.

Assume that $\Omega = \overline{\mathbf{C}}$ and let G, in addition, be piecewise holomorphic in $\mathbf{C}\backslash\mathbf{R}$. Then $\overline{\mathbf{R}}$ is of positive type with respect to G if and only if G is a Nevanlinna function, i.e., $\operatorname{Im}\,[G(\lambda)x, x] \geq 0$ for every $\lambda \in \mathbf{C}^+$ and every $x \in \mathcal{H}$. This is a consequence of the fact that a harmonic function does not attain its minimum in the interior of its domain. If $\Omega = \overline{\mathbf{C}}$ and F is piecewise holomorphic in $\overline{\mathbf{C}} \backslash \mathbf{T}$, then $\mathbf{T}$ is of positive type with respect to F if and only if F is a Carathéodory function, i.e., $\operatorname{Re}\,[F(z)x, x] \geq 0$ for every $z \in \mathbf{D}$ and every $x \in \mathcal{H}$.

Proposition 2.6. *Let* Δ *be an open subset of* $\Omega \cap \overline{\mathbf{R}}$. *Then the following statements are equivalent.*

(i) Δ *is of positive type with respect to* G.
(i′) $\psi(\Delta)$ *is of positive type with respect to* F.

(ii) *The order of growth of G near Δ is $\leq m$ for some positive integer m, and $[(S_G.g)x, x] \geq 0$ for every nonnegative function $g \in C_0^\infty(\Delta)$ and any $x \in \mathcal{H}$.*

(ii′) *The order of growth of F near $\psi(\Delta)$ is $\leq m$ for some integer m, and $[(T_F.f)x, x] \geq 0$ for every nonnegative function $f \in C_0^\infty(\psi(\Delta))$ and any $x \in \mathcal{H}$.*

(iii) *The order of growth of G near Δ is $\leq m$ for some positive integer m, and $S_G(\cdot,\cdot).\mathbf{1}$ restricted to $C_0^\infty(\Delta, \mathcal{H})$ is positive semidefinite.*

(iii′) *The order of growth of F near $\psi(\Delta)$ is $\leq m$ for some positive integer m, and $T_F(\cdot,\cdot).\mathbf{1}$ restricted to $C_0^\infty(\psi(\Delta), \mathcal{H})$ is positive semidefinite.*

(iv) *For every open set Δ_0 which is the union of a finite number of pairwise disjoint connected open subsets of Δ such that $\overline{\Delta_0} \subset \Delta$, G can be written as a sum $G = G_0 + G_{(0)}$, where G_0 is an $\mathcal{L}(\mathcal{H})$-valued Nevanlinna function and $G_{(0)}$ is locally holomorphic on $\overline{\Delta}_0$.*

(iv′) *For every open set Γ_0 which is the union of a finite number of pairwise disjoint connected open subsets of $\psi(\Delta)$ such that $\overline{\Gamma}_0 \subset \psi(\Delta)$, F can be written as a sum $F = F_0 + F_{(0)}$, where F_0 is an $\mathcal{L}(\mathcal{H})$-valued Carathéodory function and $F_{(0)}$ is locally holomorphic on $\overline{\Gamma}_0$.*

Proof. By relation (2.3) the statements (i) and (i′) are equivalent. Since (2.7) and (2.8) remain true for the extended functionals S_G, T_F, $S_G(\cdot,\cdot).\mathbf{1}$, $T_F(\cdot,\cdot).\mathbf{1}$, (ii) and (ii′) as well as (iii) and (iii′) are equivalent. On account of [6, Lemmas 3.10 and 3.12] (i′) and (ii′) are equivalent.

Assume that (i′) and (ii′) hold. We show that (iv′) holds. We construct a decomposition of F as in the proof of Proposition 2.1 with $\psi(\Delta_0) = \Gamma_0$ and assume, in addition, that the function α in (2.10) is nonnegative. It follows from (2.6) and (2.5) that

$$T_{F_0}.f = T_F.\alpha f \text{ for all } f \in C^\infty(\mathbf{T}).$$

This relation shows that, for every $x \in \mathcal{H}$, $[T_{F_0}.(\cdot)x, x]$ is a nonnegative functional on $C^\infty(\mathbf{T})$, i.e., F_0 is a Carathéodory function. It is easy to see that (iv′) implies (i′). (iv) and (iv′) are equivalent.

Assume that (iv′) holds. Let $u, v \in C_0^\infty(\Gamma_0, \mathcal{H})$ and let f_0 be a real function in $C_0^\infty(\Gamma_0)$ which is equal to one on the supports of u and v. Then, by Lemma 2.4,

$$T_F(u,v).\mathbf{1} = T_F(u,v).f_0^2 = T_{F_0}(u,v).f_0^2 = T_{F_0}(u,v).\mathbf{1}.$$

Since F_0 is a Carathéodory function the form $T_{F_0}(\cdot,\cdot).\mathbf{1}$ is positive semidefinite (see [4, Lemma 1.7]). Therefore, $T_F(\cdot,\cdot).\mathbf{1}$ is positive semidefinite on $C_0^\infty(\Gamma_0, \mathcal{H})$. This implies (iii′).

If (iii′) holds, then for every nonnegative $f_1 \in C_0^\infty(\psi(\Delta))$ and every $x \in \mathcal{H}$ we have

$$0 \leq T_F(f_1 x, f_1 x).\mathbf{1} = [(T_F.f_1^2)x, x].$$

Since every nonnegative function $f \in C_0^\infty(\psi(\Delta))$ restricted to $\mathbf{T}$ can be approximated in $C^k(\mathbf{T})$ for every $k = 0, 1, \ldots$, by functions of the form f_1^2, $f_1 \in C_0^\infty(\psi(\Delta))$, we obtain (ii′), and Proposition 2.6 is proved. □

2.4. Open sets of type π_+ and π_- with respect to an operator function

If $\mathcal{L}$ is a linear space equipped with a Hermitian sesquilinear form $[\cdot,\cdot]$, we denote by $\kappa_+((\mathcal{L},[\cdot,\cdot]))$ ($\kappa_-((\mathcal{L},[\cdot,\cdot]))$) the least upper bound ($\leq \infty$) of the dimensions of $[\cdot,\cdot]$-positive definite (resp. $[\cdot,\cdot]$-negative definite) subspaces of $\mathcal{L}$. These quantities are called the *rank of positivity* and the *rank of negativity of* $[\cdot,\cdot]$ *on* $\mathcal{L}$.

Let G and F be as in Sections 2.1 and 2.2 and let the order of growth of G near to an open set $\Delta \subset \Omega \cap \overline{\mathbf{R}}$ be $\leq m$ for some positive integer m. Then we define

$$\kappa_\pm(\Delta, G) := \kappa_\pm((C_0^\infty(\Delta,\mathcal{H}), S_G(\cdot,\cdot).\mathbf{1})).$$

Analogously, for an open set $\Gamma \subset \psi(\Omega) \cap \mathbf{T}$ we put

$$\kappa_\pm(\Gamma, F) := \kappa_\pm((C_0^\infty(\Gamma,\mathcal{H}), T_F(\cdot,\cdot).\mathbf{1})).$$

By Proposition 2.6, Δ (Γ) is of positive type with respect to G (resp. F) if and only if $\kappa_-(\Delta, G) = 0$ (resp. $\kappa_-(\Gamma, F) = 0$). Analogously for Δ and Γ of negative type.

Definition 2.7. An open set $\Delta \subset \Omega \cap \overline{\mathbf{R}}$ ($\Gamma \subset \psi(\Omega) \cap \mathbf{T}$) is said to be of *type* π_+ with respect to G (resp. F) if the order of growth of G (resp. F) near to Δ (resp. Γ) is finite and for every open subset Δ_0, $\overline{\Delta_0} \subset \Delta$, (resp. Γ_0, $\overline{\Gamma_0} \subset \Gamma$) we have $\kappa_-(\Delta_0, G) < \infty$ (resp. $\kappa_-(\Gamma_0, F) < \infty$).

Analogously, with κ_- replaced by κ_+, sets of *type* π_- with respect to G and F are defined.

Assume that $\Omega = \overline{\mathbf{C}}$. Then the set of all piecewise meromorphic $\mathcal{L}(\mathcal{H})$-valued functions $G_0 = G_0^*$ in $\mathbf{C} \setminus \mathbf{R}$ such that the total multiplicity of the poles of G_0 in $\mathbf{C}^+$ is finite, the growth of G_0 near $\overline{\mathbf{R}}$ is of finite order and $\kappa_-(\overline{\mathbf{R}}, G_0) < \infty$ holds, coincides with the set of all generalized Nevanlinna functions, i.e., with the union of all Krein-Langer classes $N_k(\mathcal{L}(\mathcal{H}))$, $k = 0, 1, \ldots$. Similarly, the set of all piecewise meromorphic $\mathcal{L}(\mathcal{H})$-valued functions $F_0 = -\hat{F}_0$ in $\overline{\mathbf{C}} \setminus \mathbf{T}$ such that the total multiplicity of the poles of F_0 in $\mathbf{D}$ is finite, the growth of F_0 near $\mathbf{T}$ is of finite order and $\kappa_-(\mathbf{T}, F_0) < \infty$ coincides with the set of all generalized Carathéodory functions, i.e., with the union of all Krein-Langer classes $C_k(\mathcal{L}(\mathcal{H}))$, $k = 0, 1, \ldots$ (see [4]). This means that, roughly speaking, an open set Δ is of type π_+ with respect to G if and only if in a neighborhood of Δ, G behaves similarly to a generalized Nevanlinna function. Analogously for F. In the following proposition this fact is more precisely expressed with the help of decompositions.

Proposition 2.8. *Let Δ be an open subset of $\Omega \cap \overline{\mathbf{R}}$. Then the following statements are equivalent.*

- (i) *Δ is of type π_+ with respect to G.*
- (i′) *$\psi(\Delta)$ is of type π_+ with respect to F.*
- (ii) *For every open set Δ_0 which is the union of a finite number of pairwise disjoint connected open subsets of Δ such that $\overline{\Delta_0} \subset \Delta$, G can be written as a sum $G = G_0 + G_{(0)}$, where $G_0 \in N_k(\mathcal{L}(\mathcal{H}))$ for some k and $G_{(0)}$ is locally holomorphic on $\overline{\Delta_0}$.*

(ii′) *For every open set* Γ_0 *which is the union of a finite number of pairwise disjoint connected open subsets of* $\psi(\Delta)$ *such that* $\overline{\Gamma}_0 \subset \psi(\Delta)$, F *can be written as a sum* $F = F_0 + F_{(0)}$, *where* $F_0 \in C_k(\mathcal{L}(\mathcal{H}))$ *for some* k *and* $F_{(0)}$ *is locally holomorphic on* $\overline{\Gamma}_0$.

Proof. That (ii′) implies (i′) is proved as in the proof of Proposition 2.6, (iv′) $\Longrightarrow$ (iii′).

Assume that (i′) holds. Then we again construct a decomposition of F as in the proof of Proposition 2.1 and assume, in addition, that α in (2.10) is the square of a nonnegative function $\beta \in C_0^\infty(\psi(\Delta))$. Then by (2.6) and (2.5) we have $T_{F_0} \,.\, f = T_F \,.\, \beta^2 f$ for all $f \in C^\infty(\mathbf{T})$ and, by approximating functions from $C^\infty(\mathbf{T}, \mathcal{H})$ by $\mathcal{H}$-valued trigonometric polynomials,

$$T_{F_0}(u, v).\mathbf{1} = T_F(\beta u, \beta v).\mathbf{1}$$

for all $u, v \in C^\infty(\mathbf{T}, \mathcal{H})$. By condition (i′) the form $T_F(\beta\cdot, \beta\cdot).\mathbf{1}$ has a finite number of negative squares and F_0 is a generalized Carathéodory function, i.e., (ii′) is true. The rest of the proof is an immediate consequence of the above considerations. □

2.5. Locally definitizable operator functions

In the following definitions we define classes of operator functions which contain those considered in Proposition 2.8.

Definition 2.9. G is called *definitizable in* Ω if the following holds.

(α) For every finite union Δ_0 of open connected subsets of $\Omega \cap \overline{\mathbf{R}}$ with $\overline{\Delta_0} \subset \Omega \cap \overline{\mathbf{R}}$ there exists a positive integer m such that the order of growth of G near Δ_0 is $\leq m$.

(β) Every point $\lambda \in \Omega \cap \overline{\mathbf{R}}$ has an open connected neighborhood I_λ in $\overline{\mathbf{R}}$ such that both components of $I_\lambda \setminus \{\lambda\}$ are of definite type with respect to G.

Definition 2.9′. F is called *definitizable in* $\psi(\Omega)$ if the following holds.

(α') For every finite union Γ_0 of open arcs of $\psi(\Omega) \cap \mathbf{T}$ with $\overline{\Gamma}_0 \subset \psi(\Omega) \cap \mathbf{T}$ there exists a positive integer m such that the order of growth of F near Γ_0 is $\leq m$.

(β') Every point $z \in \psi(\Omega) \cap \mathbf{T}$ has an open connected neighborhood I_z in $\mathbf{T}$ such that both components of $I_z \setminus \{z\}$ are of definite type with respect to F.

Similarly to Proposition 2.6 we obtain the following proposition. For characterizations of functions definitizable in $\overline{\mathbf{C}}$ (which occur in the assertions (2) and (2′) below) with the help of definitizing rational functions see [4], [5].

Proposition 2.10. *The following statements are equivalent.*

(1) *G is definitizable in Ω.*

(1′) *F is definitizable in $\psi(\Omega)$.*

(2) *For every open set* $\Delta \subset \Omega \cap \overline{\mathbf{R}}$ *which is the union of a finite number of pairwise disjoint connected open subsets of* $\Omega \cap \overline{\mathbf{R}}$ *such that* $\overline{\Delta} \subset \Omega \cap \overline{\mathbf{R}}$, G *can be written as a sum* $G = G_0 + G_{(0)}$, *where* G_0 *is an* $\overline{\mathbf{R}}$*-symmetric* $\mathcal{L}(\mathcal{H})$*-valued function definitizable in* $\overline{\mathbf{C}}$, *and* $G_{(0)}$ *is locally holomorphic on* $\overline{\Delta}$.

(2′) *For every open set* $\Gamma \subset \psi(\Omega) \cap \mathbf{T}$ *which is the union of a finite number of pairwise disjoint connected open subsets of* $\psi(\Omega) \cap \mathbf{T}$ *such that* $\overline{\Gamma} \subset \psi(\Omega) \cap \mathbf{T}$, F *can be written as a sum* $F = F_0 + F_{(0)}$, *where* F_0 *is an* $\mathbf{T}$*-symmetric* $\mathcal{L}(\mathcal{H})$*-valued function definitizable in* $\overline{\mathbf{C}}$, *and* $F_{(0)}$ *is locally holomorphic on* $\overline{\Gamma}$.

For a function G definitizable in Ω we can even find an essentially unique decomposition of G similar to that in Proposition 2.10, (2), if we make some further requirements. Exactly the same is true for F. We shall formulate and prove it only for G.

Proposition 2.11. *Let* G *be definitizable in* Ω *and let* Δ *be as in Proposition* 2.10, (2), *and assume, additionally, that the endpoints of the connected components of* Δ *are finite and do not belong to* $\mathrm{K}(G, \Omega)$. *Moreover, let* Ω' *be a domain in* $\overline{\mathbf{C}}$ *with the same properties as* Ω *such that* $\overline{\Omega'} \subset \Omega$. *Then* G *can be written as a sum*

$$G = G_1 + G_2 + G_3, \tag{2.15}$$

where

(a) G_1 *is an* $\overline{\mathbf{R}}$*-symmetric* $\mathcal{L}(\mathcal{H})$*-valued function definitizable in* $\overline{\mathbf{C}}$ *and locally holomorphic in* $\overline{\mathbf{C}} \setminus \overline{\Delta}$. *If* t_0 *is an endpoint of a connected component of* Δ, *then* $t_0 \notin \mathrm{K}(G_1, \Omega)$ *and for every* $x \in \mathcal{H}$ *the angular limit*

$$\widehat{\lim}_{\lambda \to t_0} (\lambda - t_0)[G_1(\lambda)x, x]$$

is zero.

(b) G_2 *is a meromorphic* $\overline{\mathbf{R}}$*-symmetric* $\mathcal{L}(\mathcal{H})$*-valued function in* $\overline{\mathbf{C}}$ *with all poles contained in* $\Omega' \setminus \overline{\mathbf{R}}$.

(c) G_3 *is an* $\overline{\mathbf{R}}$*-symmetric* $\mathcal{L}(\mathcal{H})$*-valued function which is locally holomorphic on* $(\Omega' \setminus \overline{\mathbf{R}}) \cup \Delta$.

For fixed Δ *and* Ω' *as above, the terms of the decomposition* (2.15) *are uniquely determined up to addition of bounded selfadjoint operators.*

Proof. Let Δ_0 be an open subset of $\Omega \cap \overline{\mathbf{R}}$ with the same properties as Δ in Proposition 2.10 and assume that $\overline{\Delta} \subset \Delta_0$. We consider a decomposition $G = G_0 + G_{(0)}$ as in Proposition 2.10, (2), but with Δ replaced by Δ_0. Then the endpoints of the connected components of Δ do not belong to $\mathrm{K}(G_0, \overline{\mathbf{C}})$. Let $\lambda_0 \in \Omega' \cap \mathbf{C}^+$ be a point of holomorphy of G_0 and let A_0 be a minimal representing definitizable selfadjoint relation in some Krein space $\mathcal{K}$ for G_0:

$$G_0(\lambda) = S + \Gamma^+\{\lambda - \mathrm{Re}\,\lambda_0 + (\lambda - \lambda_0)(\lambda - \overline{\lambda}_0)(A_0 - \lambda)^{-1}\}\Gamma, \quad \lambda \in \rho(A_0).$$

Here S is a bounded selfadjoint operator in $\mathcal{H}$ and $\Gamma \in \mathcal{L}(\mathcal{H}, \mathcal{K})$. The endpoints of the components of Δ are no critical points of A_0 (see [5]). Then the spectral function $E(\cdot, A_0)$ of A_0 is defined on Δ and

$$G_1(\lambda) = \Gamma^+\{\lambda - \mathrm{Re}\,\lambda_0 + (\lambda - \lambda_0)(\lambda - \overline{\lambda}_0)(A_0 - \lambda)^{-1}\}E(\Delta, A_0)\Gamma, \quad \lambda \notin \sigma(A_0) \cap \overline{\Delta},$$

is a definitizable $\mathcal{L}(\mathcal{H})$-valued function locally holomorphic in $\overline{\mathbf{C}} \setminus \overline{\Delta}$, and $G - G_1$ is locally holomorphic on Δ.

Let t_0 be an endpoint of a component of Δ. Then t_0 is no eigenvalue of $A_0 \cap (E(\Delta, A_0)\mathcal{K})^2$ and, therefore,

$$\widehat{\lim}_{\lambda \to t_0} (\lambda - t_0)[\Gamma^+(A_0 - \lambda)^{-1} E(\Delta, A_0)\Gamma x, x]$$
$$= \widehat{\lim}_{\lambda \to t_0} (\lambda - t_0)[(A_0 - \lambda)^{-1} E(\Delta, A_0)\Gamma x, \Gamma x] = 0,$$

that is, G_1 fulfils condition (a).

Let $\mathcal{C}^+$ be a smooth simple closed curve in $\overline{\Omega'} \cap \mathbf{C}^+$ oriented in such a way that its interior domain is bounded. Assume that G is holomorphic on $\mathcal{C}^+$ and the set of all poles of G in the interior of $\mathcal{C}^+$ coincides with the set of all poles of G in $(\Omega' \setminus \overline{\mathbf{R}}) \cap \mathbf{C}^+$. By $\mathcal{C}^-$ we denote the curve $(\mathcal{C}^+)^*$ with the orientation opposite to that induced by $\mathcal{C}^+$. Let $\mathcal{C} := \mathcal{C}^+ \cup \mathcal{C}^-$.

We define

$$G_2(\lambda) := G(\lambda) - G_1(\lambda) - (2\pi i)^{-1} \int_{\mathcal{C}} (G(\mu) - G_1(\mu))(\mu - \lambda)^{-1} d\mu,$$

$$G_3(\lambda) := G(\lambda) - G_1(\lambda) - G_2(\lambda) = (2\pi i)^{-1} \int_{\mathcal{C}} (G(\mu) - G_1(\mu))(\mu - \lambda)^{-1} d\mu.$$

It is easy to see that the functions G_2 and G_3 satisfy the conditions (b) and (c) of Proposition 2.11. The fact that G_2 is uniquely determined up to a bounded selfadjoint operator follows from Liouville's Theorem. Evidently, the difference $\tilde{G}_1$ of any two functions satisfying the conditions on G_1 is holomorphic in the complement of the set of the endpoints of the components of Δ. Since these endpoints are no critical points of the functions, the points of nonholomorphy of $\tilde{G}_1$ are poles of first order. Then it follows from the last condition in (a) that $\tilde{G}_1$ is a constant, and Proposition 2.11 is proved. □

3. Operator and relation representations of locally definitizable operator functions

3.1. Locally definitizable operator functions defined by locally definitizable relations

Let again Ω be a domain in $\overline{\mathbf{C}}$ with the properties mentioned in the introduction, $\lambda_0 \in \Omega \cap \mathbf{C}^+$, and let besides the Krein space $\mathcal{H}$, $\mathcal{K}$ be a further Krein space. We recall the definition of local definitizability for selfadjoint relations and unitary operators in $\mathcal{K}$ from [6, Definition 4.4]. For equivalent descriptions of locally definitizable relations see [6, Theorem 4.8].

Definition 3.1. The selfadjoint relation A with $\lambda_0 \in \rho(A)$ (the unitary operator U) is called *definitizable over* Ω (resp. *definitizable over* $\psi(\Omega)$) if $\sigma(A) \cap (\Omega \setminus \overline{\mathbf{R}})$ (resp. $\sigma(U) \cap (\psi(\Omega) \setminus \mathbf{T})$) consists of isolated points which are poles of the resolvent and the function

$$\lambda \longmapsto \lambda - \operatorname{Re} \lambda_0 + (\lambda - \lambda_0)(\lambda - \overline{\lambda_0})(A - \lambda)^{-1} \quad (\text{resp. } z \longmapsto (U + z)(U - z)^{-1})$$

is definitizable in Ω (resp. definitizable in $\psi(\Omega)$).

If A is a selfadjoint relation in $\mathcal{K}$ with $\lambda_0 \in \rho(A)$ and U is the unitary operator defined by

$$U := \psi(A) = -1 + (\lambda_0 - \bar{\lambda}_0)(A - \bar{\lambda}_0)^{-1}, \tag{3.1}$$

then

$$-i(\mathrm{Im}\lambda_0)^{-1}\{\lambda - \mathrm{Re}\,\lambda_0 + (\lambda - \lambda_0)(\lambda - \bar{\lambda}_0)(A - \lambda)^{-1}\} \\ = (U + \psi(\lambda))(U - \psi(\lambda))^{-1}. \tag{3.2}$$

Therefore, A is definitizable over Ω if and only if U is definitizable over $\psi(\Omega)$.

Now let A be definitizable over Ω and $\lambda_0 \in \rho(A) \cap \Omega \cap \mathbf{C}^+$. We denote the local spectral function of A ([6]) which is defined on a collection of subsets of $\Omega \cap \overline{\mathbf{R}}$ by $E(\cdot, A)$. If ω is a subset of $\Omega \setminus \overline{\mathbf{R}}$ such that $\omega \cap \sigma(A)$ is closed and open in $\sigma(A)$, the same notation will be used to denote the Riesz-Dunford projection corresponding to $\omega \cap \sigma(A)$: $E(\omega, A)$. Let S be a bounded selfadjoint operator in $\mathcal{H}$ and let $\Gamma \in \mathcal{L}(\mathcal{H}, \mathcal{K})$. We consider the $\mathcal{L}(\mathcal{H})$-valued function G defined by

$$G(\lambda) = S + \Gamma^+\{\lambda - \mathrm{Re}\,\lambda_0 + (\lambda - \lambda_0)(\lambda - \bar{\lambda}_0)(A - \lambda)^{-1}\}\Gamma, \quad \lambda \in \rho(A) \cap \Omega. \tag{3.3}$$

Then, by Definition 2.9, also G is definitizable in Ω. If an operator function G can be written as in (3.3), A is called a *representing relation for* G. In this case, evidently,

$$S = \frac{1}{2}(G(\lambda_0) + G(\lambda_0)^+) =: \mathrm{Re}^+\, G(\lambda_0).$$

The representation (3.3) is called *minimal* if

$$\mathcal{K} = \mathrm{clsp}\,\{(1 + (\lambda - \lambda_0)(A - \lambda)^{-1})\Gamma y : \lambda \in \rho(A) \cap \Omega,\ y \in \mathcal{H}\}.$$

Similarly, if U is a unitary operator in $\mathcal{K}$ definitizable over $\psi(\Omega)$, S_0 is a bounded selfadjoint operator in $\mathcal{H}$ and $\Gamma_0 \in \mathcal{L}(\mathcal{H}, \mathcal{K})$, then the function F defined by

$$F(z) = -iS_0 + \Gamma_0^+(U + z)(U - z)^{-1}\Gamma_0 \tag{3.4}$$

is definitizable in $\psi(\Omega)$. Observe that

$$-S_0 = (2i)^{-1}(F(0) - F(0)^+) =: \mathrm{Im}^+\, F(0).$$

If a relation of the form (3.4) holds, U is called a *representing operator for* F. The representation (3.4) is called *minimal* if

$$\mathcal{K} = \mathrm{clsp}\,\{U^m\Gamma y : m = 0, \pm 1, \pm 2, \ldots,\ y \in \mathcal{H}\}.$$

If

$$U = \psi(A), \quad S_0 = S, \quad \Gamma_0 = (\mathrm{Im}\,\lambda_0)^{\frac{1}{2}}\Gamma, \tag{3.5}$$

then, in view of (3.2), the functions G and F are connected by

$$-iG(\lambda) = F(\psi(\lambda)), \quad \lambda \in \Omega.$$

In the following definition we introduce a local version of minimality.

Definition 3.2. Let A be a selfadjoint relation definitizable over Ω in a Krein space $\mathcal{K}$ with $\lambda_0 \in \rho(A)$, $\Gamma \in \mathcal{L}(\mathcal{H}, \mathcal{K})$ and S a bounded selfadjoint operator in $\mathcal{H}$. Let G be defined by (3.3) and let U and F be as in (3.1) and (3.4).

Then (3.3) is called an *Ω-minimal representation of* G if the following holds: If Ω' is a domain with the same properties as Ω and $\overline{\Omega'} \subset \Omega$, $\lambda_0 \in \Omega'$, if Δ is a finite union of connected open subsets of $\Omega \cap \overline{\mathbf{R}}$ such that the endpoints of the components of Δ belong to Ω and possess open neighborhoods of definite type with respect to A, and if we set

$$\tilde{E} := E(\Delta, A) + E(\overline{\Omega'} \setminus \overline{\mathbf{R}}, A), \tag{3.6}$$

then

$$\tilde{E}\mathcal{K} = \operatorname{closp}\{(1 + (\lambda - \lambda_0)(A - \lambda)^{-1})\tilde{E}\Gamma y : \lambda \in \rho(A) \cap \Omega', \ y \in \mathcal{H}\}.$$

Similarly, (3.4) is called an *$\psi(\Omega)$-minimal representation of* F if for every projection $\tilde{E}$ as above (note that $\tilde{E}$ coincides with $E(\psi(\Delta), U) + E(\psi(\overline{\Omega'}) \setminus \mathbf{T}, U))$ we have

$$\tilde{E}\mathcal{K} = \operatorname{closp}\{U^m \tilde{E}\Gamma y : m = 0, \pm 1, \pm 2, \ldots, \ y \in \mathcal{H}\}.$$

Evidently, if (3.3) is minimal, it is also Ω-minimal. If (3.3) is Ω-minimal, then it is Ω_0-minimal for every domain Ω_0 with the same properties as Ω and $\Omega_0 \subset \Omega$. The following lemma is an easy consequence of Definition 3.2.

Lemma 3.3. *Let G and F be as in Definition* 3.2. *Then the following statements are equivalent.*

(i) *The representation* (3.3) *is Ω-minimal.*

(ii) *The representation* (3.4) *is $\psi(\Omega)$-minimal.*

(iii) *For every Ω' and Δ as in Definition* 3.2, *with $\tilde{E}$ as in* (3.6) *and $\tilde{A} := A|\tilde{E}\mathcal{K}$, the function*

$$\lambda \longmapsto (\tilde{E}\Gamma)^+\{\lambda - \operatorname{Re}\lambda_0 + (\lambda - \lambda_0)(\lambda - \overline{\lambda}_0)(\tilde{A} - \lambda)^{-1}\}(\tilde{E}\Gamma) \tag{3.7}$$

is minimally represented by (3.7) *with $\tilde{A}$ as representing relation.*

(iv) *For every Ω' and Δ as in Definition* 3.2, (3.6) *and $\tilde{U} := U|\tilde{E}\mathcal{K}$, the function*

$$z \longmapsto (\tilde{E}\Gamma_0)^+(\tilde{U} + z)(\tilde{U} - z)^{-1}(\tilde{E}\Gamma_0) \tag{3.8}$$

is minimally represented by (3.8) *with $\tilde{U}$ as representing operator.*

Proof. Evidently, (i) is equivalent to (iii) and (ii) is equivalent to (iv). In order to show that (i) implies (ii) let $\lambda \in \rho(A) \cap \Omega' \cap \mathbf{C}^+$ and connect the point λ by a smooth curve in $\lambda \in \rho(A) \cap \Omega' \cap \mathbf{C}^+$ with the point λ_0. Making use of Taylor expansions of the resolvent of A at a finite number of points of this curve we see that every element of the form $(A - \lambda)^{-1}\tilde{E}\Gamma y$, $y \in \mathcal{H}$, can be approximated by linear combinations of elements of the form $(A - \lambda_0)^{-j}\tilde{E}\Gamma y_j$, $y_j \in \mathcal{H}$, $j = 0, 1, \ldots$. Since by (3.1)

$$(A - \lambda_0)^{-j} = (\overline{\lambda}_0 - \lambda_0)^{-j}(1 + U^{-1})^j, \quad j = 0, 1, \ldots,$$

$(A-\lambda)^{-1}\tilde{E}\Gamma y$ can be approximated by linear combinations of elements of the form

$$U^{-k}\tilde{E}\Gamma u_k, \quad u_k \in \mathcal{H}, \quad k = 0, 1, \ldots.$$

Analogously for $\lambda \in \rho(A) \cap \Omega' \cap \mathbf{C}^-$. Therefore, (i) implies (ii).

On the other hand, every element of the form

$$U^{-m}\tilde{E}\Gamma u = (-1 + (\overline{\lambda}_0 - \lambda_0)(A - \lambda_0)^{-1})^m \tilde{E}\Gamma u, \quad m \in \mathbf{N}, \quad u \in \mathcal{H},$$

can be approximated by linear combinations of elements of the form

$$(A - \lambda_j)^{-1}\tilde{E}\Gamma y_j \quad \text{or} \quad \tilde{E}\Gamma y_j, \quad \lambda_j \in \rho(A) \cap \Omega' \cap \mathbf{C}^+, \quad y_j \in \mathcal{H}.$$

Analogously for m replaced by $-m$. This shows that (i) is equivalent to (ii). □

In the following proposition the local "sign multiplicities" of a function and a representing relation are compared.

Proposition 3.4. *If A, U, G and F are as above in this section, then the following holds.*

(1) *Let Δ_0 be the union of a finite number of pairwise disjoint connected open subsets of $\Omega \cap \overline{\mathbf{R}}$ such that $\overline{\Delta_0} \subset \Omega \cap \overline{\mathbf{R}}$ and $E(\Delta_0, A)$ and $E(\psi(\Delta_0), U)$ are defined. Then*

$$\begin{aligned}\kappa_\pm(\Delta_0, G) &= \kappa_\pm(\psi(\Delta_0), F) \\ &\le \kappa_\pm((E(\Delta_0, A)\mathcal{K}, [\cdot,\cdot])) = \kappa_\pm((E(\psi(\Delta_0), U)\mathcal{K}, [\cdot,\cdot])).\end{aligned} \tag{3.9}$$

If, in addition, the representations (3.3) and (3.4) are Ω-minimal and $\psi(\Omega)$-minimal, respectively, we have equality in (3.9).

(2) *Let $\mu \in \Omega \setminus \overline{\mathbf{R}}$ be a pole of G of multiplicity l, or equivalently, let $\psi(\mu) \in \psi(\Omega) \setminus \mathbf{T}$ be a pole of F of multiplicity l. Then*

$$l \le \dim\, E(\{\mu\}, A)\mathcal{K} = \dim\, E(\{\psi(\mu)\}, U)\mathcal{K}, \tag{3.10}$$

where $E(\{\mu\}, A)$ and $E(\{\psi(\mu)\}, U)$ denote the Riesz-Dunford projections corresponding to A and $\{\mu\}$, and to U and $\psi(\mu)$, respectively. Under the condition mentioned in (1) we have equality in (3.10).

Proof. 1. By (2.14) it is sufficient to prove (3.9) for F and U. If F_0 and $F_{(0)}$ are the functions defined by

$$F_0(z) = \Gamma_0^+(U+z)(U-z)^{-1}E(\psi(\Delta_0), U)\Gamma_0,$$

$$F_{(0)}(z) = -iS_0 + \Gamma_0^+(U+z)(U-z)^{-1}(1 - E(\psi(\Delta_0), U))\Gamma_0,$$

which are definitizable over $\psi(\Omega)$, then we have $F = F_0 + F_{(0)}$. By Lemma 2.4 the forms $T_F(\cdot,\cdot)$ and $T_{F_0}(\cdot,\cdot)$ coincide on $\psi(\Delta_0)$. Therefore,

$$\kappa_\pm(\psi(\Delta_0), F) = \kappa_\pm(\psi(\Delta_0), F_0). \tag{3.11}$$

Since F_0 is a definitizable function [5, Theorem 1.12,(iii)] can be applied. We find

$$\kappa_\pm(\psi(\Delta_0), F_0) = \kappa_\pm((E(\psi(\Delta_0), U)\mathcal{K}, [\cdot,\cdot])), \tag{3.12}$$

and equality holds if the representation (3.4) is Ω-minimal (see Lemma 3.3, (ii) $\Leftrightarrow$ (iv)). The relations (3.11) and (3.12) imply assertion (1).

2. To prove (2) it is again sufficient to verify (2) for F and U. If $E_1 := E(\{\psi(\mu)\} \cup \{\overline{\psi(\mu)}^{-1}\}, U)$ and

$$F_1(z) = \Gamma_0^+(U+z)(U-z)^{-1}E_1\Gamma_0, \qquad (3.13)$$
$$F_{(1)}(z) = -iS_0 + \Gamma_0^+(U+z)(U-z)^{-1}(1-E_1)\Gamma_0,$$

then $F = F_1 + F_{(1)}$, F_1 is a definitizable function and $\psi(\mu)$ is a pole of multiplicity l of F_1. Then [5, Theorem 1.12,(iv)] implies

$$l \leq \dim\ E(\{\psi(\mu)\}, U)\mathcal{K}.$$

If the representation (3.4) is Ω-minimal, then (3.13) is a minimal representation of F_1 and, by the result mentioned above we have equality. □

In the rest of Section 3.1 we consider two Ω-minimal representing relations A_1 and A_2 of the same operator function G. By Proposition 3.4 the local "sign properties inside Ω" of A_1 and A_2 coincide. In Theorem 3.6 below we will show that the restrictions of A_1 and A_2 to spectral subspaces which correspond to certain subsets of $\Omega \cap \overline{\mathbf{R}}$ are even unitarily equivalent. We need the following lemma.

Lemma 3.5. *Let $(\mathcal{K}_j, [\cdot,\cdot])$, $j = 1, 2$, be Krein spaces and U_j, $j = 1, 2$, unitary operators in $\mathcal{K}_j$ definitizable over $\psi(\Omega)$, $\Gamma_{0,j} \in \mathcal{L}(\mathcal{H}, \mathcal{K}_j)$ and $S_{0,j}$ bounded selfadjoint operators in $\mathcal{H}$, $j = 1, 2$.*

We denote by Ξ the linear space of all functions χ defined on the union of $\psi(\Omega) \cap \mathbf{T}$ and a neighborhood $\mathcal{U}$ (depending on χ) of $(\overline{\mathbf{C}} \setminus \psi(\Omega)) \cup \sigma(U_1) \cup \sigma(U_2)$ which are sums $\chi = \chi_{\mathbf{T}} + \chi_{(\mathbf{T})}$ of a function $\chi_{\mathbf{T}} \in C_0^\infty(\psi(\Omega) \cap \mathbf{T})$ and a function $\chi_{(\mathbf{T})}$ locally holomorphic on $(\overline{\mathbf{C}} \setminus \psi(\Omega)) \cup \sigma(U_1) \cup \sigma(U_2) \cup \mathbf{T}$ which is zero outside of some compact subset of $\psi(\Omega) \setminus \mathbf{T}$.

Assume that the difference of the functions

$$\begin{aligned} F_j(z) &:= -iS_{0,j} + \Gamma_{0,j}^+(U_1+z)(U_j-z)^{-1}\Gamma_{0,j}, \\ &j = 1, 2, \quad z \in \rho(U_1) \cap \rho(U_2) \cap (\psi(\Omega) \setminus \mathbf{T}), \end{aligned} \qquad (3.14)$$

can be analytically continued to the whole domain $\psi(\Omega)$.

Then the linear relation

$$\begin{aligned} V := &\left\{ \begin{pmatrix} \sum_{k=1}^n \chi_k(U_1)\Gamma_{0,1}x_k \\ \sum_{k=1}^n \chi_k(U_2)\Gamma_{0,2}x_k \end{pmatrix} : \ \chi_k \in \Xi,\ x_k \in \mathcal{H},\ k = 1, 2, \ldots, n \right\} \\ &\subset \mathcal{K}_1 \times \mathcal{K}_2 \end{aligned} \qquad (3.15)$$

is isometric, i.e., $\binom{u_1}{u_2}, \binom{u_1'}{u_2'} \in V$ implies

$$[u_1, u_1']_1 = [u_2, u_2']_2. \qquad (3.16)$$

Moreover, V intertwines U_1 and U_2, i.e., $\binom{u_1}{u_2} \in V$ implies $\binom{U_1 u_1}{U_2 u_2} \in V$.

Proof. If we denote by $\mathcal{R}_{0,\infty}$ the set of all functions $z \longmapsto \sum_k c_k z^k$, k integer, where the sums are finite, then, by [5, Section 1.3], for $g \in \mathcal{R}_{0,\infty}$ we have

$$T_{F_1}.g = 4\pi\Gamma_{0,1}^+ g(U_1)\Gamma_{0,1}, \quad T_{F_2}.g = 4\pi\Gamma_{0,2}^+ g(U_2)\Gamma_{0,2}.$$

By continuity properties of T_{F_1}, T_{F_2} and of the functional calculi of U_1 and U_2 these relations remain true for g replaced by an arbitrary $\chi \in \Xi$.

By the definition of V and since Ξ is an algebra contained in the domains of the functional calculi for U_1 and U_2, the left-hand side of (3.16) is a finite sum of the form

$$\sum_{i,j} ((T_{F_1} \cdot \chi_{i,j})x_i, y_j), \quad \chi_{i,j} \in \Xi, \quad x_i, y_j \in \mathcal{H}. \tag{3.17}$$

Then the right-hand side of (3.16) coincides with

$$\sum_{i,j} ((T_{F_2} \cdot \chi_{i,j})x_i, y_j). \tag{3.18}$$

Since the difference of F_1 and F_2 can be analytically continued to $\psi(\Omega)$, the expressions (3.17) and (3.18) coincide, which shows that V is isometric. That V intertwines U_1 and U_2 follows from the definition of V and the fact that for $\chi \in \Xi$ also the function $z \longmapsto z\chi(z)$ belongs to Ξ. Lemma 3.5 is proved. □

Theorem 3.6. *Let $(\mathcal{K}_j, [\cdot,\cdot]_j)$, $j = 1,2$, be Krein spaces and A_j, $j = 1,2$, selfadjoint relations in $\mathcal{K}_j$ definitizable over Ω, let $\lambda_0 \in (\Omega \cap \mathbf{C}^+) \cap \rho(A_1) \cap \rho(A_2)$, $\Gamma_j \in \mathcal{L}(\mathcal{H}, \mathcal{K}_j)$ and S_j bounded selfadjoint operators in $\mathcal{H}$, $j = 1,2$. Assume that the difference of the functions G_j defined by*

$$\begin{aligned} G_j(\lambda) :=& S_j + \Gamma_j^+\{\lambda - \mathrm{Re}\lambda_0 + (\lambda - \lambda_0)(\lambda - \overline{\lambda}_0)(A_j - \lambda)^{-1}\}\Gamma_j, \\ & j = 1,2, \quad \lambda \in \rho(A_1) \cap \rho(A_2) \cap (\Omega \setminus \mathbf{R}), \end{aligned} \tag{3.19}$$

can be analytically continued to the whole domain Ω.

If $U_j := \psi(A_j)$, $F_j := -iG_j \circ \phi$, $S_{0,j} := S_j$, $\Gamma_{0,j} := (\mathrm{Im}\ \lambda_0)^{\frac{1}{2}}\Gamma_j$, $j = 1,2$, then the above assumptions are equivalent to the assumptions of Lemma 3.5.

Assume further that the representations (3.19) of G_1 and G_2 are Ω-minimal or, equivalently, that the representations (3.14) of F_1 and F_2 are $\psi(\Omega)$-minimal Then the following holds.

(i) *An open set $\Delta \subset \Omega \cap \overline{\mathbf{R}}$ ($\Gamma \subset \psi(\Omega) \cap \mathbf{T}$) is of positive type with respect to A_1 (resp. U_1), that is, the spectral function $E(\cdot, A_1)$ (resp. $E(\cdot, U_1)$) is defined on all connected subsets of Δ (resp. Γ) with endpoints in Δ (resp. Γ) and its values are nonnegative projections in $\mathcal{K}_1$, if and only if it is of positive type with respect to A_2 (resp. U_2). Analogously for sets of negative type, that is, nonnegativity of the spectral projections replaced by nonpositivity.*

(ii) *Let Δ' be an open connected subset of $\Omega \cap \overline{\mathbf{R}}$ with $\overline{\Delta'} \subset \Omega \cap \overline{\mathbf{R}}$ such that the endpoints of Δ' are contained in intervals of positive or negative type. Then there exists a densely defined closed isometric operator V' from $E_1'\mathcal{K}_1$ into $E_2'\mathcal{K}_2$, where $E_j' := E(\Delta', A_j) = E(\psi(\Delta'), U_j)$, $j = 1,2$, with dense range which intertwines the resolvents of $A_j' := A_j \cap (E_j'\mathcal{K}_j)^2$ as well as the operators $U_j' := U_j|E_j'\mathcal{K}$, $i = 1,2$, i.e., for $\lambda \notin \Delta'$ we have $V'(A_1' - \lambda)^{-1} = (A_2' - \lambda)^{-1}V'$, $V'U_1' = U_2'V'$. In particular, we have*

$$\kappa_\pm((E_1'\mathcal{K}_1, [\cdot,\cdot]_1)) = \kappa_\pm((E_2'\mathcal{K}_2, [\cdot,\cdot]_2)).$$

(iii) *If, in addition to the assumptions of* (ii), $\kappa_+((E_1'\mathcal{K}_1, [\cdot,\cdot]_1)) < \infty$, *then* A_1' *and* A_2' (U_1' *and* U_2') *are isometrically equivalent, that is, there exists an operator* V' *as in* (ii) *which is even an isometric isomorphism of* $(E_1'\mathcal{K}_1, [\cdot,\cdot]_1)$ *onto* $(E_2'\mathcal{K}_2, [\cdot,\cdot]_2)$.

(iv) *If* $\mu \in \Omega \setminus \overline{\mathbf{R}}$ *is a pole of* G_1 *and* G_2 *or, equivalently,* $\psi(\mu) \in \psi(\Omega) \setminus \mathbf{T}$ *is a pole of* F_1 *and* F_2, *then there exists an injective densely defined closed operator* V_μ *from* $\mathcal{K}_{1,\mu} := E(\{\mu\}, A_1)\mathcal{K}_1 = E(\{\psi(\mu)\}, U_1)\mathcal{K}_1$ *into* $\mathcal{K}_{2,\mu} := E(\{\mu\}, A_2)\mathcal{K}_2 = E(\{\psi(\mu)\}, U_2)\mathcal{K}_2$ *with dense range such that* $A_1\mathcal{D}(V_\mu) \subset \mathcal{D}(V_\mu)$, $U_1\mathcal{D}(V_\mu) \subset \mathcal{D}(V_\mu)$ *and* $V_\mu A_1 x = A_2 V_\mu x$, $V_\mu U_1 x = U_2 V_\mu x$ *for all* $x \in \mathcal{D}(V_\mu)$.

Proof. It is sufficient to prove Theorem 3.6 for U_1 and U_2. Assertion (i) is an immediate consequence of Proposition 3.4, (1).

Let Δ' be as in assertion (ii). By the minimality assumptions the linear set

$$\operatorname{sp}\{h(U_j)\Gamma_{0,j}x : \ h \in C_0^\infty(\psi(\Delta')),\ x \in \mathcal{H}\}$$

is dense in $E(\psi(\Delta'), U_j)\mathcal{K}_j$, $j = 1, 2$. If V is the linear relation introduced in Lemma 3.5, the relation

$$V_0' := V \cap (E(\psi(\Delta'), U_1)\mathcal{K}_1 \times E(\psi(\Delta'), U_2)\mathcal{K}_2)$$

is densely defined in $E(\psi(\Delta'), U_1)\mathcal{K}_1$ and has dense range in $E(\psi(\Delta'), U_2)\mathcal{K}_2$. Since V_0' is isometric (see Lemma 3.5) it is even a closable operator. Let V' be the closure of V_0'. Then V' is also isometric. The intertwining properties of V imply the intertwining properties of V' mentioned in (ii). Assertion (iii) is a consequence of the fact that an isometric operator from a Pontryagin space into a Pontryagin space with dense domain and dense range is an isometric isomorphism.

If μ is as in assertion (iv), then by the minimality assumptions the relation

$$V_{\mu,\bar\mu;0} := V \cap (E(\{\psi(\mu), \psi(\bar\mu)\}, U_1)\mathcal{K}_1 \times E(\{\psi(\mu), \psi(\bar\mu)\}, U_2)\mathcal{K}_2)$$

is isometric, densely defined in $E(\{\psi(\mu), \psi(\bar\mu)\}, U_1)\mathcal{K}_1$ and has dense range in $E(\{\psi(\mu), \psi(\bar\mu)\}, U_2)\mathcal{K}_2$. Therefore, $V_{\mu,\bar\mu;0}$ is a closable operator. Let $V_{\mu,\bar\mu}$ be its closure. From the definition of V it follows that

$$\mathcal{D}(V_{\mu,\bar\mu;0}) = \mathcal{D}(V_{\mu,\bar\mu;0}) \cap E(\{\psi(\mu)\}, U_1)\mathcal{K}_1 + \mathcal{D}(V_{\mu,\bar\mu;0}) \cap E(\{\psi(\bar\mu)\}, U_1)\mathcal{K}_1, \tag{3.20}$$

and the intersections on the right-hand side of (3.20) are dense in $E(\{\psi(\mu)\}, U_1)\mathcal{K}_1$ and $E(\{\psi(\bar\mu)\}, U_1)\mathcal{K}_1$, respectively. Analogously for the range of $V_{\mu,\bar\mu;0}$ with U_1, $\mathcal{K}_1$ replaced by U_2, $\mathcal{K}_2$. Moreover, $V_{\mu,\bar\mu;0}$ maps the first intersection on the right-hand side of (3.20) into $E(\{\psi(\mu)\}, U_2)\mathcal{K}_2$ and the second into $E(\{\psi(\bar\mu)\}, U_2)\mathcal{K}_2$. Then the closure $V_{\mu,\bar\mu}$ has analogous properties, and assertion (iv) is true with V_μ being the restriction of $V_{\mu,\bar\mu}$ to $E(\{\psi(\mu)\}, U_1)\mathcal{K}_1$. Theorem 3.6 is proved. □

3.2. Existence of a locally definitizable representing relation with a local minimality property

In this section we shall construct representing relations for given locally definitizable operator functions. In the following theorem, which is a variant of a result of T.Ya. Azizov ([1]), we consider operator functions holomorphic in Ω and $\psi(\Omega)$. We show that for a given neighborhood of $\overline{\mathbf{C}}\setminus\Omega$ or $\overline{\mathbf{C}}\setminus\psi(\Omega)$ there exist representing operators or relations the spectrum of which is contained in that neighborhood. The extended spectrum of a relation T in a Krein space $\mathcal{K}$ will be denoted by $\widetilde{\sigma}(T)$, i.e., $\widetilde{\sigma}(T)=\sigma(T)$ if $T\in\mathcal{L}(\mathcal{K})$ and $\widetilde{\sigma}(T)=\sigma(T)\cup\{\infty\}$ if $T\notin\mathcal{L}(\mathcal{K})$. We set $\widetilde{\rho}(T)=\overline{\mathbf{C}}\setminus\widetilde{\sigma}(T)$.

Theorem 3.7. *Let $\mathcal{V}$ be an open neighborhood of $\overline{\mathbf{C}}\setminus\Omega$ and let $\lambda_0\in\mathbf{C}^+\cap(\Omega\setminus\overline{\mathcal{V}})$. Let G be an $\mathcal{L}(\mathcal{H})$-valued function holomorphic in Ω such that $G=G^*$.*

Then there exist a Krein space $\mathcal{K}$, a selfadjoint relation A in $\mathcal{K}$ and $\Gamma\in\mathcal{L}(\mathcal{H},\mathcal{K})$ such that

$$\widetilde{\sigma}(A)\subset\mathcal{V}$$

and

$$G(\lambda)=\mathrm{Re}^+\,G(\lambda_0)+\Gamma^+\{\lambda-\mathrm{Re}\,\lambda_0+(\lambda-\lambda_0)(\lambda-\bar{\lambda}_0)(A-\lambda)^{-1}\}\Gamma,$$

$\lambda\in\Omega\setminus\overline{\mathcal{V}}$, or, equivalently, with $\psi(\lambda):=-(\lambda-\lambda_0)(\lambda-\overline{\lambda_0})^{-1}$, $F(\psi(\lambda)):=-iG(\lambda)$, $U:=\psi(A)$, $\Gamma_0:=(\mathrm{Im}\,\lambda_0)^{\frac{1}{2}}\Gamma$,

$$\sigma(U)\subset\psi(\mathcal{V}) \tag{3.21}$$

and

$$F(z)=i\,\mathrm{Im}^+\,F(0)+\Gamma_0^+(U+z)(U-z)^{-1}\Gamma_0,\quad z\in\psi(\Omega)\setminus\psi(\overline{\mathcal{V}}). \tag{3.22}$$

Moreover, $\mathcal{K}$ can be chosen minimal, that is

$$\begin{aligned}\mathcal{K}&=\mathrm{clsp}\,\{(1+(\lambda-\lambda_0)(A-\lambda)^{-1})\Gamma x:\ \lambda\in\Omega\setminus\overline{\mathcal{V}},x\in\mathcal{H}\}\\&=\mathrm{clsp}\,\{U^m\Gamma_0 x:\ m=0,\pm1,\pm2,\dots,x\in\mathcal{H}\}.\end{aligned} \tag{3.23}$$

Proof. It is sufficient to prove the assertions for F and U. We may and will assume that the set $\psi(\mathcal{V})$ is bounded and $\mathbf{T}$-symmetric. We set

$$d:=\inf\{|z-w|:\ z\in\overline{\mathbf{C}}\setminus\psi(\Omega),\ w\in\mathbf{C}\setminus\psi(\mathcal{V})\}.$$

Then with the help of a $\mathbf{T}$-symmetric covering of the bounded set $\overline{\mathbf{C}}\setminus\psi(\Omega)$ by a finite number of open disc neighborhoods of points of $\overline{\mathbf{C}}\setminus\psi(\Omega)$ with radius $\leq\frac{1}{2}d$ it is not difficult to find an open neighborhood $\mathcal{W}$ of $\overline{\mathbf{C}}\setminus\psi(\Omega)$ with the following properties: (a) $\overline{\mathcal{W}}\subset\psi(\mathcal{V})$, (b) $\overline{\mathbf{C}}\setminus\overline{\mathcal{W}}$ is a piecewise analytic $\mathbf{T}$-symmetric domain of $\overline{\mathbf{C}}$, (c) $\mathbf{D}\setminus\overline{\mathcal{W}}$ is simply connected, (d) $\overline{\mathcal{W}}\cap\mathbf{T}$ consists of a finite number of pairwise disjoint closed arcs of $\mathbf{T}$. Observe that to find $\mathcal{W}$ with the property (c) the fact that $\mathbf{D}\cap\psi(\Omega)$ is simply connected has to be used. Then with the help of a conformal mapping of $\mathbf{D}\setminus\overline{\mathcal{W}}$ onto $\mathbf{D}$ and its $\mathbf{T}$-symmetric continuation it is easy to see that there exist bounded simply connected domains O_i, $i=1,\dots,n$, with analytic boundaries and the following properties: (a′) $O_i=\hat{O}_i$, $i=1,\dots,n$. (b′) The closures $\overline{O}_i$, $i=1,\dots,n$, are pairwise disjoint. (c′) $\overline{\mathbf{C}}\setminus\psi(\Omega)\subset O:=\bigcup_{i=1}^n O_i$. (d′) $\overline{O}\subset\psi(\mathcal{V})$. Then F is an $\mathcal{L}(\mathcal{H})$-valued function which is locally holomorphic on $\overline{\mathbf{C}}\setminus O$ such that $F=-\hat{F}$.

For $u,v\in H(\overline{O},\mathcal{H})$ we define the positive definite inner product

$$(u,v)_{H^2}:=\int_{\partial O}(u(z),v(z))|dz|,$$

and we denote by $H^2(\overline{O},\mathcal{H})$ the Hilbert space obtained by completion of $H(\overline{O},\mathcal{H})$ with respect to $\|\cdot\|_{H^2}$, where $\|u\|_{H^2} := (u,u)_{H^2}^{\frac{1}{2}}$, $u \in H(\overline{O},\mathcal{H})$. If we identify the space $H(\overline{O},\mathcal{H})$ with the product $H(\overline{O}_1,\mathcal{H})\times\cdots\times H(\overline{O}_n,\mathcal{H})$ and if Θ_i, $i=1,\ldots,n$, is a conformal mapping of O_i on the unit disc $\mathbf{D}$, then the linear mapping

$$\mathbf{\Theta} : (H(\overline{\mathbf{D}},\mathcal{H}))^n \ni (f_1,\ldots,f_n)^T \longmapsto (f_1\circ\Theta_1,\ldots,f_n\circ\Theta_n)^T \in H(\overline{O},\mathcal{H})$$

is bijective. It is easy to see that $\mathbf{\Theta}$ can be extended by continuity to an isomorphism $\widetilde{\mathbf{\Theta}}$ of the product $(H^2(\mathcal{H}))^n$ of the usual H^2-spaces $H^2(\mathcal{H})$ of $\mathcal{H}$-valued functions and $H^2(\overline{O},\mathcal{H})$. Making use of the isomorphism $\widetilde{\mathbf{\Theta}}$ and well-known results on H^2-spaces (see, e.g., [8, Section V, §1]) we see that $H^2(\overline{O},\mathcal{H})$ can be regarded as a Hilbert subspace of the linear space of all locally holomorphic $\mathcal{H}$-valued functions in O such that for every compact subset K of O we have

$$\sup\{\|u(\lambda)\| : \lambda\in K,\ u\in H^2(\overline{O},\mathcal{H}),\ \|u\|_{H^2}\le 1\} < \infty. \tag{3.24}$$

Let $O_{0,i}$, $i=1,\ldots,n$, be smooth domains such that $\overline{O}_{0,i}\subset O_i$ and F is still locally holomorphic on $\overline{\mathbf{C}}\setminus O_0$, $O_0 := \bigcup_{i=1}^n O_{0,i}$. Then we define, for $u,v\in H^2(\overline{O},\mathcal{H})$,

$$[u,v]_0 := -\int_{\partial O_0} [F(z)u(z), v(\bar{z}^{-1})](iz)^{-1}dz.$$

By (3.24) $[\cdot,\cdot]_0$ is a continuous Hermitian sesquilinear form on $H^2(\overline{O},\mathcal{H})$. Let W be the Gram operator of $[\cdot,\cdot]_0$ in $H^2(\overline{O},\mathcal{H})$ and let P_0 be the orthogonal projection in $H^2(\overline{O},\mathcal{H})$ on the orthogonal complement (in $H^2(\overline{O},\mathcal{H})$) of $\ker W$. Let $\mathcal{K}$ be the completion of $P_0H^2(\overline{O},\mathcal{H})$ with respect to the restriction of the quadratic norm $\||W|^{\frac{1}{2}}\cdot\|_{H^2}$ to $P_0H^2(\overline{O},\mathcal{H})$. Evidently, the form $[\cdot,\cdot]_0$ can be extended by continuity to a form $[\cdot,\cdot]_{\mathcal{K}}$ in $\mathcal{K}$ and $(\mathcal{K},[\cdot,\cdot]_{\mathcal{K}})$ is a Krein space.

Let U' and V' be the operators of multiplication by z and z^{-1}, respectively, in the Hilbert space $H^2(\overline{O},\mathcal{H})$. These operators are bounded and we have

$$[U'u_1,u_2]_0 = [u_1,V'u_2]_0, \qquad u_1,u_2\in H^2(\overline{O},\mathcal{H}).$$

Therefore, $U'\ker W\subset\ker W$, $V'\ker W\subset\ker W$, and, if we define bounded operators U_0, V_0 in $P_0H^2(\overline{O},\mathcal{H})$ by

$$U_0 := P_0U'|P_0H^2(\overline{O},\mathcal{H}), \quad V_0 := P_0V'|P_0H^2(\overline{O},\mathcal{H})$$

we find $U_0V_0 = V_0U_0 = 1$ and

$$[U_0u_1,u_2]_0 = [u_1,V_0u_2]_0, \quad u_1,u_2\in P_0H^2(\overline{O},\mathcal{H}).$$

Then, by a generalization of Krein's Lemma (see [2, Lemma 1.1]), U_0 and V_0 can be extended by continuity to bounded operators U and V, respectively, in $\mathcal{K}$ such that $UV=VU=1$ and

$$[Ux_1,x_2]_{\mathcal{K}} = [x_1,Vx_2]_{\mathcal{K}}, \qquad x_1,x_2\in\mathcal{K}.$$

The operator U is unitary in the Krein space $\mathcal{K}$.

Assume that $z \notin \overline{O}$. This implies $z, \bar{z}^{-1} \in \rho(U')$ and $z^{-1}, \bar{z} \in \rho(V')$. As the resolvents of U' and V' map ker W into itself, we find $z, \bar{z}^{-1} \in \rho(U_0)$, $z^{-1}, \bar{z} \in \rho(V_0)$, and

$$(U_0 - z)^{-1} = P_0(U' - z)^{-1} | P_0 H^2(\overline{O}, \mathcal{H}). \tag{3.25}$$

Moreover, by [2, Corollary 1.2],

$$z,\ \bar{z}^{-1} \in \rho(U).$$

In order to show that a relation of the form (3.22) holds we define an operator $\Gamma_0 \in \mathcal{L}(\mathcal{H}, \mathcal{K})$ by

$$\Gamma_0 :\ \mathcal{H} \ni y \longmapsto (2\sqrt{\pi})^{-1} P_0 \mathbf{1}y.$$

Here $\mathbf{1}y$ denotes the function identically equal to y on a neighborhood of $\overline{O}$. Let $z \notin \overline{O}$ and $h_z(\zeta) := (4\pi)^{-1}(\zeta + z)(\zeta - z)^{-1}$. Then making use of (2.6) and (3.25) we find, for $x, y \in \mathcal{H}$,

$$\begin{aligned}
[(F(z) - i\mathrm{Im}^+\ F(0))x, y] &= -\int_{\partial O_0} [F(\zeta)h_z(\zeta)x, y](i\zeta)^{-1} d\zeta \\
&= [h_z \mathbf{1}x, \mathbf{1}y]_0 = (4\pi)^{-1}[(U' + z)(U' - z)^{-1}\mathbf{1}x, \mathbf{1}y]_0 \\
&= [P_0(U' + z)(U' - z)^{-1}(2\sqrt{\pi})^{-1}P_0\mathbf{1}x, (2\sqrt{\pi})^{-1}P_0\mathbf{1}y]_0 \\
&= [(U_0 + z)(U_0 - z)^{-1}(2\sqrt{\pi})^{-1}P_0\mathbf{1}x, (2\sqrt{\pi})^{-1}P_0\mathbf{1}y]_{\mathcal{K}} \\
&= [\Gamma_0^+(U + z)(U - z)^{-1}\Gamma_0 x, y].
\end{aligned}$$

This proves (3.22).

In order to prove (3.23) it is sufficient to verify that the set of all functions of the form $z \mapsto z^m x$, $m = 0, \pm 1, \ldots$, $x \in \mathcal{H}$, is total in $H(\overline{O}, \mathcal{H})$. This is a consequence of Cauchy's integral formula and Runge's Theorem. □

Now with the help of Proposition 2.11, Lemma 3.3 and Theorem 3.7 it is not difficult to prove the following theorem.

Theorem 3.8. *Let G be an $\mathcal{L}(\mathcal{H})$-valued operator function definitizable in Ω, let λ_0 be a point of holomorphy of G in $\Omega \cap \mathbf{C}^+$, and let Ω' be a domain in $\overline{\mathbf{C}}$ with the same properties as Ω such that $\overline{\Omega'} \subset \Omega$ and $\lambda_0 \in \Omega'$.*

Then there exists a Krein space $\mathcal{K}$, a selfadjoint relation A in $\mathcal{K}$ definitizable in Ω' and $\Gamma \in \mathcal{L}(\mathcal{H}, \mathcal{K})$ such that the set of all points of holomorphy of G in Ω' coincides with $\tilde{\rho}(A) \cap \Omega'$,

$$\begin{aligned}
G(\lambda) =& \mathrm{Re}^+\ G(\lambda_0) + \Gamma^+\{\lambda - \mathrm{Re}\ \lambda_0 + (\lambda - \lambda_0)(\lambda - \overline{\lambda}_0)(A - \lambda)^{-1}\}\Gamma, \\
&\lambda \in \rho(A) \cap \Omega',
\end{aligned} \tag{3.26}$$

and this representation is Ω'-minimal.

If $F(\psi(\lambda)) = -i\,G(\lambda)$ for $\lambda \in \Omega$, $U := \psi(A)$, $\Gamma_0 := (\mathrm{Im}\ \lambda_0)^{\frac{1}{2}}\Gamma$, then the set of all points of holomorphy of F in $\psi(\Omega')$ coincides with $\rho(U) \cap \psi(\Omega')$,

$$F(z) = i\ \mathrm{Im}^+\ F(0) + \Gamma_0^+(U + z)(U - z)^{-1}\Gamma_0, \quad z \in \rho(U) \cap \psi(\Omega'), \tag{3.27}$$

and this representation is $\psi(\Omega')$-minimal.

Proof. Let Ω'' be a domain in $\overline{\mathbf{C}}$ with the same properties as Ω and Ω' and $\overline{\Omega'} \subset \Omega''$, $\overline{\Omega''} \subset \Omega$ and let Δ be the union of a finite number of pairwise disjoint connected open subsets of $\Omega \cap \overline{\mathbf{R}}$ such that $\overline{\Omega''} \cap \mathbf{R} \subset \Delta$ and $\overline{\Delta} \subset \Omega \cap \overline{\mathbf{R}}$. In addition, assume that the endpoints of the connected components of Δ are finite and do not belong to $\mathrm{K}(G, \Omega)$ (see Definition 2.5).

Let $G = G_1 + G_2 + G_3$ be a decomposition of G as in Proposition 2.11 with the set denoted by Ω' in Proposition 2.11 replaced by the set Ω'' defined in this proof. Then $G_1 + G_2$ is a definitizable function which is locally holomorphic in $(\overline{\mathbf{R}} \setminus \overline{\Delta}) \cup ((\overline{\mathbf{C}} \setminus \overline{\mathbf{R}}) \setminus \Omega'')$. Let $A_{1,2}$ be a minimal representing selfadjoint relation in a Krein space $\mathcal{K}_{1,2}$ for $G_1 + G_2$ (see [5]). Then the set of all points of holomorphy of $G_1 + G_2$ coincides with $\widetilde{\rho}(A_{1,2})$ and

$$\begin{aligned} G_1(\lambda) + G_2(\lambda) = \operatorname{Re}^+ & (G_1(\lambda_0) + G_2(\lambda_0)) + \\ & + \Gamma_{1,2}^+\{\lambda - \operatorname{Re}\lambda_0 + (\lambda - \lambda_0)(\lambda - \overline{\lambda}_0)(A_{1,2} - \lambda)^{-1}\}\Gamma_{1,2}, \quad \lambda \in \rho(A_{1,2}). \end{aligned} \tag{3.28}$$

The function G_3 is locally holomorphic on Ω''. Let A_3 be a minimal representing selfadjoint relation in a Krein space $\mathcal{K}_3$ for G_3 with $\widetilde{\sigma}(A_3) \subset \overline{\mathbf{C}} \setminus \overline{\Omega'}$, which exists by Theorem 3.7:

$$\begin{aligned} G_3(\lambda) = & \operatorname{Re}^+ G_3(\lambda_0) + \Gamma_3^+\{\lambda - \operatorname{Re}\lambda_0 + (\lambda - \lambda_0)(\lambda - \overline{\lambda}_0)(A_3 - \lambda)^{-1}\}\Gamma_3, \\ & \lambda \in \Omega' \setminus \{\infty\}. \end{aligned} \tag{3.29}$$

Let $\mathcal{K} := \mathcal{K}_{1,2} \times \mathcal{K}_3$, let A be the diagonal relation defined by

$$A = \left\{ \begin{pmatrix} (k_{1,2} & k_3)^{\mathrm{T}} \\ (k'_{1,2} & k'_3)^{\mathrm{T}} \end{pmatrix} : \begin{pmatrix} k_{1,2} \\ k'_{1,2} \end{pmatrix} \in A_{1,2}, \begin{pmatrix} k_3 \\ k'_3 \end{pmatrix} \in A_3 \right\} \tag{3.30}$$

and let $\Gamma := \binom{\Gamma_{1,2}}{\Gamma_3} \in \mathcal{L}(\mathcal{H}, \mathcal{K})$ w.r.t. $\mathcal{K} := \mathcal{K}_{1,2} \times \mathcal{K}_3$. Then $\widetilde{\rho}(A) \cap \Omega' = \widetilde{\rho}(A_{1,2}) \cap \Omega'$ and this set coincides with the set of all points of holomorphy in Ω' of $G_1 + G_2$ and, hence, of G. We have

$$\begin{aligned} G(\lambda) = \operatorname{Re}^+ G(\lambda_0) + \Gamma^+\{\lambda - \operatorname{Re}\lambda_0 + (\lambda - \lambda_0)(\lambda - \overline{\lambda}_0)(A - \lambda)^{-1}\}\Gamma, \\ \lambda \in \rho(A) \cap \Omega'. \end{aligned} \tag{3.31}$$

It remains to prove that (3.31) is Ω'-minimal. Let Ω_0 be a domain of $\overline{\mathbf{C}}$ with the same properties as Ω' such that $\overline{\Omega_0} \subset \Omega'$ and $\lambda_0 \in \Omega_0$, and let Δ_0 be a finite union of connected open subsets of $\Omega' \cap \overline{\mathbf{R}}$ such that the endpoints of the components of Δ_0 belong to Ω' and possess open neighborhoods of positive or of negative type with respect to A (see Theorem 3.6, (i)). If $\widetilde{E}_{1,2} := E(\Delta_0, A_{1,2}) + E(\Omega_0 \setminus \overline{\mathbf{R}}, A_{1,2})$, $\widetilde{A}_{1,2} := A_{1,2}|\widetilde{E}_{1,2}\mathcal{K}_{1,2}$, then by the minimality of the representation (3.28) the function

$$\lambda \longmapsto (\widetilde{E}_{1,2}\Gamma_{1,2})^+\{\lambda - \operatorname{Re}\lambda_0 + (\lambda - \lambda_0)(\lambda - \overline{\lambda}_0)(\widetilde{A}_{1,2} - \lambda)^{-1}\}(\widetilde{E}_{1,2}\Gamma_{1,2}) \tag{3.32}$$

is also minimally represented, that is

$$\widetilde{E}_{1,2}\mathcal{K}_{1,2}$$
$$= \operatorname{closp}\{(1+(\lambda-\lambda_0)(\widetilde{A}_{1,2}-\lambda)^{-1})\widetilde{E}_{1,2}\Gamma_{1,2}y:\ \lambda\in\rho(A_{1,2})\cap\Omega_0,\ y\in\mathcal{H}\}. \tag{3.33}$$

But in view of $\sigma(A_3)\subset\overline{\mathbf{C}}\setminus\overline{\Omega'}$ we have, for $\widetilde{E}:=E(\Delta_0,A)+E(\Omega_0\setminus\overline{\mathbf{R}},A)$,

$$\widetilde{E}=\begin{pmatrix}\widetilde{E}_{1,2} & 0\\ 0 & 0\end{pmatrix},\quad \widetilde{E}\Gamma=\begin{pmatrix}\widetilde{E}_{1,2}\Gamma_{1,2}\\ 0\end{pmatrix},\quad \text{w.r.t.}\quad \mathcal{K}:=\mathcal{K}_{1,2}\times\mathcal{K}_3.$$

Therefore, (3.33) is equivalent to

$$\widetilde{E}\mathcal{K}=\operatorname{closp}\{(1+(\lambda-\lambda_0)(A-\lambda)^{-1})\widetilde{E}\Gamma y:\ \lambda\in\rho(A)\cap\Omega_0,\ y\in\mathcal{H}\},$$

and the representation (3.26) is Ω'-minimal. Theorem 3.8 is proved. □

References

[1] Azizov, T.Ya.: Extensions of J-isometric and J-unitary operators, Funktsional. Anal. i Prilozhen. **18** (1984), 57–58 (English translation: Funct. Anal. Appl. **18** (1984), 46–48).

[2] Dijksma, A., H. Langer, H.S.V. de Snoo: Unitary colligations in Krein spaces and their role in the extension theory of isometries and symmetric linear relations in Hilbert spaces, Functional Analysis II, Proceedings Dubrovnik 1985, Lecture Notes in Mathematics **1242** (1986), 1–42.

[3] Dijksma, A., H. Langer, H.S.V. de Snoo: Eigenvalues and pole functions of Hamiltonian systems with eigenvalue depending boundary conditions, Math. Nachr. **161** (1993), 107–154.

[4] Jonas, P.: A class of operator-valued meromorphic functions on the unit disc, Ann. Acad. Sci. Fenn., Ser. A. I. Mathematica, **17** (1992), 257–284.

[5] Jonas, P.: Operator representations of definitizable functions, Ann. Acad. Sci. Fenn., Ser. A. I. Mathematica, **25** (2000), 41–72.

[6] Jonas, P.: On locally definite operators in Krein spaces, in: Spectral Theory and Its Applications, Ion Colojoară Anniversary Volume, Theta, Bucharest 2003, 95–127.

[7] Köthe, G.: Topologische lineare Räume. I, Berlin 1960.

[8] Szökefalvi-Nagy, B., Foiaş, C.: Analyse harmonique des opérateurs de l'espace de Hilbert, Paris 1967.

Peter Jonas
Institut für Mathematik, MA 6-4
Technische Universität Berlin
Str. d. 17. Juni 136
D-10623 Berlin
Germany
e-mail: jonas@math.tu-berlin.de

Operator Theory:
Advances and Applications, Vol. 162, 191–210

Symmetric Relations of Finite Negativity

M. Kaltenbäck, H. Winkler and H. Woracek

Abstract. We construct and investigate a space which is related to a symmetric linear relation S of finite negativity on an almost Pontryagin space. This space is the indefinite generalization of the completion of $\operatorname{dom} S$ with respect to $(S.,.)$ for a strictly positive S on a Hilbert space.

Mathematics Subject Classification (2000). Primary 47B25; Secondary 47B50.

Keywords. Pontryagin space, indefinite scalar product, Friedrichs Extension.

1. Introduction

It is well known that for a symmetric, semibounded and densely defined operator S on a Hilbert space $(\mathfrak{H}, (.,.))$ there exists a distinguished selfadjoint extension, the Friedrichs extension S_F of S. Besides other maximal properties (see, e.g., [9],[5]) the Friedrichs extension is distinguished among all semibounded selfadjoint extensions A of S by the fact that $\operatorname{dom}(|A|^{\frac{1}{2}})$ is minimal.

The domain $\operatorname{dom}(|S_F|^{\frac{1}{2}})$ coincides with the closure $\mathfrak{H}_S$ of $\operatorname{dom} S$ with respect to the inner product $h_m^S(.,.) = (S.,.) - m(.,.)$ where $m \in \mathbb{R}$ is sufficiently small. In fact, the usual construction of S_F is done with the help of the space $\mathfrak{H}_S$ (see Section 3).

Later on Friedrichs extensions were generalized for the case of nondensely defined operators or even for the case of symmetric linear relations ([5]). For the concept of linear relations, see for example [1].

The main subject of this note is to generalize the construction of the space $\mathfrak{H}_S$ to the almost Pontryagin space setting and to study the properties of these spaces.

An almost Pontryagin space $(\mathfrak{L}, [.,.], \mathcal{O})$ can be seen as a in general degenerated closed subspace of a Pontryagin space $(\mathfrak{P}, [,.])$, and $\mathcal{O}$ is the subspace topology induced by the Pontryagin space topology of $(\mathfrak{P}, [,.])$ on $\mathfrak{L}$. For an axiomatic treatment of such spaces see [7].

The linear relation S will be assumed to be closed and symmetric on an almost Pontryagin space $(\mathfrak{L}, [.,.], \mathcal{O})$ such that S is contained in its adjoint with finite codimension. Moreover, we assume that the form $h^S[.,.]$, which is $[S.,.]$ for operators S and which is defined accordingly if S is a proper relation, has finitely many negative squares on $\operatorname{dom} S$. Such relations S will be called to be of finite negativity and the resulting space will be denoted by $\mathfrak{L}_S$. We will also provide $\mathfrak{L}_S$ with a Hilbert space topology $\mathcal{O}_S$ such that $(\mathfrak{L}_S, \mathcal{O}_S)$ is continuously embedded in $(\mathfrak{L}, \mathcal{O})$.

In order to construct $\mathfrak{L}_S$ it is not necessary to impose special spectral assumptions on S. In particular, it can happen that S has no points of regular type.

Among other results we will see that $S - \epsilon I$ is of finite negativity for some $\epsilon > 0$ if and only if $(\mathfrak{L}_S, h^S[.,.], \mathcal{O}_S)$ is an almost Pontryagin space. This and other results about symmetries of finite negativity will be of great importance in one of our forthcoming papers about symmetric de Branges spaces ([8]).

In the short Section 2 we will introduce notations used throughout this note in the Hilbert space case as well as in the general almost Pontryagin space setting. In Section 3 we will recall well-known results in the Hilbert space situation and for convenience we will also provide short proofs. In the final section we introduce the proper analogue of the space $\mathfrak{H}_S$ in the almost Pontryagin space case so that we can generalize most of the results from Section 3 to the indefinite case.

2. Symmetric relations on almost Pontryagin spaces

We are going to consider a closed symmetric relation S on an almost Pontryagin space $(\mathfrak{L}, [.,.], \mathcal{O})$, i.e., a closed linear subspace of $\mathfrak{L}^2 = \mathfrak{L} \times \mathfrak{L}$ with the property that

$$[f_1, g_2] - [g_1, f_2] = 0, \ \ (f_1; g_1), (f_2; g_2) \in S.$$

Remark 2.1. We know from Proposition 3.2 in [7] that any almost Pontryagin space $(\mathfrak{L}, [.,.], \mathcal{O})$ can be viewed as a closed subspace of codimension $\Delta(\mathfrak{L}, [.,.])$ of a Pontryagin space $(\mathfrak{P}, [.,.])$ with degree $\kappa_-(\mathfrak{L}, [.,.]) + \Delta(\mathfrak{L}, [.,.])$ of negativity. Then a linear relation S on $(\mathfrak{L}, [.,.], \mathcal{O})$ is symmetric (closed) if and only if it is symmetric (closed) as a linear relation on $(\mathfrak{P}, [.,.])$.

If, in addition, J is a fundamental symmetry on $(\mathfrak{P}, [.,.])$, then S is symmetric (closed) on $(\mathfrak{L}, [.,.], \mathcal{O})$ if and only if the linear relation JS is a symmetric (closed) relation on the Hilbert space $(\mathfrak{P}, [J.,.])$. This fact is as easily verifiable by the following connection between the adjoint relation $S^{[*]}$ of S in $(\mathfrak{P}, [.,.])$ and the adjoint relation $(JS)^*$ of JS in the Hilbert space $(\mathfrak{P}, [J.,.])$:

$$(JS)^* = JS^{[*]}.$$

Definition 2.2. Let S be a symmetric relation on an almost Pontryagin space $(\mathfrak{L}, [.,.], \mathcal{O})$. We define a scalar product $h^S[.,.]$ on

$$\operatorname{dom} S = \{x \in \mathfrak{L} : (x; y) \in S \text{ for some } y \in \mathfrak{L}\}.$$

For $x, u \in \operatorname{dom} S$ let $y, v \in \mathfrak{L}$ be such that $(x; y), (u; v) \in S$ and set

$$h^S[x, u] = [y, u].$$

This scalar product is well defined and Hermitian. In fact, if $\tilde{y} \in \mathfrak{L}$ with $(x;\tilde{y}) \in S$, then the fact that S is symmetric yields

$$h^S[x,u] = [y,u] = [x,v] = [\tilde{y},u],$$

and

$$h^S[x,u] = [y,u] = [x,v] = \overline{[v,x]} = \overline{h^S[u,x]}.$$

Note also that $h^S[x,u] = [Sx,u]$, if S is an operator.

Remark 2.3. If $(\mathfrak{P},[.,.])$ is a Pontryagin space containing $(\mathfrak{L},[.,.],\mathcal{O})$ as a closed subspace (see Remark 2.1) and J is a fundamental symmetry on it, then it is straight forward to check that

$$h^S[.,.] = h^{JS}[J.,.]. \tag{2.1}$$

The following little lemma will be of use later on. Hereby an orthogonal projection P in an almost Pontryagin space $(\mathfrak{L},[.,.],\mathcal{O})$ is an everywhere defined linear operator on $\mathfrak{L}$ which satisfies $P^2 = P$ and $[Px,y] = [x,Py]$ for $x,y \in \mathfrak{L}$.

Lemma 2.4. *Let S be a symmetric relation on an almost Pontryagin space $(\mathfrak{L},[.,.],\mathcal{O})$. If P is an orthogonal projection in $(\mathfrak{L},[.,.],\mathcal{O})$ such that* $\operatorname{dom}(S) \subseteq P(\mathfrak{L})$, *then $h^S[.,.] = h^{PS}[.,.]$.*

Proof. For $(x_1;y_1),(x_2;y_2) \in S$ we have

$$h^S[x_1,x_2] = [y_1,x_2] = [y_1,Px_2] = [Py_1,x_2] = h^{PS}[x_1,x_2]. \qquad \square$$

3. Semibounded linear relations on Hilbert spaces

In this section we recall some results about semibounded relations on Hilbert spaces which are going to be important for us later on. A symmetric relation S on a Hilbert space is called semibounded if there exists a real number m such that

$$m(x,x) \le h^S(x,x), \text{ for all } x \in \operatorname{dom} S. \tag{3.1}$$

The maximum of all $m \in \mathbb{R}$ such that (3.1) holds true is denoted by $m(S)$ and is called the lower bound of S.

In order to avoid complicated formulas in the sequel we define the scalar product $(m \in \mathbb{R})$

$$h_m^S(.,.) = h^S(.,.) - m(.,.).$$

For $m < m(S)$ the inner product $h_m^S(.,.)$ is a positive definite inner product.

Note further that with S also its closure in $\mathfrak{H}^2$ is semibounded with the same lower bound, i.e., $m(\overline{S}) = m(S)$.

Definition 3.1. Let S be a semibounded relation on a Hilbert space $(\mathfrak{H},(.,.))$, and let $m < m(S)$. By $\mathfrak{H}_S$ we denote the completion of $\operatorname{dom} S$ with respect to $h_m^S(.,.)$.

The following remarks are more or less explicitly contained in [5].

Remark 3.2. For $m_2 \le m_1 < m(S)$ and $x \in \operatorname{dom} S$ we have

$$h^S_{m_1}(x,x) = h^S(x,x) - m_1(x,x) \le h^S(x,x) - m_2(x,x) = h^S_{m_2}(x,x),$$

and

$$\frac{m(S)-m_1}{m(S)-m_2} h^S_{m_2}(x,x) = \frac{m(S)-m_1}{m(S)-m_2}(h^S(x,x) - m(S)(x,x)) + (m(S)-m_1)(x,x).$$

As $h^S(x,x) - m(S)(x,x) \ge 0$ and $m(S) - m_1 \le m(S) - m_2$ this expression is less or equal to

$$(h^S(x,x) - m(S)(x,x)) + (m(S) - m_1)(x,x) = h^S_{m_1}(x,x).$$

Therefore, the topology induced by $h^S_m(.,.)$ on $\operatorname{dom} S$ and, hence, the Hilbert space $\mathfrak{H}_S$ does not depend on the choice of $m < m(s)$.

By Lemma 2.4 with $h^S_m(.,)$ also $\mathfrak{H}_S$ remains unaltered if we switch from S to PS where P is an orthogonal projection onto a subspace of $\mathfrak{H}$ which contains $\operatorname{dom} S$, i.e., $\mathfrak{H}_S = \mathfrak{H}_{PS}$.

Since $((a;b);(x;y)) \mapsto h^S_m(a,x)$ is continuous with resect to the graph norm, we have $\mathfrak{H}_S = \mathfrak{H}_{\overline{S}}$.

Remark 3.3. For $m < m(S)$ and $x \in \operatorname{dom} S$ we have

$$(m(S) - m)(x,x) \le h^S(x,x) - m(S)(x,x) + (m(S) - m)(x,x) = h^S_m(x,x).$$

Thus by continuity one can extend $(.,.)$ to $\mathfrak{H}_S$. Having done this we can define $h^S_l(.,.)$ on $\mathfrak{H}_S$ for all $l \in \mathbb{R}$ by

$$h^S_l(.,.) = h^S_m(.,.) + (m-l)(.,.).$$

Clearly, $h^S_l(.,.)$ is the unique extension by continuity of the originally on $\operatorname{dom} S$ defined scalar product $h^S_l(.,.)$.

Remark 3.4. From Remark 3.3 we conclude that the embedding

$$\iota : (\operatorname{dom} S, h^S_m(.,.)) \to (\mathfrak{H}, (.,.))$$

is bounded and can therefore be continued to a bounded mapping

$$\iota : (\mathfrak{H}_S, h^S_m(.,.)) \to (\mathfrak{H}, (.,.)).$$

The latter operator is in fact an embedding. For if $\iota(x) = 0$, then let $x_n \in \operatorname{dom} S$ converge to x within $\mathfrak{H}_S$. By continuity $\iota(x_n) = x_n \to 0$ within $\mathfrak{H}$. For $(a;b) \in S$ we have

$$h^S_m(a,x) = \lim_{n\to\infty} h^S_m(a,x_n) = \lim_{n\to\infty} (h^S(a,x_n) - m(a,x_n)) =$$

$$\lim_{n\to\infty} ((b,x_n) - m(a,x_n)) = 0,$$

and, hence, x is orthogonal to $\operatorname{dom} S$ within $\mathfrak{H}_S$ which yields $x = 0$.

As a consequence of the injectivity of ι we can consider $\mathfrak{H}_S$ as a linear subspace of $\mathfrak{H}$ where $x \in \mathfrak{H}$ belongs to $\mathfrak{H}_S$ if there exists a sequence $((x_n; y_n))$ in S such that

$$\lim_{n\to\infty} (x - x_n, x - x_n) = 0, \quad \lim_{k,l\to\infty} (x_k - x_l, y_k - y_l) = 0. \tag{3.2}$$

Finally, it is elementary to see that for $x \in \mathfrak{H}_S$ and $(a; b) \in S$ we have

$$h_m^S(a, x) = (b - ma, x).$$

We will use this fact without giving explicit references.

The space $\mathfrak{H}_S$ is used to define the Friedrichs extension of S as defined in [5]. The following way to introduce the Friedrichs extension is slightly different from the conventional access and is closely connected to the constructions given in [10],[11] and [12]. See also [2].

Theorem 3.5. *Let S be a symmetric and semibounded linear relation on the Hilbert space $(\mathfrak{H}, (.,.))$. Let $m < m(S)$ and consider the Hilbert space $(\mathfrak{H}_S, h_m^S(.,.))$ and the embedding*

$$\iota : (\mathfrak{H}_S, h_m^S(.,.)) \to (\mathfrak{H}, (.,.)).$$

Then the linear relation $S_F = (\iota\iota^*)^{-1} + mI$ is a selfadjoint and semibounded extension of S with $m(S_F) = m(S)$. Moreover, it does not depend on the particularly chosen $m < m(S)$. In fact,

$$S_F = \{(x; y) \in S^* : x \in \mathfrak{H}_S\}. \tag{3.3}$$

Proof. Clearly, $\iota\iota^*$ is a selfadjoint and bounded linear operator on $\mathfrak{H}$. Using standard arguments about linear relations we see that $(\iota\iota^*)^{-1}$ is a selfadjoint linear relation. Since for $y \in \operatorname{dom}(\iota\iota^*)^{-1} = \operatorname{ran}\iota\iota^*$ with $\iota\iota^* x = y$ we have

$$h^{(\iota\iota^*)^{-1}}(y, y) - (x, y) = h_m^S(\iota^* x, \iota^* x) \geq 0, \tag{3.4}$$

this relation is semibounded with a non-negative lower bound. With $(\iota\iota^*)^{-1}$ also S_F is selfadjoint and semibounded. If $(a; b) \in S - mI$ and $u \in \operatorname{dom} S$, then $(a; b + ma) \in S$ and $\iota(u) = u$ because we identify $\mathfrak{H}_S$ with a subspace of $\mathfrak{H}$. Therefore

$$h_m^S(a, u) = (b + ma, u) - m(a, u) = (b, u) = (b, \iota(u)) = h_m^S(\iota^* b, u),$$

and we obtain from the density of $\operatorname{dom} S$ in $\mathfrak{H}_S$ that $a = \iota^* b = \iota\iota^* b$. This proves $S \subseteq S_F$, and by the selfadjointness of S_F we see that S_F is contained in the right-hand side of (3.3). Conversely, if $(x; y) \in S^* - mI$ and $x \in \mathfrak{H}_S$, let (x_n) be a sequence in $\operatorname{dom} S$ which converges to x within $\mathfrak{H}_S$ and, hence, also within $\mathfrak{H}$. We calculate for $(u; v) \in S$

$$h_m^S(u, \iota^*(y)) = (\iota(u), y) = (u, y) = (v, x) - m(u, x) =$$

$$\lim_{n\to\infty} ((v, x_n) - m(u, x_n)) = \lim_{n\to\infty} h_m^S(u, x_n) = h_m^S(u, x),$$

and obtain $\iota\iota^*(y) = x$. Thus we verified (3.3) which, in turn, together with Remark 3.2 implies the independence of S_F from $m < m(S)$.

Finally, from $m((\iota\iota^*)^{-1}) \geq 0$ we get $m(S_F) \geq m$ and from the independence of S_F from $m < m(S)$ the relation $m(S_F) \geq m(S)$. The converse inequality is an immediate consequence of $S \subseteq S_F$. $\square$

Definition 3.6. The selfadjoint linear relation S_F is called the Friedrichs extension of S.

Remark 3.7. It is easy to see that $\mathfrak{H}_{S+rI} = \mathfrak{H}_S$ and $(S+rI)_F = S_F + rI$ for $r \in \mathbb{R}$.

With the notation from the proof of Theorem 3.5 we have

$$S_F(0) = (\iota\iota^*)^{-1}(0) = \ker \iota\iota^* = (\operatorname{dom} S)^\perp.$$

Remark 3.8. First note that since S has a selfadjoint extension any closed, symmetric and semibounded relation has equal defect indices, i.e., the Hilbert space dimension of $\ker(S^* - zI)$ is the same for all $z \in r(S)$ where $r(S)$ $(\supseteq \mathbb{C} \setminus \mathbb{R})$ is the set of all points of regular type for S.

For $m < m(S) = m(S_F)$ and $(x; y) \in S_F$ we have

$$\|x\| \|y - mx\| \geq (y - mx, x) \geq (m(S) - m)(x, x).$$

We conclude $m \in \rho(S_F)$ and

$$\|(S_F - mI)^{-1}\| \leq \frac{1}{m(S) - m}. \tag{3.5}$$

Therefore $\mathbb{C} \setminus [m(S), \infty) \subseteq \rho(S_F)$ and, hence, $\mathbb{C} \setminus [m(S), \infty) \subseteq r(S)$.

The fact that $(-\infty, m(S)) \subseteq \rho(S_F)$ can also be seen from the proof of Theorem 3.5. In fact, if we provide $\mathfrak{H}_S$ with $h_m^S(.,.)$, $m < m(S)$, then we constructed S_F such that $(S_F - mI)^{-1}$ is the bounded operator $\iota\iota^*$.

We are going to consider arbitrary selfadjoint and semibounded extensions H of S in $\mathfrak{H}$ and for $m < m(H)$ the relation between the Hilbert spaces $(\mathfrak{H}_H, h_m^H(.,.))$ and $(\mathfrak{H}_S, h_m^S(.,.))$. This well-known result is strongly connected with the second representation theorem from Kato, [9]. See also Chapter 10 of [3].

Theorem 3.9. *Let S be semibounded on the Hilbert space $(\mathfrak{H}, (.,.))$ and H be a selfadjoint and semibounded extension of S. Moreover, let $H = H_s \oplus H_\infty$ be the decomposition of H into the purely relational part $H_\infty = \{0\} \times H(0)$ and the operator part H_s, which is a selfadjoint operator on $H(0)^\perp$.*

Then the space $\mathfrak{H}_H$ as a subspace of $\mathfrak{H}$ coincides with $\operatorname{dom} |H_s|^{\frac{1}{2}}$, and for $m < m(H)$ the Hilbert space inner product $h_m^H(.,.)$ can be calculated as

$$h_m^H(x, y) = ((H_s - mI)^{\frac{1}{2}} x, (H_s - mI)^{\frac{1}{2}} y), \quad x, y \in \mathfrak{H}_H \tag{3.6}$$

The space $\mathfrak{H}_H$ contains $\mathfrak{H}_S$ as a closed subspace, and on this closed subspace the products $h_m^H(.,.)$ and $h_m^S(.,.)$ coincide. If $\mathfrak{H}_H$ is provided with $h_m^H(.,.)$, then

$$\mathfrak{H}_H \ominus \mathfrak{H}_S = \mathfrak{H}_H \cap \ker(S^* - mI). \tag{3.7}$$

We have $\mathfrak{H}_H = \mathfrak{H}_S$ if and only if $H = S_F$.

Proof. The assumption $S \subseteq H$ immediately yields $h_m^H(.,.) = h_m^S(.,.)$ on $\operatorname{dom} S$. Thus the completion $\mathfrak{H}_S$ of $\operatorname{dom} S$ with respect to $h_m^S(.,.)$ is a closed subspace of $\mathfrak{H}_H$.

Since H is semibounded and $m < m(H)$, the selfadjoint operator $H_s - mI$ is strictly positive on $H(0)^\perp$. Therefore, we can consider the square root of it. For $x, y \in \operatorname{dom} H_s = \operatorname{dom} H$ we have $(x; H_s x), (y; H_s y) \in H_s$, and hence

$$h_m^H(x,y) = (H_s x, y) - m(x,y) = ((H_s - mI)x, y) = ((H_s - m)^{\frac{1}{2}} x, (H_s - m)^{\frac{1}{2}} y).$$

Using the boundedness of $(H_s - mI)^{-1}$ we see that the norm induced by $h_m^H(.,.)$ is equivalent to the graph norm of $(H_s - mI)^{\frac{1}{2}}$ on $\operatorname{dom} H_s$. By the functional calculus for selfadjoint operators $\operatorname{dom}(H_s - mI)^{\frac{1}{2}} = \operatorname{dom} |H_s|^{\frac{1}{2}}$, and $\operatorname{dom} H_s$ is dense in $\operatorname{dom}(H_s - mI)^{\frac{1}{2}}$ with respect to the the graph norm of $(H_s - mI)^{\frac{1}{2}}$. Thus $\mathfrak{H}_H = \operatorname{dom} |H_s|^{\frac{1}{2}}$, and relation (3.6) extends to all $x, y \in \mathfrak{H}_H$.

If $H = S_F$, we obtain from (3.3) that $\operatorname{dom} S_F \subseteq \mathfrak{H}_S$. As we already identified $\mathfrak{H}_S$ as a subspace of $\mathfrak{H}_H$ we get $\mathfrak{H}_H = \mathfrak{H}_S$. Conversely, if we assume $\mathfrak{H}_H = \mathfrak{H}_S$, then by definition $\operatorname{dom} H \subseteq \mathfrak{H}_H$ and hence

$$H \subseteq \{(x; y) \in S^* : x \in \mathfrak{H}_H\} = \{(x; y) \in S^* : x \in \mathfrak{H}_S\} = S_F.$$

As both relations are selfadjoint we obtain $S_F = H$. To verify (3.7) note that for $x \in \mathfrak{H}_H$ and $(a; b) \in S$

$$h_m^H(a, x) = ((H_s - m)^{\frac{1}{2}} a, (H_s - m)^{\frac{1}{2}} x) = ((H_s - m)a, x) = (b - ma, x).$$

The final equality follows from $H_s a - b \in H(0)$ and the fact that

$$\mathfrak{H}_H = \operatorname{dom} |H_s|^{\frac{1}{2}} \perp H(0).$$

Thus $x \in \mathfrak{H}_H \ominus \mathfrak{H}_S$ if and only if $x \in \operatorname{ran}(S - mI)^\perp = \ker(S^* - mI)$. □

Remark 3.10. If we choose $H = S_F$ in (3.7), then we see that $\mathfrak{H}_S$ is disjoint to $\ker(S^* - mI)$ for all $m < m(S)$.

Remark 3.11. If S is closed with finite defect indices, then any selfadjoint extension H of S in $\mathfrak{H}$ is a finite-dimensional perturbation of S_F. Hence every canonical selfadjoint extension is semibounded. Hereby canonical means that H is a selfadjoint extension within $\mathfrak{H}$.

Moreover, by Theorem 3.9 any space $\mathfrak{H}_H$ contains $\mathfrak{H}_S$ and is contained in $\mathfrak{H}_S \dot{+} \ker(S^* - mI)$. We are going to show that any linear space $\mathfrak{G}$ with $\mathfrak{H}_S \subseteq \mathfrak{G} \subseteq \mathfrak{H}_S \dot{+} \ker(S^* - mI)$ equals a space $\mathfrak{H}_H$ for some H.

From now on we assume that S is a closed, symmetric and semibounded linear relation with finite defect indices.

Remark 3.12. As already mentioned the space $\mathfrak{H}_S \dot{+} \ker(S^* - mI)$ is of particular interest for $m < m(S)$. If $z \in \rho(S_F)$, we have

$$\mathfrak{H}_S \dot{+} \ker(S^* - zI) = \mathfrak{H}_S + \operatorname{dom} S^*. \tag{3.8}$$

As $(-\infty, m(S)) \subseteq \rho(S_F)$ we conclude that $\mathfrak{H}_S \dot{+} \ker(S^* - mI)$ does not depend on $m < m(S)$.

To verify (3.8) recall that for $z, w \in \rho(S_F)$ the operator

$$I + (z - w)(S_F - z)^{-1},$$

maps $\ker(S^* - wI)$ bijectively onto $\ker(S^* - zI)$. Since $\operatorname{dom} S_F \subseteq \mathfrak{H}_S$ (Theorem 3.9), we see that the space on the left-hand side of the equality sign in (3.8) is independent from $z \in \rho(S_F)$. The relation (3.8) is now an immediate consequence of the von Neumann formula (see, e.g., Theorem 6.1 in [6]).

Definition 3.13. By $\mathfrak{H}^S$ we denote the space in (3.8).

Proposition 3.14. *Assume that S is a closed, symmetric and semibounded linear relation with finite defect indices. Let $\mathfrak{G}$ be a subspace of $\mathfrak{H}^S$ which contains $\mathfrak{H}_S$. Then there exists a canonical selfadjoint extension H of S such that $\mathfrak{H}_H = \mathfrak{G}$.*

Proof. We provide $\mathfrak{G}$ with a Hilbert space inner product $h_m^{\mathfrak{G}}(.,.)$ which extends $h_m^S(.,.)$, $m < m(S)$, such that

$$\mathfrak{G} = \mathfrak{H}_S \oplus_{h_m^{\mathfrak{G}}(.,.)} (\ker(S^* - mI) \cap \mathfrak{G}). \tag{3.9}$$

As $\dim \ker(S^* - mI) < \infty$ the Hilbert space $(\mathfrak{G}, h_m^{\mathfrak{G}}(.,.))$ is continuously embedded in $\mathfrak{H}$, and we denote by $\iota_{\mathfrak{G}}$ the corresponding inclusion map.

Similar as for ι in the proof of Theorem 3.5 we see that $(\iota_{\mathfrak{G}}\iota_{\mathfrak{G}}^*)^{-1}$ is a semibounded selfadjoint linear relation with a non-negative lower bound. Then also $H := (\iota_{\mathfrak{G}}\iota_{\mathfrak{G}}^*)^{-1} + mI$ is a semibounded selfadjoint linear relation.

If $(a; b) \in S - mI$ and $u = u_1 + u_2 \in \operatorname{dom} S \dot{+} (\ker(S^* - mI) \cap \mathfrak{G})$, then $(a; b + ma) \in S$ and $\iota_{\mathfrak{G}}(u) = u$ as we identify $\mathfrak{G}$ with a linear subspace of $\mathfrak{H}$. As $\ker(S^* - mI) = \operatorname{ran}(S - mI)^\perp$

$$\begin{aligned} h_m^{\mathfrak{G}}(a, u) &= h_m^S(a, u_1) = (b + ma, u_1) - m(a, u_1) \\ &= (b, u_1) = (b, u) = (b, \iota_{\mathfrak{G}}(u)) = h_m^{\mathfrak{G}}(\iota_{\mathfrak{G}}^* b, u), \end{aligned}$$

and we obtain from the density of $\operatorname{dom} S \dot{+} (\ker(S^* - mI) \cap \mathfrak{G})$ in $\mathfrak{G}$ that $a = \iota_{\mathfrak{G}}\iota_{\mathfrak{G}}^* b$. Thus we verified $S \subseteq H$.

Since $\iota_{\mathfrak{G}}$ is injective, its adjoint has a dense range in $\mathfrak{G}$. This range clearly coincides with $\operatorname{dom} H$. Moreover,

$$\begin{aligned} h_m^H(a, x) &= (b - ma, x) = (b - ma, \iota_{\mathfrak{G}}\iota_{\mathfrak{G}}^*(y - mx)) \\ &= h_m^{\mathfrak{G}}(\iota_{\mathfrak{G}}^*(b - ma), \iota_{\mathfrak{G}}^*(y - mx)) = h_m^{\mathfrak{G}}(a, x), \end{aligned}$$

for $(a; b), (x; y) \in H$, and hence $\mathfrak{H}_H = \mathfrak{G}$. □

As an immediate consequence of the previous results we obtain

Corollary 3.15. *With the same assumptions as in Proposition 3.14 the space $\mathfrak{H}^S$ contains $\mathfrak{H}_H$ for all canonical selfadjoint extensions H of S, and for some canonical selfadjoint extensions H of S we have $\mathfrak{H}^S = \mathfrak{H}_H$.*

Let $\mathfrak{G}$ be such that $\mathfrak{H}_S \subseteq \mathfrak{G} \subseteq \mathfrak{H}^S$, and let $\mathfrak{G}$ be provided with a Hilbert space scalar product $h_m^{\mathfrak{G}}(.,.)$ which coincides with $h_m^S(.,.)$ on $\mathfrak{H}_S$ such that (3.9)

holds. We denote by P the orthogonal projection of $\mathfrak{G}$ onto $\mathfrak{H}_S$. Now we set

$$T = S \cap (\mathfrak{G} \times \mathfrak{G}).$$

Proposition 3.16. *Under the above assumptions the linear relation T considered in $(\mathfrak{G}, h_m^{\mathfrak{G}}(.,.))$ is closed, symmetric and semibounded with a lower bound larger than m.*

If (n, n) denotes the defect index of S, then T is of defect index (r, r) with $r \leq n$. If $\mathfrak{H}$ satisfies the minimality condition

$$\mathfrak{H} = \operatorname{cls}(\operatorname{dom} S \cup \operatorname{ran} S), \tag{3.10}$$

then $r = n$.

Proof. The closedness is an immediate consequence of the boundedness of the inclusion map $\iota_{\mathfrak{G}}$. For $(a; b), (x; y) \in T$ we have $Pa = a$, $Px = x$. Using $\ker(S^* - mI) \perp_{(.,.)} \operatorname{ran}(S - mI)$, the fact that T is symmetric follows from

$$h_m^{\mathfrak{G}}(a, y) = h_m^S(a, Py) = (b - ma, Py) = (b - ma, y) = (b, y - mx) = h_m^{\mathfrak{G}}(b, x).$$

For later use we point out that more generally we have for $(a; b) \in S$, $y \in \mathfrak{G}$

$$h_m^{\mathfrak{G}}(a, y) = h_m^S(a, Py) = (b - ma, Py) = (b - ma, y). \tag{3.11}$$

As

$$\begin{aligned} h_m^{\mathfrak{G}}(a, b) = (b - ma, b) &= (b - ma, b - ma) + m(b - ma, a) \\ &= (b - ma, b - ma) + m h_m^{\mathfrak{G}}(a, a), \end{aligned}$$

T is semibounded with a lower bound larger or equal to m. For $\epsilon > 0, m + \epsilon < m(S)$ we obtain from (3.5)

$$\begin{aligned} (b - ma, b - ma) &= (b - (m + \epsilon)a, b - (m + \epsilon)a) + 2\epsilon(b - (m + \epsilon)a, a) + \epsilon^2(a, a) \\ &= \|b - (m + \epsilon)a\|^2 + 2\epsilon h_m^{\mathfrak{G}}(a, a) - \epsilon^2(a, a) \\ &\geq (m(S) - (m + \epsilon) - \epsilon^2)\|a\|^2 + 2\epsilon h_m^{\mathfrak{G}}(a, a). \end{aligned}$$

For sufficiently small ϵ we get

$$h_m^{\mathfrak{G}}(a, b) \geq (m + 2\epsilon) h_m^{\mathfrak{G}}(a, a),$$

and therefore $m(T) > m$.

As $\operatorname{dom} S \subseteq \mathfrak{H}_S \subseteq \mathfrak{G}$ we have for $z \in r(T)$,

$$\begin{aligned} \operatorname{ran}(T - zI) &= \operatorname{ran}(S - zI) \cap \mathfrak{G} \\ &= \{x \in \mathfrak{G} : (\iota_{\mathfrak{G}}(x), y) = 0, \ y \in \ker(S^* - \bar{z}I)\} \\ &= (\iota_{\mathfrak{G}}^* \ker(S^* - \bar{z}I))^{\perp_{h_m^{\mathfrak{G}}(.,.)}}. \end{aligned}$$

Therefore, T has defect index (r, r) where $r \leq n$.

If $r < n$, then $\iota_{\mathfrak{G}}^*(y) = 0$ for some $y \in \ker(S^* - \bar{z}I)$, $y \neq 0$. From $y \in \ker \iota_{\mathfrak{G}}^* = (\operatorname{ran} \iota_{\mathfrak{G}})^{\perp} \subseteq S^*(0)$ we conclude $y \in \ker(S^*)$. Hence, condition (3.10) cannot be satisfied. $\square$

As a consequence of the previous proof note that

$$\iota_{\mathfrak{G}}^*(\ker(S^* - mI)) = \ker(T^* - mI),$$

where this correspondence between the defect spaces is bijective if (3.10) holds true. On $\operatorname{ran}(S - mI) = \ker(S^* - mI)^\perp$ we have ($x \in \mathfrak{G}$)

$$h_m^{\mathfrak{G}}(\iota_{\mathfrak{G}}^*(b - ma), x) = (b - ma, x) = h_m^{\mathfrak{G}}(a, x).$$

Hence, $\iota_{\mathfrak{G}}^*(b - ma) = a$.

In the following, $h^T h_m^{\mathfrak{G}}(.,.)$ is the scalar product and $\mathfrak{G}_T$ is the space constructed from $\mathfrak{G}, h_m^{\mathfrak{G}}(.,.), T$ in the same as $h^S(.,.)$ and $\mathfrak{H}_S$ were constructed from $\mathfrak{H}, (.,.), S$.

As already noted we have for $(a; b), (x; y) \in T$

$$\begin{aligned} h^T h_m^{\mathfrak{G}}(a, x) = h_m^{\mathfrak{G}}(a, y) = h_m^{\mathfrak{G}}(a, Py) &= (b - ma, Py) \\ &= (b - ma, y - mx) + m(b - ma, x) \\ &= (b - ma, y - mx) + m(h^{\mathfrak{G}}(a, x) - m(a, x)) \\ &= (b - ma, y - mx) + m h_m^{\mathfrak{G}}(a, x), \end{aligned}$$

and hence $h_m^T h_m^{\mathfrak{G}}(a, x) = (b - ma, y - mx)$.

Proposition 3.17. *With the above assumptions and notations $\iota_{\mathfrak{G}}^*$ maps $(\overline{\mathfrak{G}} \cap \operatorname{ran}(S - mI), (.,.))$ unitarily onto $(\mathfrak{G}_T, h_m^T h_m^{\mathfrak{G}}(.,.))$, where $\mathfrak{G}_T$ coincides with* $\operatorname{dom}(S \cap (\overline{\mathfrak{G}} \times \overline{\mathfrak{G}}))$ *and $h_m^T h_m^{\mathfrak{G}}(.,.)$ induces a norm on $\mathfrak{G}_T$, which is equivalent to the graph norm induced by $S \cap (\overline{\mathfrak{G}} \times \overline{\mathfrak{G}})$.*

If we denote by R the symmetry $T \cap (\mathfrak{G}_T \times \mathfrak{G}_T)$ on $(\mathfrak{G}_T, h_m^T h_m^G(.,.))$, then

$$((\iota_{\mathfrak{G}}^*)^{-1} \times (\iota_{\mathfrak{G}}^*)^{-1})(R) = S \cap ((\overline{\mathfrak{G}} \cap \operatorname{ran}(S - mI)) \times (\overline{\mathfrak{G}} \cap \operatorname{ran}(S - mI))).$$

Proof. For the proof we first mention that the fact that $\operatorname{ran}(S - mI)$ has finite codimension in $\mathfrak{H}$ ensures

$$\overline{\mathfrak{G} \cap \operatorname{ran}(S - mI)} = \overline{\mathfrak{G}} \cap \operatorname{ran}(S - mI).$$

As

$$h_m^T h_m^G(\iota_{\mathfrak{G}}^*(b - ma), \iota_{\mathfrak{G}}^*(y - mx)) = h_m^T h_m^G(a, x) = (b - ma, y - mx), \qquad (3.12)$$

we see that $\iota_{\mathfrak{G}}^*|_{\operatorname{ran}(S-mI)} = (S - mI)^{-1}$ maps $\operatorname{ran}(T - mI)$ unitarily onto $\operatorname{dom} T$. By continuity $\iota_{\mathfrak{G}}^*|_{\operatorname{ran}(S-mI)} = (S - mI)^{-1}$ then maps $(\overline{\mathfrak{G}} \cap \operatorname{ran}(S - mI), (.,.))$ unitarily onto $(\mathfrak{G}_T, h_m^T h_m^{\mathfrak{G}}(.,.))$. Thus

$$\mathfrak{G}_T = (S - mI)^{-1}(\overline{\mathfrak{G}} \cap \operatorname{ran}(S - mI)) = \operatorname{dom}(S \cap (\overline{\mathfrak{G}} \times \overline{\mathfrak{G}})).$$

The continuity of $(S - mI)^{-1}$ together with (3.12) shows that $h_m^T h_m^{\mathfrak{G}}(.,.)$ induces a norm on $\mathfrak{G}_T$, which is equivalent to the graph norm induced by $S \cap (\overline{\mathfrak{G}} \times \overline{\mathfrak{G}})$.

For $x, y \in \overline{\mathfrak{G}} \cap \operatorname{ran}(S - mI)$ we have $(x; y) \in S$ if and only if $x = (H - m)^{-1}(y - mx)$, where H is the selfadjoint extension $(\iota_{\mathfrak{G}} \iota_{\mathfrak{G}}^*)^{-1} + mI$ of S (see Proposition 3.14). As $\iota_{\mathfrak{G}}(\mathfrak{G})^\perp = \ker \iota_{\mathfrak{G}}^* = H(0)$ this is equivalent to $(H - m)^{-2}(y - mx) =$

$(H-m)^{-1}x$ or because of $(H-m)^{-1}y - m(H-m)^{-1}x = x \in \operatorname{ran}(S-m)$ in turn equivalent to

$$(\iota_{\mathfrak{G}}^* x; \iota_{\mathfrak{G}}^* y) = ((H-m)^{-1}x; (H-m)^{-1}y) \in S \cap (\mathfrak{G}_T \times \mathfrak{G}_T) = R. \qquad \square$$

Thus we showed that for a closed and semibounded symmetry S with finite defect index (n, n) one can partially reconstruct $\mathfrak{H}$ and S from $\mathfrak{H}^S$ and T by focusing on $\overline{\mathfrak{G}} \cap \operatorname{ran}(S - mI)$.

4. Symmetric relations of finite negativity

Definition 4.1. Let $(\mathfrak{L}, [.,.], \mathcal{O})$ be an almost Pontryagin space, and let S be a closed symmetric relation on $\mathfrak{L}$ such that S has finite codimension in

$$S^{[*]} = \{(a; b) \in \mathfrak{L} \times \mathfrak{L} : [a, y] = [b, x] \text{ for all } (x; y) \in S\}.$$

Then S is called to be of finite negativity κ_S in $(\mathfrak{L}, [.,.], \mathcal{O})$ if the inner product $h^S[.,.]$ has κ_S negative squares on $\operatorname{dom} S$. If $\kappa_S = 0$, we shall call S non-negative.

By well-known results in the theory of inner product spaces (see, e.g., [4]) $h^S[.,.]$ has finitely many negative squares if and only if there exists a linear subspace of $\operatorname{dom} S$ of finite codimension such that $h^S[.,.]$ restricted to this subspace is positive semidefinite. Moreover, $h^S[.,.]$ has κ_S negative squares on $\operatorname{dom} S$ if and only if there exists a κ_S-dimensional subspace $\mathfrak{N}$ of $\operatorname{dom} S$ such that $(\mathfrak{N}, -h^S[.,.])$ is a Hilbert space, and there is no higher-dimensional subspace of $\operatorname{dom} S$ with this property. In this case we can decompose $\operatorname{dom} S$ as

$$\operatorname{dom} S = \mathfrak{M} \dot{+} \mathfrak{N},$$

where $\mathfrak{M}$ is the orthogonal complement of $\mathfrak{N}$ with respect to $h^S[.,.]$, and $h^S[.,.]$ is non-negative on $\mathfrak{M}$.

Remark 4.2. It is easy to see that S is of finite negativity κ_S in $(\mathfrak{L}, [.,.], \mathcal{O})$, if and only if it is of finite negativity κ_S as a relation on a Pontryagin space $(\mathfrak{P}, [.,.])$ containing $(\mathfrak{L}, [.,.], \mathcal{O})$ as a closed subspace with finite codimension (see Remark 2.1).

If J is a fundamental symmetry of $(\mathfrak{P}, [.,.])$, then we see from (2.1) that S is of finite negativity κ_S in $(\mathfrak{L}, [.,.], \mathcal{O})$ if and only if JS is of finite negativity κ_S in the Hilbert space $(\mathfrak{P}, [J.,.])$.

Thus certain questions related to symmetries with finite negativity can be considered in a Hilbert space setting. There symmetries have the following important property.

Lemma 4.3. *Every symmetric relation of finite negativity on a Hilbert space is semibounded. Moreover,* $\operatorname{ran}(S - mI)$ *is closed and of finite codimension for all* $m < 0$.

Proof. Let S be a symmetry in a Hilbert space $(\mathfrak{H}, (.,.))$ of finite negativity κ_S. Now we consider $\mathfrak{G} = \mathfrak{H} \oplus \mathfrak{H}$ with the symmetric relation $T = S \oplus S^{-1}$ on it. As $T^* = S^* \oplus S^{-1*}$ it is straightforward to check that T is of finite negativity $2\kappa_S$ and that T has finite and equal defect indices.

Let A be a canonical selfadjoint extension of T in $\mathfrak{G}$. Since $\operatorname{dom} T \subseteq \operatorname{dom} A$ with finite codimension, also A is of finite negativity. Using the functional calculus for selfadjoint relations we derive from this fact that $\sigma(A) \cap (-\infty, 0)$ consists of finitely many eigenvalues of finite multiplicity. The proof for this assertion is very similar to the proof of Proposition 2.3 in [7] and is therefore omitted.

So we see that A and with A also its restriction S is semibounded. From the mentioned spectral properties for A we also see that $\operatorname{ran}(A - mI)$ is closed and of finite codimension for $m < 0$. The mapping $(x; y) \mapsto y - mx$ from A onto $\operatorname{ran}(A - mI)$ is continuous and has a finite-dimensional kernel. Hence the closed subspace T of A is mapped onto a closed subspace of $\operatorname{ran}(A - mI)$ of finite codimension. The structure of T shows that $\operatorname{ran}(S - mI)$ is closed and of finite codimension. □

Due to the previous lemma we can define a space associated to a symmetry of finite negativity.

Definition 4.4. Let $(\mathfrak{L}, [.,.], \mathcal{O})$ be an almost Pontryagin space, and let S be a symmetric relation of finite negativity on $(\mathfrak{L}, [.,.], \mathcal{O})$. Moreover, let $(\mathfrak{P}, [.,.])$ be a Pontryagin space which contains $(\mathfrak{L}, [.,.], \mathcal{O})$ as a closed subspace of finite codimension, and let J be a fundamental symmetry on this Pontryagin space. Then we define the space $\mathfrak{L}_S$ by

$$\mathfrak{L}_S = \mathfrak{P}_{JS},$$

where $\mathfrak{P}_{JS}$ is the space corresponding to the symmetry JS on the Hilbert space $(\mathfrak{P}, [J.,.])$ defined as in Definition 3.1.

We provide $\mathfrak{L}_S$ with the inner product $h^{JS}[J.,.]$ and denote it by $h^S[.,.]$ (see Remark 3.3). Moreover, let $\mathcal{O}_S$ denote the Hilbert space topology induced by $h_m^{JS}[J.,.]$, $m < m(JS)$ on $\mathfrak{L}_S$.

Remark 4.5. By Remark 3.4 $\mathfrak{P}_{JS}$ is continuously embedded in $\mathfrak{P}$. Denoting the inclusion mapping by ι its continuity yields

$$\iota(\mathfrak{P}_{JS}) = \iota(\overline{\operatorname{dom} JS}) = \iota(\overline{\operatorname{dom} S}) \subseteq \overline{\operatorname{dom} S} \subseteq \mathfrak{L}.$$

Hereby the latter closure is taken with respect to the topology $\mathcal{O}$ (which coincides with the topology induced by $[J.,.]$, see [7]) and the others are taken with respect to $\mathcal{O}_S$.

Thus $\mathfrak{L}_S$ is a linear subspace of $\mathfrak{L}$. Moreover, it is independent from the fundamental symmetry J and even from the space $\mathfrak{P}$. For by (3.2) a vector $x \in \mathfrak{L}$ belongs to $\mathfrak{L}_S$ if and only if there exists a sequence $((x_n; y_n))$ in S such that $x_n \to x$ with respect to $\mathcal{O}$ and

$$\lim_{k,l\to\infty} [x_k - x_l, y_k - y_l] = 0.$$

This characterization also shows that $\mathfrak{L}_S = \mathfrak{L}_{S-mI}$ whenever $S - mI$ is of finite negativity.

By the closed graph theorem and by the fact that ι is continuous the topology $\mathcal{O}_S$ is also independent from J and from $\mathfrak{P}$.

Finally, the $\mathcal{O}_S$-continuous scalar product $h^S[.,.]$ (on $\mathfrak{L}_S$) restricted to the the $\mathcal{O}_S$-dense linear subspace $\operatorname{dom} S$ coincides with $h^S[.,.]$ as it was defined in Definition 2.2. Hence $h^S[.,.]$ on $\mathfrak{L}_S$ is the unique continuation of $h^S[.,.]$ on $\operatorname{dom} S$ by continuity. Therefore, also $h^S[.,.]$ is independent from J and from $\mathfrak{P}$.

Remark 4.6. With the same assumptions as in Definition 4.4 let $\mathfrak{M}$ be a closed subspace of $(\mathfrak{L}, [.,.], \mathcal{O})$ such that $S \subseteq \mathfrak{M} \times \mathfrak{M}$. Then $(\mathfrak{M}, [.,.], \mathcal{O} \cap \mathfrak{M})$ is also an almost Pontryagin space (see [7]). By similar arguments as in the previous remark it is easy to verify that the triple $(\mathfrak{L}_S, h^S[.,.], \mathcal{O}_S)$ coincides with $(\mathfrak{M}_S, h^S[.,.], (\mathcal{O} \cap \mathfrak{M})_S)$. The latter is defined as above but just with the use of $(\mathfrak{M}, [.,.], \mathcal{O} \cap \mathfrak{M})$ instead of $(\mathfrak{L}, [.,.], \mathcal{O})$.

Proposition 4.7. *The triple $(\mathfrak{L}_S, h^S[.,.], \mathcal{O}_S)$ is an almost Pontryagin space if and only if there exists an $\epsilon > 0$ such that $S - \epsilon I$ is of finite negativity.*

Proof. Let $(\mathfrak{P}, [.,.])$ be a Pontryagin space which contains $(\mathfrak{L}, [.,.], \mathcal{O})$ as a closed subspace, and let J be a fundamental symmetry on this Pontryagin space. By definition $(\mathfrak{L}_S, h^S[.,.], \mathcal{O}_S) = (\mathfrak{P}_{JS}, h^{JS}[J.,.], \mathcal{O}_{JS})$, where $\mathcal{O}_{JS}$ denotes the Hilbert space topology induced by $h_m^{JS}[J.,.]$, $m < m(JS)$, on $\mathfrak{P}_{JS}$.

By Remark 4.2 the symmetric relation $S - \epsilon I$ is of finite negativity on $(\mathfrak{L}, [.,.], \mathcal{O})$ if and only if $JS - \epsilon J$ is of finite negativity on the Hilbert space $(\mathfrak{P}, [J.,.])$. Since the fundamental symmetry operator J is a finite-dimensional perturbation of I, the scalar product $h^{JS-\epsilon J}[J.,.]$ is a finite-dimensional perturbation of $h^{JS-\epsilon I}[J.,.]$ on $\operatorname{dom} S$. Hence $JS - \epsilon J$ is of finite negativity if and only if $JS - \epsilon I$ has this property.

We just showed that in order to prove the present proposition we may assume that $(\mathfrak{L}, [.,.])$ is a Hilbert space. Under this additional assumption let $(\mathfrak{L}_S, h^S[.,.], \mathcal{O}_S)$ be an almost Pontryagin space. By the definition of almost Pontryagin spaces (see [7]) there exists a closed subspace $\mathfrak{M}_S$ of finite codimension of $(\mathfrak{L}_S, h^S[.,.], \mathcal{O}_S)$ such that $(\mathfrak{M}_S, h^S[.,.])$ is a Hilbert space. Hence, if we choose $m < m(S)$, then there exist $c, d > 0$ such that for all $x \in \mathfrak{M}_S$

$$c h^S[x,x] \le h_m^S[x,x] \le d h^S[x,x]. \tag{4.1}$$

The space $\mathfrak{M}_S \cap \operatorname{dom} S$ has finite codimension in $\operatorname{dom} S$, and for $x \in \mathfrak{M}_S \cap \operatorname{dom} S$ we have

$$d h^{S - \frac{m(S)-m}{d} I}[x,x] \ge h_m^S[x,x] - (m(S) - m)[x,x] = h^S[x,x] - m(S)[x,x] \ge 0.$$

If we set

$$\epsilon = \frac{m(S) - m}{d},$$

then $\epsilon > 0$ and $h^{S-\epsilon I}[.,.]$ has finitely many negative squares, i.e., $S - \epsilon I$ is of finite negativity.

Conversely, if $S-\epsilon I$ is of finite negativity, then we can find a linear subspace $\mathfrak{M}$ of $\operatorname{dom} S$ of finite codimension such that

$$0 \leq h^{S-\epsilon I}[x,x] = h^S[x,x] - \epsilon[x,x],$$

for all $x \in \mathfrak{M}$. Since $h^S[.,.]$ and $[.,.]$ are continuous with respect to $\mathcal{O}_S$ on $\mathfrak{L}_S$, we see that $h^S[x,x] \geq \epsilon[x,x]$ for all x belonging to the closure $\mathfrak{M}_S$ of $\mathfrak{M}$ with respect to $\mathcal{O}_S$. Thus $h^S[.,.]$ induces a topology on $\mathfrak{M}_S$ with respect to which $[.,.]$, and hence also $h^S_m[.,.]$, $m \in \mathbb{R}$, is continuous. If $m < 0$ and $m < m(S)$, we see that (4.1) holds for $x \in \mathfrak{M}_S$ and for some $c, d > 0$. This means that $\mathcal{O}_S$ is also induced by $h^S[.,.]$ on $\mathfrak{M}_S$, and as this closed subspace has finite codimension in $\mathfrak{L}_S$ the triple $(\mathfrak{L}_S, h^S[.,.], \mathcal{O}_S)$ is an almost Pontryagin space. □

Remark 4.8. As the sum of Hermitian scalar products with finitely many negative squares also has this property we see that if $S - \epsilon I$, $\epsilon > 0$ is of finite negativity, then $S - \eta I$ is of finite negativity for all $\eta \leq \epsilon$.

Remark 4.9. If the condition from the previous proposition is satisfied, then $\operatorname{ran} S$ is closed and of finite codimension. In fact, this assertion is equivalent to the fact that $\operatorname{ran} JS$ is closed and of finite codimension in the Hilbert space $(\mathfrak{P}, [J.,.])$. We saw in the previous proof that $JS - \epsilon I$ is of finite negativity. Therefore, by Lemma 4.3, $\operatorname{ran} JS$ is closed and of finite codimension.

As $\operatorname{ran} S \perp_{[.,.]} \ker S$ we in particular obtain $\dim \ker S < \infty$.

The following lemma has an interesting consequence.

Lemma 4.10. *Let $(\mathfrak{L}, [.,.], \mathcal{O})$ be an almost Pontryagin space, and let S be a symmetric relation of finite negativity. Moreover, assume that*

$$\operatorname{dom} S = \operatorname{dom} T + \mathfrak{N},$$

where T is a closed restriction of S such that the adjoint of T contains T with finite codimension. Moreover, assume $\dim \mathfrak{N} < \infty$. Then

$$\mathfrak{L}_S = \mathfrak{L}_T + \mathfrak{N}.$$

Proof. Let $\mathfrak{P}$ and J be as in Definition 4.4. As $JT \subseteq JS$ it follows from Definition 3.1 that $\mathfrak{P}_{JT}(=\mathfrak{L}_T)$ is a closed subspace of $\mathfrak{P}_{JS}(=\mathfrak{L}_S)$. Since $\mathfrak{N}$ is finite-dimensional, $\mathfrak{L}_T + \mathfrak{N}$ is also a closed subspace of $\mathfrak{L}_S$. On the other hand $\operatorname{dom} S = \operatorname{dom} T + \mathfrak{N} (\subseteq \mathfrak{L}_T + \mathfrak{N})$ is dense in $\mathfrak{L}_S$. □

In the following we will consider two scalar products $[.,.]_1$ and $[.,.]$ on $\mathfrak{L}$. Then $[.,.]_1$ is said to be finite-dimensional perturbation of $[.,.]$, if for some linear subsapce $\mathfrak{M}$ of $\mathfrak{L}$ of finite codimension one has $[x,y]_1 - [x,y] = 0$ for all $x \in \mathfrak{M}$, $y \in \mathfrak{L}$.

Corollary 4.11. *Let $(\mathfrak{L}, [.,.], \mathcal{O})$ be an almost Pontryagin space, and let $[.,.]_1$ be another scalar product on $\mathfrak{L}$ which is continuous with respect to $\mathcal{O}$ and which is a finite-dimensional perturbation of $[.,.]$. Moreover, let S be a symmetric relation of finite negativity on $(\mathfrak{L}, [.,.], \mathcal{O})$ such that S is also symmetric with respect to $[.,.]_1$.*

Under these assumptions $(\mathfrak{L}, [.,.]_1, \mathcal{O})$ is an almost Pontryagin space. The symmetry S is of finite negativity on $(\mathfrak{L}, [.,.]_1, \mathcal{O})$. Moreover, the space $\mathfrak{L}_S$ and *the topology* $\mathcal{O}_S$ *remain the same if they are defined with* $(\mathfrak{L}, [.,.]_1, \mathcal{O})$ *instead of* $(\mathfrak{L}, [.,.], \mathcal{O})$. *Finally,* $(\mathfrak{L}_S, h^S[.,.], \mathcal{O}_S)$ *is an almost Pontryagin space if and only if* $(\mathfrak{L}_S, h^S[.,.]_1, \mathcal{O}_S)$ *is an almost Pontryagin space.*

Proof. By our assumptions there exists a closed subspace $\mathfrak{M}$ of $\mathfrak{L}$ of finite codimension such that

$$[x, y]_1 - [x, y] = 0, \ x \in \mathfrak{M}, \ y \in \mathfrak{L}.$$

By the definition of almost Pontryagin spaces there exists a closed subspace $\mathfrak{N}$ of $\mathfrak{L}$ of finite codimension such that $[.,.]$ restricted to $\mathfrak{N}$ is a Hilbert space inner product which induces $\mathcal{O} \cap \mathfrak{N}$ on $\mathfrak{N}$. Hence, $\mathfrak{M} \cap \mathfrak{N}$ is a closed subspace of $\mathfrak{L}$ of finite codimension such that $[.,.]_1$ restricted to $\mathfrak{M} \cap \mathfrak{N}$ is a Hilbert space inner product which induces $\mathcal{O} \cap (\mathfrak{M} \cap \mathfrak{N})$ on $\mathfrak{M} \cap \mathfrak{N}$. This in turn means that $(\mathfrak{L}, [.,.]_1, \mathcal{O})$ is an almost Pontryagin space.

By what was mentioned after Definition 4.1 the finite negativity of S on $(\mathfrak{L}, [.,.], \mathcal{O})$ is equivalent to the fact that $h^S[.,.]$ is positive semidefinite on a linear subspace $\mathfrak{Q}$ of finite codimension of $\operatorname{dom} S$. With $\mathfrak{Q}$ also $\mathfrak{Q} \cap \mathfrak{M}$ is a subspace of finite codimension of $\operatorname{dom} S$, and $h^S[.,.]$ coincides with $h^S[.,.]_1$ on $\mathfrak{Q} \cap \mathfrak{M}$. Hence S is of finite negativity on $(\mathfrak{L}, [.,.]_1, \mathcal{O})$.

Clearly, the almost Pontryagin spaces $(\mathfrak{M}, [.,.], \mathcal{O} \cap \mathfrak{M})$ and $(\mathfrak{M}, [.,.]_1, \mathcal{O} \cap \mathfrak{M})$ coincide. If we set $T = S \cap (\mathfrak{M} \times \mathfrak{M})$, then we obtain from Remark 4.6 that the space $\mathfrak{L}_T$ remains unchanged if we used $(\mathfrak{L}, [.,.]_1, \mathcal{O})$ instead of $(\mathfrak{L}, [.,.], \mathcal{O})$ for its construction. Since $\operatorname{dom} T$ is of finite codimension in $\operatorname{dom} S$, we can apply Lemma 4.10 and see that also $\mathfrak{L}_S$ remains unchanged. Using the fact that the inclusion mapping from $\mathfrak{L}_S$ into $\mathfrak{L}$ is injective and continuous the closed graph theorem implies that the topology $\mathcal{O}_S$ is also independent from the scalar product, which was used for its construction , i.e., $[.,.]$ or $[.,.]_1$.

By what was proved above $S - \epsilon I$ is of finite negativity on $(\mathfrak{L}, [.,.], \mathcal{O})$ if and only if it has this property on $(\mathfrak{L}, [.,.]_1, \mathcal{O})$. Thus the final assertion is an immediate consequence of Proposition 4.7. □

Definition 4.12. Let $(\mathfrak{L}, [.,.], \mathcal{O})$ be an almost Pontryagin space, and let S be a closed symmetric linear relation of finite negativity on $(\mathfrak{L}, [.,.], \mathcal{O})$. Moreover, let $(\mathfrak{P}, [.,.])$ be a Pontryagin space which contains $(\mathfrak{L}, [.,.], \mathcal{O})$ as a closed subspace of finite codimension, and let J be a fundamental symmetry on this Pontryagin space. Then we define the space $\mathfrak{L}^S$ as

$$\mathfrak{L}^S = \mathfrak{P}^{JS} \cap \mathfrak{L},$$

where $\mathfrak{P}^{JS}$ is the space corresponding to the symmetry JS on the Hilbert space $(\mathfrak{P}, [J.,.])$ defined as in Definition 3.13.

Remark 4.13. As $J(JS)^{(*)} = S^{[*]}$ we obtain from (3.8) and Remark 4.5

$$\mathfrak{L}^S = \mathfrak{L}_S + (\operatorname{dom} S^{[*]} \cap \mathfrak{L}).$$

By $S^{[*]}$ we mean here the adjoint relation within $(\mathfrak{P}, [.,.])$.

We can describe $\operatorname{dom} S^{[*]} \cap \mathfrak{L}$ as the set of all $a \in \mathfrak{L}$ such that for $(x;y) \in S$

$$x \mapsto [y,a],$$

is a well-defined and $\mathcal{O}$ continuous linear functional on $\operatorname{dom} S$. Hence $\mathfrak{L}^S$ neither depends on J nor on $\mathfrak{P}$.

If $S - mI$ is also of finite negativity, then we immediately see that $\mathfrak{L}^S = \mathfrak{L}^{S-mI}$.

Since we always assume that $\operatorname{codim}_{S^{[*]}} S < \infty$, $\mathfrak{L}^S$ contains $\mathfrak{L}_S$ as a subspace of finite codimension. It therefore carries a unique Hilbert space topology such that $(\mathfrak{L}_S, \mathcal{O}_S)$ is a closed subspace of it. We are going to denote this topology by $\mathcal{O}^S$.

In analogy to Corollary 4.11 we have

Proposition 4.14. *Let $(\mathfrak{L}, [.,.], \mathcal{O})$, $[.,.]_1$ and S be as in Corollary 4.11. Moreover, assume that for all $a \in \mathfrak{L}$ the mapping*

$$x \mapsto [y,a] - [y,a]_1, \ \textit{for } (x;y) \in S,$$

is a well-defined and $\mathcal{O}$ continuous linear functional on $\operatorname{dom} S$. Then the space $\mathfrak{L}^S$ is the same whether it is defined via $(\mathfrak{L}, [.,.]_1, \mathcal{O})$ or via $(\mathfrak{L}, [.,.], \mathcal{O})$.

Proof. This result immediately follows from the corresponding invariance property for $\mathfrak{L}_S$ (Corollary 4.11) and from the characterization of $\operatorname{dom} S^{[*]} \cap \mathfrak{L}$ given in Remark 4.13. $\square$

The rest of the paper is devoted to indefinite generalizations of the results in the part of Section 3 which comes after Corollary 3.15. These results will be an essential tool in our forthcoming paper [8].

From now on we will study the case that $(\mathfrak{L}_S, h^S[.,.], \mathcal{O}_S)$ is an almost Pontryagin space. We introduce a linear relation T on any subspace $\mathfrak{G} \subseteq \mathfrak{L}^S$ which contains $\mathfrak{L}_S$:

$$T = S \cap (\mathfrak{G} \times \mathfrak{G}).$$

By $\mathcal{O}_{\mathfrak{G}}$ we denote the Hilbert space topology $\mathcal{O}^S \cap \mathfrak{G}$.

Definition 4.15. An admissible scalar product $h^{\mathfrak{G}}[.,.]$ on $\mathfrak{G}$ is a Hermitian continuation of $h^S[.,.]$ such that $(\mathfrak{L}_S, h^S[.,.], \mathcal{O}_S)$ is an almost Pontryagin subspace of $(\mathfrak{G}, h^{\mathfrak{G}}[.,.], \mathcal{O}_{\mathfrak{G}})$ and such that

$$h^{\mathfrak{G}}[b,x] = [b,y],$$

for all $b \in \mathfrak{G}, (x;y) \in S$.

Such an admissible product always exists. To see this note that $\mathfrak{L}_S = \mathfrak{P}_{JS} \subseteq \mathfrak{G} \subseteq \mathfrak{L}^S \subseteq \mathfrak{P}^{JS}$. If $(.,.) = [J.,.]$, $m < m(JS)$, and $h_m^{\mathfrak{G}}(.,.)$ is defined as in Proposition 3.16 with S replaced by JS, then we set

$$h^{\mathfrak{G}}[.,.] = h_m^{\mathfrak{G}}(.,.) + m(.,.).$$

This Hermitian product is a continuation of $h^{JS}(.,.) = h^S[.,.]$ and for $b \in \mathfrak{G}$, $(x;y) \in S$ we obtain from (3.11) that

$$h^{\mathfrak{G}}[b,x] = h_m^{\mathfrak{G}}(b,x) + m(b,x) = (b, Jy - mx) + m(b,x) = (b, Jy) = [b,y].$$

Proposition 4.16. *Assume that $(\mathfrak{L}_S, h^S[.,.], \mathcal{O}_S)$ is an almost Pontryagin space and let $h^{\mathfrak{G}}[.,.]$ be an admissible Hermitian inner product on $\mathfrak{G}$ such that $(\mathfrak{L}_S, h^S[.,.], \mathcal{O}_S)$ is an almost Pontryagin subspace of $(\mathfrak{G}, h^{\mathfrak{G}}[.,.], \mathcal{O}_{\mathfrak{G}})$.*

Then T considered in $(\mathfrak{G}, h^{\mathfrak{G}}[.,.], \mathcal{O}_{\mathfrak{G}})$ is closed, symmetric, of finite codimension in $T^{h^{\mathfrak{G}}[*]}$ and it is of finite negativity κ_T, where κ_T coincides with the degree of negativity $\kappa_-(\mathrm{ran}(T), [.,.])$ of $(\mathrm{ran}(T), [.,.])$.

Finally, for sufficiently small $\epsilon > 0$ also $T - \epsilon I$ is of finite negativity.

Proof. For $(a; b), (x; y) \in T$ we see from

$$h^{\mathfrak{G}}[b, x] = [b, y] = h^{\mathfrak{G}}[a, y], \tag{4.2}$$

that T is symmetric. Moreover, this relation proves that $h^T h^{\mathfrak{G}}[.,.]$ has as many negative squares as $[.,.]$ on $\mathrm{ran}(T)$.

We see from Proposition 3.16 that $R = (JS) \cap (\mathfrak{G} \times \mathfrak{G})$ is a symmetry with finite defect indices, or equivalently it is contained in its adjoint (with respect to $h_m^{\mathfrak{G}}(.,.)$) with finite codimension. Let $\mathfrak{M}$ be a $\mathcal{O}_{\mathfrak{G}}$-closed subspace of $\mathfrak{G}$ on which $J = I$ and such that $h^{\mathfrak{G}}[.,.]$ is a Hilbert space inner product on $\mathfrak{M}$. With R also $R \cap (\mathfrak{M} \times \mathfrak{M})$ has finite defect index. Clearly,

$$R \cap (\mathfrak{M} \times \mathfrak{M}) = S \cap (\mathfrak{M} \times \mathfrak{M}) = T \cap (\mathfrak{M} \times \mathfrak{M}).$$

It is straightforward to show that also the adjoint of $R \cap (\mathfrak{M} \times \mathfrak{M})$ within $\mathfrak{M}$ with respect to $h^{\mathfrak{G}}[.,.]$ contains $R \cap (\mathfrak{M} \times \mathfrak{M})$ with finite codimension. The same is true for the adjoint of $R \cap (\mathfrak{M} \times \mathfrak{M})$ within $\mathfrak{G}$. Hence also the symmetric extension T of $R \cap (\mathfrak{M} \times \mathfrak{M})$ is contained in $T^{h^{\mathfrak{G}}[*]}$ with finite codimension. Thus according to Definition 4.1 the symmetry T is of finite negativity in $(\mathfrak{G}, h^{\mathfrak{G}}[.,.], \mathcal{O}_{\mathfrak{G}})$.

By Proposition 4.7 $S - \epsilon I$ is of finite negativity for sufficiently small $\epsilon > 0$. For $(a; b), (x; y) \in T$ we have

$$h^{\mathfrak{G}}[b - \epsilon a, x] = [b, y] - h^S[\epsilon a, x] = [b - \epsilon a, y] =$$

$$[b - \epsilon a, y - \epsilon x] + \epsilon h^{S - \epsilon I}[a, x].$$

So we identify $h^{T - \epsilon I}(h^{\mathfrak{G}}[.,.])$ as the sum of two Hermitian scalar products with finitely many negative squares. Therefore, it also has finitely many negative squares and $T - \epsilon I$ is of finite negativity. □

By Proposition 4.7 $(\mathfrak{G}_T, h^T h^{\mathfrak{G}}[.,.], (\mathcal{O}_{\mathfrak{G}})_T))$ is an almost Pontryagin space.

Proposition 4.17. *The space $\mathfrak{G}_T$ coincides with the domain of the relation*

$$X = S \cap (\overline{\mathfrak{G}} \times \overline{\mathfrak{G}}),$$

where the closure is taken in $\mathfrak{L}$ with respect to $\mathcal{O}$.

The topology $(\mathcal{O}_{\mathfrak{G}})_T$ coincides with the topology induced by the graph norm of the closed operator

$$\{(x; y + X(0)) : (x; y) \in X\} \subseteq \overline{\mathfrak{G}} \times (\overline{\mathfrak{G}}/X(0)),$$

where $\overline{\mathfrak{G}}$ is provided with $\mathcal{O} \cap \overline{\mathfrak{G}}$ and $\overline{\mathfrak{G}}/X(0)$ with the factor topology $(\mathcal{O} \cap \overline{\mathfrak{G}})/X(0)$.

Proof. From Remark 4.5 we know that $\mathfrak{G}_T$ is the set of all $x \in \mathfrak{G}$ such that there exists a sequence $((x_n; y_n))$ in T which satisfies

$$x_n \to x \text{ w.r.t. } \mathcal{O}_{\mathfrak{G}} \text{ and } [y_n - y_m, y_n - y_m] = h^{\mathfrak{G}}[y_n - y_m, x_n - x_m] \to 0. \quad (4.3)$$

The convergence of x_n with respect to $\mathcal{O}_{\mathfrak{G}}$ implies

$$[y_n - y_m, y] = h^{\mathfrak{G}}[x_n - x_m, y] \to 0,$$

for all $y \in \mathfrak{G}$. Therefore (4.3) is equivalent to $x_n \to x$ with respect to $\mathcal{O}_{\mathfrak{G}}$ and the fact that $(y_n + \overline{\mathfrak{G}}^{[o]})$ is a Cauchy sequence within the Pontryagin space $(\overline{\mathfrak{G}}/\overline{\mathfrak{G}}^{[o]}, [.,.])$ with respect to its Pontryagin space topology.

Using Remark 4.5 once more we see that $x \in \mathfrak{G}_T$ if and only if there exists a sequence $((x_n; y_n))$ in T such that $x_n \to x$ with respect to $\mathcal{O}$,

$$[y_n - y_m, x_n - x_m] \to 0,$$

and $(y_n + \overline{\mathfrak{G}}^{[o]})$ is a Cauchy sequence within the Pontryagin space $(\overline{\mathfrak{G}}/\overline{\mathfrak{G}}^{[o]}, [.,.])$. By the Cauchy-Schwartz inequality here the second condition is a consequence of the remaining two.

Hence, $\mathfrak{G}_T$ is the domain of the linear relation $Q \subseteq \overline{\mathfrak{G}} \times \overline{\mathfrak{G}}/\overline{\mathfrak{G}}^{[o]}$ where Q is the closure of $T + (\{0\} \times \overline{\mathfrak{G}}^{[o]})/(\{0\} \times \overline{\mathfrak{G}}^{[o]})$. As $\overline{\mathfrak{G}}^{[o]}$ is finite-dimensional

$$Q = \overline{T} + (\{0\} \times \overline{\mathfrak{G}}^{[o]})/(\{0\} \times \overline{\mathfrak{G}}^{[o]}).$$

On the other hand as $\operatorname{ran} S$ is closed and of finite codimension (see Remark 4.9) we obtain

$$\overline{\operatorname{ran} T} = \operatorname{ran} S \cap \overline{\mathfrak{G}}.$$

Since the mapping $(x; y) \mapsto y$ from S onto $\operatorname{ran} S$ has a finite-dimensional kernel (see Remark 4.9),

$$\overline{\operatorname{ran} T} = \operatorname{ran} \overline{T},$$

and we see that

$$\overline{T} + (\{0\} \times (S(0) \cap \overline{\mathfrak{G}})) = S \cap (\overline{\mathfrak{G}} \times \overline{\mathfrak{G}}),$$

and hence

$$\mathfrak{G}_T = \operatorname{dom}(Q) = \operatorname{dom}(\overline{T}) = \operatorname{dom}(S \cap (\overline{\mathfrak{G}} \times \overline{\mathfrak{G}})).$$

The assertion about the topology follows from the closed graph theorem since all involved topologies are Hilbert space topologies. $\square$

Corollary 4.18. *In addition to the assumptions in Proposition* 4.16 *suppose that* S *is an invertible operator. Then* $S^{-1}|_{\operatorname{ran} S \cap \overline{\mathfrak{G}}}$ *sets up an isomorphism from the almost Pontryagin space* $(\operatorname{ran} S \cap \overline{\mathfrak{G}}, [.,.], \mathcal{O} \cap \operatorname{ran} S \cap \overline{\mathfrak{G}})$ *onto* $(\mathfrak{G}_T, h^T h^{\mathfrak{G}}[.,.], (\mathcal{O}_{\mathfrak{G}})_T)$.

If we denote by R *the symmetry* $T \cap (\mathfrak{G}_T \times \mathfrak{G}_T)$ *on* $(\mathfrak{G}_T, h^T h^{\mathfrak{G}}[.,.], (\mathcal{O}_{\mathfrak{G}})_T)$, *then*

$$\{(Sx; Sy) : (x; y) \in R\} = S \cap ((\operatorname{ran} S \cap \overline{\mathfrak{G}}) \times (\operatorname{ran} S \cap \overline{\mathfrak{G}})).$$

Proof. Using the notation from Proposition 4.17 and its proof with S also X is an invertible operator. By the proof of Proposition 4.17

$$\operatorname{dom} X = \mathfrak{G}_T, \ \operatorname{ran} X = \operatorname{ran} S \cap \overline{\mathfrak{G}}.$$

Since $\operatorname{ran} X$ is closed, the closed graph theorem implies that X^{-1} is even continuous. Hence, by Proposition 4.17 the topology $(\mathcal{O}_{\mathfrak{G}})_T$ is just the initial topology induced by X.

Because of (4.2) we have

$$[b, y] = h^T h^{\mathfrak{G}}[X^{-1}b, X^{-1}y],$$

for $y \in \operatorname{ran} S \cap \mathfrak{G}$. By continuity we can extend this relation to $\operatorname{ran} S \cap \overline{\mathfrak{G}}$.

For $x, y \in \operatorname{ran} S \cap \overline{\mathfrak{G}}$ we conclude from $(x; y) \in S$ that $S^{-1}y = x = SS^{-1}x$ and $y = SS^{-1}y \in \overline{\mathfrak{G}}$. Hence $(S^{-1}x; S^{-1}y), (S^{-1}y, y) \in X$ (see Proposition 4.17), and further

$$(S^{-1}x; S^{-1}y) \in S \cap (\operatorname{dom}(X) \times \operatorname{dom}(X)) = T \cap (\operatorname{dom}(X) \times \operatorname{dom}(X)) = R.$$

Conversely, if $(S^{-1}x; S^{-1}y) \in R$, then $S^{-1}x \in \mathfrak{G}_T \subseteq \operatorname{dom} S$ and $x = SS^{-1}x = S^{-1}y$, or $(x; y) \in S$. □

References

[1] R. Arens, *Operational calculus of linear relations*, Pacific J. Math., **11** (1961), 9–23

[2] Yu.M. Arlinskiĭ, S. Hassi, Z. Sebestyén, H.S.V. de Snoo, *On the class of extremal extensions of a nonnegative operator*, Recent advances in operator theory and related topics (Szeged, 1999), Operator Theory Adv. Appl. **127**, 41–81, Birkhäuser, Basel, 2001

[3] M.Sh. Birman, M.Z. Solomjak, *Spectral theory of selfadjoint operators in Hilbert space*, Mathematics and its Applications (Soviet Series), D. Reidel Publishing Co., Dordrecht 1987

[4] J. Bognar, *Indefinite inner product spaces*, Springer Verlag, Berlin 1974

[5] E.A. Coddington, H. de Snoo, *Positive selfadjoint extensions of positive symmetric subspaces*, Mathematische Zeitschrift **159** (1978), 203–214

[6] A. Dijksma, H. de Snoo, *Self-adjoint extensions of symmetric subspaces*, Pacific J. Math. **54** (1974), 71–100

[7] M. Kaltenbäck, H. Winkler, H. Woracek, *Almost Pontryagin spaces*, Proceedings of the IWOTA 2003 in Cagliari, Operator Theory: Adv. Appl. **160**, 253–271

[8] M. Kaltenbäck, H. Winkler, H. Woracek, *Symmetric dB-spaces*, submitted

[9] T. Kato, *Perturbation Theory for Linear Operator*, Springer Verlag, Berlin 1980

[10] V. Prokaj, Z. Sebestyén, *On Friedrichs extensions of operators*, Acta Sci. Math. (Szeged) **62** (1996), 243–246

[11] V. Prokaj, Z. Sebestyén, *On extremal positive operator extensions*, Acta Sci. Math. (Szeged) **62** (1996), 485–491

[12] Z. Sebestyén, J. Stochel, *Restrictions of positive selfadjoint operators*, Acta Sci. Math. (Szeged) **55** (1991), 149–154

M. Kaltenbäck
Institut für Analysis und Scientific Computing
Technische Universität Wien
Wiedner Hauptstraße 8-10
A-1040 Wien
Austria
e-mail: `michael.kaltenbaeck@tuwien.ac.at`

H. Winkler
Faculteit der Wiskunde en Natuurwetenschappen
Rijksuniversiteit Groningen
NL-9700 AV Groningen
The Netherlands
e-mail: `winkler@math.rug.nl`

H. Woracek
Institut für Analysis und Scientific Computing
Technische Universität Wien
Wiedner Hauptstraße 8-10
A-1040 Wien
Austria
e-mail: `harald.woracek@tuwien.ac.at`

Operator Theory:
Advances and Applications, Vol. 162, 211–229

An Operator-theoretic Approach to a Multiple Point Nevanlinna-Pick Problem for Generalized Carathéodory Functions

Lutz Klotz and Andreas Lasarow

Abstract. We study a Nevanlinna-Pick type interpolation problem for matrix-valued generalized Carathéodory functions, where the values of the function and the values of its derivatives up to a certain order are prescribed at finitely many points of the open unit disk. Under the assumption that the generalized Schwarz-Pick-Potapov block matrix, which is associated to the given data, is non-singular we establish a correspondence between the set of solutions of the problem and the set of minimal unitary extensions of a certain isometry in a Pontryagin space, which is one-to-one modulo unitary equivalence.

Mathematics Subject Classification (2000). Primary 47A57; Secondary 47B50.

Keywords. Nevanlinna-Pick interpolation problem, generalized Carathéodory functions, generalized Schwarz-Pick-Potapov block matrix.

1. Introduction

Starting with Carathéodory's paper [7] interpolation problems for classes of analytic functions have been studied widely. There is an extensive literature on several types of such problems (see, e.g., the survey article [17] and the books [11], [14], [3], [10]). The present paper is another contribution to this topic and deals with an interpolation problem for matrix-valued generalized Carathéodory functions on the open unit disk $\mathbb{D} := \{z \in \mathbb{C} : |z| < 1\}$ of the complex plane $\mathbb{C}$, i.e., matrix-valued meromorphic functions on $\mathbb{D}$, for which the corresponding Carathéodory kernel can have a finite number κ of negative squares.

The problem we are going to study is a multiple point interpolation problem, i.e., a problem, where along with the values of the function the values of its deriva-

The work of the second author of the present paper was supported by the German Academy of Natural Scientists Leopoldina by means of the Federal Ministry of Education and Research under grant number BMBF-LPD 9901/8-88.

tives up to a certain order are prescribed at some points, see problem **(MNP)** of Section 2 for the exact formulation. The definite case $\kappa = 0$ of such a type of interpolation problems was studied, e.g., in [13] and [8]. Woracek [28] discussed a problem for generalized Nevanlinna functions analogous to ours and we acknowledge the strong influence of Woracek's paper to our investigations. Woracek's (and our) method is operator-theoretic. Perhaps, the first paper, where operator theory was applied to interpolation problems, was [26]. The operator approach concerning an indefinite metric was developed in [20]. In fact, the considerations below bank on a synthesis of [26], [20], and Potapov's approach to interpolation problems (see, e.g., [12], [17], and [18]).

The first task in applying operator theory to interpolation problems is to construct from the interpolation data W an inner product space $\mathfrak{H}_W$. This construction will be carried out in Section 4. It requires the introduction of a generalization of the Schwarz-Pick-Potapov block matrix. Section 3 contains some considerations, which are to motivate our choice of the generalized Schwarz-Pick-Potapov block matrix $\mathbf{P}_W$. We obtain our main result under the additional assumption that $\mathbf{P}_W$ is non-singular. In this case the corresponding space $\mathfrak{H}_W$ is a Pontryagin space and we shall use several facts on Pontryagin spaces. We refer to [15] for a nice introduction to the subject and to [4] or [2] for more comprehensive treatises on more general indefinite inner product spaces.

In Section 4 we also define an isometric operator V in $\mathfrak{H}_W$ corresponding to $\mathbf{P}_W$. As is pointed out in Section 5, the solutions of problem **(MNP)** can be described by unitary extensions of V. Our main result establishes a one-to-one correspondence (modulo unitary equivalence) between the set of minimal unitary extensions of V in a Pontryagin space Π_κ and the set of all solutions of problem **(MNP)** with κ negative squares, see Theorem 5.4.

The concluding Section 6 deals with a realization of the abstract space $\mathfrak{H}_W$ as a space of rational functions. This construction was motivated by the work of Bultheel, González-Vera, Hendriksen, and Njåstad [6] (see also [22]), where the connection between orthogonal rational functions and definite interpolation problems for scalar Carathéodory functions was investigated. We shall show that the operator V becomes an operator of multiplication in this concrete model.

Let us mention some further notations and conventions we will use. The symbols $\mathbb{N}$ and $\mathbb{N}_0$ signify the sets of positive and non-negative integers, respectively. We write 0 for the zero element of any linear space. The domain of definition, range, null space, and resolvent set of a linear operator X are denoted by $\mathcal{D}(X)$, $\mathcal{R}(X)$, $\mathcal{N}(X)$, and $\rho(X)$, respectively. Throughout the paper, the symbol q stands for a positive integer. By $\mathbb{C}^q$ we denote the Hilbert space of column vectors with q complex entries, by $(\cdot,\cdot)$ and I_q its canonical inner product and the identity operator, respectively, and by $\mathbb{C}^{q\times q}$ the algebra of complex $q\times q$-matrices. For all other inner products and identity operators we use the same notations $\langle\cdot,\cdot\rangle$ and I, respectively. (We hope that this will not lead to confusion.) Similarly, the symbol X^* stands for any adjoint operator of a linear operator X, where the underlying inner products should be clear from the context.

2. Formulation of the problem

Let us recall some facts on $\mathbb{C}^{q\times q}$-valued generalized Carathéodory functions. They and even stronger and more general results can be found in [19], see [20] (if $q=1$) and [21] for additional information.

Let $\kappa \in \mathbb{N}_0$ and let Λ be a non-empty set. A $\mathbb{C}^{q\times q}$-valued kernel K on $\Lambda \times \Lambda$ is said to *have κ negative squares* (on Λ) if it has the following two properties:

(1) $K(\lambda,\mu) = K(\mu,\lambda)^*$, $\lambda,\mu \in \Lambda$;

(2) for any choice of $r \in \mathbb{N}$, $\lambda_1,\lambda_2,\dots,\lambda_r \in \Lambda$, and $x_1,x_2,\dots,x_r \in \mathbb{C}^q$, the complex $r\times r$-matrix

$$\Big(\big(K(\lambda_j,\lambda_k)x_j,x_k\big)\Big)_{j,k=1,\dots,r}$$

has at most κ negative eigenvalues and for at least one such choice it has exactly κ negative eigenvalues.

A $\mathbb{C}^{q\times q}$-valued meromorphic function F on $\mathbb{D}$ is called a *generalized Carathéodory function with κ negative squares* if the kernel

$$K_F(u,v) := \frac{1}{1-u\overline{v}}\Big(F(u)+F(v)^*\Big), \quad u,v \in \varrho(F),$$

has κ negative squares. Here $\varrho(F)$ denotes the domain of holomorphy of F. The class of all $\mathbb{C}^{q\times q}$-valued generalized Carathéodory functions F with κ negative squares such that 0 belongs to $\varrho(F)$ will be denoted by $\mathcal{C}_\kappa^{q\times q}$.

The following theorem summarizes basic results on the representation of generalized Carathéodory functions, cf. [19, §2]. We mention that assertion (ii) of Theorem 2.1 (b) is stated without proof in [19], but a proof of a similar fact for generalized Nevanlinna functions was given in [9, proof of Theorem 1.1].

Theorem 2.1. (a) *Let $V_{\kappa'}$ be an isometric operator in a Pontryagin space $\Pi_{\kappa'}$ such that $\mathcal{R}(V_{\kappa'}) = \Pi_{\kappa'}$, let Γ be a linear operator from $\mathbb{C}^q$ into $\Pi_{\kappa'}$, and let H be an Hermitian $q\times q$-matrix. Then the function F:*

$$F(z) := \mathrm{i}\,H + \Gamma^*(V_{\kappa'}+zI)(V_{\kappa'}-zI)^{-1}\Gamma, \quad z \in \rho(V_{\kappa'})\cap\mathbb{D}, \tag{2.1}$$

belongs to the class $\mathcal{C}_{\kappa''}^{q\times q}$ for some $\kappa'' \le \kappa'$. If $V_{\kappa'}$ is unitary and if $V_{\kappa'}$ and Γ are minimal, *i.e., the linear span of*

$$\big\{V_{\kappa'}-zI)^{-1}\Gamma x:\ z\in\rho(V_{\kappa'}), x\in\mathbb{C}^q\big\}$$

is dense in $\Pi_{\kappa'}$, then $\kappa''=\kappa'$.

(b) *Let $F \in \mathcal{C}_\kappa^{q\times q}$. Then there exist a Pontryagin space Π_κ, a unitary operator U in Π_κ, and a linear operator Γ from $\mathbb{C}^q$ into Π_κ such that*

(i) *U and Γ are minimal;*

(ii) *$\varrho(F) = \rho(U)\cap\mathbb{D}$;*

(iii) *$F(z) = \mathrm{i}\,\Im\mathfrak{m}\,F(0) + \Gamma^*(U+zI)(U-zI)^{-1}\Gamma$, $z \in \varrho(F)$.*

The operator U is unique to within unitary equivalence.

For functions of the class $\mathcal{C}_\kappa^{q\times q}$ we wish to study the following multiple point interpolation problem.

(MNP) *Let $\kappa \in \mathbb{N}_0$, $n \in \mathbb{N}$, let $z_1 := 0, z_2, \dots, z_n$ be n distinct points of $\mathbb{D}$, and let $l_j \in \mathbb{N}$, $j = 1, 2, \dots, n$. For the set $\mathcal{D}_W$ of ordered pairs*

$$\mathcal{D}_W := \{(z_j, s) : \ s = 0, 1, \dots, l_j - 1, \ j = 1, 2, \dots, n\}$$

and a function $W : \mathcal{D}_W \to \mathbb{C}^{q \times q}$, describe the set $\mathcal{C}_\kappa^{q\times q}(W)$ of all functions F belonging to $\mathcal{C}_\kappa^{q\times q}$ such that

$$\frac{1}{s!} F^{(s)}(z_j) = W\big((z_j, s)\big) =: W_{js}, \quad (z_j, s) \in \mathcal{D}_W. \tag{2.2}$$

Some remarks are in order.

Remark 2.2. Problem **(MNP)** is a generalization of the trigonometric moment problem or of the Carathéodory coefficient problem (see [7] and [1]) as well as of the classical Nevanlinna-Pick problem (see [24] and [23]).

Remark 2.3. Relation (2.2) shows that the domain of definition and the values of the matrix-valued function W can be interpreted as interpolation data. Accordingly, we shall speak of the interpolation data W.

Remark 2.4. Applying a suitable linear fractional transformation of $\mathbb{D}$, one can see that the assumptions $0 \in \varrho(F)$ and $(0, 0) \in \mathcal{D}_W$ can be considered as certain normalization conditions and do not detract the generality of $\mathcal{C}_\kappa^{q\times q}$ and problem **(MNP)**, respectively.

Remark 2.5. In what follows, the sets

$$\Delta := \{(j, s) : \ s = 0, 1, \dots, l_j - 1, \ j = 1, 2, \dots, n\}$$

and $\Delta \times \Delta$ will appear as index sets of vectors and matrices, respectively. In these cases we shall always assume that Δ is ordered lexicographically, i.e., (j, s) precedes (k, t) if and only if either $j < k$ or $j = k$ and $s < t$.

3. A generalized Schwarz-Pick-Potapov block matrix

If $l_1 = l_2 = \dots = l_n = 1$ in problem **(MNP)**, the Schwarz-Pick-Potapov block matrix

$$\left(\frac{1}{1 - z_j \overline{z_k}}\Big(W_{j0} + W_{k0}^*\Big)\right)_{j,k=1,\dots,n}$$

plays a crucial role (see, e.g., [17]). It arises the question by what matrix it should be replaced in the more general setting. To motivate our choice, assume for a moment that $F \in \mathcal{C}_\kappa^{q\times q}$. From (iii) of Theorem 2.1 (b) we get

$$F(z) = \mathrm{i}\,\Im\mathfrak{m}\, F(0) + \Gamma^*\Gamma + 2z\Gamma^* U^* R(z), \quad z \in \varrho(F), \tag{3.1}$$

where $R(z) := (I - zU^*)^{-1}\Gamma$. By induction it follows

$$F^{(r)}(z) = 2\Gamma^* U^* \big(rR^{(r-1)}(z) + zR^{(r)}(z)\big), \quad z \in \varrho(F),\ r \in \mathbb{N},$$

which shows that a large part of information on the derivatives of F is contained in the derivatives of R. On the other hand, it is not hard to see that

$$(1-uv)K_F(u,\overline{v}) = F(u) + F(\overline{v})^* = 2(1-uv)R(\overline{v})^*R(u), \quad u,\overline{v} \in \varrho(F),$$

which yields

$$2\frac{\partial^{s+t}}{\partial u^s \partial v^t} R(\overline{v})^* R(u) = \frac{\partial^{s+t}}{\partial u^s \partial v^t} \frac{1}{1-uv}\Big(F(u) + F(\overline{v})^*\Big), \quad s,t \in \mathbb{N}_0. \tag{3.2}$$

These facts suggest that the matrix $\mathbf{P}_W := \big(P_{(j,s),(k,t)}\big)_{(j,s),(k,t)\in\Delta}$ with entries

$$P_{(j,s),(k,t)} := \frac{1}{s!t!}\frac{\partial^{s+t}}{\partial u^s \partial v^t}\frac{1}{1-uv}\Big(F(u)+F(\overline{v})^*\Big)\Bigg|_{\substack{u=z_j\\ v=\overline{z_k}}}, \quad (j,s),(k,t) \in \Delta, \tag{3.3}$$

could serve as a substitute for the Schwarz-Pick-Potapov block matrix. Indeed, in the definite case such kind of block matrices was used in [8]. In order to simplify the notation slightly we set

$$P_{(j,s),(k,t)} =: P_{st}^{(jk)}, \quad (j,s),(k,t) \in \Delta.$$

Thus, taking into account the lexicographic ordering of Δ, we see that $\mathbf{P}_W$ can be written as

$$\mathbf{P}_W = \big(P_{jk}\big)_{j,k=1,\ldots,n}, \tag{3.4}$$

where $P_{jk} := \big(P_{st}^{(jk)}\big)_{\substack{s=0,1,\ldots,l_j-1\\ t=0,1,\ldots,l_k-1}}$ is a complex $l_jq \times l_kq$-matrix, $(j,s),(k,t) \in \Delta$. What we still need is an explicit expression of the entries of $\mathbf{P}_W$ by the interpolation data W.

Lemma 3.1. *If $P_{(j,s),(k,t)} = P_{st}^{(jk)}$ is defined by (3.3), then*

$$\begin{aligned} P_{st}^{(jk)} = &\sum_{h=0}^{s}\sum_{r=0}^{\min\{t,h\}} \frac{(h+t-r)!}{(t-r)!r!(h-r)!}\frac{z_j^{t-r}\overline{z_k}^{h-r}}{(1-z_j\overline{z_k})^{h+t-r+1}}W_{j,s-h} \\ &+\sum_{h=0}^{t}\sum_{r=0}^{\min\{s,h\}} \frac{(h+s-r)!}{(s-r)!r!(h-r)!}\frac{z_j^{h-r}\overline{z_k}^{s-r}}{(1-z_j\overline{z_k})^{h+s-r+1}}W_{k,t-h}^*. \end{aligned} \tag{3.5}$$

Proof. The result follows by an all in all fourfold application of Leibniz' product rule to the right-hand side of (3.3) and by (2.2). □

The considerations above urge the following definition.

Definition 3.2. For the interpolation data W of problem **(MNP)**, the matrix

$$\mathbf{P}_W := \big(P_{st}^{(jk)}\big)_{(j,s),(k,t)\in\Delta},$$

whose entries $P_{st}^{(jk)}$ are defined by (3.5), is called the *generalized Schwarz-Pick-Potapov block matrix associated to* **(MNP)**.

Note that $\mathbf{P}_W$ is an Hermitian matrix. From Lemma 3.1 one can derive some useful recurrence formulae concerning the entries of $\mathbf{P}_W$.

Lemma 3.3. *The entries $P_{st}^{(jk)}$, $(j,s),(k,t)\in\Delta$, of the generalized Schwarz-Pick-Potapov block matrix associated to* **(MNP)**, *satisfy the following identities:*

(i) $(1-z_j\overline{z_k})P_{00}^{(jk)} = W_{j0}+W_{k0}^*$, $j,k=1,\dots,n$;

(ii) $(1-z_j\overline{z_k})P_{s0}^{(jk)} = \overline{z_k}P_{s-1,0}^{(jk)}+W_{js}$, $s=1,\dots,l_j-1$, *and*

$(1-z_j\overline{z_k})P_{0t}^{(jk)} = z_jP_{0,t-1}^{(jk)}+W_{kt}^*$, $t=1,\dots,l_k-1$;

(iii) $(1-z_j\overline{z_k})P_{st}^{(jk)} = P_{s-1,t-1}^{(jk)}+z_jP_{s,t-1}^{(jk)}+\overline{z_k}P_{s-1,t}^{(jk)}$, $s\geq 1$, $t\geq 1$.

Proof. Relation (i) immediately follows by setting $s=t=0$ in (3.5). For $t=0$, (3.5) implies that

$$(1-z_j\overline{z_k})P_{s0}^{(jk)} = \sum_{h=0}^{s}\frac{\overline{z_k}^h}{(1-z_j\overline{z_k})^h}W_{j,s-h}+\frac{\overline{z_k}^s}{(1-z_j\overline{z_k})^s}W_{k0}^*. \tag{3.6}$$

On the other hand, if we set $t=0$, substitute $s-1$ for s and replace the summation index h of the first outer sum on the right-hand side of (3.5) by $h-1$, we compute

$$\overline{z_k}P_{s-1,0}^{(jk)} = \sum_{h=1}^{s}\frac{\overline{z_k}^h}{(1-z_j\overline{z_k})^h}W_{j,s-h}+\frac{\overline{z_k}^s}{(1-z_j\overline{z_k})^s}W_{k0}^*.$$

A comparison with (3.6) yields the first identity of (ii). The second one follows similarly. To prove (iii), it is enough to show that the coefficients of $W_{j,s-h}$ as well as the coefficients of $W_{k,t-h}^*$ on both sides coincide. This can be done by some boring manipulations with indices in the style of the proof of (ii) and by invoking some well-known identities for binomial coefficients. Let us sketch the way in the case of the coefficients of $W_{j,s-h}$. According to (3.5) the coefficient of $W_{j,s-h}$ on the left-hand side of (iii) is equal to

$$\sum_{r=0}^{\min\{t,h\}}\frac{(h+t-r)!}{(t-r)!r!(h-r)!}\frac{z_j^{t-r}\overline{z_k}^{h-r}}{(1-z_j\overline{z_k})^{h+t-r}} = \sum_{r=0}^{\min\{t,h\}}\binom{h+t-r}{t-r}\binom{h}{r}\frac{z_j^{t-r}\overline{z_k}^{h-r}}{(1-z_j\overline{z_k})^{h+t-r}}.$$

Replacing s by $s-1$, t by $t-1$, and also the summation index h by $h-1$ in (3.5), we see that the coefficient of $W_{j,s-h}$ in the term of $P_{s-1,t-1}^{(jk)}$ equals

$$\sum_{r=0}^{\min\{t-1,h-1\}}\frac{(h+t-r-2)!}{(t-r-1)!r!(h-r-1)!}\frac{z_j^{t-r-1}\overline{z_k}^{h-r-1}}{(1-z_j\overline{z_k})^{h+t-r-1}} = \sum_{r=1}^{\min\{t,h\}}\binom{h+t-r-1}{t-r}\binom{h-1}{r-1}\frac{z_j^{t-r}\overline{z_k}^{h-r}}{(1-z_j\overline{z_k})^{h+t-r}}.$$

Similarly, one can obtain the coefficient

$$\sum_{r=0}^{\min\{t-1,h\}}\frac{(h+t-r-1)!}{(t-r-1)!r!(h-r)!}\frac{z_j^{t-r}\overline{z_k}^{h-r}}{(1-z_j\overline{z_k})^{h+t-r}} = \sum_{r=0}^{\min\{t-1,h\}}\binom{h+t-r-1}{t-r-1}\binom{h}{r}\frac{z_j^{t-r}\overline{z_k}^{h-r}}{(1-z_j\overline{z_k})^{h+t-r}}$$

of $W_{j,s-h}$ in the expression of $z_jP_{s,t-1}^{(jk)}$ and the coefficient

$$\sum_{r=0}^{\min\{t,h-1\}}\frac{(h+t-r-1)!}{(t-r)!r!(h-r-1)!}\frac{z_j^{t-r}\overline{z_k}^{h-r}}{(1-z_j\overline{z_k})^{h+t-r}} = \sum_{r=0}^{\min\{t,h-1\}}\binom{h+t-r-1}{t-r}\binom{h-1}{r}\frac{z_j^{t-r}\overline{z_k}^{h-r}}{(1-z_j\overline{z_k})^{h+t-r}}$$

of $W_{j,s-h}$ in the expression of $\overline{z_k}P^{(jk)}_{s-1,t}$. Thus, we have to verify the identity

$$\begin{aligned}\sum_{h=0}^{s}\sum_{r=0}^{\min\{t,h\}}&\binom{h+t-r}{t-r}\binom{h}{r}\frac{z_j^{t-r}\overline{z_k}^{h-r}}{(1-z_j\overline{z_k})^{h+t-r}}W_{j,s-h}\\ &=\sum_{h=1}^{s}\sum_{r=1}^{\min\{t,h\}}\binom{h+t-r-1}{t-r}\binom{h-1}{r-1}\frac{z_j^{t-r}\overline{z_k}^{h-r}}{(1-z_j\overline{z_k})^{h+t-r}}W_{j,s-h}\\ &\quad+\sum_{h=0}^{s}\sum_{r=0}^{\min\{t-1,h\}}\binom{h+t-r-1}{t-r-1}\binom{h}{r}\frac{z_j^{t-r}\overline{z_k}^{h-r}}{(1-z_j\overline{z_k})^{h+t-r}}W_{j,s-h}\\ &\quad+\sum_{h=1}^{s}\sum_{r=0}^{\min\{t,h-1\}}\binom{h+t-r-1}{t-r}\binom{h-1}{r}\frac{z_j^{t-r}\overline{z_k}^{h-r}}{(1-z_j\overline{z_k})^{h+t-r}}W_{j,s-h}.\end{aligned}\tag{3.7}$$

Since for $h=0$ we get $\frac{z_j^t}{(1-z_j\overline{z_k})^t}W_{js}$ on both sides of (3.7), it is enough to show

$$\begin{aligned}\sum_{r=0}^{\min\{t,h\}}&\binom{h+t-r}{t-r}\binom{h}{r}\frac{z_j^{t-r}\overline{z_k}^{h-r}}{(1-z_j\overline{z_k})^{h+t-r}}\\ &=\sum_{r=1}^{\min\{t,h\}}\binom{h+t-r-1}{t-r}\binom{h-1}{r-1}\frac{z_j^{t-r}\overline{z_k}^{h-r}}{(1-z_j\overline{z_k})^{h+t-r}}\\ &\quad+\sum_{r=0}^{\min\{t-1,h\}}\binom{h+t-r-1}{t-r-1}\binom{h}{r}\frac{z_j^{t-r}\overline{z_k}^{h-r}}{(1-z_j\overline{z_k})^{h+t-r}}+\sum_{r=0}^{\min\{t,h-1\}}\binom{h+t-r-1}{t-r}\binom{h-1}{r}\frac{z_j^{t-r}\overline{z_k}^{h-r}}{(1-z_j\overline{z_k})^{h+t-r}}.\end{aligned}\tag{3.8}$$

The identity

$$\binom{h+t-r-1}{t-r}\binom{h-1}{r-1}+\binom{h+t-r-1}{t-r-1}\binom{h}{r}+\binom{h+t-r-1}{t-r}\binom{h-1}{r}=\binom{h+t-r}{t-r}\binom{h}{r}$$

implies that if in all sums of (3.8) the summation index r would only run from 1 to $\min\{t-1,h-1\}$, one would have equality. Consequently, it remains to compare the rest of summands on both sides of (3.8), which occur if $r=0$, $r=\min\{t,h\}$, $r=\min\{t,h-1\}$, or $r=\min\{t-1,h\}$. One can do this by direct calculation considering the three cases $t\geq h+1$, $t=h$, and $t\leq h-1$. □

Before continuing our main direction of investigations we would like to emphasize the importance of the preceding lemma. We shall show that the Fundamental Identity of problem **(MNP)** is a simple consequence of the relations given in Lemma 3.3. We mention that the basic role of Fundamental Identities in Potapov's approach to interpolation problems (see, e.g., [12], [17], [5], and [16]) was pointed out by Sakhnovich (cf. [25]). For $j,k=1,\dots,n$, let S_{jk} be the zero matrix belonging to $\mathbb{C}^{l_jq\times l_kq}$ if $j\neq k$, S_{jj} be the complex $l_jq\times l_jq$-matrix

$$S_{jj}:=\begin{pmatrix} z_jI_q & 0 & \cdots & \cdots & 0\\ I_q & z_jI_q & \ddots & & \vdots\\ 0 & \ddots & \ddots & \ddots & \vdots\\ \vdots & \ddots & \ddots & \ddots & 0\\ 0 & \cdots & 0 & I_q & z_jI_q\end{pmatrix},$$

and U_j and W_j be the complex $l_j q \times q$-matrix

$$U_j := \begin{pmatrix} I_q \\ 0 \\ \vdots \\ 0 \end{pmatrix} \quad \text{and} \quad W_j := \begin{pmatrix} W_{j0} \\ W_{j1} \\ \vdots \\ W_{j,l_j-1} \end{pmatrix},$$

respectively. It is not hard to see that the identities of Lemma 3.3 can be written in the matrix form

$$P_{jk} - S_{jj} P_{jk} S_{kk}^* = U_j W_k^* + W_j U_k^*, \quad j,k = 1,\ldots,n,$$

which yields the Fundamental Identity

$$\mathbf{P}_W - \mathbf{S}\mathbf{P}_W\mathbf{S}^* = \mathbf{U}\mathbf{W}^* + \mathbf{W}\mathbf{U}^*,$$

where

$$\mathbf{S} := \big(S_{jk}\big)_{j,k=1,\ldots,n}, \quad \mathbf{U} := \begin{pmatrix} U_1 \\ U_2 \\ \vdots \\ U_n \end{pmatrix}, \quad \text{and} \quad \mathbf{W} := \begin{pmatrix} W_1 \\ W_2 \\ \vdots \\ W_n \end{pmatrix}.$$

4. An isometric operator corresponding to $\mathbf{P}_W$

In a standard way, the matrix $\mathbf{P}_W$ gives rise to an inner product space $\mathfrak{H}_W$. The (finite-dimensional) space $\mathfrak{H}_W$ is the linear space of all formal sums

$$\mathfrak{x} := \sum_{(j,s)\in\Delta} x_{js} e_{js},$$

where $x_{js} \in \mathbb{C}^q$ and where e_{js} is a symbol, $(j,s) \in \Delta$. It is equipped with the inner product

$$\big\langle \mathfrak{x}, \mathfrak{y} \big\rangle := \frac{1}{2} \sum_{(j,s),(k,t)\in\Delta} \Big(P_{st}^{(jk)} x_{js}, y_{kt} \Big), \tag{4.1}$$

where $\mathfrak{y} := \sum_{(k,t)\in\Delta} y_{kt} e_{kt}$, $y_{kt} \in \mathbb{C}^q$, $(k,t) \in \Delta$. In particular,

$$\Big\langle x_{js} e_{js}, y_{kt} e_{kt} \Big\rangle = \frac{1}{2} \Big(P_{st}^{(jk)} x_{js}, y_{kt} \Big), \quad x_{js}, y_{kt} \in \mathbb{C}^q, \ (j,s),(k,t) \in \Delta. \tag{4.2}$$

To the end of the paper we shall assume that the following condition (A) is satisfied.

(A) The matrix $\mathbf{P}_W$ is non-singular.

Condition (A) is equivalent to the fact that the space $\mathfrak{H}_W$ is a Pontryagin space. Denote by κ_W the dimension of its maximal negative subspaces. Recall that κ_W is equal to the number of negative squares of the kernel $K_W \colon \Delta \times \Delta \to \mathbb{C}^{q\times q}$ given by

$$K_W\big((j,s),(k,t)\big) := P_{st}^{(jk)}, \quad (j,s),(k,t) \in \Delta.$$

Moreover, note that in the case $q = 1$ the number κ_W is equal to the number of negative eigenvalues of $\mathbf{P}_W$.

Let T be the linear operator of $\mathfrak{H}_W$ such that

$$T\Big(\sum_{j=1}^{n} x_{j0}e_{j0}\Big) = \sum_{j=1}^{n} z_j x_{j0}e_{j0}, \quad x_{j0} \in \mathbb{C}^q,\ j = 1,\dots,n, \tag{4.3}$$

and

$$T\Big(x_{js}e_{js}\Big) = z_j x_{js}e_{js} + x_{js}e_{j,s-1}, \quad x_{js} \in \mathbb{C}^q,\ (j,s) \in \Delta, s > 0. \tag{4.4}$$

Note that to partition (3.4) of $\mathbf{P}_W$ there corresponds a matrix representation $T = \big(T_{jk}\big)_{j,k=1,\dots,n}$, where $T_{jk} = S_{jk}$ if $j \neq k$ and T_{jj} is the transpose of S_{jj}, $j,k = 1,\dots,n$.

Let V be the restriction of T to the linear space

$$\mathcal{D}(V) := \Big\{\mathfrak{x} \in \mathfrak{H}_W : \sum_{j=1}^{n} x_{j0} = 0\Big\}. \tag{4.5}$$

From Lemma 3.3 one can derive some important properties of V.

Proposition 4.1. *The operator V is an isometry in $\mathfrak{H}_W$ and does not have any eigenvalues.*

Proof. We shall prove that V has the isometric property on a generating system of $\mathcal{D}(V)$. If $\mathfrak{x}, \mathfrak{y} \in \mathcal{D}(V)$ are of the form $\mathfrak{x} = \sum_{j=1}^{n} x_{j0}e_{j0}$ and

$$\mathfrak{y} = \sum_{k=1}^{n} y_{k0}e_{k0}, \tag{4.6}$$

then we get

$$\begin{aligned}\Big\langle V\mathfrak{x}, V\mathfrak{y}\Big\rangle &= \frac{1}{2}\sum_{j,k=1}^{n} z_j\overline{z_k}\Big(P_{00}^{(jk)}x_{j0}, y_{k0}\Big) \\ &= \frac{1}{2}\sum_{j,k=1}^{n}\Big(\Big(P_{00}^{(jk)}x_{j0}, y_{k0}\Big) - \Big((W_{j0} + W_{k0}^*)x_{j0}, y_{k0}\Big)\Big) = \Big\langle \mathfrak{x}, \mathfrak{y}\Big\rangle\end{aligned}$$

applying (4.3), (4.1), Lemma 3.3 (i), and (4.5). If $\mathfrak{x} = x_{js}e_{js}$, $x_{js} \in \mathbb{C}^q$, $(j,s) \in \Delta$, $s > 0$ and if $\mathfrak{y} \in \mathcal{D}(V)$ is of the form (4.6), it follows

$$\begin{aligned}\Big\langle V\mathfrak{x}, V\mathfrak{y}\Big\rangle &= \frac{1}{2}\sum_{k=1}^{n}\Big(z_j\overline{z_k}\Big(P_{s0}^{(jk)}x_{js}, y_{k0}\Big) + \overline{z_k}\Big(P_{s-1,0}^{(jk)}x_{js}, y_{k0}\Big)\Big) \\ &= \frac{1}{2}\sum_{j,k=1}^{n}\Big(\Big(P_{s0}^{(jk)}x_{js}, y_{k0}\Big) - \Big(W_{js}x_{js}, y_{k0}\Big)\Big) = \Big\langle \mathfrak{x}, \mathfrak{y}\Big\rangle\end{aligned}$$

from (4.3), (4.4), (4.1), the first identity of Lemma 3.3 (ii), and (4.5). If $\mathfrak{x} = x_{js}e_{js}$ and if $\mathfrak{y} = y_{kt}e_{kt}$, $x_{js}, y_{kt} \in \mathbb{C}^q$, $(j,s),(k,t) \in \Delta$, $s > 0$, $t > 0$, then (4.4), (4.1), Lemma 3.3 (iii), and (4.2) yield

$$\begin{aligned}\Big\langle V\mathfrak{x}, V\mathfrak{y}\Big\rangle &= \frac{1}{2}\Big(\Big(z_j\overline{z_k}P_{st}^{(jk)} + z_jP_{s,t-1}^{(jk)} + \overline{z_k}P_{s-1,t}^{(jk)} + P_{s-1,t-1}^{(jk)}\Big)x_{js}, y_{kt}\Big) \\ &= \frac{1}{2}\Big(P_{st}^{(jk)}x_{js}, y_{kt}\Big) = \Big\langle \mathfrak{x}, \mathfrak{y}\Big\rangle.\end{aligned}$$

Finally, since the eigenvectors of T are the vectors $x_{j0}e_{j0}$, $x_{j0} \in \mathbb{C}^q$, $j = 1, \dots, n$, which do not belong to $\mathcal{D}(V)$, the operator V cannot have any eigenvalues. □

Some further properties of V, which are useful in the proof of our main result, are contained in the following lemma.

Lemma 4.2. (i) *If $s = 0, \dots, l_1 - 1$ and $x \in \mathbb{C}^q$, then $xe_{10} \in \mathcal{R}\big(V^s\big)$ and*

$$V^{-s}(xe_{10}) = xe_{1s}. \tag{4.7}$$

(ii) *If $(j, s) \in \Delta$, $j \neq 1$, and $x \in \mathbb{C}^q$, then $xe_{10} \in \mathcal{R}\big((V - z_jI)^{s+1}\big)$ and*

$$(V - z_jI)^{-(s+1)}(xe_{10}) = \frac{(-1)^{s+1}}{z_j^{s+1}} xe_{10} + \sum_{h=0}^{s} \frac{(-1)^{s-h}}{z_j^{s+1-h}} xe_{jh}. \tag{4.8}$$

Proof. (i) Use (4.4) and recall that $z_1 = 0$ to obtain

$$V^s(xe_{1s}) = xe_{10}, \quad x \in \mathbb{C}^q,\ s = 0, \dots, l_1 - 1,$$

which implies $xe_{10} \in \mathcal{R}\big(V^s\big)$ and (4.7).
(ii) Let $j = 2, \dots, n$ and let $x \in \mathbb{C}^q$. We shall prove (ii) by induction on s. Since $\frac{1}{z_j}xe_{j0} - \frac{1}{z_j}xe_{10} \in \mathcal{D}(V)$, from (4.3) it follows

$$(V - z_jI)\left(\frac{1}{z_j}xe_{j0} - \frac{1}{z_j}xe_{10}\right) = xe_{10},$$

which proves (ii) if $s = 0$. Now assume that (ii) is true if s is replaced by $s - 1$. It is not hard to see that the element on the right-hand side of (4.8) belongs to $\mathcal{D}(V)$. Then (4.3) and (4.4) yield that

$$\begin{aligned}
&(V - z_jI)\left(\frac{(-1)^{s+1}}{z_j^{s+1}} xe_{10} + \sum_{h=0}^{s} \frac{(-1)^{s-h}}{z_j^{s+1-h}} xe_{jh}\right)\\
&\quad = V\left(\frac{(-1)^{s+1}}{z_j^{s+1}} xe_{10} + \frac{(-1)^{s}}{z_j^{s+1}} xe_{j0} + \sum_{h=1}^{s} \frac{(-1)^{s-h}}{z_j^{s+1-h}} xe_{jh}\right)\\
&\qquad + \frac{(-1)^s}{z_j^s} xe_{10} - z_j \sum_{h=0}^{s} \frac{(-1)^{s-h}}{z_j^{s+1-h}} xe_{jh}\\
&\quad = z_j \frac{(-1)^s}{z_j^{s+1}} xe_{j0} + z_j \sum_{h=1}^{s} \frac{(-1)^{s-h}}{z_j^{s+1-h}} xe_{jh} + \sum_{h=1}^{s} \frac{(-1)^{s-h}}{z_j^{s+1-h}} xe_{j,h-1}\\
&\qquad + \frac{(-1)^s}{z_j^s} xe_{10} - z_j \sum_{h=0}^{s} \frac{(-1)^{s-h}}{z_j^{s+1-h}} xe_{jh}\\
&\quad = \frac{(-1)^s}{z_j^s} xe_{10} + \sum_{h=0}^{s-1} \frac{(-1)^{s-h-1}}{z_j^{s-h}} xe_{jh},
\end{aligned}$$

which equals $(V - z_jI)^{-s}(xe_{10})$ by the induction assumption. Thus, we get finally relation (4.8). □

5. The main result

In view of Theorem 2.1, we define a linear operator $\Gamma_W : \mathbb{C}^q \to \mathfrak{H}_W$ by

$$\Gamma_W x := xe_{10}, \quad x \in \mathbb{C}^q. \tag{5.1}$$

Proposition 5.1. *Let $\kappa' \in \mathbb{N}_0$, $\kappa' \geq \kappa_W$, and let $V_{\kappa'}$ be an isometric extension of V in a Pontryagin space $\Pi_{\kappa'} \supseteq \mathfrak{H}_W$ such that $\mathcal{R}(V_{\kappa'}) = \Pi_{\kappa'}$ and $\{z_1 = 0, z_2, \ldots, z_n\}$ is a subset of $\rho(V_{\kappa'})$. Then the function F defined by (2.1), where $H := \Im\mathfrak{m}\, W_{10}$ and $\Gamma := \Gamma_W$, belongs to $\mathcal{C}^{q\times q}_{\kappa''}(W)$ for some $\kappa'' \leq \kappa'$.*

Proof. Theorem 2.1 (a) yields $F \in \mathcal{C}^{q\times q}_{\kappa''}$ for some $\kappa'' \leq \kappa'$. It remains to show that (2.2) is satisfied. Since $F(z) = \mathrm{i}\,\Im\mathfrak{m}\, W_{10} + \Gamma_W^*\Gamma_W + 2z\Gamma_W^*(V_{\kappa'} - zI)^{-1}\Gamma_W$, from (5.1), (4.2), and Lemma 3.3 (i) it follows

$$\begin{aligned}\Big(F(z)x, y\Big) &= \mathrm{i}\Big(\Im\mathfrak{m}\, W_{10}x, y\Big) + \Big\langle xe_{10}, ye_{10}\Big\rangle + 2z\Big\langle (V_{\kappa'} - zI)^{-1} xe_{10}, ye_{10}\Big\rangle \\ &= \Big(W_{10}x, y\Big) + 2z\Big\langle (V_{\kappa'} - zI)^{-1} xe_{10}, ye_{10}\Big\rangle, \quad z \in \rho(V_{\kappa'}) \cap \mathbb{D},\ x, y \in \mathbb{C}^q.\end{aligned}$$

This immediately gives $F(0) = W_{10}$. If $j = 2, \ldots, n$, we get

$$\begin{aligned}\Big(F(z_j)x, y\Big) &= \Big(W_{10}x, y\Big) + 2z_j\Big\langle (V_{\kappa'} - z_jI)^{-1} xe_{10}, ye_{10}\Big\rangle \\ &= \Big(W_{10}x, y\Big) + 2z_j\Big\langle (V - z_jI)^{-1} xe_{10}, ye_{10}\Big\rangle \\ &= \Big(W_{10}x, y\Big) + 2z_j\Big\langle -\frac{1}{z_j}xe_{10} + \frac{1}{z_j}xe_{j0}, ye_{10}\Big\rangle = \Big(W_{j0}x, y\Big), \quad x, y \in \mathbb{C}^q,\end{aligned}$$

first applying Lemma 4.2 (ii) with $s = 0$ and then (4.1) and Lemma 3.3 (i). Thus, (2.2) is verified for $s = 0$. If $s \geq 1$, then by induction it can be shown that

$$F^{(s)}(z) = 2s!\Gamma_W^*\big(z(V_{\kappa'} - zI)^{-(s+1)} + (V_{\kappa'} - zI)^{-s}\big)\Gamma_W, \quad z \in \rho(V_{\kappa'}) \cap \mathbb{D}.$$

Hence, we obtain

$$\begin{aligned}\Big(F^{(s)}(0)x, y\Big) &= 2s!\Big\langle V_{\kappa'}^{-s} xe_{10}, ye_{10}\Big\rangle \\ &= 2s!\Big\langle V^{-s} xe_{10}, ye_{10}\Big\rangle = s!\Big(W_{1s}x, y\Big), \quad x, y \in \mathbb{C}^q,\end{aligned}$$

according to (5.1), Lemma 4.2 (i), (4.2), and the first part of Lemma 3.3 (ii). If $j = 2, \ldots, n$, then (5.1), Lemma 4.2 (ii), and the first assertion of Lemma 3.3 (ii) imply that

$$\begin{aligned}\Big(F^{(s)}(z_j)x, y\Big) &= 2s!\Big\langle \big(z_j(V_{\kappa'} - z_jI)^{-(s+1)} + (V_{\kappa'} - z_jI)^{-s}\big) xe_{10}, ye_{10}\Big\rangle \\ &= 2s!\Big(z_j\Big\langle (V - z_jI)^{-(s+1)} xe_{10}, ye_{10}\Big\rangle + \Big\langle (V - z_jI)^{-s} xe_{10}, ye_{10}\Big\rangle\Big) \\ &= s!\Big(W_{js}x, y\Big), \quad x, y \in \mathbb{C}^q.\end{aligned}$$

□

A unitary operator U in a Pontryagin space $\Pi_{\kappa'} \supseteq \mathfrak{H}_W$ is called *minimal* if the operators U and Γ_W are minimal.

Proposition 5.2. *If $F \in \mathcal{C}_\kappa^{q\times q}(W)$, then there exist a minimal unitary extension U_W of V in a Pontryagin space $\widetilde{\Pi}_\kappa \supseteq \mathfrak{H}_W$ such that*

$$F(z) = \mathrm{i}\,\Im\mathfrak{m}\, W_{10} + \Gamma_W^*(U_W + zI)(U_W - zI)^{-1}\Gamma_W, \quad z \in \varrho(F). \tag{5.2}$$

Proof. Since F belongs to $\mathcal{C}_\kappa^{q\times q}$, there exists a representation of F according to Theorem 2.1 (b). Obviously, $\Im\mathfrak{m}\, F(0) = H$, and since $F(0) = W_{10}$, one has

$$H = \Im\mathfrak{m}\, W_{10}.$$

Assertion (ii) of Theorem 2.1 (b) implies that

$$\{z_1 = 0, z_2, \dots, z_n\} \subseteq \rho(U).$$

Consider the subspace $\Pi_{\kappa'}$ of Π_κ spanned by the elements

$$\left\{\frac{1}{s!}R^{(s)}(z_j)x\colon s = 0, \dots, l_j - 1, j = 1, \dots, n, x \in \mathbb{C}^q\right\},$$

where $R(z) := (I - zU^*)^{-1}\Gamma$, $z \in \varrho(F)$, compare (3.1). From (3.2), (3.3), and (4.2) it follows that there is a unitary operator $V'\colon \Pi_{\kappa'} \to \mathfrak{H}_W$ such that

$$V'\Big(\frac{1}{s!}R^{(s)}(z_j)x\Big) = xe_{js}, \quad (j,s) \in \Delta,\ x \in \mathbb{C}^q. \tag{5.3}$$

Particularly, κ' coincides with κ_W and $\Pi_{\kappa'} = \Pi_{\kappa W}$ is a Pontryagin space. Let $\Pi_{\kappa W}^\perp$ be its orthogonal complement. Define the Pontryagin space $\widetilde{\Pi}_\kappa := \mathfrak{H}_W \oplus \Pi_{\kappa W}^\perp$ and a unitary operator $\widetilde{V}\colon \Pi_\kappa \to \widetilde{\Pi}_\kappa$ in such a way that $\widetilde{V}$ coincides with V' on $\Pi_{\kappa W}$ and $\widetilde{V}$ is the identity on $\Pi_{\kappa W}^\perp$. Further, let $U_W := \widetilde{V}U\widetilde{V}^{-1}$ and $\Gamma' := \widetilde{V}\Gamma$. Since

$$\Gamma' x = \widetilde{V}(\Gamma x) = \widetilde{V}\big(R(z_1)x\big) = xe_{10}, \quad x \in \mathbb{C}^q,$$

it follows $\Gamma' = \Gamma_W$. Hence, equality (5.2) is an easy consequence of (iii) in Theorem 2.1 (b). To prove that U_W is an extension of V, according to (5.3), (4.3), (4.4), and the definition of V it is enough to verify the following two assertions:

(a) If $x_j \in \mathbb{C}^q$, $j = 1, \dots, n$, are such that $\sum_{j=1}^n x_j = 0$, then

$$U\Big(\sum_{j=1}^n R(z_j)x_j\Big) = \sum_{j=1}^n z_j R(z_j)x_j.$$

(b) If $(j,s) \in \Delta$, $s > 0$, then

$$U\Big(\frac{1}{s!}R^{(s)}(z_j)\Big) = z_j\frac{1}{s!}R^{(s)}(z_j) + \frac{1}{(s-1)!}R^{(s-1)}(z_j).$$

Assertion (a) follows from

$$\begin{aligned}\sum_{j=1}^{n} R(z_j)x_j &= U\Big(\sum_{j=1}^{n}(U-z_jI)^{-1}\Gamma x_j\Big)\\ &= U\Big(\sum_{j=1}^{n}(U-z_jI)^{-1}\Gamma x_j\Big) - \sum_{j=1}^{n}(U-z_jI)(U-z_jI)^{-1}\Gamma x_j\\ &= \sum_{j=1}^{n} z_j(U-z_jI)^{-1}\Gamma x_j = U^*\Big(\sum_{j=1}^{n} z_jR(z_j)x_j\Big).\end{aligned}$$

Assertion (b) is seen to be true because of

$$\begin{aligned}\frac{1}{s!}R^{(s)}(z_j) &= (U^*)^s(I-z_jU^*)^{-(s+1)}\Gamma\\ &= (U^*)^s\Big(z_jU^*(I-z_jU^*)^{-(s+1)} + (I-z_jU^*)^{-s}\Big)\Gamma\\ &= U^*\Big((U^*)^s z_j(I-z_jU^*)^{-(s+1)} + (U^*)^{s-1}(I-z_jU^*)^{-s}\Big)\Gamma\\ &= U^*\Big(z_j\frac{1}{s!}R^{(s)}(z_j) + \frac{1}{(s-1)!}R^{(s-1)}(z_j)\Big).\end{aligned}$$

□

Proposition 5.2 has the following consequence.

Corollary 5.3. *If $\kappa < \kappa_W$, then the set $\mathcal{C}_\kappa^{q\times q}(W)$ is empty.*

Our main result is a description of $\mathcal{C}_\kappa^{q\times q}(W)$ if $\kappa \ge \kappa_W$. It can be derived from Propositions 5.1 and 5.2.

Theorem 5.4. *Let $\kappa \ge \kappa_W$. The map $U_W \to F$ defined by (5.2) establishes a correspondence between the set of all minimal unitary extensions U_W of V in a Pontryagin space $\Pi_\kappa \supseteq \mathfrak{H}_W$ and the set $\mathcal{C}_\kappa^{q\times q}(W)$. If unitarily equivalent extensions are identified, this correspondence is one-to-one.*

Proof. Let U_W be a minimal unitary extension of V in a Pontryagin space Π_κ such that $\Pi_\kappa \supseteq \mathfrak{H}_W$. Since U_W is unitary, $z_1 = 0 \in \rho(U_W)$. Let $j = 2, \dots, n$ and assume that $\mathfrak{u} \in \mathcal{N}\big(U_W - \frac{1}{\overline{z_j}}I\big)$. Then for $x \in \mathbb{C}^q$,

$$\begin{aligned}\Big\langle \mathfrak{u}, xe_{j0}\Big\rangle - \Big\langle \mathfrak{u}, xe_{10}\Big\rangle &= \Big\langle U_W\mathfrak{u}, U_W(xe_{j0}-xe_{10})\Big\rangle\\ &= \Big\langle U_W\mathfrak{u}, V(xe_{j0}-xe_{10})\Big\rangle = \frac{1}{\overline{z_j}}\overline{z_j}\Big\langle \mathfrak{u}, xe_{j0}\Big\rangle\end{aligned}$$

and, therefore, $\mathfrak{u}$ is orthogonal to xe_{10}. Since for any $z \in \rho(U_W)$ also the relation $(U_W^* - \overline{z}I)^{-1}\mathfrak{u} \in \mathcal{N}\big(U_W - \frac{1}{\overline{z_j}}I\big)$ is satisfied, we get

$$\begin{aligned}0 &= \Big\langle (U_W^* - \overline{z}I)^{-1}\mathfrak{u}, xe_{10}\Big\rangle = \Big\langle \mathfrak{u}, (U_W - zI)^{-1}xe_{10}\Big\rangle\\ &= \Big\langle \mathfrak{u}, (U_W - zI)^{-1}\Gamma_W x\Big\rangle, \quad x \in \mathbb{C}^q,\end{aligned}$$

which yields $\mathfrak{u} = 0$ by the minimality of U_W. It follows that $\frac{1}{\overline{z_j}} \in \rho(U_W)$ and, hence, $z_j \in \rho(U_W)$. An application of Proposition 5.1 gives that the function F defined by (5.2) belongs to $\mathcal{C}_{\kappa'}^{q\times q}(W)$ for some $\kappa' \leq \kappa$ and the second part of Theorem 2.1 (a) implies that $\kappa' = \kappa$. From Proposition 5.2 one can obtain that the map $U_W \to F$ defined by (5.2) is surjective and from the uniqueness assertion of Theorem 2.1 (b) it follows that it is a one-to-one correspondence if unitarily equivalent extensions of V are identified. □

6. $\mathfrak{H}_W$ as a space of rational functions

Now we give a concrete model of the abstract linear space $\mathfrak{H}_W$ by identifying e_{js}, $(j,s)\in\Delta$, with a certain rational function. In fact, we consider $\mathfrak{H}_W$ as the space

$$\mathfrak{H}_W = \left\{ \mathfrak{x} := \frac{1}{\mathfrak{q}_W}\mathfrak{p} : \mathfrak{p} \in \mathfrak{P}_m^{(q)} \right\}, \tag{6.1}$$

where $m+1$ stands for the cardinality of the set Δ, i.e.,

$$m := \sum_{j=1}^{n} l_j - 1,$$

$\mathfrak{P}_m^{(q)}$ denotes the linear space of all $\mathbb{C}^q$-valued polynomials, whose degree does not exceed m, and $\mathfrak{q}_W$ is the $\mathbb{C}$-valued polynomial

$$\mathfrak{q}_W(u) := \prod_{j=1}^{n} (1 - \overline{z_j} u)^{l_j}, \quad u \in \mathbb{T}. \tag{6.2}$$

Here $\mathbb{T}$ denotes the unit circle. Preferably, we shall consider rational functions defined on $\mathbb{T}$, but occasionally we shall extend their domains of definition by analytic continuation without introducing new notations for them.

Let $f_{js} : \mathbb{T} \to \mathbb{C}$ be defined by

$$f_{js}(u) := \frac{u^s}{(1 - \overline{z_j} u)^{s+1}}, \quad u \in \mathbb{T},\ (j,s) \in \Delta,$$

and let $\varepsilon_1, \ldots, \varepsilon_q$ be the canonical orthonormal basis of $\mathbb{C}^q$. It is not hard to see that the set

$$\{ f_{js}\varepsilon_r : (j,s) \in \Delta, r = 1, \ldots, q \}$$

forms a basis of the $(m+1)q$-dimensional space $\mathfrak{H}_W$.

For $z \in \mathbb{D}$, we define the elementary Blaschke factor b_z by

$$b_z(u) := \frac{\overline{z}}{|z|} \frac{z - u}{1 - \overline{z} u}, \quad u \in \mathbb{T}.$$

Here and in the following we use the convention $\frac{\overline{0}}{|0|} := -1$, so that $b_0(u) = u$. Furthermore, denote by B_W the Blaschke product

$$B_W(u) := \prod_{j=1}^{n} \left(b_{z_j}(u)\right)^{l_j}, \quad u \in \mathbb{T}.$$

An important tool for studying spaces of rational functions is the notion of the adjoint rational function, cf. [6, Section 2.2]. If $q = 1$, we define the *adjoint rational function* $\mathfrak{x}^{[W]}$ of $\mathfrak{x} \in \mathfrak{H}_W$ by

$$\mathfrak{x}^{[W]}(u) := B_W(u)\, \overline{\mathfrak{x}\Big(\frac{1}{\overline{u}}\Big)}, \quad u \in \mathbb{T}.$$

If $q > 1$, we write $\mathfrak{x} \in \mathfrak{H}_W$ as a sum $\mathfrak{x} = \sum\limits_{r=1}^{q} \mathfrak{x}_r \varepsilon_r$ and set

$$\mathfrak{x}^{[W]} := \sum_{r=1}^{q} \mathfrak{x}_r^{[W]} \varepsilon_r .$$

The notion of the adjoint rational function is a generalization of and closely related to the notion of the reciprocal polynomial, cf. [27, Equation (1.12.4)]. We recall that if $\mathfrak{p} \in \mathfrak{P}_m^{(1)}$, then the *reciprocal polynomial* $\widetilde{\mathfrak{p}}^{[m]}$ is defined by

$$\widetilde{\mathfrak{p}}^{[m]}(u) := u^m\, \overline{\mathfrak{p}\Big(\frac{1}{\overline{u}}\Big)}, \quad u \in \mathbb{T}.$$

If $q > 1$ and if $\mathfrak{p} \in \mathfrak{P}_m^{(q)}$, then write $\mathfrak{p} = \sum\limits_{r=1}^{q} \mathfrak{p}_r \varepsilon_r$ and set

$$\widetilde{\mathfrak{p}}^{[m]} := \sum_{r=1}^{q} \widetilde{\mathfrak{p}}_r^{[m]} \varepsilon_r .$$

The following properties of the mapping $\mathfrak{p} \to \widetilde{\mathfrak{p}}^{[m]}$ are well known and can be easily derived from the definition, cf. [27, Section 1.12].

(I) $\widetilde{\left(\widetilde{\mathfrak{p}}^{[m]}\right)}^{[m]} = \mathfrak{p}$ if $\mathfrak{p} \in \mathfrak{P}_m^{(q)}$;
(II) $\widetilde{\mathfrak{p}}^{[m]}(0) = 0$ if and only if $\mathfrak{p} \in \mathfrak{P}_{m-1}^{(q)}$;
(III) $\widetilde{\mathfrak{p}}^{[m]} \in \mathfrak{P}_{m-1}^{(q)}$ if and only if $\mathfrak{p}(0) = 0$.

A simple calculation shows that if $\mathfrak{x} = \frac{1}{\mathfrak{q}_W}\mathfrak{p} \in \mathfrak{H}_W$, then

$$\mathfrak{x}^{[W]} = \eta \frac{1}{\mathfrak{q}_W} \widetilde{\mathfrak{p}}^{[m]}, \tag{6.3}$$

where the polynomial $\mathfrak{q}_W$ is given as in (6.2) and where

$$\eta := \prod_{j=2}^{n} \left(-\frac{\overline{z_j}}{|z_j|} \right)^{l_j} \in \mathbb{T}.$$

Note that, because of (6.3), the properties (I), (II), and (III) lead to similar properties of the mapping $\mathfrak{x} \to \mathfrak{x}^{[W]}$.

If $\mathfrak{H}_W$ is realized as the space of rational functions in accordance with (6.1) and if we choose

$$e_{js} := f_{js}^{[W]}, \quad (j,s) \in \Delta,$$

then the operator V, which was defined in Section 4, becomes an operator of multiplication. The formulation of the following proposition as well as (II) and (III) are correct even in the case $m = 0$ if one interprets the symbol $\mathfrak{P}_{-1}^{(q)}$ as the space whose only element is the $\mathbb{C}^q$-valued zero polynomial.

Proposition 6.1. *If $\mathfrak{H}_W$ is given as in (6.1), then*

$$\begin{aligned}\mathcal{D}(V) &= \left\{\mathfrak{x} \in \mathfrak{H}_W : \mathfrak{x}^{[W]}(0) = 0\right\} = \left\{\mathfrak{x} = \frac{1}{\mathfrak{q}_W}\mathfrak{p} \in \mathfrak{H}_W : \widetilde{\mathfrak{p}}^{[m]}(0) = 0\right\} \\ &= \left\{\mathfrak{x} = \frac{1}{\mathfrak{q}_W}\mathfrak{p} \in \mathfrak{H}_W : \mathfrak{p} \in \mathfrak{P}_{m-1}^{(q)}\right\},\end{aligned} \tag{6.4}$$

$$(V\mathfrak{x})(u) = u\,\mathfrak{x}(u), \quad u \in \mathbb{T},\ \mathfrak{x} \in \mathcal{D}(V), \tag{6.5}$$

$$\begin{aligned}\mathcal{R}(V) &= \left\{\mathfrak{x} \in \mathfrak{H}_W : \mathfrak{x}(0) = 0\right\} = \left\{\mathfrak{x} = \frac{1}{\mathfrak{q}_W}\mathfrak{p} \in \mathfrak{H}_W : \mathfrak{p}(0) = 0\right\} \\ &= \left\{\mathfrak{x} = \frac{1}{\mathfrak{q}_W}\mathfrak{p} \in \mathfrak{H}_W : \widetilde{\mathfrak{p}}^{[m]} \in \mathfrak{P}_{m-1}^{(q)}\right\}.\end{aligned} \tag{6.6}$$

Proof. To avoid trivialities we assume $m > 0$. If

$$\mathfrak{x} = \sum_{(j,s)\in\Delta} x_{js} e_{js}, \quad x_{js} \in \mathbb{C}^q,\ (j,s) \in \Delta,$$

then

$$\mathfrak{x}^{[W]}(0) = \sum_{(j,s)\in\Delta} \overline{x_{js}} f_{js}(0) = \overline{\left(\sum_{j=1}^{n} x_{j0}\right)},$$

where for a $x \in \mathbb{C}^q$ the notion $\overline{x}$ means $(x^*)^\top$. Thus, the first equality of (6.4) follows from the definition of $\mathcal{D}(V)$, see (4.5). The second equality of (6.4) follows from (6.3) and the third one from (II). To prove (6.5) note first that

$$\overline{f_{js}\left(\frac{1}{\overline{u}}\right)} = \frac{u}{(u - z_j)^{s+1}}, \quad u \in \mathbb{T},\ (j,s) \in \Delta,$$

and if

$$\sum_{j=1}^{n} x_{j0} = 0, \quad x_{j0} \in \mathbb{C}^q,\ j = 1, \dots, n,$$

then

$$\begin{aligned}\sum_{j=1}^{n} \frac{z_j u}{u - z_j} x_{j0} &= u \sum_{j=1}^{n} \left(\frac{z_j}{u - z_j} x_{j0} - \frac{u}{u - z_j} x_{j0} + \frac{u}{u - z_j} x_{j0}\right) \\ &= u\left(-\sum_{j=1}^{n} x_{j0} + \sum_{j=1}^{n} \frac{u}{u - z_j} x_{j0}\right) = u \sum_{j=1}^{n} \frac{u}{u - z_j} x_{j0}, \quad u \in \mathbb{T}.\end{aligned}$$

Hence, for

$$\mathfrak{x} = \sum_{(j,s)\in\Delta} x_{js} e_{js} \left(= \sum_{(j,s)\in\Delta} f_{js}^{[W]} x_{js} \right) \in \mathcal{D}(V),$$

we get

$$\begin{aligned}(V\mathfrak{x})(u) &= \sum_{j=1}^{n} z_j f_{j0}^{[W]}(u) x_{j0} + \sum_{\substack{(j,s)\in\Delta \\ s>0}} \left(z_j f_{js}^{[W]}(u) + f_{j,s-1}^{[W]}(u) \right) x_{js} \\ &= B_W(u) \Bigg(\sum_{j=1}^{n} \frac{z_j u}{u - z_j} x_{j0} + \sum_{\substack{(j,s)\in\Delta \\ s>0}} \left(\frac{z_j u}{(u-z_j)^{s+1}} + \frac{u}{(u-z_j)^s} \right) x_{js} \Bigg) \\ &= B_W(u) \Bigg(u \sum_{j=1}^{n} \frac{u}{u - z_j} x_{j0} + u \sum_{\substack{(j,s)\in\Delta \\ s>0}} \frac{u}{(u-z_j)^{s+1}} x_{js} \Bigg) \\ &= u \sum_{(j,s)\in\Delta} f_{js}^{[W]}(u) x_{js} = u\,\mathfrak{x}(u), \quad u \in \mathbb{T}.\end{aligned}$$

From (6.4) and (6.5) one can immediately obtain the first equality of (6.6). The second equality of (6.6) is trivial and the third one is a consequence of (III). □

Remark 6.2. Since $\mathfrak{H}_W$ is finite-dimensional, it is a reproducing kernel space. Using some properties of the mapping $\mathfrak{x} \to \mathfrak{x}^{[W]}$, one can obtain several results on the reproducing kernel, cf. [6, Section 2.2] for the case $\kappa = 0$ and $q = 1$ (see also [22, Proposition 3.1]). We omit the details since the object of the present paper was the operator approach to interpolation problems.

Acknowledgement

The second author is very grateful to Professor H. Langer and Professor H. Woracek for many useful and stimulating discussions on the subject he had during his one year stay at the Technical University of Vienna.

References

[1] N.I. Akhiezer, *The Classical Moment Problem and Some Related Questions in Analysis.* Oliver and Boyd, London 1965.

[2] T.Ya. Azizov and I.S. Iohvidov, *Foundations of the Theory of Linear Operators in Spaces with an Indefinite Metric* (Russian). Nauka, Moscow 1986; Engl. transl.: *Linear Operators in Spaces with Indefinite Metric.* Wiley, New York 1989.

[3] J.A. Ball, I. Gohberg, and L. Rodman, *Interpolation of Rational Matrix Functions.* Operator Theory: Adv. and Appl. **45**, Birkhäuser, Basel 1990.

[4] J. Bognar, *Indefinite Inner Product Spaces.* Springer-Verlag, Berlin 1974.

[5] P. Bruinsma, *Interpolation for Schur and Nevanlinna pairs.* Dissertation, University of Groningen 1991.

[6] A. Bultheel, P. González-Vera, E. Hendriksen, and O. Njåstad, *Orthogonal Rational Functions.* Cambridge Monographs on Applied and Comput. Math. **5**, Cambridge University Press, Cambridge 1999.

[7] C. Carathéodory, *Über den Variabilitätsbereich der Koeffizienten von Potenzreihen, die gegebene Werte nicht annehmen.* Math. Ann. **64** (1907), 95–115.

[8] G. Chen and Y. Hu, *On the multiple Nevanlinna-Pick matrix interpolation problem in the class $\mathcal{C}_p$ and the Carathéodory matrix coefficient problem.* Linear Algebra Appl. **283** (1998), 179–203.

[9] A. Dijksma, H. Langer, and H. de Snoo, *Eigenvalues and pole functions of Hamiltonian systems with eigenvalue depending boundary conditions.* Math. Nachr. **161** (1993), 107–154.

[10] V.K. Dubovoj, B. Fritzsche, and B. Kirstein, *Matricial Version of the Classical Schur Problem.* Teubner-Texte zur Mathematik Bd. **129**, Teubner, Leipzig 1992.

[11] H. Dym, *J Contractive Matrix Functions, Reproducing Kernel Hilbert Spaces and Interpolation.* CBMS Regional Conf. Ser. Math. **71**, Amer. Math. Soc., Providence, R. I. 1989.

[12] A.V. Efimov and V.P. Potapov, *J-expansive matrix-valued functions and their role in the analytic theory of electrical circuits* (Russian). Uspekhi Mat. Nauk **28** 1 (1973), 65–130; English transl. in: Russian Math. Surveys **28** 1 (1973), 69–140.

[13] I.P. Fedčina, *The tangential Nevanlinna-Pick problem with multiple points* (Russian). Dokl. Akad. Nauk Armjan. SSR **61** (1975), 214–218.

[14] C. Foiaş and A.E. Frazho, *The Commutant Lifting Approach to Interpolation Problems.* Operator Theory: Adv. and Appl. **44**, Birkhäuser, Basel 1990.

[15] I.S. Iohvidov, M.G. Kreĭn, and H. Langer, *Introduction to the Spectral Theory of Operators in Spaces with an Indefinite Metric.* Math. Forschung, Akademie-Verlag Bd. **9**, Berlin 1982.

[16] V.E. Katsnelson, *On transformations of Potapov's fundamental matrix inequality.* in: Operator Theory: Adv. and Appl. **95**, pp. 253–281, Birkhäuser, Basel 1997.

[17] I.V. Kovalishina, *Analytic theory of a class of interpolation problems* (Russian). Izv. Akad. Nauk SSSR Ser. Mat. **47** 3 (1983), 455–497.

[18] I.V. Kovalishina and V.P. Potapov, *An indefinite metric in the Nevanlinna-Pick problem* (Russian). Dokl. Akad. Nauk Armjan. SSR **59** 1 (1974), 17–22.

[19] M.G. Kreĭn and H. Langer, *Über die verallgemeinerten Resolventen und die charakteristische Funktion eines isometrischen Operators im Raum Π_κ.* in: Hilbert space operators and operator algebras (Proc. Internat. Conf., Tihany 1970), pp. 353–399, 5. Colloq. Math. Soc. János Bolyai, North-Holland, Amsterdam 1972.

[20] M.G. Kreĭn and H. Langer, *Über einige Fortsetzungsprobleme, die eng mit der Theorie hermitescher Operatoren im Raum Π_κ zusammenhängen* Teil I: - *Einige Funktionenklassen und ihre Darstellungen.* Math. Nachr. **77** (1977), 187–236; Teil II: - *Verallgemeinerte Resolventen, u-Resolventen und ganze Operatoren.* J. Funct. Anal. **30** (1978), 390–447; Teil III: - *Indefinite analogues of the Hamburger and Stieltjes moment problems.* Beitr. Anal. **14** (1979), 25–40; Beitr. Anal. **15** (1980), 27–45.

[21] M.G. Kreĭn and H. Langer, *Some propositions on analytic matrix functions related to the theory of operators in the space Π_κ.* Acta Sci. Math. (Szeged) **43** (1981), 181–205.

[22] H. Langer and A. Lasarow, *Solution of a multiple Nevanlinna-Pick problem via orthogonal rational functions.* J. Math. Anal. Appl. **293** (2004), 605–632.

[23] R. Nevanlinna, *Über beschränkte analytische Funktionen.* Ann. Acad. Sci. Fenn. **A 32** 7 (1929), 1–75.

[24] G. Pick, *Über die Beschränkungen analytischer Funktionen, welche durch vorgegebene Funktionswerte bewirkt werden.* Math. Ann. **77** (1916), 7–23.

[25] L.A. Sakhnovich, *Similarity of operators* (Russian). Sibir. Mat. Z. **13** 4 (1972), 868–883; English transl. in: Sibirian Math. J. **13** 4 (1972), 604–615.

[26] B. Sz.-Nagy and A. Korányi, *Operatortheoretische Behandlung und Verallgemeinerung eines Problemkreises in der komplexen Funktionentheorie.* Acta Math. **100** (1958), 171–202.

[27] G. Szegö, *Orthogonal Polynomials.* Amer. Math. Soc. Coll. Publ. **23**, Providence, R. I. 1939.

[28] H. Woracek, *Multiple point interpolation in Nevanlinna classes.* Integr. Equ. Oper. Theory **28** (1997), 97–109.

Lutz Klotz
Fakultät für Mathematik und Informatik
Universität Leipzig
Augustusplatz 10
D-04109 Leipzig
Germany
e-mail: klotz@math.uni-leipzig.de

Andreas Lasarow
Departement Computerwetenschappen
K.U. Leuven
Celestijnenlaan 200A
B-3001 Heverlee (Leuven)
Belgium
e-mail: Andreas.Lasarow@cs.kuleuven.be

Operator Theory:
Advances and Applications, Vol. 162, 231–251

Bounded Normal Operators in Pontryagin Spaces

Heinz Langer and Franciszek Hugon Szafraniec

Abstract. We establish some spectral properties of normal operators in a Pontryagin space Π_κ. If $\kappa = 1$ a classification of the normal operators is given according to the structure of the eigenspaces of N and N^+ which contain a non-positive eigenelement.

1. Introduction

Whereas the spectral properties of self-adjoint and unitary operators in Pontryagin spaces Π_κ are well understood, see, e.g., [4], [1], [7], [10], not so much is known about normal operators. Recall that a bounded operator N in Π_κ is normal if $NN^+ = N^+N$, where N^+ denotes the adjoint operator of N. The starting point of our considerations is a result of M.A. Naimark, see [12], which implies that for a normal operator N in Π_κ there exists a κ-dimensional non positive common invariant subspace for N and its adjoint N^+. The aim is to say something about the properties of the spectrum, the eigenspaces and the spectral function of N and N^+. In [5], the irreducible normal operators in a *finite-dimensional* space Π_1 were described, which led to spaces of dimension ≤ 4. In the present note, for a space Π_1 of arbitrary dimension, we give a classification of all bounded normal operators with respect to their critical spectral point(s); here the interesting case is when λ_0 is the unique eigenvalue of N with a neutral eigenelement and λ_0^* is an eigenvalue of N^+ with the same eigenelement [1].

Whereas for a normal operator N in a Hilbert space we have

$$(N - \lambda_0)x_0 = 0 \iff (N^* - \lambda_0^*)x_0 = 0,$$

an eigenelement x_0 of N at λ_0 need not be an eigenelement of N^+, see Remark 5.7, or it can be an eigenelement of N^+ at an eigenvalue $\neq \lambda_0^*$. The latter is shown by the following simple example in a two-dimensional space: choose $\Pi_1 = \mathbb{C}^2$ with

[1] The * stands for the complex conjugate of a number and for the adjoint of a Hilbert space operator.

indefinite inner product generated by the Gram matrix $G = \begin{pmatrix} 0 & 1 \\ 1 & 0 \end{pmatrix}$ and let N be the operator given by $N = \begin{pmatrix} 0 & \mathrm{i} \\ 0 & 1 \end{pmatrix}$. Then $N^+ = \begin{pmatrix} 1 & -\mathrm{i} \\ 0 & 0 \end{pmatrix}$ and $N^+N = NN^+ = 0$, hence N is normal. The eigenvalues and eigenvectors of N and N^+ can be seen from the following table:

	N	N^+
$\lambda = 0$	$\binom{1}{0}$	$\binom{1}{-\mathrm{i}}$
$\lambda = 1$	$\binom{1}{-\mathrm{i}}$	$\binom{1}{0}$

In the next section we introduce some notation and recall the classification of the self-adjoint operators in a space Π_1 from, e.g., [6]. In Section 3 we give matrix representations of the operators N and N^+ corresponding to the common invariant κ-dimensional non-positive subspace of N and N^+ (for self-adjoint operators corresponding results can be found in [8]) and show the existence of a spectral function. In the last two sections we restrict our considerations to spaces Π_1: in Section 4 some properties of the eigenvalues with a non-positive eigenelement are proved, in Section 5 the above-mentioned classification is given.

Some results of Section 3 are close to results in [2] and [3], where special classes of normal operators in Krein spaces are considered.

2. Preliminaries

We start with some notation. The linear span of elements $x_1, x_2, \ldots, x_n \in \Pi_\kappa$ is denoted by $\mathrm{lin}\{x_1, x_2, \ldots, x_n\}$ and, if $x_1, x_2, \ldots, x_n$ are linearly independent, it is identified with the space $\mathbb{C}^n$ according to $\xi_1 x_1 + \cdots + \xi_n x_n \sim \begin{pmatrix} \xi_1 \\ \vdots \\ \xi_n \end{pmatrix}$. If the space Π_κ is decomposed as

$$\Pi_\kappa = \mathrm{lin}\{x_1, \ldots, x_\kappa\} \dotplus \mathcal{H}_0 \tag{2.1}$$

with some Hilbert space $\mathcal{H}_0$ then the element $x = \xi_1 x_1 + \cdots + \xi_\kappa x_\kappa + u \in \Pi_\kappa$ with $\xi_1, \ldots, \xi_\kappa \in \mathbb{C}$, $u \in \mathcal{H}_0$ is identified with $\begin{pmatrix} \xi_1 \\ \vdots \\ \xi_\kappa \\ u \end{pmatrix} \in \mathbb{C}^\kappa \oplus \mathcal{H}_0$, and the Gram operator G relates the inner product $[\,\cdot\,,\,\cdot\,]$ of Π_κ to the inner product of $\mathbb{C}^\kappa \oplus \mathcal{H}_0$:

$$[\xi_1 x_1 + \cdots + \xi_\kappa x_\kappa + u, \xi_1' x_1 + \cdots + \xi_\kappa' x_\kappa + u'] = \left(G \begin{pmatrix} \xi_1 \\ \vdots \\ \xi_\kappa \\ u \end{pmatrix}, \begin{pmatrix} \xi_1' \\ \vdots \\ \xi_\kappa' \\ u' \end{pmatrix} \right)_{\mathbb{C}^\kappa \oplus \mathcal{H}_0}.$$

If $\text{lin}\{x_1, x_2, \ldots, x_n\}$ and $\mathcal{H}_0$ are orthogonal in Π_κ then the sum on the right-hand side of (2.1) is written as $\text{lin}\{x_1, x_2, \ldots, x_n\} \oplus \mathcal{H}_0$.

Recall that a *dual pair of subspaces* of a Pontryagin space Π_κ (or, more generally, of a Krein space) is a pair $\{\mathcal{L}_+, \mathcal{L}_-\}$, such that $\mathcal{L}_+$ is a non-negative subspace, $\mathcal{L}_-$ is a non-positive subspace, and $\mathcal{L}_+[\perp]\mathcal{L}_-$. A *maximal dual pair of subspaces* is a dual pair $\{\mathcal{L}_+^{\max}, \mathcal{L}_-^{\max}\}$ for which $\mathcal{L}_+^{\max}$ is a maximal non-negative and $\mathcal{L}_-^{\max}$ is a maximal non-positive subspace. According to a theorem of R.S. Phillips, for each dual pair of subspaces $\{\mathcal{L}_+, \mathcal{L}_-\}$ there exists a maximal dual pair $\{\mathcal{L}_+^{\max}, \mathcal{L}_-^{\max}\}$ such that $\mathcal{L}_+ \subset \mathcal{L}_+^{\max}$ and $\mathcal{L}_- \subset \mathcal{L}_-^{\max}$.

We recall the notion of the spectral function with critical points of a self-adjoint operator A in the space Π_κ, see [10]. We suppose without loss of generality that A has real spectrum. By $\sigma_0(A)$ we denote the set of all eigenvalues of A with a non-positive eigenelement. The semiring of all bounded intervals of the real axis $\mathbb{R}^1$ with endpoints not in $\sigma_0(A)$ and their complements in $\mathbb{R}^1$ is denoted by $\mathcal{R}_A$. Then there exists a mapping $\Delta \mapsto E(\Delta)$ from $\mathcal{R}_A$ into the set of orthogonal projections in Π_κ with the following properties ($\Delta, \Delta' \in \mathcal{R}_A$):

(i) $E(\varnothing) = 0,\ E(\Delta) = I$ if $\sigma(A) \subset \Delta$.

(ii) $E(\Delta \cap \Delta') = E(\Delta)E(\Delta')$.

(iii) $E(\Delta \cup \Delta') = E(\Delta) + E(\Delta')$ if $\Delta \cup \Delta' \in \mathcal{R}_A, \quad \Delta \cap \Delta' = \varnothing$.

(iv) if $\Delta \cap \sigma_0(A) = \varnothing$ then $E(\Delta)\Pi_\kappa$ is a positive subspace.

(v) $E(\Delta)A = AE(\Delta)$.

(vi) $\sigma(A|E(\Delta)\Pi_\kappa) \subset \overline{\Delta}$.

If $\lambda_0 \in \sigma_0(A)$ we set

$$\mathcal{S}_{\lambda_0} := \bigcap_{\Delta \in \mathcal{R}(A),\, \lambda_0 \in \Delta} E(\Delta)\Pi_\kappa.$$

Then $\mathcal{S}_{\lambda_0}$ coincides with the algebraic eigenspace $\mathcal{L}_{\lambda_0}(A)$ of A at λ_0, see [10]. The next lemma is an immediate consequence of [10, Proposition 5.6].

Lemma 2.1. *If $\lambda_0 \in \sigma_0(A)$ then the following statements are equivalent:*

(a) *$\mathcal{L}_{\lambda_0}(A)$ is non-degenerated.*

(b) *There exists a positive number k such that $\|E(\Delta)\| \leq k$ for all $\Delta \in \mathcal{R}_A$ from a neighborhood of λ_0.*

It is easy to see that the spectral function E of A can be extended to intervals Δ with one or both endpoints in $\sigma_0(A)$ but such that the algebraic eigenspaces at these endpoints are non-degenerated. A point $\lambda_0 \in \sigma_0(A)$, for which $E(\Delta)\Pi_\kappa$ contains positive as well as negative elements for all $\Delta \in \mathcal{R}_A$, $\lambda_0 \in \Delta$, is called a *critical point* of A. The critical point λ_0 of A is called a *regular critical point* if for it the equivalent statements of Lemma 2.1 hold, otherwise it is a *singular critical point* of A.

Next we recall some results for a self-adjoint operator A in a Pontryagin space Π_1 with negative index one, cf. [6]. According to the theorem of Pontryagin, A has (at least one) non-positive eigenvector $x_0 : \ (A - \lambda_0)x_0 = 0$, $[x_0, x_0] \le 0$. Then exactly one of the following cases appears; here A_0 always stands for a self-adjoint operator in a Hilbert space $(\mathcal{H}_0, (\,\cdot\,,\,\cdot\,))$.

a) $\lambda_0 \neq \lambda_0^*$. Then $[x_0, x_0] = 0$, there exists an element y_0 with $[y_0, y_0] = 0$, $[x_0, y_0] = 1$ and $(A - \lambda_0^*)y_0 = 0$, and with respect to the decomposition $\Pi_1 = \mathrm{lin}\{x_0, y_0\} \oplus \mathcal{H}_0$ we have

$$G = \begin{pmatrix} 0 & 1 & 0 \\ 1 & 0 & 0 \\ 0 & 0 & I \end{pmatrix}, \qquad A = \begin{pmatrix} \lambda_0 & 0 & 0 \\ 0 & \lambda_0^* & 0 \\ 0 & 0 & A_0 \end{pmatrix},$$

b) $\lambda_0 = \lambda_0^*$ and $[x_0, x_0] < 0$. Then $\Pi_1 = \mathrm{lin}\{x_0\} \oplus \mathcal{H}_0$ and

$$G = \begin{pmatrix} 1 & 0 \\ 0 & I \end{pmatrix}, \qquad A = \begin{pmatrix} \lambda_0 & 0 \\ 0 & A_0 \end{pmatrix}.$$

c$_1$) $\lambda_0 = \lambda_0^*$, $[x_0, x_0] = 0$ and $x_0 \notin \mathrm{ran}(A - \lambda_0)$. We choose an $y_0 \in \Pi_1$ such that $[y_0, y_0] = 0$ and $[x_0, y_0] = 1$ and represent Π_1 as $\Pi_1 = \mathrm{lin}\{x_0, y_0\} \oplus \mathcal{H}_0$. Then

$$G = \begin{pmatrix} 0 & 1 & 0 \\ 1 & 0 & 0 \\ 0 & 0 & I \end{pmatrix}, \qquad A = \begin{pmatrix} \lambda_0 & \alpha & (\,\cdot\,, a) \\ 0 & \lambda_0 & 0 \\ 0 & a & A_0 \end{pmatrix},$$

with $\alpha \in \mathbb{R}$ and $a \in \mathcal{H}_0$, such that either $a \in \overline{\mathrm{ran}(A_0 - \lambda_0)} \setminus \mathrm{ran}(A_0 - \lambda_0)$, or $a \in \mathrm{ran}(A_0 - \lambda_0)$, say $a = (A_0 - \lambda_0)\widehat{a}$, and $\alpha = ((A_0 - \lambda_0)\widehat{a}, \widehat{a})$.

c$_2$) $\lambda_0 = \lambda_0^*$, $[x_0, x_0] = 0$, $x_0 = (A - \lambda_0)x_1$ with some $x_1 \in \Pi_1$, and $\mathrm{lin}\{x_0, x_1\}$ is non-degenerated, hence $[x_1, x_1] > 0$. Choosing, if necessary, another Jordan chain $x_0' = \alpha x_0$, $x_1' = \beta x_0 + \alpha x_1$, α and β can be determined such that $[x_0', x_1'] = \delta$ with $\delta = \pm 1$ and $[x_1', x_1'] = 0$. Therefore without loss of generality we can suppose that $[x_0, x_1] = \delta$, $\delta = \pm 1$, and $[x_1, x_1] = 0$. If Π_1 is represented as $\Pi_1 = \mathrm{lin}\{x_0, x_1\} \oplus \mathcal{H}_0$, then

$$G = \begin{pmatrix} 0 & \delta & 0 \\ \delta & 0 & 0 \\ 0 & 0 & I \end{pmatrix}, \qquad A = \begin{pmatrix} \lambda_0 & 1 & 0 \\ 0 & \lambda_0 & 0 \\ 0 & 0 & A_0 \end{pmatrix}.$$

c$_2'$) $\lambda_0 = \lambda_0^*$, $[x_0, x_0] = 0$, $x_0 = (A - \lambda_0)x_1$ with some $x_1 \notin \mathrm{ran}(A - \lambda_0)$, and $\mathrm{lin}\{x_0, x_1\}$ is degenerated. Then $[x_1, x_1] > 0$, and without loss of generality we can suppose that $[x_1, x_1] = 1$. If we choose an element $y_0 \in \Pi_1$ such that

$[y_0, y_0] = 0$, $[x_0, y_0] = 1$, and Π_1 is represented as $\Pi_1 = \mathrm{lin}\{x_0, x_1, y_0\} \oplus \mathcal{H}_0$, then

$$G = \begin{pmatrix} 0 & 0 & 1 & 0 \\ 0 & 1 & 0 & 0 \\ 1 & 0 & 0 & 0 \\ 0 & 0 & 0 & I \end{pmatrix}, \qquad A = \begin{pmatrix} \lambda_0 & 1 & \alpha & (\,\cdot\,, a) \\ 0 & \lambda_0 & 1 & 0 \\ 0 & 0 & \lambda_0 & 0 \\ 0 & 0 & a & A_0 \end{pmatrix}$$

with $\alpha \in \mathbb{R}$ and $a \in \mathcal{H}_0$, such that $a \neq \mathrm{ran}(A_0 - \lambda_0)$.

c$_3$) $\lambda_0 = \lambda_0^*$, $[x_0, x_0] = 0$ and A has a Jordan chain x_0, x_1, x_2 of length three at λ_0: $x_0 = (A - \lambda_0)x_1$, $x_1 = (A - \lambda_0)x_2$. Then this chain can be chosen such that $[x_0, x_0] = [x_0, x_1] = 0$, $[x_0, x_2] = [x_1, x_1] = 1$, $[x_1, x_2] = [x_2, x_2] = 0$. If Π_1 is represented as $\Pi_1 = \mathrm{lin}\{x_0, x_1, x_2\} \oplus \mathcal{H}_0$ we have

$$G = \begin{pmatrix} 0 & 0 & 1 & 0 \\ 0 & 1 & 0 & 0 \\ 1 & 0 & 0 & 0 \\ 0 & 0 & 0 & I \end{pmatrix}, \qquad A = \begin{pmatrix} \lambda_0 & 1 & 0 & 0 \\ 0 & \lambda_0 & 1 & 0 \\ 0 & 0 & \lambda_0 & 0 \\ 0 & 0 & 0 & A_0 \end{pmatrix}.$$

In all cases, all the real spectral points $\neq \lambda_0$ of A are of positive type (see [10]). Moreover, in the cases a), b), c_2) and c$_3$) the operator A is the orthogonal sum of a Hilbert space self-adjoint operator and an at most three-dimensional operator, whereas in the cases c_1) and c_2') such a decomposition is impossible. That is, exactly in the latter two cases λ_0 is a singular critical point of A.

We also need the following lemma about normal operators in a Hilbert space.

Lemma 2.2. *Let N_0 be a normal operator in the Hilbert space $\mathcal{H}_0$ and let a, b be nonzero elements of $\mathcal{H}_0$ satisfying the relations*

$$\|a\| \leq \|b\|, \quad N_0 a = N_0^* b. \tag{2.2}$$

Then $b = \lim_{n\to\infty} N_0 \widehat{b}_n$ with some sequence $(\widehat{b}_n) \subset \mathcal{H}_0$ implies $a = \lim_{n\to\infty} N_0^* \widehat{b}_n$, and $b = N_0 \widehat{b}$ with some $\widehat{b} \in \mathcal{H}_0$ implies $a = N_0^* \widehat{b}$. Consequently,

$$b \in \overline{\mathrm{ran} N_0} \quad \Longrightarrow \quad a \in \mathrm{ran} N_0^*, \tag{2.3}$$

and

$$b \in \mathrm{ran} N_0 \quad \Longrightarrow \quad a \in \mathrm{ran} N_0^*. \tag{2.4}$$

Proof. If $b \in \overline{\mathrm{ran} N_0}$, say $b = \lim_{n\to\infty} N_0 \widehat{b}_n$, then

$$\|N_0^* (\widehat{b}_n - \widehat{b}_m)\| = \|N_0 (\widehat{b}_n - \widehat{b}_m)\| \to 0, \qquad n, m \to \infty,$$

hence the sequence $(N_0^* \widehat{b}_n)$ converges in norm to some element $g \in \mathcal{H}$. It follows that

$$N_0 a = N_0^* b = \lim N_0^* N_0 \widehat{b}_n = \lim N_0 N_0^* \widehat{b}_n = N_0 g,$$

hence $N_0(a - g) = 0$. If P_0 denotes the orthogonal projection onto $\overline{\mathrm{ran} N_0} = \overline{\mathrm{ran} N_0^*}$ then $P_0 a = g$. On the other hand,

$$\|a\| \leq \|b\| = \lim \|N_0 \widehat{b}_n\| = \lim \|N_0^* \widehat{b}_n\| = \|g\|,$$

and $P_0 a = a$ follows.

For the proof of (2.4) the sequence $(\widehat{b}_n)$ can be chosen as constant: $\widehat{b}_n = \widehat{b}$ for $n = 1, 2, \ldots$. □

Corollary 2.3. *With the notations of Lemma 2.2, if the relations*

$$\|a\| = \|b\|, \qquad N_0 a = N_0^* b.$$

hold, then

(i) $b \in \overline{\operatorname{ran} N_0} \Longleftrightarrow a \in \overline{\operatorname{ran} N_0^*}$, *and* $b = \lim N_0 \widehat{b}_n$ *for some sequence* $(\widehat{b}_n) \subset \mathcal{H}$ *is equivalent to* $a = \lim N_0^* \widehat{b}_n$;

(i) $b \in \operatorname{ran} N_0 \Longleftrightarrow a \in \operatorname{ran} N_0^*$, *and* $b = N_0 \widehat{b}$ *is equivalent to* $a = N_0^* \widehat{b}$.

3. Normal operators in Π_κ

1. According to a theorem of M.A. Naimark, any commutative family $\mathcal{A}$ of bounded self-adjoint operators in a space Π_κ has a common κ-dimensional invariant non-positive subspace, see [12]. More generally, any dual pair of subspaces, which is invariant under a commutative family $\mathcal{A}$ of bounded self-adjoint operators in a space Π_κ, can be extended to a maximal dual pair which is also invariant under the operators of $\mathcal{A}$, see [9, Folgerung 4.1]. If we apply this result to the real and imaginary part A and B of a given normal operator N:

$$A := \frac{N + N^+}{2}, \qquad B := \frac{N - N^+}{2\,\mathrm{i}}, \tag{3.1}$$

it follows that in a space Π_κ a normal operator N and its adjoint N^+ have a common non-positive κ-dimensional invariant subspace, and, more generally, any dual pair of subspaces, invariant under the normal operator N and its adjoint N^+, can be extended to a maximal dual pair which is also invariant under N and N^+. In particular, any non-positive subspace which is invariant under N and N^+ can be extended to a κ-dimensional non-positive subspace invariant under N and N^+. By $\sigma_0(N)$ we denote the set of eigenvalues of N with a non-positive eigenelement. According to the above, $\sigma_0(N) \neq \varnothing$.

Let $\mathcal{L}$ be a κ-dimensional invariant non-positive subspace for N and N^+, and denote by $\mathcal{N}$ its isotropic part: $\mathcal{N} := \mathcal{L} \cap \mathcal{L}^{[\perp]}$. We choose a complementary subspace $\mathcal{L}'$ of $\mathcal{N}$ in $\mathcal{L}$ and a complementary subspace $\mathcal{H}_0$ of $\mathcal{N}$ in $\mathcal{L}^{[\perp]}$, that is,

$$\mathcal{L} = \mathcal{L}' \,[\dotplus]\, \mathcal{N}, \quad \mathcal{L}^{[\perp]} = \mathcal{N} \,[\dotplus]\, \mathcal{H}_0.$$

Evidently, $\mathcal{L}'$ is a negative subspace and $\mathcal{H}_0$ is a Hilbert space. Then the space Π_κ can be decomposed as follows:

$$\Pi_\kappa = \mathcal{L}' \,[\dotplus]\, (\mathcal{N} \dotplus \mathcal{M}) \,[\dotplus]\, \mathcal{H}_0, \tag{3.2}$$

where $\mathcal{M}$ is a neutral subspace of Π_κ which is skewly linked with $\mathcal{N}$; the latter means that no element of $\mathcal{M}$ is orthogonal to all of $\mathcal{N}$ and no element of $\mathcal{N}$ is orthogonal to all of $\mathcal{M}$. If in $\mathcal{M}$ a basis $e_1, e_2, \dots, e_k$ and in $\mathcal{N}$ a basis $f_1, f_2, \dots, f_k$ are chosen such that

$$[e_i, f_j] = \delta_{ij}, \quad i, j = 1, 2, \dots, k,$$

then the Gram operator for the inner product $[\cdot\,,\cdot]$ and the decomposition (3.2) of Π_κ becomes

$$G = \begin{pmatrix} I & 0 & 0 & 0 \\ 0 & 0 & I & 0 \\ 0 & I & 0 & 0 \\ 0 & 0 & 0 & I \end{pmatrix}.$$

The following theorem is an easy consequence of the fact that the subspace $\mathcal{L}' \, [+] \, \mathcal{N}$ is invariant under N and N^+ and of the normality of N.

Theorem 3.1. *With respect to the decomposition* (3.2) *of the space* Π_κ *the operators* N *and* N^+ *admit the matrix representations*

$$N = \begin{pmatrix} N_{11} & 0 & N_{13} & 0 \\ N_{21} & N_{22} & N_{23} & N_{24} \\ 0 & 0 & N_{33} & 0 \\ 0 & 0 & N_{43} & N_0 \end{pmatrix}, \quad N^+ = \begin{pmatrix} N_{11}^+ & 0 & N_{21}^+ & 0 \\ N_{13}^+ & N_{33}^+ & N_{23}^+ & N_{43}^+ \\ 0 & 0 & N_{22}^+ & 0 \\ 0 & 0 & N_{24}^+ & N_0^* \end{pmatrix}, \tag{3.3}$$

where the entries satisfy the following relations:

$$N_0 N_0^* = N_0^* N_0, \quad N_{11} N_{11}^+ = N_{11}^+ N_{11}, \quad N_{22} N_{33}^+ = N_{33}^+ N_{22},$$

$$N_{11} N_{21}^+ + N_{13} N_{22}^+ = N_{11}^+ N_{13} + N_{21}^+ N_{33}, \quad N_{22} N_{43}^+ + N_{24} N_0^* = N_{33}^+ N_{24} + N_{43}^+ N_0,$$

$$N_{21} N_{21}^+ + N_{22} N_{23}^+ + N_{23} N_{22}^+ + N_{24} N_{24}^+ = N_{13}^+ N_{13} + N_{33}^+ N_{33} + N_{23}^+ N_{33} + N_{43}^+ N_{43}.$$

Here we write $^+$ for the adjoint operators with respect to the inner product $[\cdot\,,\cdot]$ in $\mathcal{L}'$ and also in and between the other subspaces on the right-hand side of (3.2), only for N_0 we write * that it becomes apparent that this is a Hilbert space adjoint. Observe that in (3.3) N_0 is a normal operator in the Hilbert space $\mathcal{H}_0$, all the other operators in the matrices on the right-hand sides are finite-dimensional. Evidently,

$$\sigma(N) = \sigma(N_0) \cup \sigma(N_{11}) \cup \sigma(N_{22}) \cup \sigma(N_{33}), \quad \sigma_0(N) = \sigma(N_{11}) \cup \sigma(N_{22}) \cup \sigma(N_{33}).$$

2. Some invariant subspaces of N are automatically invariant under N^+. This holds for example for a one-dimensional geometric eigenspace of N, since $Nx_0 = \lambda_0 x_0$ implies $N^+ N x_0 = N N^+ x_0 = \lambda_0 N^+ x_0$ and, because the eigenspace is one-dimensional, $N^+ x_0$ must be a multiple of x_0. The following theorem is another result in this direction.

Theorem 3.2. *If the normal operator* N *in* Π_κ *has a* κ*-dimensional invariant negative subspace* $\mathcal{L}$ *then* $\mathcal{L}$ *is also invariant under* N^+*, and* N *is the orthogonal sum of a normal operator in the Hilbert space* $(\mathcal{L}^{[\perp]}, [\cdot\,,\cdot])$ *and a normal operator in the finite-dimensional Hilbert space* $(\mathcal{L}, -[\cdot\,,\cdot])$.

Proof. Since $\dim \mathcal{L} = \kappa$, $\mathcal{L}^{[\perp]}$ is a Hilbert space; we denote it again by $\mathcal{H}_0$. Choosing a proper orthonormal basis in $\mathcal{L}$ we can suppose that with respect to the decomposition $\Pi_\kappa = \mathcal{L} \oplus \mathcal{H}_0$ the operator N and the Gram operator G have the matrix representations

$$N = \begin{pmatrix} \lambda_1 & \alpha_{12} & \cdots & \alpha_{1\kappa} & (\,\cdot\,, a_1) \\ 0 & \lambda_2 & \cdots & \alpha_{2\kappa} & (\,\cdot\,, a_2) \\ \vdots & \vdots & \ddots & \vdots & \vdots \\ 0 & 0 & \cdots & \lambda_\kappa & (\,\cdot\,, a_\kappa) \\ 0 & 0 & \cdots & 0 & N_0 \end{pmatrix}, \qquad G = \begin{pmatrix} -1 & 0 & \cdots & 0 & 0 \\ 0 & -1 & \cdots & 0 & 0 \\ \vdots & \vdots & \ddots & \vdots & \vdots \\ 0 & 0 & \cdots & -1 & 0 \\ 0 & 0 & \cdots & 0 & I \end{pmatrix}.$$

It follows easily that

$$N^+ = \begin{pmatrix} \lambda_1^* & 0 & \cdots & 0 & 0 \\ \alpha_{12}^* & \lambda_2^* & \cdots & 0 & 0 \\ \vdots & \vdots & \ddots & \vdots & \vdots \\ \alpha_{1\kappa}^* & \alpha_{2\kappa}^* & \cdots & \lambda_\kappa^* & \\ -a_1 & -a_2 & \cdots & -a_\kappa & N_0^* \end{pmatrix}.$$

Comparing in the relation

$$NN^+ = N^+N \tag{3.4}$$

the entries with index κ, κ we obtain

$$|\lambda_\kappa|^2 - \|a_\kappa\|^2 = \sum_1^{\kappa-1} |\alpha_{j\kappa}|^2 + |\lambda_\kappa|^2,$$

which yields

$$a_\kappa = 0, \quad \alpha_{1\kappa} = \cdots = \alpha_{\kappa-1,\kappa} = 0.$$

Considering in the same way the entries with last index κ–1 etc, it follows that all the α_{jk} and all the elements a_j are zero. □

Since in the sequel we are mainly interested in the case $\kappa = 1$ we also formulate the following special case.

Corollary 3.3. *If the normal operator N in Π_1 has an eigenvalue λ_0 with a negative eigenelement x_0:*

$$Nx_0 = \lambda_0 x_0, \qquad [x_0, x_0] < 0,$$

then also $N^+x_0 = \lambda_0^* x_0$, and with respect to the decomposition $\Pi_1 = \mathrm{lin}\{x_0\} \oplus \mathcal{H}_0$ we have

$$G = \begin{pmatrix} -1 & 0 \\ 0 & I \end{pmatrix}$$

and

$$N = \begin{pmatrix} \lambda_0 & 0 \\ 0 & N_0 \end{pmatrix}, \qquad N^+ = \begin{pmatrix} \lambda_0^* & 0 \\ 0 & N_0^* \end{pmatrix} \tag{3.5}$$

with a normal operator N_0 in the Hilbert space $\mathcal{H}_0$.

We mention that also a neutral eigenelement of N in Π_1 at λ_0 is an eigenelement of N^+ but not necessarily at λ_0^*, see the first step in the proof of Theorem 4.1 or Corollary 3.5 below. However, a positive eigenelement of N is not necessarily an eigenelement of N^+, see Remark 5.7 below.

For a space Π_κ with general $\kappa \geq 1$ we have the following result.[2]

Theorem 3.4. *If $\mathcal{L}$ is a neutral subspace of Π_κ which is invariant under N then there exists a neutral subspace $\widetilde{\mathcal{L}} \supset \mathcal{L}$ which is invariant under N and N^+.*

Proof. Consider the subspace

$$\widetilde{\mathcal{L}} := \operatorname{lin}\{\mathcal{L},\, N^+\mathcal{L},\, (N^+)^2\mathcal{L},\, \dots\}.$$

For $x, y \in \mathcal{L},\ j, \ell \in \mathbb{N}_0$ and $x' := N^j x,\ y' := N^j y \in \mathcal{L}$ we have

$$\left[(N^+)^{j+\ell}x, (N^+)^j y\right] = \left[N^j (N^+)^\ell x, N^j y\right] = \left[(N^+)^\ell x', y'\right] = \left[x', N^\ell y'\right] = 0,$$

hence $\widetilde{\mathcal{L}}$ is neutral. Clearly, $\dim \widetilde{\mathcal{L}} \leq \kappa$ and $\widetilde{\mathcal{L}}$ is invariant under N and N^+. □

With the subspace $\widetilde{\mathcal{L}}$ from Theorem 3.4, the dual pair $\{\widetilde{\mathcal{L}}, \widetilde{\mathcal{L}}\}$ is invariant under N and N^+. According to [9, Folgerung 4.1], it can be extended to a maximal dual pair which is also invariant under N and N^+, and we have proved the following corollary.

Corollary 3.5. *If $\mathcal{L}$ is a neutral subspace of Π_κ which is invariant under N, then there exists a maximal dual pair $\{\mathcal{L}_+^{max}, \mathcal{L}_-^{max}\}$ which is invariant under N and N^+ and such that $\mathcal{L} \subset \left(\mathcal{L}_+^{max} \cap \mathcal{L}_-^{max}\right)$.*

3. The non-real spectrum of a self-adjoint operator A in Π_κ consists of at most κ pairs of eigenvalues which lie symmetrically with respect to the real axis, the algebraic eigenspaces corresponding to such a pair are skewly linked and hence the sum of these eigenspaces is a non-degenerate subspace of Π_κ. We denote for a self-adjoint operator A by $\mathcal{L}(A)$ the linear span of all the algebraic eigenspaces of A corresponding to the non-real eigenvalues of A. With the self-adjoint operators A and B in (3.1), we set

$$\mathcal{L}(N) := \operatorname{lin}\{\mathcal{L}(A), \mathcal{L}(B)\}.$$

The subspace $\mathcal{L}(N)$ can also be obtained as follows: With $\mathcal{L}(A)$ we represent the space Π_κ as

$$\Pi_\kappa = \mathcal{L}(A)\,[\dot{+}]\,\Pi'_{\kappa'}.$$

Then also the operator B decomposes accordingly: $B = B_0\,[\dot{+}]\,B'$ with a self-adjoint operator B' in $\Pi'_{\kappa'}$, and $\mathcal{L}(N) = \mathcal{L}(A)\,[\dot{+}]\,\mathcal{L}(B')$, which, as the orthogonal sum of non-degenerated subspaces is a non-degenerated subspace.

We decompose the space Π_κ as

$$\Pi_\kappa = \mathcal{L}(N)\,[\dot{+}]\,\Pi^1_{\kappa_1}. \tag{3.6}$$

[2]Theorem 3.4 and Corollary 3.5 were added in August 2004, after L. Rodman pointed out to one of the authors the results from [11].

If the index of negativity of $\mathcal{L}(N)$ is denoted by κ_0, then $\dim \mathcal{L}(N) = 2\kappa_0$ and $\kappa_1 = \kappa - \kappa_0$, the decomposition (3.6) reduces the operators N and N^+:

$$N = N^0 \,[\dot{+}]\, N_1, \qquad N^+ = (N^0)^+ \,[\dot{+}]\, N_1^+,$$

and the real and imaginary parts A_1 and B_1 of N_1 do not have non-real spectrum. We shall call the normal operator N in Π_κ *reduced* if its real and imaginary parts A and B do not have non-real spectrum. The above consideration shows that any normal operator in Π_κ is the direct sum of a reduced normal operator and a normal operator in a $2\kappa_1$-dimensional space with index of negativity κ_1.

We mention that the decomposition (3.2) can be chosen such that $\mathcal{L}(N) \subset \mathcal{N} \dot{+} \mathcal{M}$.

Now let N be a reduced normal operator in Π_κ. By A and B we denote its real and imaginary part, see (3.1), by E_A and E_B the spectral functions of A and B, respectively, and $\mathcal{R}_N$ is the semi-ring generated by all finite closed rectangles Δ in the complex plane with sides parallel to the coordinate axes and such that their boundaries do not intersect $\sigma_0(N)$. If Δ is such a rectangle,

$$\Delta = \{z : z = x + \mathrm{i}\, y,\ \alpha \le x \le \beta,\ \gamma \le y \le \delta\},$$

we define

$$E(\Delta) := E_A([\alpha,\beta])\, E_B([\gamma,\delta])\,. \tag{3.7}$$

Evidently, since the spectral functions E_A and E_B commute, $E(\Delta)$ is a self-adjoint projection in Π_κ.

Theorem 3.6. *If N is a reduced normal operator in a Pontryagin space Π_κ then the mapping E from the semiring $\mathcal{R}_N$ into the set of self-adjoint projections in Π_κ defined by* (3.7) *has the following properties* ($\Delta, \Delta' \in \mathcal{R}_N$):

(i) $E(\varnothing) = 0,\ E(\Delta) = I$ if $\sigma(N) \subset \Delta$.
(ii) $E(\Delta \cap \Delta') = E(\Delta)E(\Delta')$.
(iii) $E(\Delta \cup \Delta') = E(\Delta) + E(\Delta')$ if $\Delta \cup \Delta' \in \mathcal{R}_N,\quad \Delta \cap \Delta' = \varnothing$.
(iv) *if* $\Delta \cap \sigma_0(N) = \varnothing$ *then* $E(\Delta)\Pi_\kappa$ *is a positive subspace.*
(v) $E(\Delta)N = NE(\Delta)$.
(vi) $\sigma(N|E(\Delta)\Pi_\kappa) \subset \overline{\Delta}$.

Denote by $\mathcal{S}$ the σ-algebra generated by $\mathcal{R}_N$. It is not hard to show that the homomorphism $\Delta \mapsto E(\Delta)$ in Theorem 3.6 can be extended to all elements Δ of $\mathcal{S}$ such that the boundary of Δ does not contain points of $\sigma_0(N)$. As for a self-adjoint operator in Π_κ, a point $\lambda_0 \in \sigma_0(N)$ is called a *critical point* of N if for each $\Delta \in \mathcal{R}_N$ with $\lambda_0 \in \Delta$ the range $E(\Delta)\Pi_\kappa$ contains positive as well as negative elements. The critical point λ_0 of N is called a *regular critical point* if the norms of all the projections $E(\Delta)$ are uniformly bounded for all $\Delta \in \mathcal{R}_N$ in some neighborhood of λ_0, otherwise λ_0 is called a *singular critical point* of N. These definitions imply immediately that λ_0 is a critical point (a singular critical point, respectively) of N if and only if λ_0^* is a critical point (a singular critical point, respectively) of N^+. They yield also the following proposition.

Proposition 3.7. *Let $N = A + \mathrm{i}B$ be a reduced normal operator in a Pontryagin space Π_κ. Then $\lambda_0 = \mu_0 + \mathrm{i}\,\nu_0$, $\mu_0, \nu_0 \in \mathbb{R}$, is a critical point of N if and only if μ_0 is a critical point of A and ν_0 is a critical point of B; λ_0 is a singular critical point on N if and only if μ_0 is a singular critical point of A or ν_0 is a singular critical point of B.*

We mention that as in the self-adjoint case the algebraic eigenspace $\mathcal{L}_{\lambda_0}(N)$ of N at λ_0 can still be characterized by the spectral function of N:

$$\mathcal{L}_{\lambda_0}(N) = \bigcap_{\Delta \in \mathcal{R}(N), \lambda_0 \in \Delta} E(\Delta)\Pi_\kappa. \tag{3.8}$$

Moreover, also the analogue of Lemma 2.1 remains valid, that is the critical point λ_0 of N is a regular regular critical point if and only if $\mathcal{L}_{\lambda_0}(N)$ is non-degenerated.

Evidently, with the normal operator N also the normal operator N^+ is reduced. If we denote the spectral function of N^+ by E_+, then we have

$$E_+(\Delta) = E(\Delta^*),$$

where $\Delta^* := \{z^* | z \in \Delta\}$, and the representation of $\mathcal{L}_{\lambda_0}(N)$ by (3.8) implies that

$$\mathcal{L}_{\lambda_0}(N) = \mathcal{L}_{\lambda_0^*}(N^+).$$

4. Normal operators in Π_1: first reduction

In the following we consider only the case $\kappa = 1$.

Theorem 4.1. *If the normal operator N in Π_1 has an eigenvalue λ_0 with a neutral eigenelement x_0:*

$$Nx_0 = \lambda_0 x_0, \qquad [x_0, x_0] = 0, \tag{4.1}$$

then the following alternative holds:
either

(i) $N^+ x_0 = \lambda_0^* x_0$

or

(ii) *there exist $\mu_0 \in \mathbb{C}$ and $y_0 \in \Pi_1$ such that*

$$N^+ y_0 = \mu_0 y_0, \qquad [y_0, y_0] = 0, \quad [x_0, y_0] = 1.$$

With respect to the decomposition $\Pi_1 = \mathrm{lin}\{x_0,\, y_0\} \oplus \mathcal{H}_0$ the Gram operator is

$$G = \begin{pmatrix} 0 & 1 & 0 \\ 1 & 0 & 0 \\ 0 & 0 & I \end{pmatrix},$$

and N and N^+ admit the matrix representations

$$N = \begin{pmatrix} \lambda_0 & 0 & 0 \\ 0 & \mu_0^* & 0 \\ 0 & 0 & N_0 \end{pmatrix}, \qquad N^+ = \begin{pmatrix} \mu_0 & 0 & 0 \\ 0 & \lambda_0^* & 0 \\ 0 & 0 & N_0^* \end{pmatrix} \tag{4.2}$$

with a normal operator N_0 in the Hilbert space $\mathcal{H}_0$.

Proof. Since $[x_0, x_0] = 0$, $[N^+x_0, x_0] = [x_0, Nx_0] = \lambda_0^*[x_0, x_0] = 0$, and

$$[N^+x_0, N^+x_0] = [Nx_0, Nx_0] = |\lambda_0|^2[x_0, x_0] = 0,$$

the subspace $\operatorname{lin}\{x_0, N^+x_0\}$ is neutral. The assumption $\kappa = 1$ implies that this linear span is one-dimensional, hence

$$N^+x_0 = \gamma x_0 \tag{4.3}$$

for some $\gamma \in \mathbb{C}$. If $\gamma = \lambda_0^*$ we arrive at case (i).

If $\gamma \neq \lambda_0^*$ we proceed as follows. For the self-adjoint operators A and B in (3.1) we obtain from (4.1) and (4.3)

$$Ax_0 = \alpha x_0, \quad Bx_0 = \beta x_0 \tag{4.4}$$

with $\alpha = \frac{1}{2}(\lambda_0 + \gamma)$ and $\beta = \frac{1}{2\,\mathrm{i}}(\lambda_0 - \gamma)$. Notice that either α or β is not real since $\gamma \neq \lambda_0^*$. If, for example, $\mathfrak{Im}\,\alpha \neq 0$, then also α^* is an eigenvalue of A: $Ay_0 = \alpha^* y_0$ for some y_0 with $[y_0, y_0] = 0$ and $[x_0, y_0] = 1$. Since A commutes with B and the eigenspace of A corresponding to α^* is one-dimensional, it follows that $By_0 = \delta y_0$ for some $\delta \in \mathbb{C}$. Consequently, $Ny_0 = (\alpha^* + \mathrm{i}\,\delta)y_0$ and $N^+y_0 = \mu_0 y_0$ with $\mu_0 = \alpha^* - \mathrm{i}\,\delta$, which implies the first statement of (ii). Moreover, $N^+x_0 = (\alpha - \mathrm{i}\,\beta)x_0$, and since $Nx_0 = \lambda_0 x_0$ by assumption, we conclude that $\operatorname{lin}\{x_0, y_0\}$ is invariant for both N and N^+. Now the second part of (ii) follows easily. □

Under the assumptions of Corollary 3.3 and Theorem 4.1, (ii), the operator N has been described completely in (3.5) and (4.2). It remains to study the situation of Theorem 4.1, (i). Considering $N - \lambda_0$ instead of N, without loss of generality we can suppose that $\lambda_0 = 0$, that is, zero is an eigenvalue of N and N^+ with a common neutral eigenelement e_0:

$$Ne_0 = 0, \qquad N^+e_0 = 0, \qquad [e_0, e_0] = 0. \tag{4.5}$$

Moreover, we can assume that e_0 is (up to constant multiples) the only non-positive eigenelement of N at zero, since otherwise there would exist also a negative eigenelement for N at zero, and we would be in the situation of Corollary 3.3.

We shall study this case (4.5) in the following section. In the rest of the present section we establish some more general properties of a normal operator in Π_1.

Theorem 4.2. *Let N be a bounded normal operator in Π_1. If there are elements e_0, $e_1 \in \Pi_1$ such that $Ne_0 = 0$, $Ne_1 = e_0$ and $N^+e_0 = 0$, then $[e_0, e_0] = 0$ and $N^+e_1 = \alpha e_0$ for some $\alpha \in \mathbb{C}$.*

Proof. Indeed, $[e_0, e_0] = [Ne_1, e_0] = [e_1, N^+e_0] = 0$. Furthermore, we have

$$[N^+e_1, e_0] = [e_1, Ne_0] = 0, \quad [N^+e_1, N^+e_1] = [Ne_1, Ne_1] = [e_0, e_0] = 0.$$

Thus the subspace $\text{lin}\{e_0, N^+e_1\}$ is neutral. Since Π_1 has one negative square, it has to be of dimension 1, so $N^+e_1 = \alpha e_0$ for some $\alpha \in \mathbb{C}$. □

Examples show that $\alpha = 0$ is possible (see Section 5), which means that the Jordan chain e_0, e_1 of N belongs to $\ker N^+$.

Theorem 4.3. *A normal operator N in a Pontryagin space Π_1 cannot have a Jordan chain of length > 3. If N has a Jordan chain of length 3 at zero, say $Ne_0 = 0$, $Ne_1 = e_0$, $Ne_2 = e_1$, then the number α in Theorem 4.2 is $\neq 0$ and*

$$[e_0, e_0] = [e_0, e_1] = 0.$$

Moreover, this Jordan chain can be chosen such that also

$$[e_1, e_2] = [e_2, e_2] = 0, \quad [e_1, e_1] = 1, \ [e_0, e_2] =: \gamma \neq 0. \tag{4.6}$$

Proof. Evidently, in the situation of (ii) of Theorem 4.1 the operator N cannot have a Jordan chain of length greater than one. Therefore we can suppose that the eigenvalue with the considered Jordan chain is $\lambda_0 = 0$, and that (4.5) holds. If the elements e_0, e_1, e_2 belong to a chain of N of length ≥ 3 then $[e_0, e_1] = [e_0, Ne_2] = [N^+e_0, e_2] = 0$. If the chain could be continued with an element e_3 such that $Ne_3 = e_2$ we would find, using Theorem 4.2,

$$[e_1, e_1] = [e_1, N^2e_3] = [(N^+)^2e_1, e_3] = [N^+\alpha e_0, e_3] = 0.$$

Consequently, $\text{lin}\{e_0, e_1\}$ is a 2-dimensional neutral subspace, a contradiction.

Now suppose that we have a chain e_0, e_1, e_2 of length three. Since $[e_0,e_0] = [e_0,e_1] = 0$, $[e_1,e_1]$ must be positive and we can suppose that $[e_1,e_1] = 1$. The following relation implies that the number α in Lemma 4.2 and also $[e_0, e_2]$ are $\neq 0$:

$$1 = [e_1, e_1] = [e_1, Ne_2] = [N^+e_1, e_2] = \alpha[e_0, e_2].$$

If not all the relations in (4.6) are satisfied then we determine η and ζ such that the new chain

$$\widehat{e}_0 = e_0, \ \widehat{e}_1 = e_1 + \eta e_0, \ \widehat{e}_2 = e_2 + \eta e_1 + \zeta e_0$$

has all the desired properties. □

5. The case of a common neutral eigenelement of N and N^+

1. In this subsection we consider the case that (4.5) holds and that there are no associated elements to the eigenelement e_0. The inner product $[\,\cdot\,,\cdot\,]$ on $\mathcal{H}_0$, since it is a Hilbert inner product, is denoted by $(\,\cdot\,,\cdot\,)$.

Theorem 5.1. *Suppose that $e_0 \neq 0$ satisfies (4.5), that it (with its nonzero scalar multiples) is the unique element with these properties and that $e_0 \notin \text{ran}\, N$. Then also $e_0 \notin \text{ran}\, N^+$. If we choose an arbitrary element $f_0 \in \Pi_1$ such that $[e_0, f_0] = 1$,*

$[f_0, f_0] = 0$ and decompose the space Π_1 as $\Pi_1 = \text{lin}\{e_0, f_0\} \oplus \mathcal{H}_0$ with a Hilbert space $\mathcal{H}_0$, then the corresponding Gram operator is

$$G = \begin{pmatrix} 0 & 1 & 0 \\ 1 & 0 & 0 \\ 0 & 0 & I \end{pmatrix}, \tag{5.1}$$

and

$$N = \begin{pmatrix} 0 & \beta & (\cdot\,, a) \\ 0 & 0 & 0 \\ 0 & b & N_0 \end{pmatrix}, \qquad N^+ = \begin{pmatrix} 0 & \beta^* & (\cdot\,, b) \\ 0 & 0 & 0 \\ 0 & a & N_0^* \end{pmatrix}. \tag{5.2}$$

Here $\beta \in \mathbb{C}$, N_0 is a normal operator in the Hilbert space $\mathcal{H}_0$, and a, b are nonzero elements of $\mathcal{H}_0$ such that

$$\|a\| = \|b\|, \qquad N_0 a = N_0^* b, \tag{5.3}$$

and the following holds:
either

1. *$b \notin \text{ran} N_0$ and $a \in \overline{\text{ran} N_0^*}$,*

which is equivalent to

(1′) *$a \notin \text{ran} N_0^*$ and $b \in \overline{\text{ran} N_0}$,*

or

(2) *$b \in \text{ran} N_0$, say $b = N_0 \widehat{b}$, and $\beta = (N_0^* \widehat{b}, \widehat{b})$,*

which is equivalent to

(2′) *$a \in \text{ran} N_0^*$, $a = N_0^* \widehat{b}$, and $\beta = (N_0^* \widehat{b}, \widehat{b})$.*

Proof. With the element f_0 chosen as in the theorem the form of G (5.1) is clear. Since $e_0 \in \ker N$, the matrix representation of N is of the form

$$N = \begin{pmatrix} 0 & \beta & (\cdot\,, a) \\ 0 & \gamma & (\cdot\,, c) \\ 0 & b & N_0 \end{pmatrix}.$$

Further, $N^+ = G^{-1} N^* G = G N^* G$ and hence

$$N^+ = \begin{pmatrix} \gamma^* & \beta^* & (\cdot\,, b) \\ 0 & 0 & 0 \\ c & a & N_0^* \end{pmatrix}.$$

Since $e_0 \in \ker N^+$ we find $\gamma = 0$, $c = 0$, and (5.2) is proved. Now $NN^+ = N^+ N$ is equivalent to the normality of N_0 in $\mathcal{H}_0$ and the relations (2.2) to hold. The fact that $e_0 \neq \text{ran} N$ means that there do not exist ξ, $\eta \in \mathbb{C}$, $x \in \mathcal{H}_0$, such that

$$N \begin{pmatrix} \xi \\ \eta \\ x \end{pmatrix} = \begin{pmatrix} 0 & \beta & (\cdot\,, a) \\ 0 & 0 & 0 \\ 0 & b & N_0 \end{pmatrix} \begin{pmatrix} \xi \\ \eta \\ x \end{pmatrix} = \begin{pmatrix} 1 \\ 0 \\ 0 \end{pmatrix},$$

or that

$$\begin{aligned} \beta\eta \;+\; (x, a) &= 1, \\ \eta b \;+\; N_0 x &= 0. \end{aligned}$$

This takes place if and only if either $b \notin \operatorname{ran} N_0$, $a \in \overline{\operatorname{ran} N_0^*}$, or, if $b = N_0\widehat{b}$ with some $\widehat{b} \in \mathcal{H}_0$, for $x = -\eta\widehat{b} + a'$ with $a' \in \ker N_0$ the equation

$$\beta\,\eta + (-\eta\widehat{b} + a', b) = 1$$

does not have a solution. Since $(a', b) = 0$ the latter is equivalent to

$$\beta - (\widehat{b}, b) = \beta - (\widehat{b}, N_0\widehat{b}) = 0.$$

It remains to apply Lemma 2.2. □

Remark 5.2. Under the conditions of Theorem 5.1 the operators A, B from (3.1) have the following matrix representations:

$$A = \begin{pmatrix} 0 & \mathfrak{Re}\,\beta & (\,\cdot\,, \frac{a+b}{2}) \\ 0 & 0 & 0 \\ 0 & \frac{a+b}{2} & A_0 \end{pmatrix}, \qquad B = \begin{pmatrix} 0 & \mathfrak{Im}\,\beta & (\,\cdot\,, \frac{a-b}{2\,\mathrm{i}}) \\ 0 & 0 & 0 \\ 0 & \frac{a-b}{2\,\mathrm{i}} & B_0 \end{pmatrix}.$$

For at least one of these operators A, B, zero is an eigenvalue with a neutral eigenelement which does not belong to the range of A or B, respectively, that is, at least one of these operators is in the case c_1) of Section 2. Moreover, neither A nor B can have a degenerated chain of length 2 at zero.

Indeed, in the case (1) of Theorem 5.1 we have

$$a \in \overline{\operatorname{ran} N_0^*} \setminus \operatorname{ran} N_0^*, \quad b \in \overline{\operatorname{ran} N_0} \setminus \operatorname{ran} N_0. \tag{5.4}$$

According to Lemma 2.2 and Corollary 2.3 there exists a sequence $(\widehat{b}_n)$ such that

$$b = \lim N_0\widehat{b}_n, \quad a = \lim N_0^*\widehat{b}_n, \tag{5.5}$$

and hence

$$a \pm b = \lim\,(N_0^* \pm N_0)\widehat{b}_n,$$

such that

$$a + b \in \overline{\operatorname{ran} A_0}, \ a - b \in \overline{\operatorname{ran} B_0}.$$

If we would have $a + b \in \operatorname{ran} A_0$ and $a - b \in \operatorname{ran} B_0$, say $a + b = 2A_0u$, $a - b = -2\,\mathrm{i}\,B_0v$ then we would also have $A_0u = \lim 2A_0\widehat{b}_n$, $B_0v = \lim(-2)B_0\widehat{b}_n$, and hence $A_0B_0(u - v) = 0$.

We decompose the space $\mathcal{H}_0$ as $\mathcal{H}_0 = \mathcal{R}_0 \oplus \mathcal{A}_0 \oplus \mathcal{B}_0 \oplus \mathcal{N}_0$ where

$\mathcal{R}_0 := \overline{\operatorname{ran} A_0} \cap \overline{\operatorname{ran} B_0}$, $\mathcal{N}_0 := \ker A_0 \cap \ker B_0$, $\ker A_0 := \mathcal{A}_0 \oplus \mathcal{N}_0$, $\ker B_0 := \mathcal{B}_0 \oplus \mathcal{N}_0$.

With respect to this decomposition we can choose

$$u = w + u_0, \quad v = w + v_0, \quad \text{with } w \in \mathcal{R}_0,\ u_0 \in \mathcal{B}_0,\ v_0 \in \mathcal{A}_0,$$

which implies for $\widetilde{w} := w + u_0 + v_0$:

$$A_0\widetilde{w} = A_0u, \qquad B_0\widetilde{w} = B_0v,$$

and hence $a \in \operatorname{ran} N_0^*$, $b \in \operatorname{ran} N_0$, which is not the case because of (5.4).

Similarly, in the case (2) of Theorem 5.1 it follows, e.g., that

$$\frac{a+b}{2} = A_0\widehat{b}, \quad \mathfrak{Re}\,\beta = (A_0\widehat{b}, \widehat{b}),$$

hence at least one of the operators A or B is in the situation of c_1). If, e.g., for A there would exist an element $e_1 \in \Pi_1$ such that $[e_1, e_0] = 0$ and $Ae_1 = e_0$ then e_1 could be chosen of the form $e_1 = \begin{pmatrix} 0 \\ 0 \\ y \end{pmatrix}$ with $y \in \ker A_0$. Since on the other hand $a + b \in \overline{\operatorname{ran} A_0}$ the relation

$$Ae_1 = \begin{pmatrix} 0 & \mathfrak{Re}\,\beta & (\cdot\,, \frac{a+b}{2}) \\ 0 & 0 & 0 \\ 0 & \frac{a+b}{2} & A_0 \end{pmatrix} \begin{pmatrix} 0 \\ 0 \\ y \end{pmatrix} = \begin{pmatrix} 1 \\ 0 \\ 0 \end{pmatrix}$$

cannot hold. However, one of the operators A or B can have a non-degenerated chain of length 2 at zero. This can easily seen to hold for the operator

$$N = \begin{pmatrix} 0 & \beta & (\cdot\,, a) \\ 0 & 0 & 0 \\ 0 & a & A_0 \end{pmatrix}$$

with a self-adjoint operator A_0 in $\mathcal{H}_0$, $a \in \mathcal{H}_0$ with $a \in \overline{\operatorname{ran} A_0} \setminus \operatorname{ran} A_0$ and $\beta \neq \beta^*$.

2. In the following theorem we consider the case that to e_0 there corresponds a non-degenerated Jordan chain of length two.

Theorem 5.3. *Suppose that $e_0 \neq 0$ satisfies (4.5), that it (with its nonzero scalar multiples) is the unique element with these properties, that there exists an associated element e_1 such that $Ne_1 = e_0$ and $\delta := [e_0, e_1] \neq 0$. Then $e_1 \notin \operatorname{ran} N$, and without loss of generality we can assume that $|\delta| = 1$ and $[e_1, e_1] = 0$. The space Π_1 can be decomposed as $\Pi_1 = \operatorname{lin}\{e_0, e_1\} \oplus \mathcal{H}_0$ with a Hilbert space $\mathcal{H}_0$, the corresponding Gram operator is*

$$G = \begin{pmatrix} 0 & \delta^* & 0 \\ \delta & 0 & 0 \\ 0 & 0 & I \end{pmatrix},$$

and

$$N = \begin{pmatrix} 0 & 1 & 0 \\ 0 & 0 & 0 \\ 0 & 0 & N_0 \end{pmatrix}, \qquad N^+ = \begin{pmatrix} 0 & \delta^{*2} & 0 \\ 0 & 0 & 0 \\ 0 & 0 & N_0^* \end{pmatrix} \tag{5.6}$$

with a normal operator N_0 in the Hilbert space $\mathcal{H}_0$. In particular, the elements e_0, $\delta^2 e_1$ form a Jordan chain of N^+ at zero.

Proof. If e_1 would belong to $\operatorname{ran} N$: $e_1 = Ne_2$, then from Theorem 4.3 it would follow that $[e_0, e_1] = 0$, which is impossible according to the assumption. The reduction to the case $|\delta| = 1$ and $[e_1, e_1] = 0$ is by changing the Jordan chain as in the reasoning in c_2) of Section 2.

Since $Ne_0 = 0$, $Ne_1 = e_0$, it follows that

$$N = \begin{pmatrix} 0 & 1 & (\cdot\, a) \\ 0 & 0 & (\cdot\,, b) \\ 0 & 0 & N_0 \end{pmatrix}$$

with elements $a, b \in \mathcal{H}_0$. Therefore

$$N^+ = GN^*G = \begin{pmatrix} 0 & \delta^{*2} & 0 \\ 0 & 0 & 0 \\ b\delta^* & a\delta^* & N_0 \end{pmatrix},$$

and $N^+e_0 = 0$, $N^+e_1 = \alpha e_0$ (see Theorem 4.2) imply $a = b = 0$. □

Remark 5.4. Under the assumptions of Theorem 5.3 the operators A and B of (3.1) have the matrix representation

$$A = \begin{pmatrix} 0 & \frac{1+\delta^{*2}}{2} & 0 \\ 0 & 0 & 0 \\ 0 & 0 & A_0 \end{pmatrix}, \qquad B = \begin{pmatrix} 0 & \frac{1-\delta^{*2}}{2} & 0 \\ 0 & 0 & 0 \\ 0 & 0 & B_0 \end{pmatrix}.$$

Thus, A and B have non-degenerated algebraic eigenspaces at zero with Jordan chains of length at most 2, and at least one of the operators A, B has a non-degenerated chain of length 2 at zero, that is, it is in the situation c_2) of Section 2.

3. It remains to consider the case that for the element e_0 in (4.5) there exists an associated element e_1: $Ne_1 = e_0$, such that $[e_0, e_1] = 0$. Then, since the subspace $\operatorname{lin}\{e_0, e_1\}$ can have only one non-negative square, $[e_1, e_1] > 0$, and without loss of generality we suppose that $[e_1, e_1] = 1$.

We choose an element f_0 such that

$$[e_0, f_0] = 1, \quad [e_1, f_0] = 0, \quad [f_0, f_0] = 0, \tag{5.7}$$

and decompose the space Π_1 as

$$\Pi_1 = \operatorname{lin}\{e_0, e_1, f_0\} \oplus \mathcal{H}_0 \tag{5.8}$$

with a Hilbert space $\mathcal{H}_0$. The corresponding Gram operator is

$$G = \begin{pmatrix} 0 & 0 & 1 & 0 \\ 0 & 1 & 0 & 0 \\ 1 & 0 & 0 & 0 \\ 0 & 0 & 0 & I \end{pmatrix}. \tag{5.9}$$

The fact that e_0, e_1 form a Jordan chain of N and αe_0, e_1 form a Jordan chain of N^+ at zero and the relation $NN^+ = N^+N$ easily lead to the following representations of the operators N and N^+:

$$N = \begin{pmatrix} 0 & 1 & \beta & (\,\cdot\,, a) \\ 0 & 0 & \alpha^* & 0 \\ 0 & 0 & 0 & 0 \\ 0 & 0 & b & N_0 \end{pmatrix}, \qquad N^+ = \begin{pmatrix} 0 & \alpha & \beta^* & (\,\cdot\,, b) \\ 0 & 0 & 1 & 0 \\ 0 & 0 & 0 & 0 \\ 0 & 0 & a & N_0^* \end{pmatrix}; \tag{5.10}$$

here $\beta \in \mathbb{C}$, $a, b \in \Pi_1$, N_0 is a normal operator in $\mathcal{H}_0$, and the following relations hold:

$$1 + \|a\|^2 = |\alpha|^2 + \|b\|^2, \qquad N_0 a = N_0^* b. \tag{5.11}$$

Observe that for $\alpha \neq 0$ both operators N and N^+ have a Jordan chain of length two at zero, formed by e_0 and e_1 for N, and αe_0 and e_1 for N^+. If $\alpha = 0$ the operator N has still the chain e_0, e_1, but e_0, $e_1 \in \ker N^+$.

In order describe the algebraic eigenspaces $\mathcal{L}_0(N)$ of N and $\mathcal{L}_0(N^+)$ of N^+ at zero we consider the equations

$$N\begin{pmatrix}\xi_1\\ \xi_2\\ \xi_3\\ x\end{pmatrix}=\begin{pmatrix}\mu\\ \nu\\ 0\\ 0\end{pmatrix},\qquad N^+\begin{pmatrix}\eta_1\\ \eta_2\\ \eta_3\\ y\end{pmatrix}=\begin{pmatrix}\mu\\ \nu\\ 0\\ 0\end{pmatrix},$$

for arbitrary $\mu,\nu\in\mathbb{C}$. They are equivalent to the systems of equations

$$\begin{array}{rcl} \xi_2+\beta\,\xi_3+(x,a) &=& \mu,\\ \alpha^*\,\xi_3 &=& \nu,\\ \xi_3\,b+N_0x &=& 0, \end{array}\qquad \begin{array}{rcl} \alpha\,\eta_2+\beta^*\,\eta_3+(y,b) &=& \mu,\\ \eta_3 &=& \nu,\\ \eta_3\,a+N_0^*y &=& 0. \end{array}\tag{5.12}$$

In the following, for an element $x\in\mathcal{H}_0$ we denote $\mathbf{x}:=\begin{pmatrix}0\\0\\0\\x\end{pmatrix}$ and set

$$\mathcal{K}_0:=\{\mathbf{x}:x\in\ker N_0\}=\ker(N_0P),$$

where P denotes the projection onto the last component $\mathcal{H}_0$ in the decomposition (5.8).

Theorem 5.5. *Suppose that $e_0\neq 0$ satisfies (4.5), that it (with its nonzero scalar multiples) is the unique element with these properties, that there exists an associated element e_1 such that $Ne_1=e_0$ and $[e_0,e_1]=0$, and that the algebraic eigenspace $\mathcal{L}_0(N)$ of N at zero is a degenerated subspace; the latter is equivalent to the fact that*

$$a\notin \operatorname{ran}N_0^*,\qquad b\notin\operatorname{ran}N_0.\tag{5.13}$$

holds in the representation (5.10). Then

$$\mathcal{L}_0(N^+)=\mathcal{L}_0(N)=\operatorname{lin}\{e_0,e_1,\mathcal{K}_0\},$$

and

$$\mathcal{L}_0(N)=\operatorname{lin}\{e_1\}\dotplus\ker N,$$

$$\mathcal{L}_0(N^+)=\begin{cases}\operatorname{lin}\{e_1\}\dotplus\ker N^+ & \text{if }\ \alpha\neq 0,\\ \ker N & \text{if }\ \alpha=0,\ (b,\ker N_0)=\{0\},\\ \operatorname{lin}\{\mathbf{b}\}\dotplus\ker N & \text{if }\ \alpha=0,\ (b,\ker N_0)\neq\{0\}.\end{cases}$$

For the proof we mention that the space $\mathcal{L}_0(N)=\mathcal{L}_0(N^+)$ is non-degenerated if and only if the systems in (5.12) for arbitrary μ,ν do not have nontrivial solutions with $\xi_3\neq 0$ or $\eta_3\neq 0$. It can easily be seen that this is equivalent to the statement concerning (5.13). The other claims of the theorem follow by straightforward computations.

Theorem 5.6. *Suppose that $e_0\neq 0$ satisfies (4.5), that it (with its nonzero scalar multiples) is the unique element with these properties, that there exists an associated element e_1 such that $Ne_1=e_0$ and $[e_0,e_1]=0$, and that the algebraic eigenspace $\mathcal{L}_0(N)$ of N at zero is non-degenerated. Then exactly one of the* *following cases prevails:*

(I) $a \in \operatorname{ran} N_0^*$ *and* $b \notin \operatorname{ran} N_0$: *Then, if* $a = N_0^* \widehat{a}$,

$$\mathcal{L}_0(N^+) = \mathcal{L}_0(N) = \operatorname{lin}\{e_0, e_1, f_0 - \widehat{\mathbf{a}}, \mathcal{K}_0\}.$$

The subspace $\operatorname{lin}\{e_0, e_1, f_0 - \widehat{\mathbf{a}}, N(f_0 - \widehat{\mathbf{a}})\}$ *in invariant under* N *and* N^+, *the elements*

$$\alpha^* e_0,\ (\beta - (\widehat{a}, a))\, e_0 + \alpha^* e_1 + \mathbf{b} - \mathbf{N_0}\widehat{\mathbf{a}},\ f_0 - \widehat{\mathbf{a}} \tag{5.14}$$

form a Jordan chain of N *at zero, the elements*

$$\alpha e_0,\ (\beta^* - (\widehat{a}, b))\, e_0 + e_1,\ f_0 - \widehat{\mathbf{a}} \tag{5.15}$$

form a Jordan chain of N^+ *at zero; these chains are of length three if* $\alpha \neq 0$ *and of length two if* $\alpha = 0$.

(II) $a \notin \operatorname{ran} N_0^*$ *and* $b \in \operatorname{ran} N_0$: *In this case* $|\alpha| \geq 1$. *If* $b = N_0 \widehat{b}$, *then*

$$\mathcal{L}_0(N^+) = \mathcal{L}_0(N) = \operatorname{lin}\{e_0, e_1, f_0 - \widehat{\mathbf{b}}, \mathcal{K}_0\}.$$

The subspace $\operatorname{lin}\{e_0, e_1, f_0 - \widehat{\mathbf{b}}, N^+(f_0 - \widehat{\mathbf{b}})\}$ *in invariant under* N *and* N^+, *the elements*

$$\alpha^* e_0,\ \left(\beta - (\widehat{b}, a)\right) e_0 + \alpha^* e_1,\ f_0 - \widehat{\mathbf{b}}$$

form a Jordan chain of N *at zero, the elements*

$$\alpha e_0,\ \left(\beta^* - (\widehat{b}, b)\right) e_0 + e_1 + \mathbf{a} - \mathbf{N_0^*}\widehat{\mathbf{b}},\ \mathbf{f_0} - \widehat{\mathbf{b}}$$

form a Jordan chain of N^+ *at zero, both chains are of length three.*

(III) $a \in \operatorname{ran} N_0^*$ *and* $b \in \operatorname{ran} N_0$: *Then, if* $a = N_0^* \widehat{a}$, $b = N_0 \widehat{b}$,

$$\mathcal{L}_0(N^+) = \mathcal{L}_0(N) = \operatorname{lin}\{e_0, e_1, f_0 - \widehat{\mathbf{a}}, \mathcal{K}\} = \operatorname{lin}\{e_0, e_1, f_0 - \widehat{\mathbf{b}}, \mathcal{K}_0\}.$$

The Jordan chains of N *and* N^+ *at zero are given by* (5.14) *and* (5.15), *respectively. They are of length three if* $\alpha \neq 0$ *and of length two if* $\alpha = 0$.

The proof follows by straightforward computations, using the equations (5.12) and observing that

$$\ker N_0 = \ker N_0^*,$$

and that under the assumptions of (I) $b - N_0 \widehat{a} \in \ker N_0$ and under the assumptions of (II) $a - N_0^* \widehat{b} \in \ker N_0$.

Remark 5.7. Suppose that (as, e.g., in case (II)) the operator N has a chain e_0, e_1, e_2 of length three at the eigenvalue $\lambda_0 = 0$:

$$N e_0 = 0,\ \ N e_1 = e_0,\ \ N e_2 = e_1. \tag{5.16}$$

Then $\mathcal{L}_0 := \operatorname{lin}\{e_0, e_1, e_2\}$ is non-degenerated and (trivially) invariant under N, but it need not be invariant under N^+ since $N^+ e_2$ does not necessarily belong to $\mathcal{L}_0$. However, the subspace $\mathcal{L}_1 := \operatorname{lin}\{e_0, e_1, e_2, N^+ e_2\}$ is non-degenerated and invariant under N and N^+, and

$$\sigma(N|_{\mathcal{L}_1}) = \sigma(N^+|_{\mathcal{L}_1}) = \{0\}.$$

In addition to the chain (5.16), N has at zero the eigenelement $N^+e_2 - \alpha e_1$. The operator N^+ has at zero the Jordan chain αe_0, $e_1 + \mathbf{a}$, $\alpha^* e_2$, and the eigenvector $\mathbf{a}$. This shows, that the geometric eigenspaces of N and N^+ need not coincide.

Remark 5.8. The operators A and B from (3.1) have the following matrix representations:

$$A = \begin{pmatrix} 0 & \frac{1+\alpha}{2} & \mathfrak{Re}\,\beta & \left(\cdot, \frac{a+b}{2}\right) \\ 0 & 0 & \frac{\alpha^*+1}{2} & 0 \\ 0 & 0 & 0 & 0 \\ 0 & 0 & \frac{a+b}{2} & A_0 \end{pmatrix}, \quad B = \begin{pmatrix} 0 & \frac{1-\alpha}{2\,\mathrm{i}} & \mathfrak{Im}\,\beta & \left(\cdot, \frac{b-a}{2\,\mathrm{i}}\right) \\ 0 & 0 & \frac{\alpha^*-1}{2} & 0 \\ 0 & 0 & 0 & 0 \\ 0 & 0 & \frac{b-a}{2\,\mathrm{i}} & B_0 \end{pmatrix} \tag{5.17}$$

In the situation of Theorem 5.5 at least one of the operators A or B has a degenerated Jordan chain of length two at zero which cannot be continued. In the situation of Theorem 5.6 the Jordan chains of A and B at zero are non-degenerated; they can easily be found from the matrix-representations (5.17).

Remark 5.9. The above representations imply that a normal operator in Π_1 has at most two eigenvalues with a non-positive eigenelement. Moreover, in the situations of Corollary 3.3, Theorem 4.1 (ii), Theorem 5.3, and Theorem 5.6, the operator N (and also its adjoint N^+) can be decomposed as orthogonal sum of a normal operator N_0 (or its adjoint N_0^*) in a Hilbert space $\mathcal{H}_0$ and an operator in an at most four-dimensional space with one negative square, whereas in the situations of Theorem 5.1 and Theorem 5.5 such a decomposition is impossible. In the latter case, the corresponding unique eigenvalue λ_0 of N with a non-positive eigenelement is a singular critical point of N and λ_0^* is a singular critical point of N^+.

References

[1] Azizov, Tomas Ya.; Iokhvidov, Iosif S.: Linear Operators in Spaces with an Indefinite Metric. John Wiley and Sons, Chichester, 1989.

[2] Azizov, Tomas Ya.; Jonas, Peter: On compact perturbations of normal operators in a Krein space. Ukrainskiĭ Matem. Žurnal 42 (1990), 1299–1306.

[3] Azizov, Tomas Ya.; Strauss, Vladimir A.: Spectral decompositions for special classes of self-adjoint and normal operators on Krein spaces. Theta Series in Advanced Mathematics: Spectral Analysis and its Applications, 45–67, The Theta Foundation, Bucharest, 2003.

[4] Bognár, Janos: Indefinite Inner Product Spaces. Springer, New York-Heidelberg, 1974.

[5] Gohberg, Israel; Reichstein, Boris: On classification of normal matrices in an indefinite scalar product. Integral Equations Operator Theory 13 (1990), 364–394.

[6] Jonas, Peter; Langer, Heinz: A model for π-selfadjoint operators in π_1-spaces and a special linear pencil. Integral Equations Operator Theory 8 (1985), 13–35.

[7] Krein, Mark G.; Langer, Heinz: On the spectral function of a self-adjoint operator in a space with indefinite metric. Dokl. Akad. Nauk SSSR 152 (1963) 39–42 (Russian).

[8] Langer, Heinz: Habilitationsschrift Technische Universität Dresden, 1965.

[9] Langer, Heinz: Invariante Teilräume definisierbarer J-selbstadjungierter Operatoren. Ann. Acad. Sci. Fenn. Ser. A I No. 475, 1971.

[10] Langer, Heinz: Spectral functions of definitizable operators in Kreĭn spaces. Functional analysis (Dubrovnik, 1981), Lecture Notes in Math., 948, 1–46, Springer, Berlin-New York, 1982.

[11] Mehl, Christian; Ran, Andre; Rodman, Leiba: Semidefinite invariant subspaces: degenerate inner products, Operator Theory: Advances and Applications, 149 (2004), 467–486.

[12] Naĭmark, Mark A.: On commuting unitary operators in spaces with indefinite metric. Acta Sci. Math. (Szeged) 24 (1963), 177–189.

Heinz Langer
Institute of Analysis and Scientific Computing
Vienna University of Technology,
Wiedner Hauptstrasse 8–10
A-1040 Vienna
Austria
e-mail: hlanger@mail.zserv.tuwien.ac.at

Franciszek Hugon Szafraniec
Instytut Matematyki
Uniwersytet Jagielloński,
ul. Reymonta 4
PL-30059 Kraków
Poland
e-mail: fhszafra@im.uj.edu.pl

Operator Theory:
Advances and Applications, Vol. 162, 253–276

Scalar Generalized Nevanlinna Functions: Realizations with Block Operator Matrices

Matthias Langer and Annemarie Luger

Abstract. In this paper a concrete realization for a scalar generalized Nevanlinna function $q \in \mathcal{N}_\kappa$ is given using the realizations of the factors in the basic factorization of q. Some cases are discussed in more detail and the representing operators are given as block operator matrices.

Mathematics Subject Classification (2000). Primary 47A06; Secondary 47A50, 47B32, 47B50.

Keywords. Scalar generalized Nevanlinna function, Realization, Block operator matrix.

1. Introduction

It is well known that a generalized Nevanlinna function $q \in \mathcal{N}_\kappa$ (for the definition of $\mathcal{N}_\kappa$ see Section 2 below) possesses a realization in a Pontryagin space $(\mathcal{K}, [\,\cdot\,,\,\cdot\,])$, and hence can be written as

$$q(z) = q(z_0)^* + (z - z_0^*)\big[\big(I + (z - z_0)(A - z)^{-1}\big)v, v\big] \qquad z \in \varrho(A)$$

where A is a self adjoint relation in $\mathcal{K}$, and with $v \in \mathcal{K}$ and $z_0 \in \varrho(A)$. Recently (see [DLLuSh2] and also [DeH]), such realizations were constructed based on the basic factorization of q, that is,

$$q(z) = r^\#(z)\, q_0(z)\, r(z),$$

where $q_0 \in \mathcal{N}_0$ is a usual Nevanlinna function, the rational function r collects the generalized poles and zeros of q that are not of positive type, and $r^\#(z) := r(z^*)^*$. In these papers the realizations were constructed with the help of a matrix function which was defined using the basic factorization of q and using reproducing kernel space methods. In the present paper, however, we construct a realization in the space $\mathcal{K} = \mathcal{K}_0 \dot{+} \mathbb{C}^{2\kappa}$, where $\mathcal{K}_0$ is a Hilbert space in which a minimal realization of q_0

The authors gratefully acknowledge the support of the "Fond zur Förderung der wissenschaftlichen Forschung" (FWF, Austria), grant number P15540-N05.

acts, which can be chosen arbitrarily. The self-adjoint relation A in the realization of q is constructed using the realization of q_0 and Jordan blocks connected with the generalized poles not of positive type of q.

In Section 2 preliminaries are put together, in the first two subsections we recall the definitions of realizations and boundary mappings together with its basic properties, as far as we need it. In Subsection 2.3 the canonical realization, which is a particular realization in a reproducing kernel Pontryagin space, is discussed briefly. For a usual Nevanlinna function the connection between its integral representation and the realization is detailed in Subsection 2.4. Furthermore, in Subsection 2.5 we recall the basic factorization of a scalar generalized Nevanlinna function.

Section 3 is devoted to the main result, Theorem 3.1, where the realization is given. The proof is carried out by constructing an isomorphism to the model given in [DLLuSh2]. At the end of Section 3 we describe how to obtain a minimal realization. In the case that the representing relation A is in fact an operator, block operator representations can be given, which are discussed in Subsections 4.1 and 4.2. An example where A is not an operator is given in Subsection 4.3.

2. Preliminaries

By definition a scalar function $q : \mathcal{D} \subseteq \mathbb{C} \to \mathbb{C}$ belongs to the *generalized Nevanlinna class* $\mathcal{N}_\kappa$ if it is meromorphic in $\mathbb{C} \setminus \mathbb{R}$, symmetric with respect to the real line, i.e., $q(z^*) = q(z)^*$, and if the so-called *Nevanlinna kernel*

$$K_q(z, w) := \frac{q(z) - q(w^*)}{z - w^*} \qquad z, w \in \mathcal{D}$$

has κ negative squares. This means that for arbitrary numbers $N \in \mathbb{N}$, and points $z_1, \ldots, z_N \in \mathcal{D} \cap \mathbb{C}^+$ the matrices

$$\big(K_q(z_i, z_j)\big)_{i,j=1}^{N}$$

have at most κ negative eigenvalues, and κ is minimal with this property. Here $\mathcal{D}$ is the domain of holomorphy of q, $\mathbb{C}^+$ is the upper half plane, and by * we denote the complex conjugate of a complex number as well as the adjoint of an operator or a relation.

2.1. Realizations of $\mathcal{N}_\kappa$-functions

Generalized Nevanlinna functions can also be characterized by their realizations. It is well known (see, e.g., [KL]) that a function q belongs to the class $\mathcal{N}_\kappa$ if and only if it admits a minimal realization in some Pontryagin space with negative index κ. A *realization* (A, φ) for a function q is given by a self-adjoint linear relation A in a Pontryagin space $\big(\mathcal{K}, [\,\cdot\,,\,\cdot\,]\big)$ and a corresponding *defect function* $\varphi(z)$, that is, a function $\varphi : \varrho(A) \to \mathcal{K}$ with the property

$$\varphi(w) = \big(I + (w - z)(A - w)^{-1}\big)\varphi(z),$$

such that for $z, w \in \varrho(A)$, $z \neq w^*$, the following identity holds

$$\frac{q(z) - q(w^*)}{z - w^*} = [\varphi(z), \varphi(w)].$$

In particular, this implies the following representation for the function q:

$$q(z) = q(z_0)^* + (z - z_0^*)\big[\big(I + (z - z_0)(A - z)^{-1}\big)v, v\big] \qquad z \in \varrho(A),$$

where $z_0 \in \varrho(A) \setminus \mathbb{R}$ is some fixed point and $v := \varphi(z_0)$. The realization is called *minimal* if the defect elements form a total set in $\mathcal{K}$, i.e.,

$$\mathcal{K} = \overline{\operatorname{span}}\big\{\big(I + (z - z_0)(A - z)^{-1}\big)v \,\big|\, z \in \varrho(A)\big\}.$$

In this case $\mathcal{D} = \varrho(A)$ and the realization is unique up to unitary equivalence. Two realizations (A, φ) and $(\widetilde{A}, \widetilde{\varphi})$ of a function q in spaces $\mathcal{K}$ and $\widetilde{\mathcal{K}}$, respectively, are said to be *unitarily equivalent* if there exists a unitary operator $\Phi : \mathcal{K} \to \widetilde{\mathcal{K}}$ such that

$$\{f; g\} \in A \Longleftrightarrow \{\Phi(f); \Phi(g)\} \in \widetilde{A}$$

and $\Phi(\varphi(z)) = \widetilde{\varphi}(z)$.

Sometimes also the triple (A, S, φ) is called realization, where S is the symmetric restriction of A given by

$$S := \big\{\{f; g\} \in A \,\big|\, [g - z_0^* f, \varphi(z_0)] = 0\big\},$$

which is independent of the particular choice of the point $z_0 \in \varrho(A)$. With this notation one has $\varphi(z) \in \ker(S^* - z)$. Note that for unitarily equivalent realizations (A, S, φ) and $(\widetilde{A}, \widetilde{S}, \widetilde{\varphi})$ it follows that

$$\{f; g\} \in S \Longleftrightarrow \{\Phi(f); \Phi(g)\} \in \widetilde{S}.$$

2.2. Boundary mappings

In this subsection we recall the notion of boundary mappings as it is contained, e.g., in [De] and [DeM]. Let S be a symmetric relation with equal defect indices in a Pontryagin space $\mathcal{K}$ and S^* its adjoint. The triple $(\mathcal{H}, B_0, B_1)$ is called a *boundary triple for* S^* if $\mathcal{H}$ is a Hilbert space with inner product $(\,\cdot\,,\,\cdot\,)$ and B_0, B_1 are linear bounded mappings from S^* into $\mathcal{H}$ such that $B_0 \times B_1$ is surjective onto $\mathcal{H} \times \mathcal{H}$ and that the following (abstract Lagrange or Green) identity holds:

$$[g, f'] - [f, g'] = \big(B_1\{f; g\}, B_0\{f'; g'\}\big) - \big(B_0\{f; g\}, B_1\{f'; g'\}\big)$$

for $\{f; g\}, \{f'; g'\} \in S^*$. The B_i are called *boundary mappings.*

It is easy to show that $S = \ker B_0 \cap \ker B_1$ and that $A := \ker B_0$ and $\ker B_1$ are self-adjoint relations. In the following we assume that $\varrho(A) \neq \emptyset$. This assumption will always be satisfied in the next sections. Define the defect subspaces by

$$\widetilde{\mathfrak{N}}_z := \{\{f; g\} \in S^* \mid g = zf\} \subseteq \mathcal{K} \times \mathcal{K},$$
$$\mathfrak{N}_z := \{f \in \mathcal{K} \mid \{f; zf\} \in S^*\} = \ker(S^* - z).$$

Since $S^* = A \dot{+} \widetilde{\mathfrak{N}}_z$ for $z \in \varrho(A)$, the mapping $B_0|_{\widetilde{\mathfrak{N}}_z}$ is bijective from $\widetilde{\mathfrak{N}}_z$ onto $\mathcal{H}$, and we can set

$$\tilde{\gamma}(z) := \left(B_0|_{\widetilde{\mathfrak{N}}_z}\right)^{-1} : \mathcal{H} \to \widetilde{\mathfrak{N}}_z,$$
$$\gamma(z) := P_1\tilde{\gamma}(z) : \mathcal{H} \to \mathfrak{N}_z \subseteq \mathcal{K},$$

where P_1 denotes the projection onto the the first component in $\mathcal{K} \times \mathcal{K}$. The map $\gamma(z)$ satisfies the following relation,

$$\gamma(w) = (I + (w - z)(A - w)^{-1})\gamma(z) \tag{2.1}$$

for $z, w \in \varrho(A)$. The *Titchmarsh–Weyl function* M is defined by

$$M(z) := B_1\tilde{\gamma}(z) \tag{2.2}$$

for $z \in \varrho(A)$, which is an operator function in $\mathcal{H}$.

If S is a densely defined operator, then S^* is also an operator and one can set $B_i f := B_i\{f; S^*f\}$ for $f \in \mathcal{D}(S^*)$ and $i = 0, 1$.

The following proposition shows the connection between boundary mappings of symmetric relations with defect $(1,1)$ and realizations of scalar $\mathcal{N}_\kappa$-functions. Note that in the case of defect $(1,1)$ the space $\mathcal{H}$ can be chosen to be $\mathbb{C}$; the Titchmarsh–Weyl function is then a scalar function.

Proposition 2.1. *Let S be a symmetric relation in a Pontryagin space with defect $(1,1)$ and $(\mathbb{C}, B_0, B_1)$ a boundary triple for S^*. Set $\varphi(z) := \gamma(z)1$ and $A :=$ ker B_0. Then (A, S, φ) is a realization of the Titchmarsh–Weyl function corresponding to $(\mathbb{C}, B_0, B_1)$.*

Conversely, let (A, S, φ) be a realization of an $\mathcal{N}_\kappa$-function q. Decompose an element $\{f; g\} \in S^*$ according to $S^* = A \dot{+} \widetilde{\mathfrak{N}}_{z_0}$ (for some $z_0 \in \varrho(A)$) as follows

$$\{f; g\} = \{f_0; g_0\} + c\{\varphi(z_0); z_0\varphi(z_0)\}$$

with $\{f_0; g_0\} \in A$ and $c \in \mathbb{C}$, and define

$$B_0\{f; g\} := c,$$
$$B_1\{f; g\} := cq(z_0) + [g_0 - z_0^* f_0, \varphi(z_0)].$$

Then $(\mathbb{C}, B_0, B_1)$ is a boundary triple for S with the properties

$$B_0\{\varphi(z); z\varphi(z)\} = 1, \quad B_1\{\varphi(z); z\varphi(z)\} = q(z), \tag{2.3}$$

and $A =$ ker B_0. Hence the Titchmarsh–Weyl function corresponding to $(\mathbb{C}, B_0, B_1)$ is equal to q.

Proof. The first part follows immediately from (2.1). It is a straightforward calculation that B_0, B_1 in the second part are boundary mappings. The relation

$$\{\varphi(z); z\varphi(z)\} = (z - z_0)\{(A - z)^{-1}\varphi(z_0); (I + z(A - z)^{-1})\varphi(z_0)\} + \{\varphi(z_0); z_0\varphi(z_0)\},$$

where the first term on the right-hand side is in A, implies $B_0\{\varphi(z); z\varphi(z)\} = 1$ and

$$\begin{aligned}
&B_1\{\varphi(z); z\varphi(z)\}\\
&= q(z_0) + (z - z_0)\big[(I + z(A - z)^{-1})\varphi(z_0) - z_0^*(A - z)^{-1}\varphi(z_0), \varphi(z_0)\big]\\
&= q(z_0) + (z - z_0)\big[\varphi(z_0), (I + (z^* - z_0)(A - z^*)^{-1})\varphi(z_0)\big]\\
&= q(z_0) + (z - z_0)[\varphi(z_0), \varphi(z^*)] = q(z),
\end{aligned}$$

which finishes the proof. □

We say that the boundary mappings B_0, B_1 are *compatible* with a realization (A, S, φ) of an $\mathcal{N}_\kappa$-function q if (2.3) holds.

If for some $z_0 \in \varrho(A)$ von Neumann's decomposition holds, i.e.,

$$\{f; g\} = \{f_{00}; g_{00}\} + c_1\{\varphi(z_0); z\varphi(z_0)\} + c_2\{\varphi(z_0^*); z^*\varphi(z_0^*)\}$$

for $\{f; g\} \in S^*$ with $\{f_{00}; g_{00}\} \in S$, $c_1, c_2 \in \mathbb{C}$, then in the second part of the proposition one could equivalently define

$$B_0\{f; g\} := c_1 + c_2, \quad B_1\{f; g\} := c_1 q(z_0) + c_2 q(z_0^*).$$

It can be shown that for the self-adjoint operator of a minimal realization of an $\mathcal{N}_\kappa$-function such a z_0 always exists.

Remark 2.2. If (A, S, φ) and $(\widetilde{A}, \widetilde{S}, \widetilde{\varphi})$ are unitarily equivalent realizations of a function $q \in \mathcal{N}_\kappa$, where the unitary operator Φ gives the equivalence, and B_0, B_1 and $\widetilde{B}_0, \widetilde{B}_1$, respectively, are boundary mappings as in Proposition 2.1, then

$$B_i\{f; g\} = \widetilde{B}_i\{\Phi(f); \Phi(g)\} \quad \text{for } i = 0, 1.$$

2.3. The canonical realization

In the proof of the main theorem we will make use of reproducing kernel spaces even for some matrix-valued generalized Nevanlinna functions and also of a particular realization in such spaces.

Let $Q : \mathcal{D} \to \mathbb{C}^{n\times n}$ be a matrix-valued generalized Nevanlinna function, $Q \in \mathcal{N}_\kappa^{n\times n}$, that is, it is meromorphic in $\mathbb{C} \setminus \mathbb{R}$, symmetric with respect to the real line, i.e., $Q(z^*) = Q(z)^*$, and the kernel

$$K_Q(z, w) := \frac{Q(z) - Q(w^*)}{z - w^*} \qquad z, w \in \mathcal{D} \tag{2.4}$$

has κ negative squares. This means that for $N \in \mathbb{N}$, points $z_1, \dots, z_N \in \mathcal{D} \cap \mathbb{C}^+$, and vectors $\boldsymbol{x}_1, \dots, \boldsymbol{x}_N \in \mathbb{C}^n$ the matrices

$$\big((K_Q(z_i, z_j)\boldsymbol{x}_i, \boldsymbol{x}_j)\big)_{i,j=1}^N$$

have at most κ negative eigenvalues, and κ is minimal with this property.

By $\mathcal{L}(Q)$ denote the reproducing kernel Pontryagin space associated with the function Q: this is the closed linear span of the kernel functions $K_Q(\,\cdot\,,z)\boldsymbol{c}$ for $z \in \mathcal{D}$ and $\boldsymbol{c} \in \mathbb{C}^n$ with respect to the norm that corresponds to the inner product

$$\big\langle K_Q(\,\cdot\,,z)\boldsymbol{c}, K_Q(\,\cdot\,,w)\boldsymbol{d}\big\rangle := \big(K_Q(w,z)\boldsymbol{c},\boldsymbol{d}\big)_{\mathbb{C}^n},$$

which has κ negative squares. The elements of this space are functions, which are holomorphic on the domain of holomorphy of Q.

Now we restrict ourselves again to the case of scalar generalized Nevanlinna functions. The following theorem describes the so-called *canonical realization*. For a detailed discussion see [DLLuSh2] and the references given there.

Proposition 2.3. *Let the function $q \in \mathcal{N}_\kappa$ be given. Define the self-adjoint linear relation A_q by*

$$A_q := \big\{\{f;g\} \in \mathcal{L}(q)^2 \,\big|\, \exists c \in \mathbb{C}:\ g(\zeta) - \zeta f(\zeta) \equiv c\big\},$$

and the symmetry S_q by

$$S_q := \big\{\{f;g\} \in \mathcal{L}(q)^2 \,\big|\, g(\zeta) - \zeta f(\zeta) \equiv 0\big\},$$

and set $\varphi_q(z) := K_q(\,\cdot\,,z^*)$. Then the triple (A_q, S_q, φ_q) forms a minimal realization for q.

Moreover, the adjoint of S_q is given by

$$S_q^* := \big\{\{f;g\} \in \mathcal{L}(q)^2 \,\big|\, \exists c,d \in \mathbb{C}:\ g(\zeta) - \zeta f(\zeta) \equiv c - d\,q(\zeta)\big\} \tag{2.5}$$

and boundary mappings that are compatible with the realization are given by

$$B_0\{f;g\} := d, \quad B_1\{f;g\} := c, \tag{2.6}$$

where c and d are as in (2.5).

Note that whenever we refer to the canonical realization, we use the function q as subscript.

2.4. Realizations of $\mathcal{N}_0$-functions

In this section we recall realizations for $\mathcal{N}_0$-functions which are connected with their integral representation

$$q(z) = a + bz + \int_{-\infty}^{\infty}\Big(\frac{1}{t-z} - \frac{t}{1+t^2}\Big)d\sigma(t),$$

where $a \in \mathbb{R}$, $b \geq 0$ and σ is a measure with $\int_{-\infty}^{\infty} d\sigma(t)/(1+t^2) < \infty$. We list the space $\mathcal{K}$, the relations A, S and S^*, and the defect function φ such that (A,S,φ) is a realization of the function q above. Moreover, it is not difficult to determine boundary mappings that are compatible with the realization, cf. [DeM].

We have to consider two cases.

1. $b = 0$:
$$\begin{aligned}
&\mathcal{K} = L^2_\sigma\\
&A = \{\{f;g\} \mid g(t) = tf(t)\}\\
&S = \{\{f;g\} \in A \mid \textstyle\int_{-\infty}^{\infty} f(t)d\sigma(t) = 0\}\\
&S^* = \{\{f;g\} \mid \exists c \in \mathbb{C} : g(t) = tf(t) - c\}\\
&\varphi(z) = \varphi(z,t) = \frac{1}{t-z}\\
&B_0\{f;g\} = c \qquad \text{(where } c \text{ is such that } g(t) = tf(t) - c)\\
&B_1\{f;g\} = ac + \int_{-\infty}^{\infty} \Big(f(t) - c\frac{t}{1+t^2}\Big)d\sigma(t)
\end{aligned}$$

2. $b > 0$:
$$\begin{aligned}
&\mathcal{K} = L^2_\sigma \oplus \mathbb{C}, \quad \big[\big(\begin{smallmatrix} f\\ \xi\end{smallmatrix}\big), \big(\begin{smallmatrix} g\\ \eta\end{smallmatrix}\big)\big]_{\mathcal{K}} = [f,g]_{L^2_\sigma} + b\xi\eta^*\\
&A = \{\{\big(\begin{smallmatrix} f\\ \xi\end{smallmatrix}\big); \big(\begin{smallmatrix} g\\ \eta\end{smallmatrix}\big)\} \mid g(t) = tf(t),\ \xi = 0\}\\
&S = \{\{\big(\begin{smallmatrix} f\\ \xi\end{smallmatrix}\big); \big(\begin{smallmatrix} g\\ \eta\end{smallmatrix}\big)\} \in A \mid b\eta + \textstyle\int_{-\infty}^{\infty} f(t)d\sigma(t) = 0\}\\
&S^* = \{\{\big(\begin{smallmatrix} f\\ \xi\end{smallmatrix}\big); \big(\begin{smallmatrix} g\\ \eta\end{smallmatrix}\big)\} \mid g(t) = tf(t) - \xi\}\\
&\varphi(z) = \varphi(z,t) = \begin{pmatrix} \frac{1}{t-z}\\ 1\end{pmatrix}\\
&B_0\{\big(\begin{smallmatrix} f\\ \xi\end{smallmatrix}\big); \big(\begin{smallmatrix} g\\ \eta\end{smallmatrix}\big)\} = \xi\\
&B_1\{\big(\begin{smallmatrix} f\\ \xi\end{smallmatrix}\big); \big(\begin{smallmatrix} g\\ \eta\end{smallmatrix}\big)\} = a\xi + b\eta + \int_{-\infty}^{\infty} \Big(f(t) - \xi\frac{t}{1+t^2}\Big)d\sigma(t)
\end{aligned}$$

If the measure σ is infinite and $b = 0$ (which is equivalent to the fact that S is densely defined), then S^* is an operator and the boundary mappings depend only on the first component.

If the measure σ is finite and $b = 0$, then the function q has also a representation as a u-resolvent, i.e.,

$$q(z) = s + [(A-z)^{-1}u, u]$$

with $s \in \mathbb{R}$, some self-adjoint operator A in a Hilbert space $\mathcal{K}$ and an element $u \in \mathcal{K}$. A corresponding realization (A, S, φ) and boundary mappings are given by

$$\begin{aligned}
&S = \{\{f;g\} \in A \mid [f,u] = 0\}\\
&S^* = \{\{f;g\} \mid f \in \mathcal{D}(A),\ \exists c \in \mathbb{C} : g = Af - cu\}\\
&\varphi(z) = (A-z)^{-1}u\\
&B_0\{f;g\} = c \qquad \text{(where } c \text{ is such that } g = Af - cu)\\
&B_1\{f;g\} = sc + [f,u]
\end{aligned}$$

Here for instance one can take $\mathcal{K} = L^2_\sigma$, A the multiplication operator by the independent variable, and $u = \mathbf{1}$.

2.5. Basic factorization

A point $\alpha \in \mathbb{C} \cup \{\infty\}$ is called a *generalized pole* of the function $q \in \mathcal{N}_\kappa$ if it is an eigenvalue of the relation A in some minimal realization of q. Of particular interest are those generalized poles that are *not of positive type*, that is the corresponding eigenvector of the self-adjoint relation is not a positive element in $\mathcal{K}$. Its *degree of non-positivity* ν_α is the dimension of a maximal non-positive invariant subspace of the root space of A at α. A point $\beta \in \mathbb{C} \cup \{\infty\}$ is called a *generalized zero* of q (not of positive type with *degree of non-positivity* κ_β) if it is a generalized pole of $\widehat{q}(z) := -\frac{1}{q(z)}$ (not of positive type with degree of non-positivity κ_β). Note that if $q \in \mathcal{N}_\kappa$ then also $\widehat{q} \in \mathcal{N}_\kappa$.

In [DLLuSh1] and also [DeHS1] it was shown that every function $q \in \mathcal{N}_\kappa$ admits a *basic factorization*: let the points α_i, $i = 1, 2, \ldots, \ell$, (β_j , $j = 1, 2, \ldots, k$, respectively) be the generalized poles (zeros, respectively) of q in $\mathbb{C}^+ \cup \mathbb{R}$ that are not of positive type, denote by ν_i (κ_j, respectively) the degree of non-positivity of α_i (β_j, respectively), and define

$$r(z) := \frac{(z-\beta_1)^{\kappa_1} \ldots (z-\beta_k)^{\kappa_k}}{(z-\alpha_1^*)^{\nu_1} \ldots (z-\alpha_\ell^*)^{\nu_\ell}} . \tag{2.7}$$

Then there exists a function $q_0 \in \mathcal{N}_0$ such that

$$q(z) = r^{\#}(z) q_0(z) r(z), \tag{2.8}$$

where $r^{\#}(z) := r(z^*)^*$. Note that if

$$\tau := \kappa_1 + \cdots + \kappa_k - (\nu_1 + \cdots + \nu_\ell) \tag{2.9}$$

is positive (negative, respectively), then ∞ is a generalized pole (zero, respectively) of q which is not of positive type and with degree of non-positivity $|\tau|$. Since ∞ cannot be a generalized zero and a generalized pole at the same time, we have

$$\kappa = \max\{\kappa_1 + \cdots + \kappa_k, \nu_1 + \cdots + \nu_\ell\}.$$

3. The realization

Let $q \in \mathcal{N}_\kappa$ be given by its basic factorization (2.8):

$$q(z) = r^{\#}(z)\, q_0(z)\, r(z).$$

Write the rational function r as partial fractional decomposition

$$r(z) = \frac{\prod_{j=1}^{k}(z-\beta_j)^{\kappa_j}}{\prod_{i=1}^{\ell}(z-\alpha_i^*)^{\nu_i}} = \sum_{i=0}^{\ell} r_i(z) \tag{3.1}$$

with the functions

$$r_0(z) = \sum_{j=0}^{\nu_0} \sigma_{0j} z^j \quad \text{and} \quad r_i(z) = \sum_{j=1}^{\nu_i} \frac{-\sigma_{ij}}{(z-\alpha_i^*)^j} \quad \text{for } i = 1, \ldots, \ell.$$

Here we assume that $\sigma_{0\nu_0} \neq 0$ if $\nu_0 > 0$. Note that $\kappa = \sum_{i=0}^{\ell} \nu_i$ and ν_0 denotes the degree of non-positivity of ∞ if ∞ is a generalized pole and $\nu_0 = 0$ otherwise.

Let (A_0, S_0, φ_0) be a minimal realization of the Nevanlinna function q_0 in a Hilbert space $(\mathcal{K}_0, [\,\cdot\,,\,\cdot\,]_0)$. By Proposition 2.1 one can find corresponding boundary mappings $B_{S_0,0}$ and $B_{S_0,1}$ that are compatible with (A_0, S_0, φ_0), i.e.,

$$B_{S_0,0}\{\varphi_0(z); z\varphi_0(z)\} = 1,$$
$$B_{S_0,1}\{\varphi_0(z); z\varphi_0(z)\} = q_0(z).$$

Using the above "ingredients" we will define a space $\mathcal{K}$, relations $\mathcal{A}$ and $\mathcal{S}$, and a function $\varphi(z)$ such that $(\mathcal{A}, \mathcal{S}, \varphi)$ is a realization of q. To this end let us first introduce some notations.

Let $\mathcal{K} := \mathcal{K}_0[+](\mathbb{C}^\kappa \dot{+} \mathbb{C}^\kappa)$ be the Pontryagin space with the inner product $[\,\cdot\,,\,\cdot\,]$ given by the Gram operator

$$G := \begin{pmatrix} I_{\mathcal{K}_0} & 0 & 0 \\ 0 & 0 & I_{\mathbb{C}^\kappa} \\ 0 & I_{\mathbb{C}^\kappa} & 0 \end{pmatrix}.$$

Here $\dot{+}$ denotes a direct sum and $[+]$ a direct sum that is even orthogonal with respect to the indefinite inner product $[\,\cdot\,,\,\cdot\,]$. In the following a vector $\boldsymbol{h} \in \mathbb{C}^\kappa$ will be decomposed according to (3.1) as $\boldsymbol{h} = \begin{pmatrix}\boldsymbol{h}_0 & \boldsymbol{h}_1 & \dots & \boldsymbol{h}_\ell\end{pmatrix}^\top$, where $\boldsymbol{h}_i = (h_{i1}, \dots, h_{i\nu_i})^\top \in \mathbb{C}^{\nu_i}$ for $i = 0, 1, \dots, \ell$. Moreover, set $\boldsymbol{e} := (1, 0, \dots, 0)^\top$, a vector of suitable size, $\boldsymbol{\sigma}_i := (\sigma_{i1}, \dots, \sigma_{i\nu_i})^\top$, and

$$\begin{aligned} \boldsymbol{e}_0(z) = \boldsymbol{e}_0^\dagger(z) &:= \begin{pmatrix} 1 \\ z \\ \vdots \\ z^{\nu_0 - 1} \end{pmatrix}, \\ \boldsymbol{e}_i(z) := \begin{pmatrix} \frac{1}{z-\alpha_i} \\ \frac{1}{(z-\alpha_i)^2} \\ \vdots \\ \frac{1}{(z-\alpha_i)^{\nu_i}} \end{pmatrix} \quad &\text{and} \quad \boldsymbol{e}_i^\dagger(z) := \begin{pmatrix} \frac{1}{z-\alpha_i^*} \\ \frac{1}{(z-\alpha_i^*)^2} \\ \vdots \\ \frac{1}{(z-\alpha_i^*)^{\nu_i}} \end{pmatrix} \quad \text{for } i = 1, \dots, \ell. \end{aligned} \tag{3.2}$$

Furthermore, denote by $J(\alpha)$ for $\alpha \in \mathbb{C}$ a lower Jordan block of suitable size,

$$J(\alpha) := \begin{pmatrix} \alpha & & & \\ 1 & \ddots & & \\ & \ddots & \ddots & \\ & & 1 & \alpha \end{pmatrix},$$

and by G_i the matrix

$$G_i := \begin{pmatrix} \sigma_{i1} & \sigma_{i2} & \dots & \sigma_{i\nu_i} \\ \sigma_{i2} & & \dots & 0 \\ \vdots & \vdots & \vdots & \vdots \\ \sigma_{i\nu_i} & 0 & \dots & 0 \end{pmatrix}^{-1},$$

which is related to $r_i(z)$.

Theorem 3.1. *Let $q \in \mathcal{N}_\kappa$ be given and let the notations be as above. Define the relation $\mathcal{S}^*$ in $\mathcal{K}$ as follows:*

$$\left\{ \begin{pmatrix} f_0 \\ \boldsymbol{h} \\ \boldsymbol{k} \end{pmatrix}; \begin{pmatrix} F_0 \\ \boldsymbol{H} \\ \boldsymbol{K} \end{pmatrix} \right\} \in \mathcal{S}^*$$

if and only if there exist constants $c_1, c_2, c_3, c_4 \in \mathbb{C}$ such that

$$\{f_0; F_0\} \in S_0^* \quad \text{with } B_{S_0,0}\{f_0; F_0\} = c_3 \text{ and } B_{S_0,1}\{f_0; F_0\} = c_1, \tag{3.3}$$

$$\boldsymbol{h}_0 = J(0)\boldsymbol{H}_0 + c_1\boldsymbol{e}, \tag{3.4}$$

$$\boldsymbol{H}_i = J(\alpha_i)\boldsymbol{h}_i + c_1\boldsymbol{e} \qquad \text{for } i = 1, \dots, \ell, \tag{3.5}$$

$$\boldsymbol{k}_0 = J(0)^*\boldsymbol{K}_0 + c_4\boldsymbol{\sigma}_0, \tag{3.6}$$

$$\boldsymbol{K}_i = J(\alpha_i)^*\boldsymbol{k}_i + c_4\boldsymbol{\sigma}_i \qquad \text{for } i = 1, \dots, \ell, \tag{3.7}$$

$$\sum_{j=1}^{\nu_0} \sigma_{0j}^* H_{0j} = \sum_{i=1}^{\ell}\sum_{j=1}^{\nu_i} \sigma_{ij}^* h_{ij} + c_2 - c_1\sigma_{00}^* \tag{3.8}$$

and

$$K_{01} = \sum_{i=1}^{\ell} k_{i1} + c_3 - c_4\sigma_{00}, \tag{3.9}$$

where $K_{01} = 0$ if $\nu_0 = 0$. Define the relations $\mathcal{A}$ and $\mathcal{S}$ as the restrictions of $\mathcal{S}^*$ to elements for which $c_4 = 0$ and $c_2 = c_4 = 0$, respectively. Furthermore, denote by $\varphi(z)$ the function

$$\varphi(z) := \begin{pmatrix} r(z)\varphi_0(z) \\ \left(r(z)q_0(z)\boldsymbol{e}_i(z)\right)_{i=0}^{\ell} \\ \left(G_i^{-1}\boldsymbol{e}_i^{\dagger}(z)\right)_{i=0}^{\ell} \end{pmatrix}.$$

Then the triple $(\mathcal{A}, \mathcal{S}, \varphi)$ is a realization for q. Boundary mappings that are compatible with this realization are given by

$$B_0\widetilde{x} = c_4, \quad B_1\widetilde{x} = c_2$$

for an element $\widetilde{x} = \{(f_0\ \boldsymbol{h}\ \boldsymbol{k})^\top; (F_0\ \boldsymbol{H}\ \boldsymbol{K})^\top\} \in \mathcal{S}^*$, where c_2 and c_4 are as in (3.3)–(3.9).

In particular, for $\nu_0 = 0$, i.e., ∞ is not a generalized pole not of positive type of q, relations (3.4) and (3.6) are void and (3.8) and (3.9) become

$$\sum_{i=1}^{\ell}\sum_{j=1}^{\nu_i} \sigma_{ij}^* h_{ij} + c_2 - c_1\sigma_{00}^* = 0 \quad \text{and} \quad \sum_{i=1}^{\ell} k_{i1} + c_3 - c_4\sigma_{00} = 0.$$

Note that the above notation is justified since from the proof we will see that the relation $\mathcal{S}^*$ is indeed the adjoint of $\mathcal{S}$.

Remark 3.2. This realization need not be minimal. For more details see the end of this section.

Remark 3.3. Note that the self-adjoint relation $\mathcal{A}$ is independent of the numbers σ_{ij}, and hence it does only depend on the generalized poles α_j but not on the generalized zeros β_j.

In order to prove the theorem we will show that $(\mathcal{A}, \mathcal{S}, \varphi)$ is unitarily equivalent to a triple $(\widetilde{\mathcal{A}}, \widetilde{\mathcal{S}}, \widetilde{\varphi})$ in a Pontryagin space $\widetilde{\mathcal{K}}$ which was introduced in [DLLuSh2] and shown to be a realization for q. For the convenience of the reader we recall these notations here. The unitary operator that yields the equivalence is then defined in (3.14) below.

Let the matrix functions $\mathcal{M} \in \mathcal{N}_\kappa^{2\times 2}$ and $\mathcal{Q} \in \mathcal{N}_\kappa^{3\times 3}$ be defined as

$$\mathcal{M}(z) := \begin{pmatrix} 0 & r^{\#}(z) \\ r(z) & 0 \end{pmatrix}, \quad \mathcal{Q}(z) := \begin{pmatrix} q_0(z) & 0 & 0 \\ 0 & 0 & r^{\#}(z) \\ 0 & r(z) & 0 \end{pmatrix}.$$

Then the corresponding reproducing kernel Pontryagin space $\mathcal{L}(\mathcal{Q})$ decomposes as $\mathcal{L}(\mathcal{Q}) = \mathcal{L}(q_0) \oplus \mathcal{L}(\mathcal{M})$. The self-adjoint relation $\widetilde{A}$ is defined as

$$\widetilde{A} := \{\{\widetilde{f}, \widetilde{g}\} \in (\mathcal{L}(\mathcal{Q}))^2 \mid \exists \boldsymbol{c} \in \mathbb{C}^3 \,.\, \widetilde{g}(\zeta) - \zeta \widetilde{f}(\zeta) = (\mathcal{I} + \mathcal{Q}(\zeta)\mathcal{B})\boldsymbol{c}\},$$

where $\mathcal{I}$ is the 3×3 identity matrix and

$$\mathcal{B} := \begin{pmatrix} 0 & 0 & -1 \\ 0 & 0 & 0 \\ -1 & 0 & 0 \end{pmatrix}.$$

By $\widetilde{\varphi}(\,\cdot\,, z)$ denote the functions

$$\widetilde{\varphi}(\zeta, z) := \frac{\mathcal{Q}(\zeta) - \mathcal{Q}(z)}{\zeta - z}\boldsymbol{v}(z),$$

where

$$\boldsymbol{v}(z) := \begin{pmatrix} r(z) \\ 1 \\ r(z) q_0(z) \end{pmatrix}.$$

In [DLLuSh2] it was proved that $(\widetilde{A}, \widetilde{\varphi})$ is a – not necessarily minimal – realization for q.

In the following lemma we identify the symmetry $\widetilde{S}$ and its adjoint $\widetilde{S}^*$ such that $\widetilde{\varphi}(\,\cdot\,,z)\in\mathcal{L}(\mathcal{Q})$ is a corresponding defect function, i.e., $\widetilde{\varphi}(\,\cdot\,,z)\in\ker(\widetilde{S}^*-z)$ for all $z\in\varrho(\widetilde{A})$. Define

$$\widetilde{S}:=\{\{\widetilde{f};\widetilde{g}\}\in\widetilde{A}\mid \boldsymbol{v}^{\#}(\zeta)(\widetilde{g}(\zeta)-\zeta\widetilde{f}(\zeta))\equiv 0\}.$$

Lemma 3.4. *The triple $(\widetilde{A},\widetilde{S},\widetilde{\varphi})$ is a realization of the function q in the space $\mathcal{L}(\mathcal{Q})$. Moreover, the symmetry $\widetilde{S}$ can be written as*

$$\widetilde{S}=\left\{\{\widetilde{f};\widetilde{g}\}\in(\mathcal{L}(\mathcal{Q}))^2\;\middle|\;\exists\, c_1,c_3\in\mathbb{C}:\widetilde{g}(\zeta)-\zeta\widetilde{f}(\zeta)\equiv\begin{pmatrix}c_1\\0\\c_3\end{pmatrix}-\mathcal{Q}(\zeta)\begin{pmatrix}c_3\\0\\c_1\end{pmatrix}\right\}\tag{3.10}$$

and its adjoint is given by

$$\widetilde{S}^*=\left\{\{\widetilde{f};\widetilde{g}\}\;\middle|\;\exists\, c_1,\dots,c_4\in\mathbb{C}:\widetilde{g}(\zeta)-\zeta\widetilde{f}(\zeta)\equiv\begin{pmatrix}c_1\\c_2\\c_3\end{pmatrix}-\mathcal{Q}(\zeta)\begin{pmatrix}c_3\\c_4\\c_1\end{pmatrix}\right\}.\tag{3.11}$$

Boundary mappings that are compatible with this realization are given by

$$B_{\widetilde{S},0}\{\widetilde{f};\widetilde{g}\}=c_4,\quad B_{\widetilde{S},1}\{\widetilde{f};\widetilde{g}\}=c_2$$

for elements as in (3.11).

Proof. We first show (3.10). By definition the pair $\{\widetilde{f};\widetilde{g}\}\in(\mathcal{L}(\mathcal{Q}))^2$ belongs to $\widetilde{S}$ if and only if there exists a vector $\boldsymbol{c}=\begin{pmatrix}c_1 & c_2 & c_3\end{pmatrix}^\top$ such that $\widetilde{g}(\zeta)-\zeta\widetilde{f}(\zeta)\equiv \boldsymbol{c}+\mathcal{Q}(\zeta)\mathcal{B}\boldsymbol{c}$ with $0=\boldsymbol{v}^{\#}(\zeta)\big(\widetilde{g}(\zeta)-\zeta\widetilde{f}(\zeta)\big)$. The latter expression equals

$$\boldsymbol{v}^{\#}(\zeta)\big(\widetilde{g}(\zeta)-\zeta\widetilde{f}(\zeta)\big)=\begin{pmatrix}r^{\#}(\zeta) & 1 & r^{\#}(\zeta)q_0(\zeta)\end{pmatrix}\begin{pmatrix}c_1-c_3q_0(\zeta)\\c_2-c_1r^{\#}(\zeta)\\c_3\end{pmatrix}=c_2,$$

which shows (3.10). Note that hence $\widetilde{S}$ is the restriction of $S_{\mathcal{Q}}^*$ (the symmetry $S_{\mathcal{Q}}$ corresponds to the canonical model for the matrix function $\mathcal{Q}$, cf. [DLLuSh2, Theorem 2.1 (iii)]),

$$S_{\mathcal{Q}}^*:=\{\{\widetilde{f};\widetilde{g}\}\in(\mathcal{L}(\mathcal{Q}))^2\mid\exists\,\boldsymbol{c},\boldsymbol{d}\in\mathbb{C}:\widetilde{g}(\zeta)-\zeta\widetilde{f}(\zeta)\equiv\boldsymbol{c}-\mathcal{Q}(\zeta)\boldsymbol{d}\},\tag{3.12}$$

to elements $\{\widetilde{f};\widetilde{g}\}$ for which $\boldsymbol{c}^\top=\begin{pmatrix}c_1 & 0 & c_3\end{pmatrix}^\top$ and $\boldsymbol{d}=-\mathcal{B}\boldsymbol{c}$. In the following we use the boundary mappings for $S_{\mathcal{Q}}$, which are given by $B_{S_{\mathcal{Q}},0}\{\widetilde{f};\widetilde{g}\}=\boldsymbol{d}$ and $B_{S_{\mathcal{Q}},1}\{\widetilde{f};\widetilde{g}\}=\boldsymbol{c}$, see, e.g., [DLLuSh2, Theorem 2.4]. Let $\{\widetilde{f};\widetilde{g}\}\in S_{\mathcal{Q}}^*$ as in (3.12) and $\{\widetilde{f}';\widetilde{g}'\}\in\widetilde{S}$ with $B_{S_{\mathcal{Q}},0}\{\widetilde{f}',\widetilde{g}'\}=\boldsymbol{d}'$ and $B_{S_{\mathcal{Q}},1}\{\widetilde{f}',\widetilde{g}'\}=\boldsymbol{c}'$. Then

$$\langle\widetilde{g},\widetilde{f}'\rangle_{\mathcal{L}(\mathcal{Q})}-\langle\widetilde{f},\widetilde{g}'\rangle_{\mathcal{L}(\mathcal{Q})}=(\boldsymbol{c},\boldsymbol{d}')_{\mathbb{C}^3}-(\boldsymbol{d},\boldsymbol{c}')_{\mathbb{C}^3}=(c_3-d_1){c_1'}^*+(c_1-d_3){c_3'}^*.$$

For fixed $\{\widetilde{f};\widetilde{g}\}$ the latter expression vanishes for all $\{\widetilde{f}';\widetilde{g}'\}\in\widetilde{S}$ if and only if $d_1=c_3$ and $d_3=c_1$, which shows (3.11).

It is easy to see that $B_{\widetilde{S},0}$, $B_{\widetilde{S},1}$ are possible boundary mappings for $\widetilde{S}$. Since

$$z\widetilde{\varphi}(\zeta,z)-\zeta\widetilde{\varphi}(\zeta,z)=\begin{pmatrix} r(z)q_0(z)\\ r^{\#}(z)q_0(z)r(z)\\ r(z)\end{pmatrix}-\mathcal{Q}(\zeta)\begin{pmatrix} r(z)\\ 1\\ r(z)q_0(z)\end{pmatrix}$$

it follows that $\{\widetilde{\varphi}(\,\cdot\,,z);z\widetilde{\varphi}(\,\cdot\,,z)\}\in\widetilde{S}^*$, i.e., $\widetilde{\varphi}(\,\cdot\,,z)\in\ker(\widetilde{S}^*-z)$, and that

$$B_{\widetilde{S},0}\{\widetilde{\varphi}(\,\cdot\,,z);z\widetilde{\varphi}(\,\cdot\,,z)\}=1,\quad B_{\widetilde{S},1}\{\widetilde{\varphi}(\,\cdot\,,z);z\widetilde{\varphi}(\,\cdot\,,z)\}=r^{\#}(z)q_0(z)r(z)=q(z),$$

which shows that $B_{\widetilde{S},0}$, $B_{\widetilde{S},1}$ are compatible with the realization $(\widetilde{A},\widetilde{S},\widetilde{\varphi})$. □

We now define a mapping $\Phi:\mathcal{L}(\mathcal{Q})\to\mathcal{K}_0[+](\mathbb{C}^\kappa\dot{+}\mathbb{C}^\kappa)$ that will give the unitary equivalence of $(\widetilde{A},\widetilde{S},\widetilde{\varphi})$ and $(\mathcal{A},\mathcal{S},\varphi)$. Note that since it was assumed that the given realization of q_0 is minimal, it is unitarily equivalent to the canonical realization in the reproducing kernel space $\mathcal{L}(q_0)$. Let this unitary equivalence be given by the mapping $\Phi_0\colon\mathcal{L}(q_0)\to\mathcal{K}_0$.

According to [DLLuSh2, Theorem 3.4] an element in $\mathcal{L}(\mathcal{Q})$ is of the form

$$\widetilde{f}=\Big(\widetilde{f}_0\,,\sum_{j=1}^{\nu_0}\widetilde{h}_{0j}\zeta^{j-1}+\sum_{i=1}^{\ell}\sum_{j=1}^{\nu_i}\frac{\widetilde{h}_{ij}}{(\zeta-\alpha_i)^j}\,,\sum_{j=1}^{\nu_0}\widetilde{k}_{0j}\zeta^{j-1}+\sum_{i=1}^{\ell}\sum_{j=1}^{\nu_i}\frac{\widetilde{k}_{ij}}{(\zeta-\alpha_i^*)^j}\Big)\tag{3.13}$$

with $\widetilde{f}_0\in\mathcal{L}(q_0)$ and $\widetilde{h}_{ij},\widetilde{k}_{ij}\in\mathbb{C}$ for $i=0,1,\dots,\ell$ and $j=1,\dots,\nu_i$. Then define

$$\Phi(\widetilde{f}):=\begin{pmatrix}\Phi_0(\widetilde{f}_0)\\ (G_i^*\widetilde{\boldsymbol{h}}_i)_{i=0}^{\ell}\\ (\widetilde{\boldsymbol{k}}_i)_{i=0}^{\ell}\end{pmatrix},\tag{3.14}$$

here we again used the notation $\widetilde{\boldsymbol{h}}_i:=\begin{pmatrix}\widetilde{h}_{i1}&\dots&\widetilde{h}_{i\nu_i}\end{pmatrix}^\top$ and accordingly for the vector $\widetilde{\boldsymbol{k}}_i$. Obviously Φ is bijective and since the inner product in $\mathcal{L}(\mathcal{Q})$ is given by

$$\left\langle\begin{pmatrix}\widetilde{f}_0\\ \widetilde{\boldsymbol{h}}\\ \widetilde{\boldsymbol{k}}\end{pmatrix},\begin{pmatrix}\widetilde{F}_0\\ \widetilde{\boldsymbol{H}}\\ \widetilde{\boldsymbol{K}}\end{pmatrix}\right\rangle_{\mathcal{L}(\mathcal{Q})}=\langle\widetilde{f}_0,\widetilde{F}_0\rangle_{\mathcal{L}(q_0)}+\sum_{i=0}^{\ell}\big[(G_i\widetilde{\boldsymbol{k}}_i,\widetilde{\boldsymbol{H}}_i)_{\mathbb{C}^{\nu_i}}+(G_i^*\widetilde{\boldsymbol{h}}_i,\widetilde{\boldsymbol{K}}_i)_{\mathbb{C}^{\nu_i}}\big],$$

it follows that Φ is unitary. In the following lemma the isomorphism is applied to kernel elements. Recall that the kernel of a matrix function was defined in (2.4).

Lemma 3.5. *With the notation in* (3.2) *the following relations hold,*

$$\Phi\begin{pmatrix}0\\ K_{\mathcal{M}}(\,\cdot\,,z^*)\begin{pmatrix}0\\1\end{pmatrix}\end{pmatrix}=\begin{pmatrix}0\\ \big(\boldsymbol{e}_i(z)\big)_{i=0}^{\ell}\\ 0\end{pmatrix}$$

and

$$\Phi\begin{pmatrix} 0 \\ K_{\mathcal{M}}(\,\cdot\,,z^*)\begin{pmatrix}1\\0\end{pmatrix}\end{pmatrix} = \begin{pmatrix} 0 \\ 0 \\ \bigl(G_i^{-1}\boldsymbol{e}_i^\dagger(z)\bigr)_{i=0}^{\ell}\end{pmatrix}.$$

Proof. For $i = 1,\dots,\ell$ we have

$$\begin{aligned}\frac{r_i(\zeta)-r_i(z)}{\zeta - z} &= \sum_{j=1}^{\nu_i}\sigma_{ij}\frac{(\zeta-\alpha_i^*)^j-(z-\alpha_i^*)^j}{(\zeta-z)(\zeta-\alpha_i^*)^j(z-\alpha_i^*)^j}\\ &= \sum_{j=1}^{\nu_i}\sigma_{ij}\sum_{k=0}^{j-1}\frac{(z-\alpha_i^*)^k(\zeta-\alpha_i^*)^{j-k-1}}{(z-\alpha_i^*)^j(\zeta-\alpha_i^*)^j}\\ &= \sum_{k=1}^{\nu_i}\frac{1}{(\zeta-\alpha_i^*)^k}\sum_{j=1}^{\nu_i-k+1}\frac{\sigma_{i,k+j-1}}{(z-\alpha_i^*)^j}\\ &= \sum_{k=1}^{\nu_i}\frac{1}{(\zeta-\alpha_i^*)^k}\Bigl(G_i^{-1}\boldsymbol{e}_i^\dagger(z)\Bigr)_k\end{aligned}$$

and likewise

$$\begin{aligned}\frac{r_0(\zeta)-r_0(z)}{\zeta - z} &= \sum_{k=1}^{\nu_0}\zeta^{k-1}\sum_{j=1}^{\nu_0-k+1}\sigma_{0,k+j-1}z^{j-1}\\ &= \sum_{k=1}^{\nu_0}\zeta^{k-1}\Bigl(G_0^{-1}\boldsymbol{e}_0^\dagger(z)\Bigr)_k.\end{aligned}$$

Hence we find

$$\Phi\begin{pmatrix} 0 \\ K_{\mathcal{M}}(\,\cdot\,,z^*)\begin{pmatrix}1\\0\end{pmatrix}\end{pmatrix} = \begin{pmatrix} 0 \\ 0 \\ \bigl(G_i^{-1}\boldsymbol{e}_i^\dagger(z)\bigr)_{i=0}^{\ell}\end{pmatrix}$$

and similarly

$$\begin{aligned}\Phi\begin{pmatrix} 0 \\ K_{\mathcal{M}}(\,\cdot\,,z^*)\begin{pmatrix}0\\1\end{pmatrix}\end{pmatrix} &= \begin{pmatrix} 0 \\ \bigl(G_i^*G_i^{-*}\boldsymbol{e}_i(z)\bigr)_{i=0}^{\ell} \\ 0\end{pmatrix}\\ &= \begin{pmatrix} 0 \\ \bigl(\boldsymbol{e}_i(z)\bigr)_{i=0}^{\ell} \\ 0\end{pmatrix}.\end{aligned}$$

□

Proof of Theorem 3.1. Let again the element $\widetilde{f} \in \mathcal{L}(\mathcal{Q})$ be given in the form (3.13) with $\widetilde{f}_0 \in \mathcal{L}(q_0)$ and $\widetilde{h}_{ij}$, $\widetilde{k}_{ij} \in \mathbb{C}$, and similarly an element $\widetilde{F} \in \mathcal{L}(\mathcal{Q})$ with $\widetilde{F}_0 \in \mathcal{L}(q_0)$ and $\widetilde{H}_{ij}$, $\widetilde{K}_{ij} \in \mathbb{C}$ for $i = 0,1,\dots,\ell$ and $j = 1,\dots,\nu_i$. Moreover, we write $\Phi(\widetilde{f}) =: \bigl(f_0, (\boldsymbol{h}_i)_{i=0}^{\ell}, (\boldsymbol{k}_i)_{i=0}^{\ell}\bigr)^\top$ and $\Phi(\widetilde{F}) =: \bigl(F_0, (\boldsymbol{H}_i)_{i=0}^{\ell}, (\boldsymbol{K}_i)_{i=0}^{\ell}\bigr)^\top$.

Inserting $\widetilde{f}$ and $\widetilde{F}$ in the description of $\widetilde{S}^*$ in (3.11) yields that $\{\widetilde{f};\widetilde{F}\}\in\widetilde{S}^*$ if and only if there exist complex numbers c_1, c_2, c_3 and c_4 such that

$$\widetilde{F}_0(\zeta)-\zeta\widetilde{f}_0(\zeta) = c_1 - c_3 q_0(\zeta), \quad (3.15)$$

$$\sum_{j=1}^{\nu_0}\widetilde{H}_{0j}\zeta^{j-1}+\sum_{i=1}^{\ell}\sum_{j=1}^{\nu_i}\frac{\widetilde{H}_{ij}}{(\zeta-\alpha_i)^j} - \zeta\sum_{j=1}^{\nu_0}\widetilde{h}_{0j}\zeta^{j-1}-\zeta\sum_{i=1}^{\ell}\sum_{j=1}^{\nu_i}\frac{\widetilde{h}_{ij}}{(\zeta-\alpha_i)^j} = c_2 - c_1 r^{\#}(\zeta), \quad (3.16)$$

$$\sum_{j=1}^{\nu_0}\widetilde{K}_{0j}\zeta^{j-1}+\sum_{i=1}^{\ell}\sum_{j=1}^{\nu_i}\frac{\widetilde{K}_{ij}}{(\zeta-\alpha_i^*)^j} - \zeta\sum_{j=1}^{\nu_0}\widetilde{k}_{0j}\zeta^{j-1}-\zeta\sum_{i=1}^{\ell}\sum_{j=1}^{\nu_i}\frac{\widetilde{k}_{ij}}{(\zeta-\alpha_i^*)^j} = c_3 - c_4 r(\zeta). \quad (3.17)$$

According to (2.5), (2.6) and Remark 2.2 equation (3.15) can be written as (3.3). Comparing coefficients in the partial fractional decomposition (3.16) yields

$$\widetilde{\boldsymbol{H}}_i = \begin{pmatrix}\alpha_i & 1 & & \\ & \ddots & \ddots & \\ & & \ddots & 1\\ & & & \alpha_i\end{pmatrix}\widetilde{\boldsymbol{h}}_i + c_1\begin{pmatrix}\sigma_{i,1}^*\\ \sigma_{i,2}^*\\ \vdots\\ \sigma_{i,\nu_i}^*\end{pmatrix} \quad \text{for } i=1,\dots,\ell,$$

$$\widetilde{\boldsymbol{h}}_0 = \begin{pmatrix}0 & 1 & & \\ & \ddots & \ddots & \\ & & \ddots & 1\\ & & & 0\end{pmatrix}\widetilde{\boldsymbol{H}}_0 + c_1\begin{pmatrix}\sigma_{0,1}^*\\ \sigma_{0,2}^*\\ \vdots\\ \sigma_{0,\nu_i}^*\end{pmatrix} \quad \text{and} \quad \widetilde{H}_{01}-\sum_{i=1}^{\ell}\widetilde{h}_{i1} = c_2 - c_1\sigma_{00}^*.$$

Observing that for $i=0,1,\dots,\ell$,

$$\begin{pmatrix}\alpha & 1 & & \\ & \ddots & \ddots & \\ & & \ddots & 1\\ & & & \alpha\end{pmatrix}G_i^{-*} = G_i^{-*}\begin{pmatrix}\alpha & & & \\ 1 & \ddots & & \\ & \ddots & \ddots & \\ & & 1 & \alpha\end{pmatrix} \quad \text{for any } \alpha\in\mathbb{C}$$

and

$$G_i^*\begin{pmatrix}\sigma_{i1}^*\\ \sigma_{i2}^*\\ \vdots\\ \sigma_{i\nu_i}^*\end{pmatrix} = \begin{pmatrix}1\\ 0\\ \vdots\\ 0\end{pmatrix}$$

and using the relations between $\widetilde{\boldsymbol{h}}_i$, $\widetilde{\boldsymbol{H}}_i$ and $\boldsymbol{h}_i$, $\boldsymbol{H}_i$, we obtain (3.4), (3.5) and (3.8). In the same way one finds that (3.17) can be written as (3.6), (3.7) and (3.9). Hence we have found that $\{\widetilde{f};\widetilde{F}\}\in\widetilde{S}^*$ if and only if $\{\Phi(\widetilde{f});\Phi(\widetilde{F})\}\in\mathcal{S}^*$.

Since $B_i\{\Phi(\widetilde{f});\Phi(\widetilde{F})\}=B_{\widetilde{S},i}\{\widetilde{f};\widetilde{F}\}$ for $i=0,1$ (cf. Lemma 3.4), B_0,B_1 are possible boundary mappings and

$$\{\widetilde{f};\widetilde{F}\}\in\widetilde{A}\Longleftrightarrow\{\Phi(\widetilde{f});\Phi(\widetilde{F})\}\in\mathcal{A},$$
$$\{\widetilde{f};\widetilde{F}\}\in\widetilde{S}\Longleftrightarrow\{\Phi(\widetilde{f});\Phi(\widetilde{F})\}\in\mathcal{S}.$$

The defect function $\widetilde{\varphi}(\,\cdot\,,z)$ for $\widetilde{S}^*$ is given by

$$\widetilde{\varphi}(\zeta,z)=\frac{\mathcal{Q}(\zeta)-\mathcal{Q}(z)}{\zeta-z}\boldsymbol{v}(z)=\begin{pmatrix} r(z)\frac{q_0(\zeta)-q_0(z)}{\zeta-z}\\ r(z)q_0(z)\frac{r^\#(\zeta)-r^\#(z)}{\zeta-z}\\ \frac{r(\zeta)-r(z)}{\zeta-z}\end{pmatrix},$$

and hence

$$\Phi(\widetilde{\varphi}(\,\cdot\,,z))=\begin{pmatrix} r(z)\varphi_0(z)\\ \big(r(z)q_0(z)\boldsymbol{e}_i(z)\big)_{i=0}^{\ell}\\ \big(G_i^{-1}\boldsymbol{e}_i^\dagger(z)\big)_{i=0}^{\ell}\end{pmatrix}$$

because of Lemma 3.5, and the relation $\varphi_0(z)=\Phi_0\big(K_{q_0}(\,\cdot\,,z^*)\big)$. It follows from Lemma 3.4 that $(\mathcal{A},\mathcal{S},\varphi)$ is a realization of the function q and that B_0,B_1 are boundary mappings that are compatible with this realization. □

The compression of the resolvent of $\mathcal{A}$ to the Hilbert space $\mathcal{K}_0$ is given in the next proposition.

Proposition 3.6. *For $z\in\varrho(\mathcal{A})$ the following relation holds,*

$$P_{\mathcal{K}_0}(\mathcal{A}-z)^{-1}|_{\mathcal{K}_0}=(A_0-z)^{-1},$$

where $P_{\mathcal{K}_0}$ denotes the orthogonal projection onto the subspace $\mathcal{K}_0$.

Proof. We calculate the element F_0 from $\begin{pmatrix}F_0\\ \boldsymbol{h}\\ \boldsymbol{k}\end{pmatrix}=(\mathcal{A}-z)^{-1}\begin{pmatrix}f_0\\ \boldsymbol{0}\\ \boldsymbol{0}\end{pmatrix}$, that is,

$$\left\{\begin{pmatrix}F_0\\ \boldsymbol{h}\\ \boldsymbol{k}\end{pmatrix};\begin{pmatrix}f_0+zF_0\\ z\boldsymbol{h}\\ z\boldsymbol{k}\end{pmatrix}\right\}\in\mathcal{A}.$$

From Theorem 3.1 it follows in particular that

$$\{F_0;f_0+zF_0\}\in S_0^*,\qquad B_{S_0,0}(\{F_0;f_0+zF_0\})=c_3 \tag{3.18}$$

and

$$z\boldsymbol{k}_i = \begin{pmatrix} \alpha_i^* & 1 & & \\ & \ddots & \ddots & \\ & & \ddots & 1 \\ & & & \alpha_i^* \end{pmatrix} \boldsymbol{k}_i \quad \text{for } i = 1, \dots, \ell.$$

Since $z \neq \alpha_i^*$, this implies $\boldsymbol{k}_i = 0$, and if $\nu_0 > 0$, we also find $\boldsymbol{k}_0 = 0$. In both cases relation (3.9) gives $c_3 = 0$. But then (3.18) implies $\{F_0; f_0 + zF_0\} \in A_0$ and hence $F_0 = (A_0 - z)^{-1} f_0$. □

Minimality. In [DLLuSh2] it is discussed that the realization given there and hence also that given in Theorem 3.1 need not be minimal. This happens exactly if a zero (pole) of r is a generalized pole (zero) of q_0. But in this case there exists a finite-dimensional positive subspace $\mathcal{K}_{\mathrm{red}} \subseteq \mathcal{K}$ such that

$$\left(\mathcal{A}|_{\mathcal{K}_{\mathrm{red}}^{[\perp]}}, \varphi\right)$$

forms a minimal realization of q in $\mathcal{K}_{\mathrm{red}}^{[\perp]}$. Here $\mathcal{K}_{\mathrm{red}}^{[\perp]}$ denotes the orthogonal complement of $\mathcal{K}_{\mathrm{red}}$ in $\mathcal{K}$ with respect to $[\,\cdot\,,\cdot\,]$. Note that in [DLLuSh2] it was shown that $\mathcal{K}_{\mathrm{red}}^{[\perp]}$ is an invariant subspace for the resolvent of $\mathcal{A}$ and hence the restriction should be understood in this sense; moreover, $\varphi(z) \in \mathcal{K}_{\mathrm{red}}^{[\perp]}$ for $z \in \varrho(\mathcal{A})$.

In order to describe the space $\mathcal{K}_{\mathrm{red}}$ we need some more notations. Put $\alpha_0 := \infty$, $\beta_0 := \infty$. In the sets I^+, J^+ we collect those indices i (j, respectively) such that the generalized pole α_i (zero β_j, respectively) of q is also a generalized zero (pole) of q_0:

$$I^+ := \left\{0 \leq i \leq \ell \mid \alpha_i \in \sigma_p\big(A_{\widehat{q}_0}\big)\right\}, \quad J^+ := \left\{0 \leq j \leq k \mid \beta_j \in \sigma_p\left(A_{q_0}\right)\right\},$$

where again $\widehat{q}_0(z) = -1/q_0(z)$. Note that the α_i and β_j are real numbers or ∞. Moreover, define the elements $\widehat{y}_{\alpha_i}$ and y_{β_j} in $\mathcal{K}_0$ such that

$$\begin{array}{lll} \text{for } i \in I^+ \setminus \{0\}: & \{\widehat{y}_{\alpha_i}; \alpha_i \widehat{y}_{\alpha_i}\} \in \ker B_{S_0,1} & \text{with } B_{S_0,0}\{\widehat{y}_{\alpha_i}; \alpha_i \widehat{y}_{\alpha_i}\} = -1, \\ \text{if } 0 \in I^+: & \{0; \widehat{y}_{\alpha_0}\} \in \ker B_{S_0,1} & \text{with } B_{S_0,0}\{0; \widehat{y}_{\alpha_0}\} = 1, \\ \text{for } j \in J^+ \setminus \{0\}: & \{y_{\beta_j}; \beta_j y_{\beta_j}\} \in \ker B_{S_0,0} & \text{with } B_{S_0,1}\{y_{\beta_j}; \beta_j y_{\beta_j}\} = 1, \\ \text{if } 0 \in J^+: & \{0; y_{\beta_0}\} \in \ker B_{S_0,0} & \text{with } B_{S_0,1}\{0; y_{\beta_0}\} = -1. \end{array}$$

Note that y_{β_j} and $\widehat{y}_{\alpha_i}$ are eigenvectors of the self-adjoint relations $A_0 = \ker B_{S_0,0}$ and $\ker B_{S_0,1}$, respectively, where $\ker B_{S_0,1}$ is unitarily equivalent to $A_{\widehat{q}_0}$. Since the eigenspaces of $\ker B_{S_0,1}$ and $\ker B_{S_0,0}$ are one-dimensional, the elements $\widehat{y}_{\alpha_i}$ and y_{β_j} are uniquely determined by the above characterization. Define the following elements in $\mathcal{K}$:

$$\widehat{\boldsymbol{y}}_{\alpha_i} = \begin{pmatrix} \widehat{y}_{\alpha_i} \\ 0 \\ \left(\delta_{ik}\boldsymbol{e}\right)_{k=0}^{\ell} \end{pmatrix} \qquad \text{for } i \in I^+,$$

$$\boldsymbol{y}_{\beta_j} = \begin{pmatrix} y_{\beta_j} \\ \left(\boldsymbol{e}_k(\beta_j)\right)_{k=0}^{\ell} \\ 0 \end{pmatrix} \qquad \text{for } j \in J^+ \setminus \{0\},$$

$$\boldsymbol{y}_{\beta_0} = \begin{pmatrix} y_{\beta_j} \\ (-\boldsymbol{e})_{k=1}^{\ell} \\ 0 \end{pmatrix} \qquad \text{if } 0 \in J^+.$$

We can now give an explicit description of the space $\mathcal{K}_{\text{red}}$.

Proposition 3.7. *With the notations as above and as in the beginning of this section the Hilbert space $\mathcal{K}_{\text{red}}$ (such that $(\mathcal{A}|_{\mathcal{K}_{\text{red}}^{[\perp]}}, \varphi)$ is a minimal realization of q) is given by*

$$\mathcal{K}_{\text{red}} = \operatorname{span}\{\widehat{\boldsymbol{y}}_{\alpha_i}, \boldsymbol{y}_{\beta_j} \mid i \in I^+, j \in J^+\}.$$

Proof. In [DLLuSh2] a space $\widetilde{\mathcal{K}}_{\text{red}}$ was constructed such that $(\widetilde{A}|_{\widetilde{\mathcal{K}}_{\text{red}}^{[\perp]}}, \widetilde{\varphi})$ is a minimal realization of q in $\widetilde{\mathcal{K}}_{\text{red}}^{[\perp]}$. Here $\widetilde{\mathcal{K}}_{\text{red}}$ is spanned by the vectors

$$\begin{pmatrix} \widehat{x}_{\alpha_i} \\ 0 \\ \widehat{q}_0\widehat{x}_{\alpha_i} \end{pmatrix}, \; i \in I^+, \quad \text{and} \quad \begin{pmatrix} x_{\beta_j} \\ -r^{\#}x_{\beta_j} \\ 0 \end{pmatrix}, \; j \in J^+,$$

where

$$\widehat{x}_{\alpha_i}(\zeta) := \begin{cases} \dfrac{q_0(\zeta)}{\zeta - \alpha_i} & \text{for } i \in I^+ \setminus \{0\}, \\ q_0(\zeta) & \text{for } i = 0 \in I^+ \end{cases}$$

and

$$x_{\beta_j}(\zeta) := \begin{cases} \dfrac{1}{\zeta - \beta_j} & \text{for } j \in J^+ \setminus \{0\}, \\ 1 & \text{for } j = 0 \in J^+ \end{cases}.$$

It remains to show that

$$\widehat{\boldsymbol{y}}_{\alpha_i} = -\Phi\big((\widehat{x}_{\alpha_i} \quad 0 \quad \widehat{q}_0\widehat{x}_{\alpha_i})^\top\big) \quad \text{and} \quad \boldsymbol{y}_{\beta_j} = -\Phi\big((x_{\beta_j} \quad -r^{\#}x_{\beta_j} \quad 0)^\top\big). \tag{3.19}$$

Indeed it follows from the facts that the element x_{β_j} is an eigenvector of A_{q_0} and $B_{S_0,m}\{y_{\beta_j}; \beta_j y_{\beta_j}\} = -B_{\widetilde{S}_0,m}\{x_{\beta_j}; \beta_j x_{\beta_j}\}$ for $m = 0, 1$ that $\Phi_0(-x_{\beta_j}) = y_{\beta_j}$ since the eigenspaces are one-dimensional and the boundary mappings yield the correct scaling. For $j > 0$ Lemma 3.5 implies that

$$\Phi\begin{pmatrix} 0 \\ \frac{r^{\#}(\zeta)}{\zeta-\beta_j} \\ 0 \end{pmatrix} = \Phi\begin{pmatrix} 0 \\ \frac{r^{\#}(\zeta)-r^{\#}(\beta_j)}{\zeta-\beta_j} \\ 0 \end{pmatrix} = \begin{pmatrix} 0 \\ \left(\boldsymbol{e}_k(\beta_j)\right)_{k=0}^{\ell} \\ 0 \end{pmatrix}. \tag{3.20}$$

If $0 \in J^+$, then ∞ is a zero of r and hence $r_0 = 0$. It follows from the definition of Φ that

$$\Phi\begin{pmatrix} 0 \\ r^{\#} \\ 0 \end{pmatrix} = \begin{pmatrix} 0 \\ \left(-G_k^* \boldsymbol{\sigma}_k^*\right)_{k=1}^{\ell} \\ 0 \end{pmatrix} = \begin{pmatrix} 0 \\ (-\boldsymbol{e})_{k=1}^{\ell} \\ 0 \end{pmatrix}. \tag{3.21}$$

The relations $\Phi_0(-x_{\beta_j}) = y_{\beta_j}$ and (3.20), (3.21) show the validity of the second equality in (3.19). The first equality in (3.19) is similar but even easier to show. □

Remark 3.8. In [DLLuSh2] it was shown that those generalized poles and zeros of the function q that contribute to the space $\mathcal{K}_{\text{red}}$ can also be characterized analytically. For $1 < i, j \leq \ell$ the following is true:

$$j \in J^+ \quad \Longleftrightarrow \quad \exists\, \mu \in \mathbb{N}: \quad \lim_{z \hat{\to} \beta_j} \frac{q(z)}{(z-\beta_j)^{2\mu-1}} < 0,$$

and

$$i \in I^+ \quad \Longleftrightarrow \quad \exists\, \nu \in \mathbb{N}: \quad \lim_{z \hat{\to} \alpha_i} (z-\alpha_i)^{2\nu-1} q(z) > 0,$$

where $\hat{\to}$ denotes a non-tangential limit.

4. Block operator matrix representations of $\mathcal{A}$

In this section we consider some cases, where the realization $(\mathcal{A}, \varphi)$ and in particular the description of the relation $\mathcal{A}$ simplify. In the first two subsections we consider the case that $\mathcal{A}$ is an operator, which is the case if neither q_0 nor r have a generalized pole at infinity. There are two cases to consider, whether S_0 (the corresponding symmetric operator to q_0) is densely defined or not. Finally, we give an example where $\mathcal{A}$ is a relation.

4.1. The case that S_0 is not densely defined

We consider the case that S_0 is not densely defined and A_0 is an operator, that is, the function q_0 possesses a minimal u-resolvent representation

$$q_0(z) = s_0 + \left[(A_0 - z)^{-1} u_0, u_0\right]_0 \tag{4.1}$$

with $s_0 \in \mathbb{R}$ and $u_0 \in \mathcal{K}_0$. Moreover, we assume that $r(z)$ has no pole at infinity, i.e., $\nu_0 = 0$.

Theorem 4.1. *Assume that q_0 has the representation* (4.1) *and that $\nu_0 = 0$. Then the operator $\mathcal{A}$ in the realization of q has the following block operator matrix representation*

$$\mathcal{A} = \left(\begin{array}{c|ccc|ccc} A_0 & & & & (\cdot,\boldsymbol{e})u_0 & \cdots & (\cdot,\boldsymbol{e})u_0 \\ \hline [\cdot,u_0]_0\boldsymbol{e} & J(\alpha_1) & & & -s_0(\cdot,\boldsymbol{e})\boldsymbol{e} & \cdots & -s_0(\cdot,\boldsymbol{e})\boldsymbol{e} \\ \vdots & & \ddots & & \vdots & & \vdots \\ [\cdot,u_0]_0\boldsymbol{e} & & & J(\alpha_\ell) & -s_0(\cdot,\boldsymbol{e})\boldsymbol{e} & \cdots & -s_0(\cdot,\boldsymbol{e})\boldsymbol{e} \\ \hline & & & & J(\alpha_1)^* & & \\ & & & & & \ddots & \\ & & & & & & J(\alpha_\ell)^* \end{array}\right), \tag{4.2}$$

where empty blocks are zero. Moreover, the function q has the u-resolvent representation

$$q(z) = s_0\sigma_{00}^2 + \big[(\mathcal{A}-z)^{-1}u, u\big],$$

where

$$u := (\sigma_{00}u_0 \quad -s_0\sigma_{00}\boldsymbol{e} \quad \cdots \quad -s_0\sigma_{00}\boldsymbol{e} \quad -\boldsymbol{\sigma}_1 \quad \cdots \quad -\boldsymbol{\sigma}_\ell)^\top. \tag{4.3}$$

Proof. According to Section 2.4 we have

$$S_0^* = \big\{\{f_0; F_0\} \in \mathcal{D}(A_0) \times \mathcal{H}_0 \mid \exists c \in \mathbb{C} : F_0 = A_0 f_0 - cu_0\big\}, \tag{4.4}$$

and

$$B_{S_0,0}\{f_0; F_0\} = c, \quad B_{S_0,1}\{f_0; F_0\} = s_0 c + [f_0, u_0]_0,$$

where c is the constant appearing in (4.4), are boundary mappings which are compatible with the realization of q_0. Now let

$$\left\{\begin{pmatrix} f_0 \\ \boldsymbol{h} \\ \boldsymbol{k}\end{pmatrix}; \begin{pmatrix} F_0 \\ \boldsymbol{H} \\ \boldsymbol{K}\end{pmatrix}\right\} \in \mathcal{A}.$$

Relations (3.3) and (3.9) yield (note that $c_4 = 0$)

$$\begin{aligned} F_0 &= A_0 f_0 - B_{S_0,0}\{f_0; F_0\}u_0 = A_0 f_0 - c_3 u_0 = A_0 f_0 + \sum_{i=1}^{\ell} k_{i1}u_0 \\ &= A_0 f_0 + \sum_{i=1}^{\ell} (\boldsymbol{k}_i, \boldsymbol{e})u_0, \end{aligned}$$

and (3.5) gives

$$\begin{aligned} \boldsymbol{H}_i &= J(\alpha_i)\boldsymbol{h}_i + c_1\boldsymbol{e} = J(\alpha_i)\boldsymbol{h}_i + s_0c_3 + [f_0, u_0]_0 \\ &= J(\alpha_i)\boldsymbol{h}_i - s_0\sum_{i=1}^{\ell}(\boldsymbol{k}_i, \boldsymbol{e}) + [f_0, u_0]_0. \end{aligned}$$

Relation (3.7) reduces to $\boldsymbol{K}_i = J(\alpha_i)^*\boldsymbol{k}_i$ and (3.8) gives no constraint since c_2 is arbitrary, which proves (4.2). Since $\varphi(z)$ is in the domain of $\mathcal{A}$, the function $q(z)$ admits again a u-resolvent representation with $u = (\mathcal{A}-z)\varphi(z)$. The constant $s_0\sigma_{00}^2$ is obtained by taking the limit $z \to \infty$. So it remains to prove (4.3). Let $u = (\mathcal{A}-z)\varphi(z) = (F_0 \quad \boldsymbol{H} \quad \boldsymbol{K})^\top$. Then

$$\begin{aligned} F_0 &= r(z)(A_0 - z)\varphi_0(z) + \sum_{i=1}^{\ell}\big(G_i^{-1}\boldsymbol{e}_i^\dagger(z), \boldsymbol{e}\big)u_0 \\ &= r(z)u_0 + \sum_{i=1}^{\ell}\sum_{j=1}^{\nu_i}\frac{\sigma_{ij}}{(z-\alpha_i^*)^j}u_0 = \sigma_{00}u_0, \end{aligned}$$

$$\begin{aligned} \boldsymbol{H}_i &= r(z)\big[\varphi_0(z), u_0\big]_0\boldsymbol{e} + r(z)q_0(z)J(\alpha_i)\boldsymbol{e}_i(z) - s_0\sum_{k=1}^{\ell}\big(G_k^{-1}\boldsymbol{e}_k^\dagger(z), \boldsymbol{e}\big)\boldsymbol{e} \\ &= r(z)\big(q_0(z) - s_0\big)\boldsymbol{e} + r(z)q_0(z)\boldsymbol{e} - s_0\sum_{k=1}^{\ell}\sum_{j=1}^{\nu_k}\frac{\sigma_{kj}}{(z-\alpha^*)^j}\boldsymbol{e} \\ &= -s_0 r(z)\boldsymbol{e} - s_0\big(-r(z) + \sigma_{00}\big)\boldsymbol{e} = -s_0\sigma_{00}\boldsymbol{e}, \end{aligned}$$

$$\boldsymbol{K}_i = J(\alpha_i)^*G_i^{-1}\boldsymbol{e}_i^\dagger(z) = G_i^{-1}J(\alpha_i^*)\boldsymbol{e}_i^\dagger(z) = -G_i^{-1}\boldsymbol{e} = -\boldsymbol{\sigma}_i,$$

which finishes the proof. □

To make the representation even more explicit, one can choose A_0 to be the multiplication operator by the independent variable in the space L^2_σ and $u_0 = \mathbf{1}$, where σ is the measure in the integral representation of q_0.

4.2. The case that S_0 is densely defined

Now we consider the case that S_0 is a densely defined operator, that is, q_0 has the integral representation

$$q_0(z) = a + \int_{\mathbb{R}}\Big(\frac{1}{t-z} - \frac{t}{1+t^2}\Big)d\sigma(t), \tag{4.5}$$

where $a \in \mathbb{R}$ and σ is a measure with

$$\int_{\mathbb{R}}\frac{1}{1+t^2}d\sigma(t) < \infty \quad \text{but} \quad \int_{\mathbb{R}}d\sigma(t) = \infty.$$

As explained in Section 2.4 the operator A_0 can be chosen to be the multiplication operator by the independent variable in L^2_σ with maximal domain, S_0^* is given by

$$\begin{aligned} &\mathcal{D}(S_0^*) = \{f \in L^2_\sigma \mid \exists\, c_f : tf(t) - c_f \in L^2_\sigma\} \\ &(S_0^*f)(t) = tf(t) - c_f \end{aligned}$$

and

$$B_{S_0,0}f := c_f, \qquad B_{S_0,1}f := c_f + \int_{\mathbb{R}}\Big(f(t) - c_f\frac{t}{1+t^2}\Big)d\sigma(t)$$

are boundary mappings that are compatible with the realization.

Theorem 4.2. *Suppose that q_0 has the integral representation (4.5) with an infinite measure σ. Moreover, assume that $\nu_0 = 0$. Then the operator $\mathcal{A}$ has the following block operator matrix representation*

$$\mathcal{A} = \left(\begin{array}{c|ccc|ccc} t\cdot & & & & (\cdot,\boldsymbol{e})\mathbf{1} & \cdots & (\cdot,\boldsymbol{e})\mathbf{1} \\ \hline (B_{S_0,1}\cdot)\boldsymbol{e} & J(\alpha_1) & & & & & \\ \vdots & & \ddots & & & & \\ (B_{S_0,1}\cdot)\boldsymbol{e} & & & J(\alpha_\ell) & & & \\ \hline & & & & J(\alpha_1)^* & & \\ & & & & & \ddots & \\ & & & & & & J(\alpha_\ell)^* \end{array}\right) \tag{4.6}$$

with domain

$$\mathcal{D}(\mathcal{A}) = \left\{ (f_0 \quad \boldsymbol{h} \quad \boldsymbol{k})^\top \in \mathcal{K} \,\middle|\, t f_0(t) + \sum_{i=1}^{\ell} (\boldsymbol{k}_i, \boldsymbol{e}) \in L^2_\sigma \right\}.$$

Remark 4.3. Note that the components of vectors in the domain of $\mathcal{A}$ are coupled by an extra condition.

Proof. Let $\{(f_0 \quad \boldsymbol{h} \quad \boldsymbol{k})^\top; (F_0 \quad \boldsymbol{H} \quad \boldsymbol{K})^\top\} \in \mathcal{A}$. Since S_0^* is an operator, f_0 determines F_0, c_1 and c_3 by (3.3), which in particular implies $F_0(t) = tf(t) - c_3$. Relation (3.7) reduces to $\boldsymbol{K}_i = J(\alpha_i)^* \boldsymbol{k}_i$, (3.8) is void, and (3.9) reduces to $\sum_{i=1}^{\ell} (\boldsymbol{k}_i, \boldsymbol{e}) + c_3 = 0$, which is a constraint for elements in the domain of $\mathcal{A}$. □

4.3. An example

In this subsection we consider the example

$$q_\mu(z) = -\frac{\pi}{2\sin \pi\mu}(-z)^\mu$$

for $\mu > -1$, $\mu \notin \mathbb{Z}$. This function appears as a Titchmarsh–Weyl coefficient in connection with the Bessel operator on the half line, see [DSh]. We choose the branch such that for $z \in \mathbb{C}^+$ it holds $(-z)^\mu = \rho^\mu e^{i\mu(\phi-\pi)}$ where $z = \rho e^{i\phi}$. Write $\mu = \mu_0 + 2\kappa$ with $-1 < \mu_0 < 1$ and $\kappa \in \mathbb{Z}$. Then the basic factorization of q_μ is given by $q_\mu(z) = r^\#(z) q_{\mu_0}(z) r(z)$ with $r(z) = z^\kappa$. It is easy to see that q_{μ_0} is an $\mathcal{N}_0$-function, which has the following integral representation:

$$q_{\mu_0}(z) = \frac{\pi}{4\sin\frac{\pi\mu_0}{2}} + \frac{1}{2}\int_0^\infty \left(\frac{1}{t-z} - \frac{t}{1+t^2}\right) t^{\mu_0}\, dt,$$

which for $\mu_0 < 0$ reduces to

$$q_{\mu_0}(z) = \frac{1}{2}\int_0^\infty \frac{1}{t-z}\, t^{\mu_0}\, dt.$$

Since the measure σ is infinite, the operator S_0^* is given by $S_0^* f = tf(t) - c_f$, where c_f is such that this expression is in L^2_σ. Corresponding boundary mappings (cf. Section 2.4) are given by

$$B_{S_0,0} f = c_f \quad \text{(where } c_f \text{ is as above)},$$

$$B_{S_0,1} f = -\frac{\pi}{4 \sin \frac{\pi \mu_0}{2}} c_f + \frac{1}{2} \int_0^\infty \Big(f(t) - c_f \frac{t}{1+t^2} \Big) t^{\mu_0} dt.$$

In the case $\mu_0 < 0$ we just have $B_{S_0,1} f = \frac{1}{2} \int_0^\infty f(t) t^{\mu_0} dt$.

With the notation of Section 3 we have $\ell = 0$, $\nu_0 = \kappa$, $\sigma_{00} = \cdots = \sigma_{0,\kappa-1} = 0$, and $\sigma_{0\kappa} = 1$. Therefore there exists a realization of q_μ in the space $\mathcal{K} = \mathcal{K}_0[+](\mathbb{C}^\kappa \dot{+} \mathbb{C}^\kappa)$, where $\mathcal{K}_0 = L^2_\sigma$ with $d\sigma = \frac{1}{2} t^{\mu_0} dt$. The self-adjoint relation $\mathcal{A}$ in this realization is described in the following theorem.

Theorem 4.4. *Let the notation be as above and $\mathcal{A}$ be the relation in the representation of the function q_μ. The pair $\{(f_0\ \boldsymbol{h}_0\ \boldsymbol{k}_0)^\top; (F_0\ \boldsymbol{H}_0\ \boldsymbol{K}_0)^\top\}$ is in $\mathcal{A}$, where $f_0, F_0 \in \mathcal{K}_0$, $\boldsymbol{h}_0, \boldsymbol{k}_0, \boldsymbol{H}_0, \boldsymbol{K}_0 \in \mathbb{C}^\kappa$, if and only if*

$$\begin{pmatrix} F_0 \\ \boldsymbol{H}_0 \\ \boldsymbol{K}_0 \end{pmatrix} = \begin{pmatrix} S_0^* & 0 & 0 \\ 0 & J(0)^* & 0 \\ (B_{S_0,0}\cdot)\boldsymbol{e} & 0 & J(0) \end{pmatrix} \begin{pmatrix} f_0 \\ \boldsymbol{h}_0 \\ \boldsymbol{k}_0 \end{pmatrix} + \gamma \begin{pmatrix} 0 \\ \hline 0 \\ \vdots \\ 0 \\ 1 \\ \hline 0 \\ \vdots \\ 0 \end{pmatrix},$$

with $k_{0\kappa} = 0$, $h_{01} = B_{S_0,1} f_0$, and γ is an arbitrary complex number. The defect function φ is given by

$$\varphi(z) = \Big(\frac{z^\kappa}{t-z};\ q_{\mu_0}(z) z^\kappa, \ldots, q_{\mu_0}(z) z^{2\kappa-1};\ z^{\kappa-1}, \ldots, z, 1 \Big)^\top.$$

Proof. Since S_0^* is an operator, the constants c_1 and c_3 in Theorem 3.1 are determined by f_0. Equations (3.4) and (3.8) yield $\boldsymbol{H}_0 = J(0)^* \boldsymbol{h}_0 + c_2 (0 \ldots 0\,1)^\top$ with $c_2 = \gamma \in \mathbb{C}$ arbitrary and $h_{01} = B_{S_0,1} f_0$. Equations (3.6) and (3.9) give $\boldsymbol{K}_0 = J(0)\boldsymbol{k}_0 + (B_{S_0,0} f_0)\boldsymbol{e}$ and $k_{0\kappa} = 0$. The calculation of φ is straightforward. □

References

[De] V.A. Derkach, *On Weyl function and generalized resolvents of a Hermitian operator in a Krein space.* Integral Equations Operator Theory, **23** (1995), 387–415.

[DeH] V.A. Derkach and S. Hassi, *A reproducing kernel space model for N_κ-functions.* Proc. Amer. Math. Soc. **131** (2003), 3795–3806.

[DeHS1] V.A. Derkach, S. Hassi and H. de Snoo, *Operator models associated with Kac subclasses of generalized Nevanlinna functions.* Methods Funct. Anal. Topology **5** (1999), 65–87.

[DeM] V.A. Derkach and M.M. Malamud, *The extension theory of Hermitian operators and the moment problem.* J. Math. Sci. **73** (1995), 141–242.

[DLLuSh1] A. Dijksma, H. Langer, A. Luger and Yu. Shondin, *A factorization result for generalized Nevanlinna functions of the class $\mathcal{N}_\kappa$*. Integral Equations Operator Theory **36** (2000), 121–125.

[DLLuSh2] A. Dijksma, H. Langer, A. Luger and Yu. Shondin, *Minimal realizations of scalar generalized Nevanlinna functions related to their basic factorization.* Oper. Theory Adv. Appl. **154**, 69–90, Birkhäuser, Basel, 2004.

[DSh] A. Dijksma and Yu. Shondin, *Singular point-like perturbations of the Bessel operator in a Pontryagin space.* J. Differential Equations **164** (2000), 49–91.

[KL] M.G. Krein and H. Langer, *Über einige Fortsetzungsprobleme, die eng mit der Theorie hermitescher Operatoren im Raume Π_κ zusammenhängen. I. Einige Funktionenklassen und ihre Darstellungen.* Math. Nachr. **77** (1977), 187–236.

Matthias Langer
Department of Mathematics
University of Strathclyde
26 Richmond Street
Glasgow G1 1XH
United Kingdom
e-mail: ml@maths.strath.ac.uk

Annemarie Luger
Institute for Analysis and Scientific Computing
Vienna University of Technology
Wiedner Hauptstraße 8/101
A-1040 Wien
Austria
e-mail: aluger@mail.zserv.tuwien.ac.at

Operator Theory:
Advances and Applications, Vol. 162, 277–292

Polar Decompositions of Normal Operators in Indefinite Inner Product Spaces

Christian Mehl, André C.M. Ran and Leiba Rodman

Abstract. Polar decompositions of normal matrices in indefinite inner product spaces are studied. The main result of this paper provides sufficient conditions for a normal operator in a Krein space to admit a polar decomposition. As an application of this result, we show that any normal matrix in a finite-dimensional indefinite inner product space admits a polar decomposition which answers affirmatively an open question formulated in [2]. Furthermore, necessary and sufficient conditions are given for a matrix to admit a polar decomposition and for a normal matrix to admit a polar decomposition with commuting factors.

Mathematics Subject Classification (2000). 47B50, 46C20, 15A23.

Keywords. Krein spaces, polar decompositions, normal operators, indefinite inner products.

1. Introduction

Let $\mathcal{H}$ be a (complex) Hilbert space, and let H be a (bounded) selfadjoint operator on $\mathcal{H}$, which is boundedly invertible. The operator H defines a Krein space structure on $\mathcal{H}$, via the indefinite inner product

$$[x, y] = \langle Hx, y\rangle, \qquad x, y \in \mathcal{H},$$

where $\langle\cdot,\cdot\rangle$ is the Hilbert inner product in $\mathcal{H}$. All operators in the paper are assumed to be linear and bounded. We denote by $\mathcal{L}(\mathcal{H})$ the Banach algebra of bounded linear operators on $\mathcal{H}$. The adjoint of an operator $X \in \mathcal{L}(\mathcal{H})$ with respect to $\langle\cdot,\cdot\rangle$ will be denoted by X^*.

A large part of this work was performed while the first author was visiting the Vrije Universiteit Amsterdam and while he was supported by the Research Training Network HPRN-CT-2000-00116 of the European Union, "Classical Analysis, Operator Theory, Geometry of Banach Spaces, their interplay and their applications".
Research of the third author was supported in part by NSF grant DMS-9988579 and by a Faculty Research Assignment grant from the College of William and Mary.

An operator $X \in \mathcal{L}(\mathcal{H})$ is said to be an *H-isometry* if $[Xx, Xy] = [x, y]$ for all $x, y \in \mathcal{H}$, and is called *H-selfadjoint* if $[Xx, y] = [x, Xy]$ for all $x, y \in \mathcal{H}$. An operator $X \in \mathcal{L}(\mathcal{H})$ is called *H-normal* if

$$XX^{[*]} = X^{[*]}X,$$

where $X^{[*]}$ is the adjoint of X with respect to the indefinite inner product $[\cdot, \cdot]$.

Given a (linear bounded) operator X on $\mathcal{H}$, a decomposition of the form

$$X = UA,$$

where U is an invertible H-isometry (in other words, U is *H-unitary*) and A is H-selfadjoint, is called an *H-polar decomposition* of X. An analogous decomposition of the form $X = AU$ will be called a *right* H-polar decomposition for X.

In the context of positive definite inner products, polar decompositions (which are usually taken with the additional requirement that A be positive semidefinite and the relaxation that U need be a partial isometry only instead of an invertible one) are a basic tool of operator theory. In context of indefinite inner products, they have been studied extensively in recent years (see, e.g., [4, 2, 3, 16, 13]), in particular, in connection with matrix computations [7, 8].

Remark 1. An operator $X \in \mathcal{L}(\mathcal{H})$ admits an H-polar decomposition if and only if it admits a right H-polar decomposition. This follows easily from the fact that $X = UA = (UAU^{-1})U$.

Our main result, Theorem 4, is stated and proved in the next section. In particular, it follows from Theorem 4 that for a finite-dimensional $\mathcal{H}$ every H-normal operator admits an H-polar decomposition, thereby settling in the affirmative an open question formulated in [2]. In Sections 3 and 4 we apply the main result to other properties that H-normal operators may have in connection with H-polar decompositions, assuming that $\mathcal{H}$ is finite-dimensional. In particular, we provide necessary and sufficient conditions for a matrix to admit a polar decomposition and for a normal matrix to admit a polar decomposition with commuting factors.

2. The main result

In this section, we will provide sufficient conditions for an H-normal operator to admit an H-polar decomposition. The proof of the main result will be based on the following decomposition that is of interest in itself.

Lemma 2. *Let $X \in \mathcal{L}(\mathcal{H})$, and let $Q_{\operatorname{Ker} X}$ be the orthogonal (in the Hilbert space sense) projection onto* $\operatorname{Ker} X$. *Assume that the operator*

$$Q_{\operatorname{Ker} X} H Q_{\operatorname{Ker} X}|_{\operatorname{Ker} X} \,:\, \operatorname{Ker} X \longrightarrow \operatorname{Ker} X \tag{1}$$

has closed range. Then there exists an invertible operator $P \in \mathcal{L}(\mathcal{H})$, a Hilbert space orthogonal decomposition

$$\mathcal{H} = \mathcal{H}_0 \oplus \mathcal{H}_1 \oplus \mathcal{H}_2 \oplus \widetilde{\mathcal{H}}_0 \tag{2}$$

and a Hilbert space isomorphism $H_{14} : \mathcal{H}_0 \to \widetilde{\mathcal{H}}_0$, *such that*

$$\operatorname{Ker}(P^{-1}XP) = \mathcal{H}_0 \oplus \mathcal{H}_1, \tag{3}$$

and with respect to decomposition (2), $P^{-1}XP$, P^*HP, *and* $P^{-1}X^{[*]}P$ *have the following block operator matrix forms:*

$$P^{-1}XP = \begin{bmatrix} 0 & 0 & X_{13} & X_{14} \\ 0 & 0 & X_{23} & X_{24} \\ 0 & 0 & X_{33} & X_{34} \\ 0 & 0 & X_{43} & X_{44} \end{bmatrix}, \quad P^*HP = \begin{bmatrix} 0 & 0 & 0 & H_{14} \\ 0 & H_{22} & 0 & 0 \\ 0 & 0 & H_{33} & 0 \\ H_{14}^* & 0 & 0 & 0 \end{bmatrix}, \tag{4}$$

and

$$P^{-1}X^{[*]}P = \begin{bmatrix} H_{14}^{-*}X_{44}^*H_{14}^* & H_{14}^{-*}X_{24}^*H_{22} & H_{14}^{-*}X_{34}^*H_{33} & H_{14}^{-*}X_{14}^*H_{14} \\ 0 & 0 & 0 & 0 \\ H_{33}^{-1}X_{43}^*H_{14}^* & H_{33}^{-1}X_{23}^*H_{22} & H_{33}^{-1}X_{33}^*H_{33} & H_{33}^{-1}X_{13}^*H_{14} \\ 0 & 0 & 0 & 0 \end{bmatrix}, \tag{5}$$

where $H_{14}^{-*} := (H_{14}^*)^{-1}$. *Moreover, if* X *is* H*-normal, then* $X_{23} = 0$, $X_{43} = 0$, *and* X_{33} *is* H_{33}*-normal.*

Proof. Let $\mathcal{H} = \mathcal{G}_0 \oplus \mathcal{G}_1$ where $\mathcal{G}_0 = \operatorname{Ker} X$ and $\mathcal{G}_1 = (\operatorname{Ker} X)^\perp$. Then with respect to this decomposition, X and H have the forms

$$X = \begin{bmatrix} 0 & \widehat{X}_{12} \\ 0 & \widehat{X}_{22} \end{bmatrix}, \quad H = \begin{bmatrix} \widehat{H}_{11} & \widehat{H}_{12} \\ \widehat{H}_{12}^* & \widehat{H}_{22} \end{bmatrix}.$$

By the hypothesis, $\widehat{H}_{11}$ has closed range, so we may further orthogonally decompose $\mathcal{G}_0 = \mathcal{H}_0 \oplus \mathcal{H}_1$ such that with respect to the decomposition $\mathcal{H}_0 \oplus \mathcal{H}_1 \oplus \mathcal{G}_1$ the operators X and H have the forms

$$X = \begin{bmatrix} 0 & 0 & \widehat{X}_{13} \\ 0 & 0 & \widehat{X}_{23} \\ 0 & 0 & \widehat{X}_{33} \end{bmatrix}, \quad H = \begin{bmatrix} 0 & 0 & H_{13} \\ 0 & H_{22} & H_{23} \\ H_{13}^* & H_{23}^* & \widehat{H}_{33} \end{bmatrix},$$

where $H_{22} : \mathcal{H}_1 \to \mathcal{H}_1$ is invertible. Then setting

$$P_1 := \begin{bmatrix} I & 0 & 0 \\ 0 & I & -H_{22}^{-1}H_{23} \\ 0 & 0 & I \end{bmatrix}$$

implies

$$P_1^{-1}XP_1 = \begin{bmatrix} 0 & 0 & \widehat{X}_{13} \\ 0 & 0 & \widehat{X}_{23} + H_{22}^{-1}H_{23}\widehat{X}_{33} \\ 0 & 0 & \widehat{X}_{33} \end{bmatrix}, \quad P_1^*HP_1 = \begin{bmatrix} 0 & 0 & H_{13} \\ 0 & H_{22} & 0 \\ H_{13}^* & 0 & \widetilde{H}_{33} \end{bmatrix}.$$

Since H is invertible, we obtain that H_{13} is right invertible. Let $\mathcal{H}_2 = \operatorname{Ker} H_{13}$, $\widetilde{\mathcal{H}}_0 = (\operatorname{Ker} H_{13})^\perp$, and decompose $\mathcal{G}_1 = \mathcal{H}_2 \oplus \widetilde{\mathcal{H}}_0$. Then there exist invertible operators $S : \mathcal{H}_0 \to \mathcal{H}_0$ and $T : \mathcal{G}_1 \to \mathcal{G}_1$ such that $S^*H_{13}T = \begin{bmatrix} 0 & H_{14} \end{bmatrix}$, where

$H_{14} : \mathcal{H}_0 \to \widetilde{\mathcal{H}}_0$ is a Hilbert space isomorphism. Then setting $P_2 = P_1 \cdot (S \oplus I_{\mathcal{H}_1} \oplus T)$, we get

$$P_2^{-1} X P_2 = \begin{bmatrix} 0 & 0 & \widetilde{X}_{13} & \widetilde{X}_{14} \\ 0 & 0 & \widetilde{X}_{23} & \widetilde{X}_{24} \\ 0 & 0 & \widetilde{X}_{33} & \widetilde{X}_{34} \\ 0 & 0 & \widetilde{X}_{43} & \widetilde{X}_{44} \end{bmatrix}, \quad P_2^* H P_2 = \begin{bmatrix} 0 & 0 & 0 & H_{14} \\ 0 & H_{22} & 0 & 0 \\ 0 & 0 & H_{33} & H_{34} \\ H_{14}^* & 0 & H_{34}^* & H_{44} \end{bmatrix}.$$

Finally, setting

$$P := P_1 P_2 \begin{bmatrix} I & 0 & -(H_{14}^*)^{-1} H_{34}^* & -\frac{1}{2}(H_{14}^*)^{-1} H_{44} \\ 0 & I & 0 & 0 \\ 0 & 0 & I & 0 \\ 0 & 0 & 0 & I \end{bmatrix},$$

we obtain that $P^{-1}XP$ and P^*HP have the form as in (4). A straightforward computation shows that $P^{-1}X^{[*]}P$ has the form (5). Furthermore,

$$P^{-1} X^{[*]} X P = \begin{bmatrix} 0 & 0 & * & * \\ 0 & 0 & 0 & 0 \\ 0 & 0 & * & * \\ 0 & 0 & 0 & 0 \end{bmatrix}.$$

Now, let X be H-normal, i.e., $P^{-1}XX^{[*]}P = P^{-1}X^{[*]}XP$. This implies that the first two operator columns of $P^{-1}XX^{[*]}P$ are zero, i.e.,

$$\begin{bmatrix} X_{13} \\ X_{23} \\ X_{33} \\ X_{43} \end{bmatrix} \begin{bmatrix} H_{33}^{-1} X_{43}^* H_{14}^* & H_{33}^{-1} X_{23}^* H_{22} \end{bmatrix} = 0. \tag{6}$$

Observe that the first operator matrix in (6) has zero kernel, because of (3). This implies $X_{43} = 0$ and $X_{23} = 0$. Then comparing the blocks in the $(3,3)$-positions of $P^{-1}XX^{[*]}P$ and $P^{-1}X^{[*]}XP$, we obtain $X_{33}H_{33}^{-1}X_{33}^*H_{33} = H_{33}^{-1}X_{33}^*H_{33}X_{33}$, i.e., X_{33} is H_{33}-normal. □

Next, we state a lemma that is of a general nature. We say that a point $\lambda \in \sigma(X)$, $X \in \mathcal{L}(\mathcal{H})$, is an *eigenvalue of finite type* if λ is an isolated point of the spectrum $\sigma(X)$ and the spectral projection $(2\pi i)^{-1} \int_{|\xi|=\epsilon} (\xi I - X)^{-1} d\xi$, where $\epsilon > 0$ is sufficiently small, has finite rank. It is easy to see (by using the decomposition of $\mathcal{H}$ as a direct sum of two X-invariant subspaces so that $X - \lambda I$ is invertible on one of them, and $X - \lambda I$ is nilpotent on the other) that if λ is an eigenvalue of finite type of X, and if $\mathcal{M}$ is an X-invariant subspace such that $\lambda \in \sigma(X|_{\mathcal{M}})$, then λ is an eigenvalue of finite type of the restriction $X|_{\mathcal{M}}$.

Lemma 3. *Let* $X \in \mathcal{L}(\mathcal{H})$ *be such that* 0 *is an eigenvalue of finite type of* X. *Then we have that* $\dim \operatorname{Ker} X = \dim \operatorname{Ker} X^{[*]}$.

Proof. By the assumption the spectral subspace $\mathcal{H}_0$ of X corresponding to the zero eigenvalue is finite-dimensional. Write $\mathcal{H} = \mathcal{H}_0 \oplus \mathcal{H}_0^{\perp}$, and with respect to this decomposition write

$$X = \begin{bmatrix} X_{11} & X_{12} \\ 0 & X_{22} \end{bmatrix}.$$

Then $\sigma(X_{11}) = \{0\}$ and X_{22} is invertible. Now $\dim \operatorname{Ker} X^{[*]} = \dim \operatorname{Ker} X^*$. We have

$$\operatorname{Ker} X^* = \left\{ \begin{bmatrix} x_1 \\ x_2 \end{bmatrix} \middle| x_1 \in \operatorname{Ker} X_{11}^*,\ x_2 = -(X_{22}^*)^{-1} X_{12}^* x_1 \right\}.$$

Also $\dim \operatorname{Ker} X_{11}^* = \dim \operatorname{Ker} X_{11}$ as $\mathcal{H}_0$ is finite-dimensional. So

$$\dim \operatorname{Ker} X^* = \dim \operatorname{Ker} X_{11}^* = \dim \operatorname{Ker} X_{11} = \dim \operatorname{Ker} X,$$

as required. □

We are now ready to state our main result.

Theorem 4. *Assume that $X \in \mathcal{L}(\mathcal{H})$ satisfies the following properties:*

(a) *X is H-normal;*
(b) *either X is invertible, or 0 is an eigenvalue of X of finite type;*
(c) *$\sigma(X)$ does not surround zero, i.e., there exists a continuous path in the complex plane that connects a sufficiently small neighborhood of zero with infinity and lies entirely in the resolvent set $\mathbb{C} \setminus \sigma(X)$.*

Assume in addition that one of the following conditions hold:

(i) $\operatorname{Ker} X = \operatorname{Ker} X^{[*]}$;
(ii) *$\mathcal{H}$ with the indefinite inner product generated by H is a Pontryagin space, i.e., at least one of the two spectral subspaces of H corresponding to the positive part of $\sigma(H)$ and to the negative part of $\sigma(H)$ is finite-dimensional.*

Then X admits an H-polar decomposition.

Proof. The proof starts with a general construction that is independent of whether we assume the additional conditions (i) or (ii) or not.

By Lemma 2 we may assume that

$$X = \begin{bmatrix} 0 & 0 & X_{13} & X_{14} \\ 0 & 0 & 0 & X_{24} \\ 0 & 0 & X_{33} & X_{34} \\ 0 & 0 & 0 & X_{44} \end{bmatrix}, \qquad H = \begin{bmatrix} 0 & 0 & 0 & H_{14} \\ 0 & H_{22} & 0 & 0 \\ 0 & 0 & H_{33} & 0 \\ H_{14}^* & 0 & 0 & 0 \end{bmatrix}, \tag{7}$$

with respect to an orthogonal decomposition

$$\mathcal{H} = \mathcal{H}_0 \oplus \mathcal{H}_1 \oplus \mathcal{H}_2 \oplus \widetilde{\mathcal{H}}_0,$$

where $\operatorname{Ker} X = \mathcal{H}_0 \oplus \mathcal{H}_1$, where X_{33} is H_{33}-normal, and where $H_{14} : \mathcal{H}_0 \to \widetilde{\mathcal{H}}_0$ is a Hilbert space isomorphism. (Note that by the hypotheses of the theorem, clearly the operator (1) has closed range.) In the following, we will identify $\mathcal{H}_0$ and $\widetilde{\mathcal{H}}_0$ via the isomorphism H_{14}, i.e., we assume without loss of generality that $\mathcal{H}_0 = \widetilde{\mathcal{H}}_0$ and $H_{14} = I_{\mathcal{H}_0}$.

We use induction on the dimension of the spectral subspace of X corresponding to the eigenvalue 0. The base of induction, i.e., the case when X is invertible, was proved in [13] (note that the finite-dimensional proof given in [13] carries over to the infinite-dimensional case using the property (c) of X).

We have

$$\sigma(X_{33}) \cup \{0\} = \sigma(\widetilde{X}),$$

where

$$\widetilde{X} := \begin{bmatrix} 0 & 0 & X_{13} \\ 0 & 0 & 0 \\ 0 & 0 & X_{33} \end{bmatrix}.$$

Moreover, the unbounded component of $\mathbb{C} \setminus \sigma(\widetilde{X})$ contains the unbounded component of $\mathbb{C} \setminus \sigma(X)$ (this is a general property of the spectrum of a restriction of an operator to its invariant subspace). Thus, the property (c) holds true for X_{33}.

To see that X_{33} satisfies property (b), we have to show that either X_{33} is invertible, or 0 is an eigenvalue of finite type of X_{33}. Assume then that X_{33} is not invertible. Since 0 is an eigenvalue of finite type of X, it is also an eigenvalue of finite type for X restricted to its invariant subspace $\mathcal{H}_0 \oplus \mathcal{H}_1 \oplus \mathcal{H}_2$. In order to show that 0 is an eigenvalue of finite type of X_{33} all we need to show is that $\dim \operatorname{Ker} X_{33}^n$ is uniformly bounded. We have that $\dim \operatorname{Ker} \widetilde{X}^n \leq \dim \operatorname{Ker} X^n$, and so $\dim \operatorname{Ker} \widetilde{X}^n$ is uniformly bounded. Now

$$\widetilde{X}^n = \begin{bmatrix} 0 & 0 & X_{13}X_{33}^{n-1} \\ 0 & 0 & 0 \\ 0 & 0 & X_{33}^n \end{bmatrix},$$

and so

$$\operatorname{Ker} \widetilde{X}^n = \mathcal{H}_0 \oplus \mathcal{H}_1 \oplus \operatorname{Ker} X_{33}^{n-1},$$

where the latter equality follows from

$$\operatorname{Ker} \begin{bmatrix} X_{13} \\ X_{33} \end{bmatrix} = \{0\} \tag{8}$$

by construction of the form (7). Hence we have that $\dim \operatorname{Ker} X_{33}^{n-1}$ is uniformly bounded, and so 0 is an eigenvalue of finite type of X_{33} whenever X_{33} is not invertible.

If (ii) is satisfied, i.e., if $\mathcal{H}$ with the indefinite inner product generated by H is a Pontryagin space, then also $\mathcal{H}_2$ with the indefinite inner product generated by H_{33} is a Pontryagin space. On the other hand, if (i) is satisfied, i.e., $\operatorname{Ker} X =$

Ker $X^{[*]}$, then we obtain $X_{24} = 0$ and $X_{44} = 0$, and

$$X^{[*]}X = \begin{bmatrix} 0 & 0 & X_{34}^*H_{33}X_{33} & X_{34}^*H_{33}X_{34} \\ 0 & 0 & 0 & 0 \\ 0 & 0 & X_{33}^{[*]}X_{33} & X_{33}^{[*]}X_{34} \\ 0 & 0 & 0 & 0 \end{bmatrix},$$

$$XX^{[*]} = \begin{bmatrix} 0 & 0 & X_{13}X_{33}^{[*]} & X_{13}H_{33}^{-1}X_{13}^* \\ 0 & 0 & 0 & 0 \\ 0 & 0 & X_{33}X_{33}^{[*]} & X_{33}H_{33}^{-1}X_{13}^* \\ 0 & 0 & 0 & 0 \end{bmatrix}.$$

Assume that $x \in \text{Ker } X_{33}$. Then

$$XX^{[*]}\begin{bmatrix} 0 & 0 & x & 0 \end{bmatrix}^T = X^{[*]}X\begin{bmatrix} 0 & 0 & x & 0 \end{bmatrix}^T = 0$$

which implies

$$\begin{bmatrix} X_{13} \\ X_{33} \end{bmatrix} X_{33}^{[*]}x = 0.$$

Because of (8), we obtain $X_{33}^{[*]}x = 0$ and Ker $X_{33} \subseteq$ Ker $X_{33}^{[*]}$. The other inclusion follows analogously. So, Ker X = Ker $X^{[*]}$ implies that Ker X_{33} = Ker $X_{33}^{[*]}$.

Hence, X_{33} satisfies all assumptions of the theorem. By the induction hypothesis, X_{33} admits an H_{33}-polar decomposition and by Remark 1 also a right H-polar decomposition $X_{33} = A_{33}U_{33}$, where U_{33} is an invertible H_{33}-isometry, and A_{33} is H_{33}-selfadjoint. In the following, we construct an H-polar decomposition for X. This will be done in five steps.

1. First, we show that there exists α real such that the operator $L - \alpha M$ is invertible, where

$$L = H_{33}A_{33} \qquad \text{and} \qquad M = (U_{33}^{-1})^* X_{13}^* X_{13} U_{33}^{-1}$$

are selfadjoint operators. For this purpose, observe that $H_{33}^{-1}LU_{33} = X_{33}$ is Fredholm, and therefore so is L. Denote by $Q_{\text{Ker}\, L}$ the orthogonal projection onto the finite-dimensional subspace Ker L. We claim that

$$\text{Ker } (Q_{\text{Ker}\, L}M|_{\text{Ker}\, L}) = \{0\}. \tag{9}$$

To this end note that Ker $X_{13} \cap$ Ker $X_{33} = \{0\}$ by (8), and hence

$$\text{Ker}\, M \cap \text{Ker}\, L = \{0\}. \tag{10}$$

Let x be such that $Lx = 0$, $Q_{\text{Ker}\, L}Mx = 0$. Then

$$\langle Mx, x\rangle = \langle Mx, Q_{\text{Ker}\, L}x\rangle = \langle Q_{\text{Ker}\, L}Mx, x\rangle = 0,$$

thus $Mx = 0$ (because M is positive semidefinite), and $x = 0$ in view of (10). This proves the claim (9). Now, with respect to the orthogonal decomposition $\mathcal{H}_2 = \text{Ker}\, L \oplus (\text{Ker}\, L)^{\perp}$, we have

$$L - \alpha M = \begin{bmatrix} -\alpha M_1 & -\alpha M_2 \\ -\alpha M_2^* & L_1 - \alpha M_3 \end{bmatrix}, \qquad \alpha \in \mathbb{R},$$

where L_1 and M_1 (because of (9) and the Fredholmness of L) are invertible. Using Schur complements we obtain that $L - \alpha M$ is invertible if and only if $\alpha \neq 0$ and the operator

$$L_1 + \alpha(-M_3 + M_2^* M_1^{-1} M_2)$$

is invertible. Clearly, such α's exist.

2. We construct an H-selfadjoint polar factor for X. For this, let $\alpha \neq 0$, $\alpha \in \mathbb{R}$, be such that $L - \alpha M$ is invertible. Then set

$$A_{13} := X_{13}U_{33}^{-1}, \quad A_{14} := \alpha^{-1} I_q, \quad A_{34} := H_{33}^{-1} A_{13}^* = H_{33}^{-1} \left(U_{33}^{-1}\right)^* X_{13}^*,$$

and

$$A := \begin{bmatrix} 0 & 0 & A_{13} & A_{14} \\ 0 & 0 & 0 & 0 \\ 0 & 0 & A_{33} & A_{34} \\ 0 & 0 & 0 & 0 \end{bmatrix}.$$

Then a straightforward computation shows that A is H-selfadjoint.

3. Next, we show $A^2 = X^{[*]}X$. Indeed, we obtain from the identities

$$\begin{aligned} A_{13}A_{33} &= X_{13}U_{33}^{-1}A_{33} = X_{13}H_{33}^{-1}H_{33}U_{33}^{-1}A_{33} = X_{13}H_{33}^{-1}U_{33}^*H_{33}A_{33} \\ &= X_{13}H_{33}^{-1}U_{33}^*A_{33}^*H_{33} = X_{13}H_{33}^{-1}X_{33}^*H_{33}, \\ A_{13}A_{34} &= X_{13}U_{33}^{-1}H_{33}^{-1}(U_{33}^*)^{-1}X_{13}^* = X_{13}H_{33}^{-1}X_{13}^*, \\ A_{33}^2 &= A_{33}H_{33}^{-1}A_{33}^*H_{33} = X_{33}U_{33}^{-1}H_{33}^{-1}(U_{33}^*)^{-1}X_{33}^*H_{33} = X_{33}H_{33}^{-1}X_{33}^*H_{33}, \\ A_{33}A_{34} &= X_{33}U_{33}^{-1}H_{33}^{-1}(U_{33}^*)^{-1}X_{13}^* = X_{33}H_{33}^{-1}X_{13}^*, \end{aligned}$$

that

$$\begin{aligned} A^2 &= \begin{bmatrix} 0 & 0 & A_{13}A_{33} & A_{13}A_{34} \\ 0 & 0 & 0 & 0 \\ 0 & 0 & A_{33}^2 & A_{33}A_{34} \\ 0 & 0 & 0 & 0 \end{bmatrix} \\ &= \begin{bmatrix} 0 & 0 & X_{13}H_{33}^{-1}X_{33}^*H_{33} & X_{13}H_{33}^{-1}X_{13}^* \\ 0 & 0 & 0 & 0 \\ 0 & 0 & X_{33}H_{33}^{-1}X_{33}^*H_{33} & X_{33}H_{33}^{-1}X_{13}^* \\ 0 & 0 & 0 & 0 \end{bmatrix} \\ &= XX^{[*]} = X^{[*]}X. \end{aligned}$$

4. Finally, we show Ker X = Ker A. From the construction, it is clear that Ker $X \subseteq$ Ker A. For the other implication, let $v = \begin{bmatrix} a & b & c & d \end{bmatrix}^T \in$ Ker A. Then

$$0 = A_{13}c + A_{14}d = X_{13}U_{33}^{-1}c + \alpha^{-1}d \implies d = -\alpha X_{13}U_{33}^{-1}c.$$

Moreover,

$$0 = A_{33}c + A_{34}d = A_{33}c - \alpha H_{33}^{-1}\left(U_{33}^{-1}\right)^* X_{13}^* X_{13} U_{33}^{-1} c.$$

The choice of α implies $c = 0$ and thus, we also obtain $d = 0$. Hence, $v \in \operatorname{Ker} X$.

Thus, we constructed an H-selfadjoint operator A that satisfies $A^2 = X^{[*]}X$ and $\operatorname{Ker} X = \operatorname{Ker} A$. Since X is Fredholm of index zero, it is easy to see that $X^{[*]}X$ and therefore also A are Fredholm operators of index zero. Define the operator U_0 on the range of A by $U_0 x = Xy$, where y is such that $x = Ay$. It is a standard exercise to check that U_0 is a well-defined H-isometry on the range of A, and the range of U_0 coincides with the range of X. Moreover, since A and X have generalized inverses and $\operatorname{Ker} A = \operatorname{Ker} X$, it follows that U_0 is bounded and $\|U_0 x\| \geq \varepsilon \|x\|$, $x \in \operatorname{Range} A$, where the positive constant ε is independent of x.

5. Extension of U_0 to an invertible H-isometry. This is where the assumptions (i) or (ii) come in that have not been used so far. First we consider the case where $\mathcal{H}$ is a Pontryagin space. By Lemma 3 we have $\dim \operatorname{Ker} X = \dim \operatorname{Ker} X^{[*]}$, so

$$\begin{aligned} \operatorname{codim} \operatorname{Range} A &= \dim(\operatorname{Range} A)^{[\perp]} = \dim \operatorname{Ker} A^{[*]} = \dim \operatorname{Ker} A = \dim \operatorname{Ker} X \\ &= \dim \operatorname{Ker} X^{[*]} = \dim(\operatorname{Range} X)^{[\perp]} = \operatorname{codim} \operatorname{Range} X. \end{aligned}$$

Then we can use [16, Theorem 2.5] to show that in case $\mathcal{H}$ is a Pontryagin space with respect to the indefinite scalar product generated by H, U_0 can be extended to an invertible H-isometry. This proves the theorem in case (ii) holds true.

Next, we consider the case that $\operatorname{Ker} X = \operatorname{Ker} X^{[*]}$. Then we have the equalities

$$(\operatorname{Range} A)^{[\perp]} = \operatorname{Ker} A^{[*]} = \operatorname{Ker} A = \operatorname{Ker} X = \operatorname{Ker} X^{[*]} = (\operatorname{Range} X)^{[\perp]}, \qquad (11)$$

and so we have that $\operatorname{Range} A = \operatorname{Range} X$. In particular we have

$$\mathcal{H}_0 \oplus \mathcal{H}_1 = \operatorname{Ker} X = (\operatorname{Range} A)^{[\perp]} = H^{-1}(\operatorname{Range} A)^{\perp}$$

which implies $(\operatorname{Range} A)^{\perp} = \ddot{\mathcal{H}}_0 \oplus \mathcal{H}_1$ and $\operatorname{Range} A = \mathcal{H}_0 \oplus \mathcal{H}_2$. Because of (11), the isotropic part of $\operatorname{Range} A$ (which is the finite-dimensional space $\mathcal{H}_0$) is the same as the isotropic part of $\operatorname{Range} X$. Choose a $\langle \cdot, \cdot \rangle$-orthonormal set of vectors $\{e_1, \ldots, e_n\}$ that form a basis for $\mathcal{H}_0$. Moreover, the $\langle \cdot, \cdot \rangle$-orthogonal complement of $\mathcal{H}_0$ in $\operatorname{Range} A$ (which is $\mathcal{H}_2$) is an H-nondegenerate subspace. Choose a basis $\{f_1, \ldots, f_n\}$ of $\widetilde{\mathcal{H}}_0$ that is skewly linked to $\{e_1, \ldots, e_n\}$, that is, $[e_i, f_j] = \delta_{ij}$ and $[f_i, f_j] = 0$. (For details on construction of skewly linked bases see, e.g., [10, 16, 3]; although it is assumed there that the indefinite inner product space is a Pontryagin space, the construction goes through without change for finite-dimensional subspaces of Krein spaces.) Then $\operatorname{Range} A \oplus \widetilde{\mathcal{H}}_0 = \operatorname{Range} X \oplus \widetilde{\mathcal{H}}_0$ is H-nondegenerate.

We start by showing that U_0 maps $\mathcal{H}_0$ into itself. Indeed, for $x_0 \in \mathcal{H}_0$ we have that $U_0 x_0$ is H-orthogonal to the whole of $\operatorname{Range} X$, and hence is in $\mathcal{H}_0$. So, if we write U_0 with respect to the decomposition $\mathcal{H}_0 \oplus \mathcal{H}_2$ of $\operatorname{Range} A = \operatorname{Range} X$

as a two by two block operator matrix, we have

$$U_0 = \begin{bmatrix} U_{11} & U_{12} \\ 0 & U_{22} \end{bmatrix},$$

Clearly, since U_0 is one-to-one and maps onto Range X, it follows that U_0 and therefore also U_{11} and U_{22} are invertible maps.

With respect to the decomposition $\mathcal{H}_0 \oplus \mathcal{H}_2 \oplus \widetilde{\mathcal{H}}_0$ we have for H the following form (where we choose the basis in $\mathcal{H}_0$ and in $\widetilde{\mathcal{H}}_0$ as above)

$$H = \begin{bmatrix} 0 & 0 & I \\ 0 & H_{33} & 0 \\ I & 0 & 0 \end{bmatrix}.$$

We shall define $\widetilde{U}_0 : \operatorname{Range} A \oplus \widetilde{\mathcal{H}}_0 \to \operatorname{Range} X \oplus \widetilde{\mathcal{H}}_0$ as the following 3×3 block operator matrix

$$\widetilde{U}_0 = \begin{bmatrix} U_{11} & U_{12} & U_{13} \\ 0 & U_{22} & U_{23} \\ 0 & 0 & U_{33} \end{bmatrix},$$

where $U_{33} := (U_{11}^*)^{-1}$, and $U_{23} := -U_{22}H_{22}^{-1}U_{12}^*(U_{11}^*)^{-1}$, and finally $U_{13} := -\frac{1}{2}U_{12}H_{22}^{-1}U_{12}^*(U_{11}^*)^{-1}$. Computing $\widetilde{U}_0^* H \widetilde{U}_0$ on $\mathcal{H}_0 \oplus \mathcal{H}_1 \oplus \widetilde{\mathcal{H}}_0$ we have that it equals to

$$\begin{bmatrix} 0 & 0 & I \\ 0 & H_{33} & U_{12}^*(U_{11}^*)^{-1} + U_{22}^*H_{33}U_{23} \\ I & U_{23}^*H_{33}U_{22} + U_{11}^{-1}U_{12} & U_{13}^*(U_{11}^*)^{-1} + U_{11}^{-1}U_{13} + U_{23}^*H_{33}U_{23} \end{bmatrix}. \tag{12}$$

We see from the definition of U_{23} that the $(2,3)$-entry of the operator matrix (12) is zero. Next,

$$U_{23}^*H_{33}U_{23} = U_{11}^{-1}U_{12}H_{33}^{-1}U_{12}^*(U_{11}^*)^{-1}.$$

Thus, from the definition of U_{13} we see that also the $(3,3)$-entry of (12) is zero. Hence $\widetilde{U}_0$ is indeed an H-isometry. The fact that $\widetilde{U}_0$ is one-to-one and maps onto Range $X \oplus \widetilde{\mathcal{H}}_0$ follows easily from the invertibility of U_{11}, U_{22}, and U_{33}.

Now using [1, Theorem VI.4.4] we see that $\widetilde{U}_0$ can be extended to an H-unitary operator on the whole space $\mathcal{H}$. This concludes the proof of Theorem 4. □

3. Applications of the main result

For the remainder of the paper, we assume that $\mathcal{H}$ is finite-dimensional, and identify $\mathcal{L}(\mathcal{H})$ with $\mathbb{C}^{n\times n}$, the algebra of $n \times n$ complex matrices. Then Theorem 4 has some important corollaries. First of all, it answers affirmatively the question posed in [2] whether each H-normal matrix allows an H-polar decomposition.

Corollary 5. *Let $X \in \mathbb{C}^{n\times n}$ be H-normal. Then X admits an H-polar decomposition.*

Corollary 5 was known to be correct for invertible H-normal matrices and for some special cases of singular H-normal matrices (see [2, 12, 11, 13]). The result for the general case is new. The next corollary gives a criterion for the existence of H-polar decompositions in terms of well-known canonical forms of pairs (A, H), where A is H-selfadjoint, under transformations of the form $(A, H) \mapsto (P^{-1}AP, P^*HP)$, where P is invertible, see, for example, [6].

Corollary 6. *Let $X \in \mathbb{C}^{n\times n}$. Then X admits an H-polar decomposition if and only if $(X^{[*]}X, H)$ and $(XX^{[*]}, H)$ have the same canonical form.*

Proof. If $X = UA$ is a polar decomposition, then

$$XX^{[*]} = UAA^{[*]}U^{[*]} = UA^2U^{-1} \quad \text{and} \quad X^{[*]}X = A^{[*]}U^{[*]}UA = A^2,$$

i.e., $(XX^{[*]}, H)$ and $(X^{[*]}X, H)$ have the same canonical forms, because U is H-unitary. On the other hand, if $(XX^{[*]}, H)$ and $(X^{[*]}X, H)$ have the same canonical forms, then there exists an H-unitary matrix U such that $UXX^{[*]}U^{-1} = X^{[*]}X$. Then $\widetilde{X} = UX$ is H-normal, since

$$\widetilde{X}^{[*]}\widetilde{X} = X^{[*]}X = UXX^{[*]}U^{-1} = \widetilde{X}\widetilde{X}^{[*]}.$$

By Corollary 5 $\widetilde{X}$ admits an H-polar decomposition $\widetilde{X} = VA$, where V is H-unitary and A is H-selfadjoint. Then $X = (U^{-1}V)A$ is an H-polar decomposition for X. □

Thus, up to multiplication by an H-unitary matrix from the left, H-normal matrices are the only matrices that admit H-polar decompositions. Corollary 6 has been conjectured in [12, 11], where also a proof has been given for the case that X is invertible or that the eigenvalue zero of $X^{[*]}X$ has equal algebraic and geometric multiplicities.

Theorem 4 also answers a question on sums of squares of H-selfadjoint matrices that has been posed in [14]. In general, the set $\{A^2 : A \text{ is } H\text{-selfadjoint}\}$ (where H is fixed) is not convex, in contrast to the convexity of the cone of positive semidefinite matrices with respect to the Euclidean inner product, as the following example shows: Let

$$H = \begin{bmatrix} 0 & 1 \\ 1 & 0 \end{bmatrix}, \quad A_1 = \begin{bmatrix} 0 & 2 \\ -2 & 0 \end{bmatrix}, \quad A_2 = \begin{bmatrix} 1 & 1 \\ 0 & 1 \end{bmatrix}, \quad A_1^2+A_2^2 = \begin{bmatrix} -3 & 2 \\ 0 & -3 \end{bmatrix}.$$

Then $A_1^2 + A_2^2$ is not a square of any H-selfadjoint matrix, since $A_1^2 + A_2^2$ has only one Jordan block associated with the eigenvalue -3. This contradicts the conditions for the existence of an H-selfadjoint square root, see Theorem 3.1 in [15]. Instead, we have the following result.

Corollary 7. *If A_1 and A_2 are two commuting H-selfadjoint matrices, then there exists an H-selfadjoint matrix A such that $A_1^2 + A_2^2 = A^2$.*

Proof. Let $X = A_1 + iA_2$. Then X is H-normal, because X and $X^{[*]} = A_1 - iA_2$ commute. By Corollary 5, X admits an H-polar decomposition $X = UA$, where U is H-unitary and A is H-selfadjoint. This implies $A_1^2 + A_2^2 = X^{[*]}X = A^2$. □

4. Polar decompositions with commuting factors

Again, we assume that $\mathcal{H}$ is finite-dimensional, and identify $\mathcal{L}(\mathcal{H})$ with $\mathbb{C}^{n\times n}$, the algebra of $n \times n$ complex matrices. It is well known that a normal matrix X (normal with respect to the standard inner product) allows a polar decomposition $X = UA$ with commuting factors, see [5], for example. The question arises whether this is still true for indefinite inner products. In [13], it has been shown by a Lie group theoretical argument that nonsingular H-normal matrices allow an H-polar decomposition with commuting factors. (For a different proof of this fact, see [12].) On the other hand, there exist singular H-normal matrices that do not allow such H-polar decompositions. The following example is borrowed from [13].

Example 8. Let

$$X = \begin{bmatrix} 0 & 0 \\ 0 & i \end{bmatrix}, \quad H = \begin{bmatrix} 0 & 1 \\ 1 & 0 \end{bmatrix}.$$

Then X is H-normal. In fact, $X^{[*]}X = XX^{[*]} = 0$. It is straightforward to check that all H-polar decompositions $X = UA$ of X are described by the formulas

$$U = \begin{bmatrix} 0 & ix \\ ix^{-1} & y \end{bmatrix}, \quad A = \begin{bmatrix} 0 & x \\ 0 & 0 \end{bmatrix},$$

where $x \neq 0$ and y are arbitrary real numbers. Clearly, U and A do not commute for any values of the parameters x and y.

In the following, we will give necessary and sufficient conditions for the existence of H-polar decompositions with commuting factors. The proof will be based on the following result on particular square roots of H-unitary matrices.

Theorem 9. *Let $V \in \mathbb{C}^{n\times n}$ be H-unitary and let $M \in \mathbb{C}^{n\times n}$ be such that $MV = VM$. Then there exists an H-unitary matrix $U \in \mathbb{C}^{n\times n}$ such that $U^2 = V$ and $MU = UM$.*

Proof. First, assume that there are no eigenvalues of V on the negative real line (including zero). Let Γ be a simple (i.e., without self-intersections) closed rectifiable contour in the complex plane such that Γ is symmetric with respect to the real axis, the eigenvalues of V are inside Γ, and the negative real axis $(-\infty, 0]$ is outside Γ. Let $f : \mathbb{C} \setminus (-\infty, 0] \to \mathbb{C}$ be the branch of the square root that assigns to $z \in \mathbb{C} \setminus (-\infty, 0]$ the solution c of $c^2 = z$ that has positive real part. Then f is analytic on Γ and analytic in the interior of Γ and hence, the matrix $f(V)$ given by the functional calculus

$$f(V) = \frac{1}{2\pi i}\int_\Gamma f(z)(zI - V)^{-1}dz \tag{13}$$

is well defined. From the fact that V is H-unitary, we obtain the formula

$$H(zI-V)^{-1} = \Big((zI-V)H^{-1}\Big)^{-1} = \Big(H^{-1}(zI-(V^*)^{-1})\Big)^{-1} = (zI-(V^*)^{-1})^{-1}H.$$

This implies $Hf(V) = f((V^*)^{-1})H$. Since $f(z^{-1}) = f(z)^{-1}$, we obtain that $f(V^{-1}) = f(V)^{-1}$, see [9, Corollary 6.2.10].

We then obtain from $f(\overline{z}) = \overline{f(z)}$, the symmetry of Γ with respect to the real axis, and the general fact that $f(M^T) = f(M)^T$, that

$$f\Big((V^*)^{-1}\Big) = \Big(f(V)^*\Big)^{-1}.$$

This implies that $U := f(V)$ is H-unitary. Clearly, $U^2 = V$ and $UM = MU$. For the case that there are negative eigenvalues of V, there exists $0 \le \theta < 2\pi$ such that the ray $re^{i\theta}$ $(r > 0)$ does not contain an eigenvalue of V. Then $\widetilde{V} = e^{i(\pi-\theta)}V$ is still H-unitary, satisfies $M\widetilde{V} = \widetilde{V}M$, and does not have negative eigenvalues. Hence, there exists an H-unitary matrix $\widetilde{U}$ such that $\widetilde{U}^2 = \widetilde{V}$ and $M\widetilde{U} = \widetilde{U}M$. Then $U = e^{i(\theta-\pi)/2}\widetilde{U}$ is an H-unitary square root of V satisfying $MU = UM$. □

The following result provides necessary and sufficient conditions for the existence of polar decompositions with commuting factors.

Theorem 10. *Let $X \in \mathbb{C}^{n\times n}$. Then the following statements are equivalent.*

i) *X admits an H-polar decomposition with commuting factors.*
ii) *X is H-normal and* Ker (X) = Ker $(X^{[*]})$.
iii) *There exists an H-unitary matrix V such that $X = VX^{[*]}$.*

Proof. i) ⇒ ii): If X allows an H-polar decomposition $X = UA$ with commuting factors, then $X^{[*]} = (UA)^{[*]} = AU^{-1} = U^{-1}A$. But then X is H-normal, because

$$XX^{[*]} = UAAU^{-1} = AU^{-1}UA = X^{[*]}X.$$

In addition, we have Ker (X) = Ker (A) = Ker $(X^{[*]})$.

ii) ⇒ iii): This is a special case of Witt's Theorem and coincides with [4, Lemma 4.1].

iii) ⇒ i): Let V be an H-unitary matrix such that $X = VX^{[*]}$. Note that X and V commute:

$$XV = VX^{[*]}V = V(VX^{[*]})^{[*]}V = VXV^{[*]}V = VX.$$

Then Theorem 9 implies that V has an H-unitary square root U that commutes with X. Now consider $X = UA$, where $A := U^{-1}X$. Clearly, U and A commute. Furthermore, A is H-selfadjoint, because

$$(U^{-1}X)^{[*]} = X^{[*]}U = V^{-1}XU = V^{-1}UX = U^{-2}UX = U^{-1}X.$$

Thus $X = UA$ is an H-polar decomposition for X with commuting factors. □

Note that if $X = UA$ is an H-polar decomposition of X, i.e., U is H-unitary and A is H-selfadjoint, then

$$UA = AU \iff UX = XU \implies XA = AX.$$

If A is invertible, then $XA = AX \implies UA = AU$, but in general $XA = AX \not\Longrightarrow UA = AU$ as the next two examples show. Thus, the equality $XA = AX$ gives a commutativity property of H-polar decomposition which is strictly weaker than commuting factors. Example 8 shows that not every H-normal matrix admits an H-polar decomposition with this weaker commutativity property.

We conclude the paper with two examples; the second example is borrowed from [14].

Example 11. Let

$$X = \begin{bmatrix} 0 & 0 & 0 & 1 & 0 \\ 0 & 0 & 1 & 0 & 0 \\ 0 & 0 & 0 & 1 & 0 \\ 0 & 0 & 0 & 0 & 0 \\ 0 & 0 & 0 & 0 & 0 \end{bmatrix}, \quad H = \begin{bmatrix} 0 & 0 & 0 & 0 & 1 \\ 0 & 0 & 0 & 1 & 0 \\ 0 & 0 & 1 & 0 & 0 \\ 0 & 1 & 0 & 0 & 0 \\ 1 & 0 & 0 & 0 & 0 \end{bmatrix}.$$

Then X is H-normal, but Ker $(X) \neq$ Ker $(X^{[*]})$. Thus, X cannot have a polar decomposition with commuting factors by Theorem 10. On the other hand, consider the matrices

$$U = \begin{bmatrix} 1 & 0 & 1 & 0 & -\frac{1}{2} \\ 0 & 1 & 0 & 0 & 0 \\ 0 & 0 & 1 & 0 & -1 \\ 0 & 0 & 0 & 1 & 0 \\ 0 & 0 & 0 & 0 & 1 \end{bmatrix}, \quad A = \begin{bmatrix} 0 & 0 & 0 & 0 & 0 \\ 0 & 0 & 1 & 0 & 0 \\ 0 & 0 & 0 & 1 & 0 \\ 0 & 0 & 0 & 0 & 0 \\ 0 & 0 & 0 & 0 & 0 \end{bmatrix}.$$

Then U is H-unitary, A is H-selfadjoint and $X = UA$. Moreover, A and X commute, but A and U do not.

Example 12. Let

$$X = \begin{bmatrix} 0 & 0 & 1 & 0 & 0 \\ 0 & 0 & 0 & r & z \\ 0 & 0 & 0 & 1 & 0 \\ 0 & 0 & 0 & 0 & 0 \\ 0 & 0 & 0 & 0 & 0 \end{bmatrix}, \quad r > 0, \ z = \pm 1, \quad H = \begin{bmatrix} 0 & 0 & 0 & 1 & 0 \\ 0 & 0 & 0 & 0 & 1 \\ 0 & 0 & 1 & 0 & 0 \\ 1 & 0 & 0 & 0 & 0 \\ 0 & 1 & 0 & 0 & 0 \end{bmatrix}.$$

A possible H-polar decomposition $X = UA$, where U is H-unitary and A is H-selfadjoint, is the following:

$$U = \begin{bmatrix} 1 & -\frac{r}{2z} & \frac{r^2}{4z} & -\frac{r^4}{32} & 0 \\ 0 & 1 & \frac{r}{2} & -\frac{3r^3}{16z} & -\frac{r^2}{8} \\ 0 & 0 & 1 & -\frac{r^2}{2z} & -\frac{r}{2} \\ 0 & 0 & 0 & 1 & 0 \\ 0 & 0 & 0 & \frac{r}{2z} & 1 \end{bmatrix}, \quad A = \begin{bmatrix} 0 & 0 & 1 & 0 & \frac{r}{2} \\ 0 & 0 & 0 & \frac{r}{2} & z \\ 0 & 0 & 0 & 1 & 0 \\ 0 & 0 & 0 & 0 & 0 \\ 0 & 0 & 0 & 0 & 0 \end{bmatrix}. \tag{14}$$

Note that A and X commute; but A and U do not commute. A MAPLE computation even shows that there does not exist an H-unitary $\widetilde{U}$ such that $X = \widetilde{U}A = A\widetilde{U}$ for the special choice of A in (14) as an H-selfadjoint polar factor of X.

However, note that Ker X = Ker $X^{[*]}$, i.e., by Theorem 10 there exists an H-polar decomposition $X = \widehat{U}\widehat{A}$ with commuting factors. Indeed, let

$$\widehat{U} = \begin{bmatrix} 1 & -\frac{rz}{2} & \frac{r^2z}{8} & -\frac{9r^4z^2}{128} & -\frac{r^3z}{8} \\ 0 & 1 & \frac{r}{2} & 0 & -\frac{r^2}{8} \\ 0 & 0 & 1 & -\frac{3r^2z}{8} & -\frac{r}{2} \\ 0 & 0 & 0 & 1 & 0 \\ 0 & 0 & 0 & \frac{rz}{2} & 1 \end{bmatrix}, \quad \widehat{A} = \begin{bmatrix} 0 & 0 & 1 & \frac{r^2z}{8} & \frac{r}{2} \\ 0 & 0 & 0 & \frac{r}{2} & z \\ 0 & 0 & 0 & 1 & 0 \\ 0 & 0 & 0 & 0 & 0 \\ 0 & 0 & 0 & 0 & 0 \end{bmatrix}. \tag{15}$$

Then $\widehat{U}$ is H-unitary, $\widehat{A}$ is H-selfadjoint, and $X = \widehat{U}\widehat{A} = \widehat{A}\widehat{U}$. It is interesting to note that a straightforward but tedious MAPLE computation reveals that the polar factor A is unique up to a sign, i.e., all H-polar decompositions for X with commuting factors necessarily have the H-selfadjoint polar factor A (or $-A$) as in (15).

References

[1] J. Bognár. *Indefinite inner product spaces.* Springer-Verlag, New York-Heidelberg, 1974.

[2] Y. Bolshakov, C.V.M. van der Mee, A.C.M. Ran, B. Reichstein, and L. Rodman. Polar decompositions in finite-dimensional indefinite scalar product spaces: general theory. *Linear Algebra Appl.*, 261:91–141, 1997.

[3] Y. Bolshakov, C.V.M. van der Mee, A.C.M. Ran, B. Reichstein, and L. Rodman. Extension of isometries in finite-dimensional indefinite scalar product spaces and polar decompositions. *SIAM J. Matrix Anal. Appl.*, 18:752–774, 1997.

[4] Y. Bolshakov and B. Reichstein. Unitary equivalence in an indefinite scalar product: an analogue of singular-value decomposition. *Linear Algebra Appl.*, 222: 155–226, 1995.

[5] F. Gantmacher. *Theory of Matrices*, volume 1. Chelsea, New York, 1959.

[6] I. Gohberg, P. Lancaster, and L. Rodman. *Matrices and Indefinite Scalar Products.* Birkhäuser Verlag, Basel, Boston, Stuttgart, 1983.

[7] N.J. Higham, J-orthogonal matrices: properties and generation, *SIAM Review* 45:504–519, 2003.

[8] N.J. Higham, D.S. Mackey, N. Mackey, and F. Tisseur. Computing the polar decomposition and the matrix sign decomposition in matrix groups, *SIAM J. Matrix Anal. Appl.* 25(4):1178–1192, 2004.

[9] R. Horn and C.R. Johnson. *Topics in Matrix Analysis.* Cambridge University Press, Cambridge, 1991.

[10] I.S. Iohvidov, M.G. Krein, and H. Langer. *Introduction to the spectral theory of operators in spaces with an indefinite metric.* Akademie-Verlag, Berlin, 1982.

[11] U. Kintzel. Polar decompositions, factor analysis, and Procrustes problems in finite-dimensional indefinite scalar product spaces. Preprint 32-2003, Institut für Mathematik, Technische Universität Berlin, Germany, 2003.

[12] U. Kintzel. Polar decompositions and Procrustes problems in finite-dimensional indefinite scalar product spaces. Doctoral Thesis, Technische Universität Berlin, Germany, 2004.

[13] B. Lins, P. Meade, C. Mehl, and L. Rodman. Normal matrices and polar decompositions in indefinite inner products. *Linear and Multilinear Algebra*, 49:45–89, 2001.

[14] B. Lins, P. Meade, C. Mehl, and L. Rodman. Research Problem: Indefinite inner product normal matrices. *Linear and Multilinear Algebra*, 49: 261–268, 2001.

[15] C.V.M. van der Mee, A.C.M. Ran, and L. Rodman. Stability of self-adjoint square roots and polar decompositions in indefinite scalar product spaces. *Linear Algebra Appl.*, 302–303:77–104, 1999.

[16] C.V.M. van der Mee, A.C.M. Ran, and L. Rodman. Polar decompositions and related classes of operators in spaces Π_κ. *Integral Equations and Operator Theory*, 44: 50–70, 2002.

Christian Mehl
Technische Universität Berlin
Institut für Mathematik, MA 4-5
D-10623 Berlin
Germany
e-mail: mehl@math.tu-berlin.de

André C.M. Ran
Afdeling Wiskunde
Faculteit der Exacte Wetenschappen
Vrije Universiteit Amsterdam
De Boelelaan 1081a
NL-1081 HV Amsterdam
The Netherlands
e-mail: ran@cs.vu.nl

Leiba Rodman
Department of Mathematics
College of William and Mary
Williamsburg VA 23187-8795
USA
e-mail: lxrodm@math.wm.edu

Operator Theory:
Advances and Applications, Vol. 162, 293–308

Bounds for Contractive Semigroups and Second-Order Systems

Krešimir Veselić

Abstract. We derive a uniform bound for the difference of two contractive semigroups, if the difference of their generators is form-bounded by the Hermitian parts of the generators themselves. We construct a semigroup dynamics for second-order systems with fairly general operator coefficients and apply our bound to the perturbation of the damping term. The result is illustrated on a dissipative wave equation. As a consequence the exponential decay of some second-order systems is proved.

1. Introduction

The aim of this paper is to derive a new perturbation bound for strongly continuous contractive semigroups in a Hilbert space and to apply it to damped systems of second order. Let e^{At}, e^{Bt} be strongly continuous contractive semigroups in a Hilbert space $\mathcal{X}$. Their generators are maximal dissipative in the sense that $\Re(A\psi, \psi) \leq 0$ and that A is maximal with this property and similarly with B. (That is, $-A$, $-B$ are maximal accretive as defined in [3]. In this paper we will follow the notations and the terminology of the Kato's monograph.)

We consider a rather restricted kind of perturbation, it reads formally

$$|(x, (B - A)y)|^2 \leq \varepsilon^2 \Re(-Bx, x) \Re(-Ay, y), \quad \varepsilon > 0. \tag{1}$$

As a result we obtain a uniform estimate for the semigroups:

$$\|e^{Bt} - e^{At}\| \leq \frac{\varepsilon}{2}.$$

Note that here we have not the classical situation: 'unperturbed object plus a small perturbation' in which the perturbed object often has first to be constructed and then the distance between the two is measured (see, e.g., [3] Ch. XI, Th. 2.1). We impose no condition whatsoever on the size of the positive constant ε but we know that both A and B are dissipative, and both operators appear in a symmetric way.

Moreover, no requirements are made about the size of the subspace $\mathcal{D}(A) \cap \mathcal{D}(B)$, it could even be trivial. To this end, (1) is rewritten in a 'weak form' as

$$|(B^*x, y) - (x, Ay)|^2 \leq \varepsilon^2 \Re(-B^*x, x) \Re(-Ay, y).$$

This kind of perturbation will appear to be the proper setting for treating semigroups, generated by second-order systems

$$M\ddot{x} + C\dot{x} + Kx = 0. \tag{2}$$

Here M, C, K can be finite symmetric matrices, with *the mass matrix* M positive semidefinite, *the stiffness matrix* K positive definite and *the damping matrix* C accretive[1] (our result seems to be new even in the matrix case). Or, M, C, K may be differential operators with similar properties. As a next result, a contractive semigroup, naturally attached to (2), will be constructed, where M, C, K are understood as sesquilinear forms satisfying some mild natural regularity conditions. This construction seems to cover damped systems, more general than those treated in previous literature (cf., e.g., [1], [2], [4]), for instance, M is allowed to have a nontrivial null-space and C need not be symmetric. We then use our abstract semigroup bound to derive a bound for second-order systems in which the damping term C is subject to a perturbation of the same type as (1).[2] As a consequence, the exponential decay of some damped systems will be proved. In particular, under the additional assumption that C be sectorial, a second-order system is exponentially stable, if and only if the system with the 'pure symmetric damping' $\widehat{C} = (C^* + C)/2$ is such. In these applications an important property of the condition (1) will be used: it is invariant under the inversion of both operators.

The article is organised as follows. In Section 2 we prove the main result in a 'local' and a 'global' version. We also include an analogous bound for discrete semigroups, although we have no application for it as yet.

In Section 3 we apply this theory to abstract damped systems of the form (2), including the construction of the semigroup itself. In Section 4 we apply our theory to the damped wave equation in one dimension.

2. An abstract perturbation bound

Let A be the generator of a strongly continuous semigroup in a Hilbert space $\mathcal{X}$. By $\mathcal{T}(A)$ we denote the set of all differentiable semigroup trajectories

$$S = \left\{x = e^{At}x_0, t \geq 0\right\}, \text{ for some } x_0 \in \mathcal{D}(A).$$

Theorem 1. *Let A, B be the generators of strongly continuous semigroups in a Hilbert space $\mathcal{X}$ (then A^*, B^* are also such). Suppose that there exist trajectories $S \in \mathcal{T}(A)$, $T \in \mathcal{T}(B^*)$ and an $\varepsilon > 0$ such that for any $y \in S$, $x \in T$*

$$|(B^*x, y) - (x, Ay)|^2 \leq \varepsilon^2 \Re(-B^*x, x) \Re(-Ay, y). \tag{3}$$

[1]For simplicity we use the term 'damping matrix' for C although it is not necessarily symmetric and thus may include a gyroscopic component.

[2]A related perturbation result for finite matrices was proved in [5].

Then for all such x, y

$$|\left(x, (e^{Bt} - e^{At})y)\right)| \leq \frac{\varepsilon}{2}\|x\|\|y\|. \tag{4}$$

(*Note that in* (3) *it is tacitly assumed that the factors on the right-hand side are non-negative.*)

Proof. For $y \in S$, $x \in T$ we have

$$\begin{aligned}&\frac{d}{ds}\left(e^{B^*s}x, e^{A(t-s)}y\right)\\ &\quad= \left(e^{B^*s}B^*x, e^{A(t-s)}y\right) - \left(e^{B^*s}x, e^{A(t-s)}Ay\right),\end{aligned}$$

which is continuous in s, so by integrating from 0 to t we obtain the weak Duhamel formula

$$(e^{B^*t}x, y) - (x, e^{At}y) = \int_0^t \left[(B^*e^{B^*s}x, e^{A(t-s)}y) - (e^{B^*s}x, Ae^{A(t-s)}y)\right] ds.$$

By using (3) and the Cauchy-Schwarz inequality it follows

$$\begin{aligned}&|\left(x, (e^{Bt} - e^{At}y)\right)|^2\\ &\leq \left(\int_0^t |(B^*e^{B^*s}x, e^{A(t-s)}y) - (e^{B^*s}x, Ae^{A(t-s)}y)|ds\right)^2\\ &\leq \varepsilon^2\left(\int_0^t \sqrt{\Re(-B^*e^{B^*s}x, e^{B^*s}x)\Re(-Ae^{A(t-s)}y, e^{A(t-s)}y)}ds\right)^2\\ &\leq \varepsilon^2\int_0^t \Re(-B^*e^{B^*s}x, e^{B^*s}x)ds\int_0^t \Re(-Ae^{As}y, e^{As}y)ds.\end{aligned}$$

By partial integration we compute

$$\mathcal{I}(A, y, t) = \int_0^t \Re(-Ae^{As}y, e^{As}y)ds = -\|e^{As}y\|^2\Big|_0^t - \mathcal{I}(A, y, t),$$

$$\mathcal{I}(A, y, t) = \frac{1}{2}\left((y, y) - \|e^{At}y\|^2\right). \tag{5}$$

Obviously

$$0 \leq \mathcal{I}(A, y, t) \leq \frac{1}{2}(y, y)$$

and $\mathcal{I}(A, y, t)$ increases with t. Thus, there exist limits

$$\mathcal{I}(A, y, t) \nearrow \mathcal{I}(A, y, \infty) = \frac{1}{2}\left((y, y) - P(A, y)\right), \quad t \to \infty$$

$$\|e^{At}y\|^2 \searrow P(A, y), \quad t \to \infty$$

with

$$0 \leq \mathcal{I}(A, y, \infty) \leq \frac{1}{2}(y, y).$$

(and similarly for B^*). Altogether

$$\begin{aligned} |\left(x, (e^{Bt} - e^{At} y)\right)|^2 &\leq \frac{\varepsilon^2}{4}\left((x,x) - P(B^*,x)\right)\left((y,y) - P(A,y)\right) \qquad (6) \\ &\leq \frac{\varepsilon^2}{4}(x,x)(y,y). \end{aligned}$$

□

Remark 1. As a matter of fact, in the proof above neither of the operators need be densely defined. In this case the assertion of the theorem is valid only in the weak form

$$|(e^{B^*t}x, y) - (x, e^{At}y)| \leq \frac{\varepsilon}{2}\|x\|\|y\|.$$

Corollary 1. *Suppose that* (3) *holds for all* y *from some* $S \in \mathcal{T}(A)$ *and all* $x \in \mathcal{D}$, *where* $\mathcal{D}$ *is a dense subspace, invariant under* e^{B^*t}, $t \geq 0$. *Then*

$$\|\left(e^{Bt} - e^{At}\right) y\| \leq \frac{\varepsilon}{2}\|y\|. \qquad (7)$$

By setting $\varepsilon = 0$ in (7) we obtain the known uniqueness of the solution of a first-order differential equation:

$$e^{Bt}y = e^{At}y, \quad t \geq 0.$$

If there is a dense subspace $\mathcal{D} \in \mathcal{D}(A)$, which is invariant under e^{At}, and on which A is dissipative we set

$$P(A,y) = (P(A)y,y), \quad P(A) = \text{s-}\lim_{t\to\infty} e^{A^*t}e^{At}. \qquad (8)$$

The strong limit $P(A)$ above exists by the dissipativity and obviously $0 \leq P(A) \leq I$ in the sense of forms (and similarly for B^*).

Corollary 2. *If* (3) *holds for all* $x \in \mathcal{D}$, $y \in \mathcal{E}$, *where* $\mathcal{D}$, $\mathcal{E}$ *are dense subspaces, invariant under* e^{B^*t}, e^{At}, *respectively, then*

$$|\left(x, (e^{Bt} - e^{At})y\right)|^2 \leq \frac{\varepsilon^2}{4}\left((x,x) - (P(B^*)x,x)\right)\left((y,y) - (P(A)y,y)\right). \qquad (9)$$

In particular,

$$\|e^{Bt} - e^{At}\| \leq \frac{\varepsilon}{2}. \qquad (10)$$

Remark 2. The corollary above certainly holds, if (3) is fulfilled for all $x \in \mathcal{D}(B^*)$ and all $y \in \mathcal{D}(A)$ (it is enough to require the validity of (3) on respective cores) and this will be the situation in our applications. In any of these cases both B^* and A (and then also B and A^*) are, in fact, maximal dissipative.

The condition (3) has a remarkable property of being *inversion invariant*, i.e., B^* and A may be replaced by their inverses.

Proposition 1. *Suppose that both B^* and A (and then also B and A^*) are (not necessarily boundedly) invertible. Then (3), valid for all $x \in \mathcal{D}(B^*)$ and all $y \in \mathcal{D}(\mathcal{A})$ is equivalent to*

$$|(B^{-*}\xi, \eta) - (\xi, A^{-1}\eta)|^2 \le \varepsilon^2 \Re(-B^{-*}\xi, \xi)\Re(-A^{-1}\eta, \eta),\ \xi \in \mathcal{D}(B^{-*}),\ \eta \in \mathcal{D}(A^{-1}). \tag{11}$$

(here B^{-*} is an abbreviation for $(B^{-1})^* = (B^*)^{-1}$).

Proof. Just set $B^*x = \xi,\ Ay = \eta$. □

Note that in all our results above no further restriction to the constant ε was imposed. This is partly due to the fact that the perturbation is measured by both the "perturbed" and the "unperturbed" operator in a completely symmetric way. This kind of perturbation bound will prove particularly appropriate for our applications below. If ε is further restricted important new conclusions can be drawn.

A semigroup is called *exponentially stable*[3] or *exponentially decaying*, if

$$\|e^{At}\| \le ce^{-\beta t}, \quad t \ge 0 \tag{12}$$

for some $c, \beta > 0$.

Corollary 3. *If in Corollary 2 we have $\varepsilon < 2$ then the exponential decay of one of the semigroups implies the same for the other.*

Proof. Just recall that the exponential stability follows, if $\|e^{At}\| < 1$ for some $t > 0$. □

Remark 3. In all that was said thus far there is an obvious symmetry: in (3) we may replace $A,\ B^*$ by $B,\ A^*$, thus obtaining the dual estimate

$$|(Bx, y) - (x, A^*y)|^2 \le \varepsilon^2 \Re(-Bx, x)\Re(-A^*y, y). \tag{13}$$

with completely analogous results. Obviously, (3) and (13) are equivalent, if $\mathcal{D}(A) = \mathcal{D}(A^*)$ and $\mathcal{D}(B) = \mathcal{D}(B^*)$.

Discrete semigroups. Every step of the perturbation theory, developed above can be correspondingly extended to discrete semigroups. An operator T is called a contraction, if $\|T\| \le 1$. For any such operator T the strong limit

$$Q(T) = \text{s-}\lim_{n\to\infty} T^{*n}T^n$$

obviously exists and satisfies

$$0 \le Q(T) \le 1.$$

The following theorem sums up the most important facts.

[3] Some authors call this property the *uniform* exponential stability.

Theorem 2. *Let A, B be contractions and*

$$|((B-A)x,y)|^2 \le \varepsilon^2((1-B^*B)x,x)((1-AA^*)y,y) \tag{14}$$

for all x,y and some $\varepsilon \ge 0$ (note that in (14) the right-hand side is always non-negative). Then

$$|((B^n-A^n)x,y)|^2 \le \varepsilon^2((1-Q(B))x,x)((1-Q(A^*))y,y) \tag{15}$$

and, in particular,

$$\|B^n-A^n\| \le \varepsilon\sqrt{\|1-Q(A^*)\|\|1-Q(B)\|} \le \varepsilon. \tag{16}$$

Proof. For any x,y we have

$$\begin{aligned}
|((B^n-A^n)x,y)|^2 &= |(\sum_{k=0}^{n-1} A^k(B-A)B^{n-k-1}x,y)|^2 \\
&\le \left(\sum_{k=0}^{n-1} |((B-A)B^{n-k-1}x, A^{*k}y)|\right)^2 \\
&\le \varepsilon^2\left(\sum_{k=0}^{n-1}\sqrt{(A^k(1-AA^*)A^{*k}y,y)(B^{*n-k-1}(1-B^*B)B^{n-k-1}x,x)}\right)^2 \\
&\le \varepsilon^2\sum_{k=0}^{n-1}(A^k(1-AA^*)A^{*k}y,y)\sum_{k=0}^{n-1}(B^{*k}(1-B^*B)B^ky,y) \\
&= \varepsilon^2((1-A^nA^{*n})y,y)((1-B^{*n}B^n)x,x)
\end{aligned}$$

and (15) follows. Here we have used the identity

$$\sum_{n=0}^{n-1} A^k(1-AB)B^k = 1-A^nB^n. \tag{17}$$

□

It may be interesting to note that (17) appears to be a discrete analog of

$$\int_0^t e^{A\tau}(A+B)e^{B\tau}d\tau = -\left(1-e^{At}e^{Bt}\right) \tag{18}$$

on which (5) was based.

Any contraction A is exponentially stable, if and only if $\|A^n\| < 1$ for some n. This leads to a result, analogous to Corollary 3.

Corollary 4. *Let A and B be contractions satisfying (15) with $\varepsilon < 1$. Then the exponential stability of one of them implies the same for the other.*

One might wonder that the bound (10) is uniform in t although the involved semigroups need not be exponentially decaying. As a simple example consider dissipative operators A, B in a finite-dimensional space. Then each of these operators is known to be an orthogonal sum of a skew-Hermitian part and an exponentially stable part. By (3) (which is now equivalent to (1)) the skew-Hermitian parts of A and B coincide and the difference $e^{Bt} - e^{At}$ decays exponentially. The situation with discrete semigroups is similar.

In the infinite-dimensional case the uniformity of the bound (10) is a more serious fact as will be illustrated on applications from Mathematical Physics below.

3. Application to damped systems

An abstract damped linear system is governed by a formal second-order differential equation in a vector space $\mathcal{Y}_0$

$$\mu(\ddot{y}, v) + \theta(\dot{y}, v) + \kappa(y, v) = 0, \tag{19}$$

where μ, θ, κ are sesquilinear forms with the following properties:

- κ symmetric, strictly positive,
- μ symmetric, positive, κ-closable,
- θ κ-bounded, accretive.

A possible way to turn (19) into an operator equation is to take $(u, v) = \kappa(u, v)$ as the scalar product and to complete accordingly $\mathcal{Y}_0$ to a Hilbert space $\mathcal{Y}$. By the known representation theorems ([3]) we have

$$\mu(u, v) = (Mu, v), \quad \theta(u, v) = (Cu, v), \tag{20}$$

where M is (possibly unbounded) selfadjoint and positive and C is bounded accretive. We now replace (19) by

$$M\ddot{y} + C\dot{y} + y = 0, \tag{21}$$

where the time derivatives $\dot{y}$, $\ddot{y}$ are taken in $\mathcal{Y}$.[4]

To the equation (21) one naturally associates the phase space system, obtained by the formal substitution

$$x_1 = y, \quad x_2 = M^{1/2}\dot{y} \tag{22}$$

which leads to the first-order equation

$$\frac{d}{dt}\begin{pmatrix} x_1 \\ x_2 \end{pmatrix} = \mathcal{A}\begin{pmatrix} x_1 \\ x_2 \end{pmatrix}$$

with

$$\mathcal{A} = \begin{pmatrix} 0 & M^{-1/2} \\ -M^{-1/2} & -M^{-1/2}CM^{-1/2} \end{pmatrix}. \tag{23}$$

[4]Our choice of the underlying scalar product in $\mathcal{Y}_0$ is fairly natural but not the only relevant one. One could show that very different, even topologically non-equivalent, choices of the scalar product still lead to the essentially same semigroup dynamics, see [4].

which then should generate a contractive semigroup which realises the dynamics. Our conditions are far too general for this $\mathcal{A}$ to make sense as it stays (note that M may have a nontrivial null-space). However, the formal inverse

$$\mathcal{A}^{+} = \begin{pmatrix} -C & -M^{1/2} \\ M^{1/2} & 0 \end{pmatrix} \tag{24}$$

is more regular, although not necessarily bounded. Considered in the 'total energy' Hilbert space $\widehat{\mathcal{X}} = \mathcal{Y} \oplus \mathcal{Y}$, $\mathcal{A}^{+}$ has the following properties

$$\mathcal{A}^{+} \text{ is maximal dissipative,} \tag{25}$$

$$\mathcal{D}(\mathcal{A}^{+}) = \mathcal{D}((\mathcal{A}^{+})^{*}) = \mathcal{D}(M^{1/2}) \oplus \mathcal{D}(M^{1/2}), \tag{26}$$

$$\mathcal{N}(\mathcal{A}^{+}) = \mathcal{N}((\mathcal{A}^{+})^{*}). \tag{27}$$

All this follows from the fact that $\mathcal{A}^{+}$ is a sum of the skew-selfadjoint operator

$$\begin{pmatrix} 0 & -M^{1/2} \\ M^{1/2} & 0 \end{pmatrix}$$

and a bounded dissipative operator

$$-\begin{pmatrix} C & 0 \\ 0 & 0 \end{pmatrix}.$$

Thus, $\mathcal{N}(\mathcal{A}^{+})$ reduces both $\mathcal{A}^{+}$ and its adjoint, the same is the case with the space $\mathcal{X}$, defined as

$$\mathcal{X} = \mathcal{N}(\mathcal{A}^{+})^{\perp}. \tag{28}$$

More precisely, $\mathcal{A}^{+}$ is a direct sum of the null operator and a maximal dissipative invertible operator $\mathcal{A}^{-1}$ in the Hilbert space $\mathcal{X}$, defined on

$$\mathcal{X} \cap \mathcal{D}(\mathcal{A}^{+})$$

which is dense in $\mathcal{X}$. Obviously, the operator $\mathcal{A}$ is again maximal dissipative and this is by definition the generator of our semigroup. The space $\mathcal{X}$ may be called *the physical phase space* for the system (21).[5]

Denoting by Q the orthogonal projection onto the space $\mathcal{X}$ in $\widehat{\mathcal{X}}$ we have, in fact,

$$(\lambda - \mathcal{A})^{-1} Q = \mathcal{A}^{+}(\lambda \mathcal{A}^{+} - 1)^{-1} = \frac{1}{\lambda} - \frac{1}{\lambda^2}\left(\frac{1}{\lambda} - \mathcal{A}^{+}\right)^{-1}, \quad \Re\lambda > 0. \tag{29}$$

Another useful identity is valid for the case of M bounded

$$(\lambda - \mathcal{A})^{-1} Q = \begin{pmatrix} \frac{1}{\lambda} - \frac{K(\lambda)^{-1}}{\lambda} & K(\lambda)^{-1} M^{-1/2} \\ -M^{1/2} K(\lambda)^{-1} & \lambda M^{-1/2} K(\lambda)^{-1} M^{-1/2} \end{pmatrix}, \tag{30}$$

whenever $K(\lambda) = \mu^2 M + \mu C + 1$ is positive definite. Both are immediately verified.

---

[5]A different but related construction was used in [4] where both M and C are symmetric, but possibly unbounded.

Proposition 2. *The null-space $\mathcal{N}(\mathcal{A}^+)$ satisfies the inclusion*

$$\mathcal{N}(\mathcal{A}^+) \supseteq (\mathcal{N}(C) \cap \mathcal{N}(M)) \oplus \mathcal{N}(M). \tag{31}$$

If, in addition, C is sectorial then we have the equality

$$\mathcal{N}(\mathcal{A}^+) = (\mathcal{N}(C) \cap \mathcal{N}(M)) \oplus \mathcal{N}(M). \tag{32}$$

Proof. [6] Now, $\mathcal{N}(\mathcal{A}^+)$ is given by the equations

$$-Cx_1 - M^{1/2}x_2 = 0, \quad M^{1/2}x_1 = 0, \quad x_{1,2} \in \mathcal{D}(M^{1/2}).$$

From this the inclusion (31) follows. Let now C be sectorial. The above equations imply $(Cx_1, x_1) = -(M^{1/2}x_2, x_1) = 0$. By the assumed sectoriality it follows $Cx_1 = 0$, so (32) follows. □

The fact that the semigroup dynamics exists only on a closed subspace $\mathcal{X}$ of $\widehat{\mathcal{X}}$ is quite natural, even in the finite-dimensional space: one cannot prescribe velocity initial data on the parts of the space where the mass is vanishing. If M is injective – no matter how singular M^{-1} may be – our dynamics exists on the whole space $\widehat{\mathcal{X}}$.

It can be shown ([4]) that this semigroup provides an appropriate solution to the second-order system (21) via the formulae (22), at least in the special case of M, C bounded symmetric. In our, more general situation we can show that $\mathcal{A}$ yields the "true" dynamics by way of approximation. We approximate the operator M by a sequence M_n of bounded, positive operators such that

$$M_n^{1/2}x \to M^{1/2}x, \quad x \in \mathcal{D}(M^{1/2}). \tag{33}$$

If, in addition, all M_n are positive definite the operator (23)

$$\mathcal{A}_n = \begin{pmatrix} 0 & M_n^{-1/2} \\ -M_n^{-1/2} & -M_n^{-1/2}CM_n^{-1/2} \end{pmatrix} \tag{34}$$

is bounded dissipative in $\widehat{\mathcal{X}}$ and its semigroup trivially reproduces the solution of the so modified second-order system

$$M_n\ddot{y} + C\dot{y} + y = 0, \tag{35}$$

An example of such sequence is

$$M_n = f_n(M), \quad f_n(\lambda) = \begin{cases} \frac{1}{n}, & 0 \le \lambda \le \frac{1}{n} \\ \lambda, & \frac{1}{n} \le \lambda \le n \\ n, & n \le \lambda \end{cases}$$

Note that here, in addition, the operators M_n are both bounded and boundedly invertible, being positive definite.

[6] In the case of C symmetric and M bounded this formula was proved in [4].

Proposition 3. *For any* $x = \begin{pmatrix} x_1 \\ x_2 \end{pmatrix} \in \mathcal{X}$ *and any approximation sequence* (33) *we have*

$$e^{\mathcal{A}_n}x \to e^{\mathcal{A}}x, \quad n \to \infty. \tag{36}$$

uniformly on any compact interval in t. *Choose, in addition,* M_n *as positive definite and set*

$$\begin{pmatrix} y_n(t) \\ u_n(t) \end{pmatrix} = e^{\mathcal{A}_n t}\begin{pmatrix} x_1 \\ x_2 \end{pmatrix}, \quad \begin{pmatrix} y(t) \\ u(t) \end{pmatrix} = e^{\mathcal{A}t}\begin{pmatrix} x_1 \\ x_2 \end{pmatrix}.$$

Then $y_n(t)$ *solves* (35) *with* $u_n(t) = M_n^{1/2}\dot{y}_n(t)$ *and*

$$y_n(t) \to y(t), \quad M_n^{1/2}\dot{y}_n(t) \to M^{1/2}\dot{y}(t), \quad n \to \infty.$$

Proof. By (33) we have $\mathcal{A}_n^{-1} \to \mathcal{A}^+$ in the strong resolvent sense (see [3], Ch. VIII, Th, 1.5), i.e.,

$$(\lambda - \mathcal{A}_n^{-1})^{-1} \to (\lambda - \mathcal{A}^+)^{-1}, \quad \Im\lambda \neq 0.$$

Hence by (29),

$$(\lambda - \mathcal{A}_n)^{-1} = \frac{1}{\lambda} - \frac{1}{\lambda^2}\left(\frac{1}{\lambda} - \mathcal{A}_n^{-1}\right)^{-1}$$

$$\to \frac{1}{\lambda} - \frac{1}{\lambda^2}\left(\frac{1}{\lambda} - \mathcal{A}^+\right)^{-1} = (\lambda - \mathcal{A})^{-1}Q.$$

all in the strong sense. Now the Trotter-Kato convergence theory ([3]) can be applied to give

$$e^{\mathcal{A}_\eta t}x \to e^{\mathcal{A}t}x, \quad \eta \to 0 \tag{37}$$

for all $x \in \mathcal{X}$. (The original Trotter-Kato theorem requires the injectivity of the strong limit in (34), but the same proof is easily seen to accommodate our slightly more general setting.) The remaining assertions are now straightforward. □

We now apply our abstract theory from Section 2 to a second-order system with variable damping.

Theorem 3. *Let*

$$\mathcal{A}^+ = \begin{pmatrix} -C & -M^{1/2} \\ M^{1/2} & 0 \end{pmatrix}, \quad \widehat{\mathcal{A}}^+ = \begin{pmatrix} -\widehat{C} & -M^{1/2} \\ M^{1/2} & 0 \end{pmatrix}$$

where M *is bounded, positive selfadjoint and* C, $\widehat{C}$ *are bounded accretive operators satisfying*

$$\left|\left((\widehat{C} - C)x, y\right)\right|^2 \leq \varepsilon^2 \Re(Cy, y)\Re(\widehat{C}x, x) \tag{38}$$

for all $x, y \in \mathcal{Y}$ *and some* $\varepsilon > 0$. *Then* $\mathcal{A}^+$ *and* $\widehat{\mathcal{A}}^+$ *have the same null-space and the respective contractive semigroup generators* $\mathcal{A}$ *and* $\widehat{\mathcal{A}}$ *in* $\mathcal{X}$ *from* (28) *satisfy the assumptions of Corollary 2, in particular,*

$$\|e^{\widehat{\mathcal{A}}t} - e^{\mathcal{A}t}\| \leq \frac{\varepsilon}{2}. \tag{39}$$

Proof. Obviously (38) is equivalent to

$$\left|\left((\widehat{\mathcal{A}}^{+} - \mathcal{A}^{+})x, y\right)\right|^2 \le \varepsilon^2 \Re(-\mathcal{A}^{+}y, y)\Re(-\widehat{\mathcal{A}}^{+}x, x) \tag{40}$$

for all $x, y \in \mathcal{D}(\mathcal{A}^{+}) = \mathcal{D}(\widehat{\mathcal{A}}^{+})$. From this it follows that $\mathcal{A}^{+}$ and $\widehat{\mathcal{A}}^{+}$ have the same null-space. Furthermore, by (26) the domains of the four operators $\mathcal{A}^{+}, \widehat{\mathcal{A}}^{+}, (\mathcal{A}^{+})^*, (\widehat{\mathcal{A}}^{+})^*$ coincide and (40) is equivalent to both (3) and (13) for for $A = \mathcal{A}^{+}$ and $B = \widehat{\mathcal{A}}^{+}$ and then also for $A = \mathcal{A}^{-1}$ and $B = \widehat{\mathcal{A}}^{-1}$ (note that in our situation we have $\Re(-\mathcal{A}^{+}y, y) = \Re(-(\mathcal{A}^{+})^*y, y)$ and $\Re(-\widehat{\mathcal{A}}^{+}x, x) = \Re(-(\widehat{\mathcal{A}}^{+})^*x, x))$. Now apply Proposition 1 and Corollary 2. □

Note the important role of the 'inverse-invariance property' in Proposition 1 in the proof above because we have no explicit formulae for the generators $\mathcal{A}$ and $\widehat{\mathcal{A}}$ and there is no control on their domains of definition.

We now prove some stability results for second-order systems.

Theorem 4. *Let the system* (21) *be exponentially stable*[7] *with a symmetric* $C = C^{(1)}$ *and let*

$$0 \le C^{(1)} \le D \le \alpha C^{(1)}, \quad \alpha \ge 1.$$

Then the exponential stability holds with $C = D$ *and vice versa.*

Proof. Set

$$C_k = C^{(1)} + \frac{k}{n}(D - C^{(1)}), \quad k = 0, \dots, n.$$

Then $C_0 = C^{(1)}$, $C_n = D$ and

$$0 \le C_{k+1} - C_k \le \frac{\alpha - 1}{n} C^{(1)}$$

and

$$|((C_{k+1} - C_k)x, y)|^2 \le ((C_{k+1} - C_k)x, x)((C_{k+1} - C_k)y, y)$$
$$\le \left(\frac{\alpha - 1}{n}\right)^2 (C^{(1)}x, x)(C^{(1)}y, y) \le \left(\frac{\alpha - 1}{n}\right)^2 (C_{k+1}x, x)(C_k y, y).$$

Now choose $n > (\alpha - 1)/2$ and use Theorem 3 and Corollary 3. Use induction: the exponential stability carries over from C_k to C_{k+1} and vice versa. □

In particular, the exponential stability with C implies the same with αC for any positive α. A similar technique can be applied to gyroscopic systems:

Theorem 5. *Suppose that in* (21) *the operator* C *is sectorial. Then the exponential stability of this system is equivalent to the exponential stability of the 'purely damped' system*

$$M\ddot{y} + \widehat{C}\dot{y} + y = 0, \text{ with } \widehat{C} = \frac{C^* + C}{2}. \tag{41}$$

[7]By the exponential stability of a second-order system we mean the exponential stability of the generated semigroup.

Proof. By sectoriality there exists $N > 0$ such that

$$|\Im(Cy, y)| \leq N\Re(Cy, y) \text{ for all } x, y. \tag{42}$$

We have

$$\Re(Cx, x) = (\widehat{C}x, x), \quad -i\Im(Cx, x) = -i((\widehat{C} - C)x, x).$$

The operators $\widehat{C}$ and $-i(\widehat{C} - C)$ are symmetric, so the inequality (42) may be polarised to read

$$((\widehat{C} - C)x, y)|^2 \leq N^2\Re(Cy, y)\Re(Cx, x) = N^2\Re(Cy, y)\Re(\widehat{C}x, x).$$

Assume first $N < 2$. Then apply Theorem 3 and Corollary 3 to obtain the exponential stability with C. Now drop the condition $N < 2$ and proceed by induction. Introduce the sequence

$$C_k = \widehat{C} + \frac{k}{n}(C - \widehat{C}), \quad k = 0, \dots, n.$$

Then obviously

$$\begin{aligned} |((C_{k+1} - C_k)x, y)|^2 &= \frac{1}{n^2}|((\widehat{C} - C)x, y)|^2 \leq \\ \frac{N^2}{n^2}\Re(Cy, y)\Re(\widehat{C}x, x) &= \frac{N^2}{n^2}\Re(C_k y, y)\Re(C_{k+1}x, x). \end{aligned}$$

Now choose $n < 2/N$ and apply the above consideration to the consecutive pairs C_k, C_{k+1}. We may begin at the bottom with $C_0 = \widehat{C}$ or at the top with $C_n = C$. □

The foregoing theorem can be nicely combined with the spectral shift techniques from [6] to obtain further results on exponential stability.

Suppose first that in (21) both M and C are bounded and symmetric (the boundedness of both operators is immediately seen to be a necessary condition for the exponential stability). The key relation in the following will be

$$\mathcal{L}(\mu)\widehat{\mathcal{A}}_\mu^+\mathcal{L}(\mu)^{-1} = (\mathcal{A} - \mu)^{-1}Q. \tag{43}$$

with $\mu < 0$ and

$$\mathcal{L}(\mu) = \begin{pmatrix} K(\mu)^{-1/2} & 0 \\ \mu M^{1/2}K(\mu)^{-1/2} & 1 \end{pmatrix}, \quad K(\mu) = \mu^2 M + \mu C + 1, \tag{44}$$

$$\widehat{\mathcal{A}}_\mu^+ = \begin{pmatrix} K(\mu)^{-1/2}(C + 2\mu M)K(\mu)^{-1/2} & -K(\mu)^{-1/2}M^{1/2} \\ M^{1/2}K(\mu)^{-1/2} & 0 \end{pmatrix}, \tag{45}$$

where μ is chosen in such a way that $K(\mu)$ remains positive definite.[8] In this way both $\mathcal{L}(\mu)$ and $\mathcal{L}(\mu)^{-1}$ are everywhere defined and bounded. The relation (43) is immediately verified (use (30)) and it is an immediate generalisation of the one obtained in [6], (16).

[8] By the assumed boundedness of both M and C there are negative μ's with $K(\mu)$ positive definite.

Under the additional assumption that $C + 2\mu M$ be positive the operator $\widehat{\mathcal{A}}_\mu^+$ is bounded and dissipative, so it is reduced by the subspaces $\mathcal{N}(\widehat{\mathcal{A}}_\mu^+)$ and $\mathcal{N}(\widehat{\mathcal{A}}_\mu^+)^\perp$ and we may write

$$\widehat{\mathcal{A}}_\mu^+ = \widehat{\mathcal{A}}_\mu^{-1} P, \tag{46}$$

where P is the orthogonal projection onto $\mathcal{N}(\widehat{\mathcal{A}}_\mu^+)^\perp$ and $\widehat{\mathcal{A}}_\mu$ is maximal dissipative in the Hilbert space $\mathcal{N}(\widehat{\mathcal{A}}_\mu^+)^\perp$. Thus, (43) implies

$$\mathcal{S}^{-1}\mathcal{A}\mathcal{S} = \widehat{\mathcal{A}}_\mu + \mu. \tag{47}$$

Here the linear operator

$$\mathcal{S} = \mathcal{L}(\mu)\Big|_{\mathcal{N}(\widehat{\mathcal{A}}_\mu^+)^\perp} : \mathcal{N}(\widehat{\mathcal{A}}_\mu^+)^\perp \to \mathcal{N}(\mathcal{A}^+)^\perp$$

is bijective and bicontinuous. We summarise:

Theorem 6. *Let the operators M and C from (21) have additional properties that M is bounded, C sectorial and $C - \alpha M$ accretive for some $\alpha > 0$. Then the system (21) is exponentially stable. If, in addition, C is symmetric then*

$$\gamma = \sup_{x\in\mathcal{Y},(Mx,x)>0} \Re\frac{-(Cx,x)+\sqrt{(Cx,x)^2-4(Mx,x)(x,x)}}{2(Mx,x)} < 0 \tag{48}$$

and

$$\|e^{\mathcal{A}t}\| \le \|\mathcal{L}(\mu)\|\|\mathcal{L}(\mu)^{-1}\|e^{\mu t}. \tag{49}$$

for any $\mu \in (\gamma, 0)$.

Proof. Take first C as symmetric. The value γ is the infimum of all μ such that $C + 2\mu M$ is positive and $K(\mu)$ is positive definite. The proof of this fact is the same as that of [6], Proposition 1, so we omit it here. So, for any $\mu \in (\gamma, 0)$ (47) implies (49).

For non-symmetric C the above considerations will obviously be valid for its symmetric part $\widehat{C}$ from (41). By using Theorem 5 the exponential stability follows. □

Remark 4.

(i) All conditions on M, C, $\widehat{C}$ in the foregoing theorems can be readily expressed in the language of the original forms in (19). For instance, (38) is equivalent to

$$\left|(\widehat{\gamma} - \gamma)(x,y)\right|^2 \le \varepsilon^2 \Re\gamma(y,y)\Re\widehat{\gamma}(x,x). \tag{50}$$

and so on.

(ii) Explicit bounds for the condition number, appearing in (49) may be taken over from [6], Lemma 1.

4. The damped wave equation

Here we apply our general theory to the wave equation in one dimension

$$\rho(x)u_{tt} + \gamma(x)u_t - (d(x)u_{tx})_x - (k(x)u_x)_x = 0 \tag{51}$$

for the unknown function $u = u(x,t)$, $a < x < b$ and $0 < t < \infty$. The functions $\rho(x)$, $\gamma(x)$, $d(x)$, $k(x)$ are assumed to be non-negative and measurable; in addition, $\rho(x)$, $\gamma(x)$ are bounded and

$$\text{ess inf}_{a<x<b} k(x) > 0, \quad \text{ess sup}_{a<x<b} \frac{d(x)}{k(x)} < \infty. \tag{52}$$

The boundary conditions are

$$u(a,t) = 0, \quad u_x(b,t) + \zeta u_t(b,t) = 0, \quad \zeta \geq 0. \tag{53}$$

This is a formally dissipative equation which we shall understand in its weak form

$$\mu(u_{tt}, v) + \theta(u_t, v) + \kappa(u, v) = 0 \tag{54}$$

with $u(a) = v(a) = 0$ and

$$\mu(u,v) = \int_a^b \rho(x) u\bar{v} dx, \tag{55}$$

$$\theta(u,v) = \int_a^b (\gamma(x)u\bar{v} + d(x)u'\bar{v}')\, dx + \zeta u(a)\bar{v}(b), \tag{56}$$

$$\kappa(u,v) = \int_a^b k(x)u'\bar{v}' dx. \tag{57}$$

The forms μ, θ are symmetric and positive. θ is obviously κ-bounded while μ is κ-closable. As the underlying Hilbert space $\mathcal{Y}$ we take the functions with the scalar product

$$(u,v) = \kappa(u,v) = \int_a^b k(x)u'\bar{v}' dx, \quad u(a) = v(a) = 0. \tag{58}$$

Then under our conditions,

$$\mu(u,v) = (Mu, v), \quad \theta(u,v) = (Cu, v) \tag{59}$$

where M, C are positive selfadjoint operators, with bounded C and M. Thus, we end up with the second-order system (21) and (54) gives rise to a contractive semigroup on the space $\mathcal{X}$ which is determined from the null-spaces of M, C.

Note that in order for M to have a non-trivial null-space it is not sufficient that the function ρ vanishes just on a set of positive measure, rather ρ must vanish on an interval (and similarly for C). If ρ vanishes on an interval and γ does not, then (51) is of mixed type (hyperbolic – parabolic). All such cases are covered by our theory.

Now for the perturbation. We perturb the damping parameters $\gamma(x)$, $d(x)$, ζ into $\widehat{\gamma}(x)$, $\widehat{d}(x)$, $\widehat{\zeta}$, which satisfy the same conditions as $\gamma(x)$, $d(x)$, ζ above and are such that

$$|\widehat{\gamma}(x) - \gamma(x)| \leq \varepsilon\sqrt{\widehat{\gamma}(x)\gamma(x)} \tag{60}$$

$$|\widehat{d}(x) - d(x)| \leq \varepsilon\sqrt{\widehat{d}(x)d(x)} \tag{61}$$

$$|\widehat{\zeta} - \zeta| \leq \varepsilon\sqrt{\widehat{\zeta}\zeta} \tag{62}$$

This is a 'relatively small' change of the damping parameters, commonly encountered in practice. The corresponding operators C and $\widehat{C}$ are immediately seen to satisfy (38). Hence Theorem 3 applies and the corresponding semigroups satisfy (39).

One might be interested to obtain perturbation results under the more common assumptions involving only the 'unperturbed' data and the perturbation:

$$|\widehat{\gamma}(x) - \gamma(x)| \leq \eta\gamma(x) \tag{63}$$

$$|\widehat{d}(x) - d(x)| \leq \eta d(x) \tag{64}$$

$$|\widehat{\zeta} - \zeta| \leq \eta\zeta \tag{65}$$

with $\eta < 1$. This implies (60)–(62) with $\varepsilon = \frac{\eta}{\sqrt{1-\eta}}$.

But the real use of (63)–(65) consists merely in insuring the non-negativity of the perturbed damping parameters and the conditions (52); all this is usually known in advance, so there is no need to abandon the much less restrictive conditions (60)–(62).

In view of Corollary 1 we conclude that *if the equation* (51) *decays exponentially with the damping parameters* $\gamma(x)$, $d(x)$, ζ, *then the same will be the case with* $\widehat{\gamma}(x)$, $\widehat{d}(x)$, $\widehat{\zeta}$, *if the constant* ε *is less than* 2. Theorem 4 also applies accordingly.

The situation in higher dimensions is similar and the results are completely analogous and straightforward. The interval $[a, b] \subseteq R$ will be replaced by a bounded domain Ω, so the only additional issue is to insure that the boundary $\partial\Omega \subseteq R^n$ be smooth enough to accommodate any of the boundary conditions from (52).

As a second example consider the equation (51) on the infinite interval $0 < x < \infty$ with the boundary condition

$$u(0, t) = 0. \tag{66}$$

For simplicity we take

$$k(x) = \rho(x) \equiv 1, \tag{67}$$

whereas $\gamma(x) \geq 0$ is supposed to satisfy

$$D = \sup_{u \in \mathcal{Y}} \frac{\int_0^\infty \gamma(x)|u(x)|^2 dx}{\int_0^\infty |u'(x)|^2 dx} < \infty, \tag{68}$$

where $\mathcal{Y}$ is the set of all u which are absolutely continuous, vanish at zero and have a square integrable u'; this is obviously a Hilbert space with the scalar product

$$(u, v) = \int_0^\infty u'(x)\bar{v}'(x)dx.$$

The class of functions γ satisfying (68) is not void since it includes

$$\gamma(x) = \frac{1}{x^2}, \text{ with } D = 4$$

([3] Ch. VI 4.1). The form

$$\mu(u, v) = \int_0^\infty u(x)\bar{v}(x)dx$$

defined on $\mathcal{D}(\mu) = L^2(0, \infty) \cap \mathcal{Y}$ is closed in $\mathcal{Y}$, so (59) yields a positive unbounded operator M with a trivial null-space and a bounded C. Hence our semigroup construction applies and under the perturbation (60) the bounds (10), (9) hold.

Such semigroups are in general not exponentially decaying (they are usually extended to uniformly bounded groups) and will give rise to a non-trivial scattering theory on an 'absorbing obstacle' represented by the short range damping function γ. Further considerations along these lines go beyond the scope of this article.

References

[1] R. Hryniv and A. Shkalikov, Operator models in elasticity theory and hydrodynamics and associated analytic semigroups, Vestnik Moskov. Univ. Ser. I Mat. Mekh. 1999, No. 5, 5–14.

[2] R. Hryniv and A. Shkalikov, Exponential stability of semigroups associated with some operator models in mechanics. (Russian) Mat. Zametki 73 (2003), No. 5, 657–664.

[3] T. Kato, Perturbation Theory for Linear Operators, Springer Berlin 1966.

[4] I. Nakić, A. Suhadolc, K. Veselić, Uniform exponential stability of an abstract vibrational system, in preparation.

[5] K. Veselić, Bounds for exponentially stable semigroups, Linear Algebra Appl. **358** (2003) 309–333.

[6] K. Veselić, Energy decay of damped systems, ZAMM **84**, Heft 12, (2004) 856–863.

Krešimir Veselić
Fernuniversität Hagen
Lehrgebiet Mathematische Physik
Postfach 940
D 58084 Hagon
Germany
e-mail: kresimir.veselic@fernuni-hagen.de